Population Genetics

Population Genetics

Principles and Applications for Fisheries Scientists

Eric M. Hallerman
Editor

American Fisheries Society
Bethesda, Maryland, USA
2003

Suggested Citation Formats

Entire book

Hallerman, E. M., editor. 2003. Population genetics: principles and applications for fisheries scientists. American Fisheries Society, Bethesda, Maryland.

Chapter within the book

Shaklee, J. B., and K. P. Currens. 2003. Genetic stock identification and risk assessment. Pages 291–328 *in* E. M. Hallerman, editor. Population genetics: principles and applications for fisheries scientists. American Fisheries Society, Bethesda, Maryland.

Library of Congress Control Number 2002114336
ISBN 1-888569-27-1

Printed in the United States of American on acid-free paper.

American Fisheries Society
5410 Grosvenor Lane, Suite 110
Bethesda, Maryland 20814-2199
USA

Table of Contents

Practical Applications of Population Genetics

Contributors

Neil Billington (Chapter 4): Department of Biological Sciences, Troy State University, Troy, Alabama 36082, USA

Bonnie Brown (Chapters 1 and 5): Department of Biology, Virginia Commonwealth University, Richmond, Virginia 23284, USA

Kenneth P. Currens (Chapter 13): Northwest Indian Fisheries Commission, 6730 Martin Way E., Olympia, Washington 98506, USA

John Epifanio (Chapters 1, 5, and 6): Center for Aquatic Ecology, Illinois Natural History Survey, 607 E. Peabody Drive, Champaign, Illinois 61820, USA

Anthony J. Gharrett (Chapter 7): School of Fisheries and Ocean Sciences, University of Alaska-Fairbanks, Post Office Box 210082, Auke Bay, Alaska 99821, USA

Eric Hallerman (Chapters 1, 6, 8–12, and 16–18): Department of Fisheries and Wildlife Sciences, Virginia Polytechnic Institute and State University, Blacksburg, Virginia 24061, USA

Anne R. Kapuscinski (Chapter 14): Department of Fisheries and Wildlife, University of Minnesota, St. Paul, Minnesota 55108, USA

Bernie May (Chapter 2): Department of Animal Science, University of California at Davis, Davis, California 95616, USA

Loren M. Miller (Chapter 14): Department of Fisheries and Wildlife, University of Minnesota, St. Paul, Minnesota 55108, USA

Ruth B. Phillips (Chapter 3): Department of Biology, University of Wisconsin-Milwaukee, Box 413, Milwaukee, Wisconsin 53211, USA

Kent M. Reed (Chapter 3): Department of Veterinary Pathobiology, University of Minnesota, St. Paul, Minnesota 55108, USA

James B. Shaklee (Chapter 13): Washington Department of Fish and Wildlife, 600 Capitol Way N., Olympia, Washington 98501, USA

Fred Utter (Chapter 15): School of Fisheries, University of Washington, Seattle, Washington 98195, USA

Lev A. Zhivotovsky (Chapter 7): Vavilov Institute of General Genetics, Russian Academy of Sciences, 3 Gubkin Street, Moskow 117809, Russia; and Morrison Institute for Population and Resource Studies, Stanford University, Stanford, California 94305, USA

List of Fish Taxa

Acadian redfish	*Sebastes fasciatus*
Alabama sturgeon	*Scaphirhynchus suttkusi*
alewife	*Alosa pseudoharengus*
Amazon molly	*Poecilia formosa*
American eel	*Anguilla rostrata*
American shad	*Alosa sapidissima*
angular rough shark	*Oxynotus centrina*
Apache trout	*Oncorhynchus apache*
Arctic char	*Salvelinus alpinus*
Arctic cisco	*Coregonus autumnalis*
armored catfishes	Families Callichthyidae and Loricariidae
Atlantic cod	*Gadus morhua*
Atlantic herring	*Clupea harengus*
Atlantic salmon	*Salmo salar*
Atlantic whitefish	*Coregonus huntsmani*
Australian bass	*Macquaria novemaculeata*
ayu	*Plecoglossus altivelis*
ayus	Family Plecoglossidae
banded sunfish	*Enneacanthus obesus*
barramundi perch	*Lates calcarifer*
billfishes	Family Istiophoridae
black basses	*Micropterus* spp.
black crappie	*Pomoxis nigromaculatus*
black mudfish	*Neochanna diversus*
black perch	*Embiotoca jacksoni*
black porgy	*Acanthopagrus schlegeli*
blackbanded sunfish	*Enneacanthus chaetodon*
bleak	*Alburnus alburnus*
blue tilapia	*Tilapia aurea*
blueback herring	*Alosa aestivalis*
bluegill	*Lepomis macrochirus*

bluespotted sunfish	*Enneacanthus gloriosus*
bonefish	*Albula vulpes*
Bonneville cutthroat trout	*Oncorhynchus clarki utah*
bowfin	*Amia calva*
Brazilian electric ray	*Torpedo ocellata*
brook trout	*Salvelinus fontinalis*
brown bullhead	*Ameiurus nebulosus*
brown trout	*Salmo trutta*
bull trout	*Salvelinus confluentus*
Cape hake	*Merluccius capensis*
capelin	*Mallotus villosus*
carps	Family Cyprinidae
chain pickerel	*Esox niger*
channel catfish	*Ictalurus punctatus*
characins	*Astyanax* spp., *Anoptichthys antrobius, Anoptichthys hubbsi*
chinook salmon	*Oncorhynchus tshawytscha*
chubs	*Notropis* spp.
chum salmon	*Oncorhynchus keta*
cisco (lake herring)	*Coregonus artedi*
ciscoes	*Coregonus* spp.
coastal cutthroat trout	*Oncorhynchus clarki clarki*
cobbler	*Cnidoglanis macrocephalus*
cods	Family Gadidae
coho salmon	*Oncorhynchus kisutch*
Colorado River cutthroat trout	*Oncorhynchus clarki pleuriticus*
Columbia River redband trout	*Oncorhynchus mykiss gairdneri*
common carp	*Cyprinus carpio*
common shiner	*Luxilus cornutus*
common snook	*Centropomus undecimalis*
crescent shiner	*Luxilus cerasinus*
cutthroat trout	*Oncorhynchus clarki*
cyprinids	*Barbus* spp., *Ptychobarbus (Diptychus) dipogon, Rutilus* spp.
deepwater redfish	*Sebastes mentella*
Dover sole	*Microstomus pacificus*
Eurasian perch	*Perca fluviatilis*
European anchovy	*Engraulis encrasicolus*
European eel	*Anguilla anguilla*
European river lamprey	*Lampetra fluviatilis*
fathead minnow	*Pimephales promelas*
flathead catfish	*Pylodictis olivaris*
Florida largemouth bass	*Micropterus salmoides floridanus*
gars	Family Lepisosteidae
Gila topminnow	*Poeciliopsis occidentalis*
Gila trout	*Oncorhynchus gilae*
goldfish	*Carassius auratus*
gourami	*Sphaerichtys espromenarides*
green sturgeon	*Acipenser medirostris*

greenback cutthroat trout	*Oncorhynchus clarki stomias*
Guadalupe bass	*Micropterus treculi*
Gulf of Mexico Atlantic sturgeon	*Acipenser oxyrhynchus desotoi*
guppy	*Poecilia reticulata*
haddock	*Melanogrammus aeglefinus*
heptapterid catfish	*Rhamdia* sp.
herrings	*Alosa* spp.
hickory shad	*Alosa mediocris*
Japanese medaka	*Oryzias latipes*
Japanese pufferfish	*Fugu rubripes*
kelp bass	*Paralabrax clathratus*
killifishes	Family Cyprinodontidae, *Fundulus* spp.
knifefish	*Eigenmannia* sp.
kokanee	*Oncorhynchus nerka*
Lahontan cutthroat trout	*Oncorhynchusclarki henshawi*
lake herring (cisco)	*Coregonus artedi*
lake trout	*Salvelinus namaycush*
lake whitefish	*Coregonus clupeaformis*
lampreys	Family Petromyzontidae
largemouth bass	*Micropterus salmoides*
livebearer	*Poeciliopsis lucida*
livebearers	Family Poeciliidae, *Poecilia* spp.
livebearing eelpout	*Zoarces viviparus*
loaches	Family Cobitidae
longnose gar	*Lepisosteus osseus*
loricarid catfish	*Microlepidogaster* spp.
lungfishes	Families Ceratodidae, Lepidosirenidae, and Protopteridae
masu (cherry salmon)	*Oncorhynchus masou*
medakas	*Oryzias* spp.
menhadens	*Brevoortia* spp.
Mexican golden trout	*Oncorhynchus chrysogaster*
Mexican livebearer	*Poeciliopsis monacha*
Mexican tetra	*Astyanax mexicanus*
minnows	Family Cyprinidae
Monterrey platyfish	*Xiphophorus couchianus*
mummichog	*Fundulus heteroclitus*
Nile tilapia	*Tilapia nilotica*
northern largemouth bass	*Micropterus salmoides salmoides*
northern pike	*Esox lucius*
northern pikeminnow	*Ptychocheilus oregonensis*
oyster toadfish	*Opsanus tau*
Pacific herring	*Clupea pallasi*
Pacific ocean perch	*Sebastes alutus*
Pacific salmon	*Oncorhynchus* spp.
paddlefishes and sturgeons	Acipenseriformes
pallid sturgeon	*Scaphirhynchus albus*
pangasiid catfishes	Family Pangasiidae
Pecos pupfish	*Cyprinodon pecosensis*

perches	Family Percidae, *Stizostedion* spp.
pink salmon	*Oncorhynchus gorbuscha*
plaice	*Pleuronectes platessa*
plated catfishes	Family Callichthyidae
pollock	*Pollachius virens*
powan	*Coregonus lavaretus*
pumpkinseed	*Lepomis gibbosus*
pupfishes	*Cyprinodon* spp.
rainbow smelt	*Osmerus mordax*
rainbow trout	*Oncorhynchus mykiss*
red drum	*Sciaenops ocellatus*
redear sunfish	*Lepomis microlophus*
redfish	*Sebastes marinus*
Rio Grande cutthroat trout	*Oncorhynchus clarki virginalis*
rivuline	*Rivulus* spp.
round goby	*Neogobius melanostomus*
sailfin molly	*Poecilia latipinna*
sand shiner	*Notropis stramineus*
sauger	*Stizostedion canadense*
saugeye (sauger × walleye hybrid)	*Stizostedion canadense* × *S. vitreum*
scale-eating cichlid(s)	*Perissodus microlepis, Perissodus* spp.
sea lamprey	*Petromyzon marinus*
sheatfishes	Family Siluridae
sheepshead minnow	*Cyprinodon variegatus*
shiners	*Notropis* spp.
shortfin mako	*Isurus oxyrinchus*
shortfin molly	*Poecilia mexicana*
shovelnose sturgeon	*Scaphirhynchus platorynchus*
silver char	*Salvelinus alpinus oquassa*
skipjack tuna	*Katsuwonus pelamis*
smallmouth bass	*Micropterus dolomieu*
smelts	*Osmerus* spp.
snappers	Subfamily Lutjaninae
sockeye salmon	*Oncorhynchus nerka*
southern bluefin tuna	*Thunnus maccoyi*
southern platyfish	*Xiphophorus maculatus*
splake (brook trout × lake trout hybrid)	*Salvelinus fontinalis* × *S. namaycush*
spotfin shiner	*Cyprinella spiloptera*
spotted bass	*Micropterus punctulatus*
spotted seatrout	*Cynoscion nebulosus*
spotted sunfish	*Lepomis punctatus*
steelhead	*Oncorhynchus mykiss*
striped bass	*Morone saxatilis*
striped mullet	*Mugil cephalus*
striped shiner	*Luxilus chrysocephalus*
sturgeons	Family Acipenseridae
sturgeons and paddlefishes	Acipenseriformes
suckermouth catfishes	Family Loricariidae
suckers	Family Catastomidae

sunfishes	Family Centrarchidae
tarakihi (jackass morwong)	*Cheilodactylus (Nedadactylus) macropterus*
tasseled-mouth loach	*Crossostoma lacustre*
temperate basses	Family Moronidae (=Percichthyidae)
threespine stickleback	*Gasterosteus aculeatus*
tilapias	*Tilapia* spp.
trouts	Family Salmonidae
tunas	*Thunnus* spp.
walleye	*Stizostedion vitreum*
walleye pollack	*Theragra chalcogramma*
warmouth	*Lepomis gulosus*
western mosquitofish	*Gambusia affinis*
westslope cutthroat trout	*Oncorhynchus clarki lewisi*
white crappie	*Pomoxis annularis*
white perch	*Morone americana*
White Sands pupfish	*Cyprinodon tularosa*
white seabass	*Atractoscion nobilis*
white shiner	*Luxilus albeolus*
white sturgeon	*Acipenser transmontanus*
white sucker	*Catastomus commersoni*
whitefishes	*Coregonus* spp.
yellow perch	*Perca flavescens*
yellowfin tuna	*Thunnus albacares*
Yellowstone cutthroat trout	*Oncorhynchus clarki bouvieri*
zander	*Stizostedion lucioperca*
zebra danio	*Danio rerio*

Preface

The effort to produce this book was driven by a sense of need. The importance of fisheries genetics to the science-based practices of aquaculture and fisheries management is becoming more widely recognized (Kapuscinski and Philipp 1988). However, educational opportunities in fisheries genetics are limited, even in many leading fisheries science programs. This is, in part, due to lack of ready teaching resources in some aspects of fisheries genetics (Hallerman 1994). Teaching of aquacultural genetics is supported by the availability of several useful books, including Kirpichnikov (1981), Tave (1993), Purdom (1993), and Lutz (2001). However, there is no textbook specifically designed for teaching population genetics as applied to fisheries science. Some instructors have used Ryman and Utter (1987) successfully, but they have had to add much material to provide background on population genetic processes for students who have had only an introductory course in general genetics. Frequently, the outcome is that, partly for lack of ready teaching resources, fisheries population genetics is not taught. Hence, cohort after cohort of recruits into the fisheries management profession lack practical knowledge of fisheries genetics. To resolve the problem, population genetics must be placed into the fisheries curriculum in a form that students can relate to and apply throughout their careers. Hence, the need for a textbook in fisheries population genetics.

Suggestions to Instructors for Use of This Book in the Classroom

I have taught a highly rated applied population genetics course to advanced undergraduates and graduate students, providing the basis for the following suggestions on how to put this book to effective use in the classroom.

Students should come to this population genetics course after successful completion of an introductory course on the principles of genetics.

The goal of this course should be to prepare the student for independent learning in population genetics as applied to fisheries. Educational objectives for the student should be to (1) achieve an understanding of the principles of population genetics and the corresponding implications for fisheries management, (2) understand and evaluate reports on fisheries genetic issues in the scientific literature, (3) communicate effectively on fisheries management related genetic issues, and (4) demonstrate integration of these skills. Effective integration can be demonstrated through execution of a literature review and synthesis paper on a chosen topic and communication of this work in an oral presentation.

Course resources should include this textbook and a collection of reprints of research articles illustrating applications of principles discussed in lectures. Key journals from which the instructor might draw these articles include *Evolution*, *Conservation Biology*, *Molecular Ecology*, *Transactions of the American Fisheries Society*, and *Canadian Journal of Fisheries and Aquatic Sciences*. Suggestions for pertinent literature can be found in the literature cited in individual chapters in this book. To keep the course up to date, however, the instructor will need to update the collection of reprints regularly.

The syllabus for the course can follow the chapters in this book. To involve the students actively in learning, I have organized the course in a lecture–discussion format. Typically, one or two lectures are presented on a particular subject, following the material in the relevant chapter to suit the course to the instructor's curriculum and interests. To support the active participation of the student, for each topic I assign to each student a paper to read from the scientific literature. The assignment is, to the degree possible, tailored to the interests of the individual. All students prepare a one-page synopsis of the paper, including description of the problem, methods, key findings, and implications. At the next class meeting, three students make approximately 10-min presentations on key papers. Use of simple visual aids, such as overheads, is encouraged. The students then answer questions, assisted as needed by the instructor. These weekly literature exercises (1) lead students to apply general principles discussed in lecture to specific fisheries problems, (2) personalize the content of the class to an individual's interests, (3) broaden the range of material discussed in class, (4) encourage students to read the primary population genetics literature, (5) improve students' written and oral communications skills, and (6) improve students' ability to think on their feet in question and answer contexts.

Sometime during the semester, individuals typically find particular interest in some aspect of fisheries population genetics. I meet individually with students to discuss a potential topic and guide them to a topic of appropriate narrowness or breadth, and I may give them key names or citations to start their literature search. Toward the end of the semester, students turn in a written review, which is the product of reading and synthesizing the information in relevant research papers. They also present a 15-min synopsis of their project to the class, followed by a brief question and answer period. The objectives of the exercise are to (1) personalize the content of the course to their interests, (2) develop and demonstrate their knowledge in a particular area of genetics, and (3) sharpen their written and oral presentation skills. The general flow of the paper should include a definition of the problem, a discussion of what the various studies measured and contributed, and most important, an assessment of the implications of the body of research. The paper will include a literature citation section, plus any tables and figures needed to support the text. Many students have commented that the literature review and synthesis project, including the in-class discussion, was the best part of the course.

Acknowledgments

This book is the product of many people's work. There are many whom I am pleased to thank.

The authors of the various chapters gave selflessly of their time and expertise, with only my thanks and the future utility of their writings to students as their reward.

Hence, I owe a heartfelt thanks to Neil Billington, Bonnie Brown, Ken Currens, John Epifanio, Tony Gharrett, Anne Kapuscinski, Bernie May, Loren Miller, Ruth Phillips, Kent Reed, Jim Shaklee, Fred Utter, Dick Wilmot, and Lev Zhivitovsky.

Many individuals in the publications office of the American Fisheries Society offered expert advice in development of this book and exhibited great patience as it developed. I thank Janet Raysick, Bob Kendall, Deborah Lehman, Bob Rand, Eva Silverfine, Beth Staehle, and Eric Wurzbacher.

Valerie Sutherland prepared many of the graphics for chapters I authored and for the cover of the book.

Many people contributed constructive criticism on drafts of the chapters. John Epifanio deserves special thanks in this regard. Others who gave selflessly of their time include Tony Gharrett, Ellen Marsden, Harold Kincaid, Bill Smoker, and numerous anonymous reviewers.

This book is a product of the ad hoc Education Committee of the Genetics Section of the American Fisheries Society. The Section provided impetus and support for the project.

The Virginia Sea Grant College Consortium provided seed money to initiate this project. Production costs were supported by the Virginia Sea Grant College Consortium and the Fisheries Management, Genetics, and Education sections of the American Fisheries Society.

I completed much of the work on this book while on a sabbatical supported by Virginia Polytechnic Institute and State University and by the Fulbright Senior Scholars Program.

I hope this book finds utility by a wide variety of students of many ages, both in and out of a formal educational setting.

ERIC HALLERMAN

References

Hallerman, E. M. 1994. Assessment of needs for education opportunities in genetics. Fisheries 19(3):6–12.

Lutz, G. C. 2001. Practical genetics for aquaculture. Iowa State University Press, Ames, Iowa.

Kapuscinski, A. R., and D. P. Philipp. 1988. Fisheries genetics: issues and priorities for research and policy development. Fisheries 13(6):4–10.

Kirpichnikov, V. S. 1981. Genetic bases of fish selection. Translated by G. G. Gause. Springer-Verlag, New York.

Purdom, C. E. 1993. Genetics and fish breeding. Chapman and Hall, London.

Ryman, N., and F. W. Utter. 1987. Population genetics and fishery management. University of Washington Press, Seattle.

Tave, D. 1993. Genetics for fish hatchery managers, 2nd edition. Van Nostrand Reinhold, New York.

Introduction

1

An Overview of Classical and Molecular Genetics

Eric Hallerman, Bonnie Brown, and John Epifanio

1.1 Introduction

1.1.1 *An Overview of the Field of Genetics*

Genetics is the scientific study of inheritance. Genetics is a broad field that can be regarded as having five interrelated branches: classical or Mendelian genetics, cytogenetics, molecular genetics, population genetics, and quantitative genetics. This book focuses on population genetics and its relationships with other branches of genetics and with practical aspects of fisheries management.

Classical, or Mendelian, genetics concerns the transmission of discrete traits from generation to generation and the modes of their expression, following the laws deduced by Gregor Mendel and subsequent workers (section 1.2.1). Classical genetics is briefly reviewed in section 1.2 from the viewpoint of fisheries science.

Cytology is the study of structure, function, and life history of the cell; hence, **cytogenetics** is the science that combines the methods and findings of cytology and genetics (King and Stansfield 1990:83). Cytogenetics focuses on the structure, function, and evolution of chromosomes. The study of cytogenetics as applied to fisheries science is discussed by Phillips and Reed (Chapter 3).

Molecular genetics is the study of the structure, function, and dynamics of genes at the molecular level. A review of basic principles of molecular genetics is presented in section 1.3. This book contains chapters describing methods and findings for each of the major classes of molecular genetic markers, including allozymes (May, Chapter 2), mitochondrial DNA (Billington, Chapter 4), and nuclear DNA (Brown and Epifanio, Chapter 5).

Population genetics is the study of the genetic composition and evolution of populations. The scope of population genetics cuts across numerous disciplines, including other branches of genetics, ecology, systematics, natural history, and statistics. This book focuses on population genetics, its interfaces with other branches of genetics, and its applications within the field of fisheries science. Discussion of population genetic processes includes chapters on mutation (Hallerman and Epifanio, Chapter 6), migration (Gharrett and Zhivotovsky, Chapter 7), selection (Hallerman, Chapter 8), random genetic drift (Hallerman, Chapter 9), inbreeding (Hallerman, Chapter 10), and coadaptation (Hallerman, Chapter 11).

Quantitative genetics is the study of traits exhibiting phenotypes that are quantitative in nature and continuous in distribution. Chapter 12 (Hallerman) presents discussion of the types of quantitative genetic variation, estimation of the underlying components of quantitative variation, and applications of quantitative genetics in fisheries science.

1.1.2 *Applications of Population Genetics in Fisheries Management*

Population genetics finds practical application in a number of contexts in fisheries management, as discussed in the last six chapters of this volume.

— The identification of distinct genetic stocks and estimation of their relative abundance in mixed-stock fisheries (Shaklee and Currens, Chapter 13) is critical for effective fisheries management and conservation.
— Effective use of fish hatcheries as a tool for supplementation of natural fish populations depends upon formulation and implementation of genetically appropriate hatchery programs (Miller and Kapuscinski, Chapter 14).
—The issue of whether and to what degree release of cultured fishes affects natural populations proved controversial through the 1990s; assessing the effects of cultured fishes on natural fish populations is discussed by Utter (Chapter 15).
— In many contexts, fisheries scientists or fisheries managers would like to mark groups of fish, for example, those released in fisheries supplementation programs and their descendants. **Genetic marking** (Hallerman, Chapter 16) is the purposeful change of allele frequencies in a group of fish in order to identify them or their descendants within a larger collection of conspecifics.
— Fisheries officers sometimes face cases of apparent misappropriation of fisheries resources, misrepresentation of fisheries products, or illegal commerce in fisheries products. **Forensics** (Hallerman, Chapter 17) includes the use of scientific methods to reach inferences with regard to fisheries, wildlife, or conservation law enforcement.
— Maintaining a population's viability is a fundamental goal of conservation biology and increasingly so for fisheries management. Population viability analysis (Hallerman, Chapter 18) is the application of an emerging body of theory that has been developed to guide conservation planning for wild populations.

Against the background of the scope of topics covered in this volume, this introductory chapter reviews basic aspects of classical and molecular genetics, preparing the reader for all subsequent chapters.

■ 1.2 Review of Classical Genetics

1.2.1 *Mendel s Laws*

To understand the foundations of population genetics, it is worthwhile to consider its historical development. In the early nineteenth century, scholars of natural history had come to realize that species evolved. In 1809, Lamarck put forward the view that species can evolve into new species through gradual strengthening and perfection of adaptive characteristics and that these characteristics are transmitted to the offspring (King and Stansfield 1990:354). Darwin's voyage on the *Beagle* in the 1830s and Wallace's studies of the fauna of the Malay Archipelago in the 1850s led them to recognize geographic modes of speciation and to propose the theory of natural selection in the late 1850s (Darwin and Wallace 1958). Despite gaining insights into certain mechanisms of evolution, however, a scientific understanding of the basic mechanisms of inheritance was lacking.

It was against this background that, in 1856, Gregor Mendel, a monk at the Augustinian monastery of St. Thomas in what is now Brno in the Czech Republic, began breeding experiments with the garden pea. Crossing inbred lines, he noted frequencies of key phenotypic characters in the F_1 and F_2 generations. He noted that certain traits, such as plant height, flower color, and seed texture, were inherited as discrete traits, whose expression often exhibited dominant or recessive modes of expression. The results of his work led to fundamental advancement of the understanding of inheritance. Mendel's laws, as restated to reflect modern understanding, are as follows.

— The **law of segregation** states that the two members of a pair for a character in a parental organism segregate. That is, the two alleles at a locus in a parental organism segregate into different gametes at meiosis. For example, an *Aa* heterozygote produces equal numbers of *A*-bearing and *a*-bearing gametes and equal numbers of offspring bearing its *A* and *a* alleles.

— The **law of independent assortment** states that members of different pairs of characters associate randomly, or independently. That is, alleles at different loci are incorporated independently into gametes at meoisis (provided they reside on different chromosomes).

In 1866, Mendel's *Versuche uber Pflanzenhybriden (Experiments on Plant Hybridization)* was published but ignored. It was not until 1900 that de Vries, Correns, and Tschermak independently rediscovered Mendel's paper (King and Stansfield 1990:355). De Vries and Correns had performed breeding experiments similar to Mendel's with several other plants, and, reaching similar conclusions, they acknowledged the significance of his work. The principles of transmission of genetically determined traits had been established and recognized.

1.2.2 Genetic Variation

To this point, we have relied on the reader's intuitive sense of the terms gene, locus, and allele, as these terms had not yet been coined in Mendel's time. A more specific understanding of these terms is needed. **Gene** is a general term referring to the physical entity transmitted from parent to offspring (Hartl 1988:1; see also below). Often, gene is used to refer to a unit encoding a specific trait or protein. Other terms are used to convey particular meanings in different contexts. **Locus** refers to the position that a gene occupies on a chromosome or within a segment of genomic DNA (King and Stansfield 1990). An **allele** is one of a series of possible alternative forms of a given gene. Referring to Mendel's peas, for example, the locus is the site on the chromosome where the gene determining plant height is encoded, and the alleles are the genetic coding for normal and dwarf phenotypes. Genetic variation as referred to in this volume generally refers to allelic variation.

Much of the development of genetics as a science has been the struggle to observe and understand the significance of genetic variation. The four central questions regarding genetic variation are (1) how much variation is found in a population, (2) what is the origin of the variation, (3) how is the variation maintained, and (4) what is the ecological and evolutionary significance of the variation? Seeking answers to these questions forms the focus of population genetics.

Part of the struggle to measure variation is in describing it in evolutionarily meaningful terms (Lewontin 1974:19). The description must be genotypic, because evolution is based on the underlying Mendelian genetics. However, the description also should define the relation between genotype and phenotype, because phenotype determines the likelihood of survival and breeding success and is operated upon by natural selection. At a practical level, however, there is a trade-off between the genetic variation that we as geneticists *want* to measure and what we actually *can* measure. In practice, we seldom measure variation directly at fitness-conferring loci.

1.2.3 Classical Mendelian Traits

With classical **Mendelian traits,** substitution of one allele for another at a given locus yields a large change in phenotype. These traits include discrete phenotypes, visible mutations, and lethal genes.

Examples of traits exhibiting discrete phenotypes are those that Mendel studied in his peas. Pea plant height was either dwarf or tall. The surface of the pea was either wrinkled or smooth. Pea flowers were either red or yellow. Mendel also succeeded in inferring the action of dominant and recessive alleles. **Dominant alleles** are those that fully express their phenotype when present in the heterozygous state. Alleles whose phenotypic expression is masked by dominant alleles are termed **recessive alleles.** In Mendel's peas, for example, the allele conferring red flower color is dominant to that conferring yellow flower color. Mendelian inheritance of discrete phenotypes is known in many fishes. It can be exemplified by the metallic blue coloration of rainbow trout *Oncorhynchus mykiss* (Kincaid 1977; Figure 1.1), for which the dominant *B* allele is expressed as normal pigmentation and the recessive *b* allele as iridescent metallic blue coloration. Other traits of fishes exhibiting simple Mendelian inheritance include telescope eyes of goldfish *Carassius auratus* (Matsui 1934); albino coloration of channel catfish *Ictalurus punctatus* (Bondari 1984), rainbow trout (Bridges and von Limbach 1972), and Japanese medaka *Oryzias latipes* (Yamamoto 1969); and red and syrup coloration morphs of Nile tilapia *Tilapia nilotica* (McAndrew et al. 1988).

Phenotypically visible mutations provide an historically important source of genetic variation. The fruit flies, especially *Drosophila melanogaster,* provide a useful model system for studying visible mutations. Variations include extra abdominal bristles, variant eye colors, and different wing morphologies, all of which arose by mutation from the respective wild types, that is, those most frequently observed and taken as "normal." Visible mutations are not restricted to laboratory model systems. The house mouse *Mus musculus* genome has five loci—*A, B, C, D,* and *ln*— encoding coat color; the mutation rate for these loci from the wild type to a mutant has been estimated to be 11.2×10^{-6} mutations per locus per generation (Lewontin 1974:37). Among the most prominent piscine examples of visible mutations are the many color, fin, and scalation morphs in fancy goldfish, which are believed to have arisen as mutations in the ancestral (orange) comet variety.

Lethal genes also exemplify alteration of phenotype by substituting one allele for another. Lethal genes persist in a gene pool because, generally, only the homozygous recessive genotype is lethal. Lethal genes are difficult to detect in a randomly breeding population. However, the action of a lethal gene can be detected readily within the context of a cross (Figure 1.2). In the parental generation (Figure 1.2A), a heterozygous car-

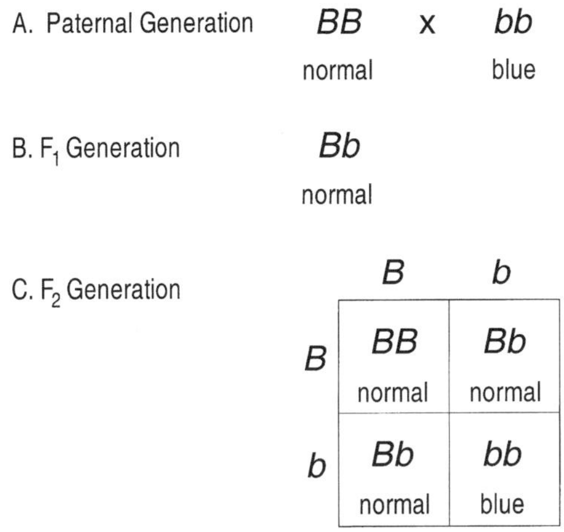

Figure 1.1 Inheritance of a recessive trait, iridescent metallic blue coloration of rainbow trout (Kincaid 1975), demonstrated by means of a Punnett square.

rier, with genotype +*m*, might be mated with an individual homozygous (++) for the wild-type allele. Among the offspring, half would be ++ and half +*m*. Should a series of full-sib matings be executed (Figure 1.2B), three types of pairings are possible, and approximately one-quarter will be matings of +*m* with +*m* individuals. Such matings will result in offspring that are homozygous wild type (++), heterozygous (+*m*), and homozygous recessive (*mm*) at a ratio of 1:2:1 (Figure 1.2C). If the recessive allele is indeed a lethal allele, then all the offspring with the homozygous recessive genotype *mm* die. By noting that in the system of matings there are families whose survival conforms to the ratio 1:2:0, a geneticist can determine the segregation of a lethal allele. Should the last term in the observed ratio prove to be less than one, the recessive allele is termed **sublethal.** Because of the need for test crosses, the fruit flies have proven a suitable model system, and hundreds of lethal and sublethal genes have been detected. It has been esti-

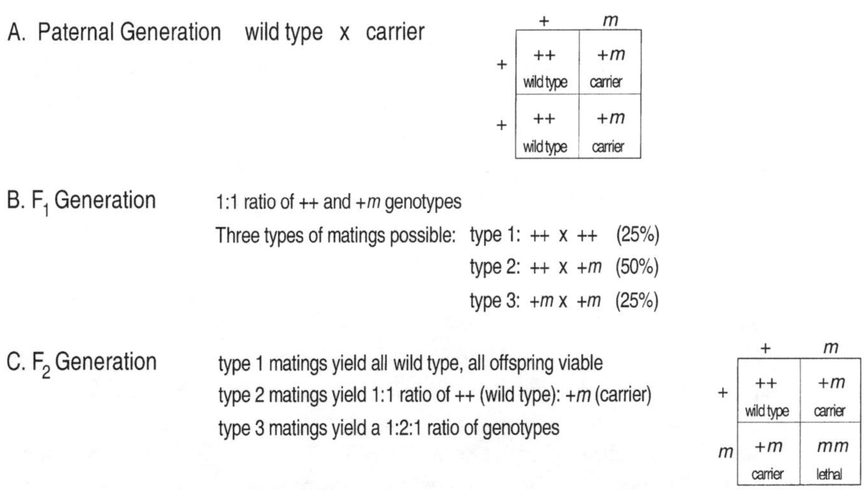

Figure 1.2 Detection of a lethal gene by means of a cross between a carrier and a wild-type individual. The Punnett square shows segregation of a type 3 mating.

mated that each human carries tens of sublethal or possibly lethal alleles. Although loci linked to scale pattern in common carp *Cyprinus carpio* (Wohlfarth et al. 1963; Kirpichnikov 1981) and to the saddleback morphology in blue tilapia *Tilapia aurea* (Tave et al. 1983) provide notable exceptions, rather few lethal alleles have been documented in fishes. The interplay of lethal genes and selection is discussed in Chapter 6 (section 6.4.2).

1.2.4 The Hardy Weinberg Model

The process of analyzing genetic variation involves scoring one or more types of genetic markers, quantifying phenotype frequencies, and then inferring the underlying genotype frequencies. A model, then, is needed to relate genotype frequencies with allele frequencies and to reach inferences about processes acting upon natural populations. The utility of such a model lies in its ability to identify critical observations that we should seek or critical experiments that we should execute to achieve new insights.

The model independently developed by G. H. Hardy (1908) and W. Weinberg (1908) is the most important model relating genotypic and allelic frequencies. The **Hardy–Weinberg model** has a number of important underlying assumptions:

— large and constant population size between generations;
— random mating, that is, the population is **panmictic,** or well-mixed;
— diploid organisms;
— nonoverlapping generations;
— sexual reproduction; and
— negligible effects of mutation, migration, or selection.

For **autosomal loci** (i.e., those not located on sex chromosomes) with two alleles, the Hardy–Weinberg model is expressed as

$$(p + q)^2 = p^2 + 2pq + q^2 = 1,$$

where p is the frequency of the more common allele, A, and q is the frequency of the less common allele, a, and $p + q = 1$. Alleles A and a are **codominant alleles** for a particular trait; that is, we can distinguish carriers of the genotypes AA, Aa, and aa on a phenotypic basis. The frequency of the more common homozygote, AA, is p^2; the frequency of the heterozygote, Aa, is $2pq$; and the frequency of the less common homozygote, aa, is q^2. The Hardy–Weinberg model also can be extended to cases of multiple alleles and to genes located on the X chromosome (Hartl 1988:34; Falconer 1989:14) and to pairs of genes located on the same chromosome (see section 1.2.5).

The Hardy–Weinberg model predicts constancy of allele frequencies between generations. The model may be used to predict genotype frequencies given present allele frequencies, subject to the satisfaction of the model's assumptions. Application of the Hardy–Weinberg model provides a basis for assessment of evolutionary forces affecting the array of genotypes in a population. Should comparison of genotype frequencies observed depart from those expected under the model, hypotheses can be put forward as to why such departures occurred, pointing toward new observations to be made or experiments to be conducted. The implication is that the model was violated because one or more of its underlying assumptions was not met. Violations of assumptions underlying the Hardy–Weinberg model may occur through various ecological mechanisms.

The Hardy–Weinberg model assumes that matings are random, but this assumption frequently is violated by certain mating behaviors. **Assortative mating** is the selection of a mate based on phenotype. **Positive assortative mating** occurs when like individuals mate more frequently than expected by chance. **Inbreeding** (Chapter 10) is the mating of related individuals, which is a particular case of positive assortative mating. **Negative assortative mating** occurs when unlike individuals mate more frequently than expected by chance, as in the rare-male advantage of *Drosophila.* **Genetic stock structure** (Box 1.1; Chapter 13) is also a type of assortative mating, in which the gene pool of a species is composed of a group of subpopulations, or stocks, that mate panmictically within themselves.

In certain instances, the effect of natural selection on genotype frequencies is not negligible. For example, a well-known example involves a selective advantage enjoyed by heterozygotes at the hemoglobin-*B* locus of humans, where the *+s* genotype (malaria resistant, not anemic) is more fit than the ++ (malaria prone) and *ss* (sickle cell anemic) genotypes. Fitness of heterozygotes is discussed in sections 8.3.1 and 10.4.

Violations of assumptions underlying the Hardy–Weinberg model might arise through a variety of other ecological mechanisms. Actually, the Hardy–Weinberg model is rather robust against minor violations of its assumptions (Hartl 1988:28), a quality that renders it more useful in practice. However, it must be recognized that should observed genotype frequencies fit Hardy–Weinberg expectations, this does not necessarily imply that all the assumptions are met. A numerical example of application of the Hardy–Weinberg model within the context of fisheries science is presented in Box 1.1.

1.2.5 Linkage and Linkage Disequilibrium

Mendel's law of independent assortment predicts random association of alleles at different loci. When this is the case, there are equal frequencies for gametes bearing each combination of the respective alleles (Figure 1.3A). These predictions hold true when the respective loci are on different chromosomes. Genes whose alleles are in random association are said to be in **linkage equilibrium.** However, random association of alleles may not occur when the loci are **linked,** that is, when they occur on the same chromosome. This condition is referred to as **linkage disequilibrium.**

The possibility that loci are linked can be examined by producing a test cross and observing genotype frequencies in an appropriate progeny group (Figure 1.3B). Consider the classical experimental design for testing the linkage of two loci, *A* and *B*. Two inbred parental lines, one expressing the marker genotype *AABB*, the other *aabb*, are crossed. In the F_1 generation, all individuals carry the *AaBb* genotype. Segregation will occur in subsequent generations. Were the loci unlinked, we would observe equal frequencies of all combinations of *A*, *a*, *B*, and *b* alleles. However, because the two loci are linked, there will be higher and equal frequencies of the two parental types, or **nonrecombinant gametes** *AB* and *ab*, and lower and equal frequencies of the **recombinant gametes,** *Ab* and *aB*. The degree to which the respective gametic frequencies differ from 0.25 is a function of the **recombination frequency,** *r;* that is, the frequency with which meiotic **crossing-over,** the exchange of material between homologous chromosomes, produces new, nonparental types of gametes. Parental-type *AB* and *ab* gametes each are produced with frequency $(1 - r)/2$. Recombinant-type Ab and aB gametes each are produced with frequencies of $r/2$. The **recombination fraction,** then, refers to the proportion of recombinant gametes produced by an individual; it is equal to *r*. For example, should an individual produce *AB* and *ab* gametes at frequencies of 0.36, and *Ab* and *aB* gametes at frequencies of 0.14, the recombination fraction is 0.14 + 0.14 = 0.28.

■ Box 1.1
Application of the Hardy Weinberg Model for Analysis of a Mixed-Stock Fishery

The Hardy–Weinberg (H–W) model finds a wide range of applications in fisheries science. One of the most important arises in the context of identifying mixed-stock fisheries and estimating their composition from nuclear-based molecular characters. If a sample from a fishery does not conform to H–W expectations, then one possibility is the mixing of differentiated subpopulations. Falconer (1989) cites an interesting case study on Atlantic cod *Gadus morhua* originally reported by Moller (1968).

Genotype frequencies at a hemoglobin locus were determined for a collection of Atlantic cod in a Norwegian fishery. Among the 2,591 Atlantic cod in the sample, 130 exhibited *AA*, 763 *Aa*, and 1,698 *aa* genotypes. The study was motivated, in part, by the desire to know whether the Atlantic cod fishery constituted one stock or more than one genetic stock. Hence, the hypotheses to be tested are H_o, the population is at H–W equilibrium, versus H_a, the population is not at H–W equilibrium. The analysis proceeds as follows.

Step 1: Calculate the allele frequencies

$$130 + 763 + 1{,}698 = 2{,}591 \text{ fish.}$$

$$2{,}591 \text{ fish} \times 2 \text{ alleles per individual} = 5{,}182 \text{ alleles.}$$

To calculate p, the frequency of the A allele, note that the collection included 130 homozygous fish × 2 alleles each + 763 heterozygous fish × 1 allele each = 260 + 763 = 1,023 A alleles. Dividing by the total number of alleles, we obtain p:

$$p = 1{,}023/5{,}182 = 0.197.$$

Calculation of q, the frequency of the a allele, goes forward in similar fashion. Observed were 763 heterozygous fish × 1 allele each + 1,698 homozygous fish × 2 alleles each = 763 + 3,396 = 4,159 a alleles.

$$q = 4{,}159/5{,}182 = 0.803.$$

Step 2: Calculate expected genotype frequencies, using the H–W model

$$(p + q)^2 = p^2 + 2pq + q^2 = 1.$$

$$= 0.121^2 + 2(0.121)(0.879) + 0.879^2$$

$$= 0.014 + 0.213 + 0.772,$$

Note that the last expression gives you the expected frequencies of the various genotypes: the frequency of *AA* is 0.039, of *Aa* is 0.316, and of *aa* is 0.645.

Step 3: Calculate the expected numbers of individuals for each genotype
The expected number of individuals for each genotype is calculated by multiplying the expected frequency of each genotype by the total number of individuals in the sample.

continued on next page

■ Box 1.1 *continued*

Number of AA expected $= 0.039 \times 2{,}591 = 101.05$ fish.

Number of Aa expected $= 0.316 \times 2{,}591 = 818.76$ fish.

Number of aa expected $= 0.645 \times 2{,}591 = 1{,}671.20$ fish.

Step 4: Apply the χ^2 test to determine whether observations conform to H–W expectations

The χ^2 test is used to determine whether the observed numbers of individuals exhibiting given genotypes conform to H–W expectations.

$$\chi^2 = (\text{observed number} - \text{expected number})^2/\text{expected number}.$$

For AA, $\chi^2 = (130 - 101.05)^2/101.05 = 8.29$.

For Aa, $\chi^2 = (763 - 818.76)^2/818.76 = 3.80$.

For aa, $\chi^2 = (1{,}698 - 1{,}671.20)^2/1671.20 = 0.43$.

$\Sigma = 12.52$

The number of degrees of freedom associated with this χ^2 value equal the number of classes of data (in this example, 3) minus 1, minus the number of parameters estimated from the data (in this example, one parameter, p, was estimated from the data), or $3 - 1 - 1 = 1$. Note that a degree of freedom is not deducted for estimating q, as once we have estimated p, q follows from the relation $q = 1 - p$. With 1 degree of freedom, this χ^2 value is highly significant ($P < 0.01$).

We conclude that observed genotype frequencies in the sample from the fishery depart from H–W expectations. The inference is that at least one of the model's assumptions was violated. Which assumption might it have been? An independent line of evidence offers a tentative explanation.

Two distinct races of Atlantic cod can be distinguished by morphological differences in their otoliths (or ear-stones, involved in hearing and balance). When otoliths from the Atlantic cod were analyzed, the catch was separated into two stocks, coastal and arctic. Genotype counts in the coastal stock were 23 AA, 250 Aa, and 946 aa, and in the arctic stock were 107 AA, 513 Aa, and 752 aa. What light do these findings shed on the composition of the overall catch?

Begin the analysis by determining whether the genotype composition in the coastal stock conforms to H–W expectations.

Step 1: Calculate the allele frequencies

$23 + 250 + 946 = 1{,}219$ fish, or 2,438 alleles.

$p = [(23 \times 2) + (250 \times 1)]/2{,}438 = 0.121.$

$q = [(250 \times 1) + (946 \times 2)]/2{,}438 = 0.879.$

continued on next page

■ Box 1.1 *continued*

Step 2: Calculate expected genotype frequencies

$$(p + q)^2 = p^2 + 2pq + q^2 = 1$$
$$= 0.121^2 + 2(0.121)(0.879) + 0.879^2$$
$$= 0.014 + 0.213 + 0.772,$$

representing the frequencies of *AA*, *Aa*, and *aa*, respectively.

Step 3: Calculate the expected numbers of individuals for each genotype

Number of *AA* expected $= 0.014 \times 1{,}219 = 17.07$ fish.

Number of *Aa* expected $= 0.213 \times 1{,}219 = 259.64$ fish.

Number of *aa* expected $= 0.772 \times 1{,}219 = 941.07$ fish.

Step 4: Apply the χ^2 test to determine whether observations conform to H–W expectations

For *AA*, $\chi^2 = (23 - 17.07)^2/17.07 = 2.06.$

For *Aa*, $\chi^2 = (250 - 259.64)^2/259.64 = 0.36.$

For *aa*, $\chi^2 = (946 - 941.07^2/941.07 = 0.03.$

$\Sigma = 2.45.$

The χ^2 test result, 2.45, is not statistically significant. Hence, genotype frequencies in the coastal stock are consistent with the H–W model.

Next, determine whether genotype frequencies in the arctic stock conform to H–W expectations.

Step 1: Calculate the allele frequencies

$107 + 513 + 752 = 1{,}372$ fish $\times 2 = 2{,}744$ alleles.

$$p = [(107 \times 2) + (513 \times 1)]/2{,}744 = 0.265.$$
$$q = [(513 \times 1) + (752 \times 2)]/2{,}744 = 0.735.$$

Step 2: Calculate expected genotype frequencies

$$(p + q)^2 = p^2 + 2pq + q^2 = 1$$
$$= 0.265^2 + 2(0.265)(0.735) + 0.735^2$$
$$= 0.070 + 0.390 + 0.540.$$

Step 3: Calculate the expected numbers of individuals for each genotype

Number of *AA* expected $= 0.070 \times 1{,}372 = 96.04$ fish.

Number of *Aa* expected $= 0.390 \times 1{,}372 = 535.08$ fish.

Number of *aa* expected $= 0.540 \times 1{,}372 = 740.88$ fish

continued on next page

Box 1.1 *continued*

Step 4: Apply the χ^2 test to determine whether observations conform to H–W expectations

$$\begin{aligned} \text{For } AA, \chi^2 &= (107 - 96.04)^2/96.04 = 1.25. \\ \text{For } Aa, \chi^2 &= (513 - 535.08)^2/535.08 = 0.91. \\ \text{For } aa, \chi^2 &= (752 - 740.88)^2/740.88 = 0.17. \\ \Sigma &= 2.33. \end{aligned}$$

The χ^2 test result, 2.33, is not statistically significant. Hence, genotype frequencies in the arctic stock are consistent with the H–W model.

What do the results of these analyses collectively tell us? The deviation from H–W equilibrium suggests that the Atlantic cod fishery is a mixed-stock fishery, comprising a coastal and an arctic stock. Genotype frequencies within each of these stocks are consistent with H–W expectations, suggesting that each stock is, within itself, panmictic. This approach is the basis for most mixed-stock analyses, such as those such as those conducted on lake trout *Salvelinus namaycush* (Krueger et al. 1989; Grewe et al. 1994), Pacific salmonids *Oncorhynchus* spp. (Pella and Milner, 1987; Smouse et al. 1990; Waples et al. 1990), and American shad *Alosa sapidissima* (Epifanio et al. 1995; Brown et al. 1996; Brown et al. 1999). Recognition of genetic stock structure poses important implications for fisheries management (Chapter 13).

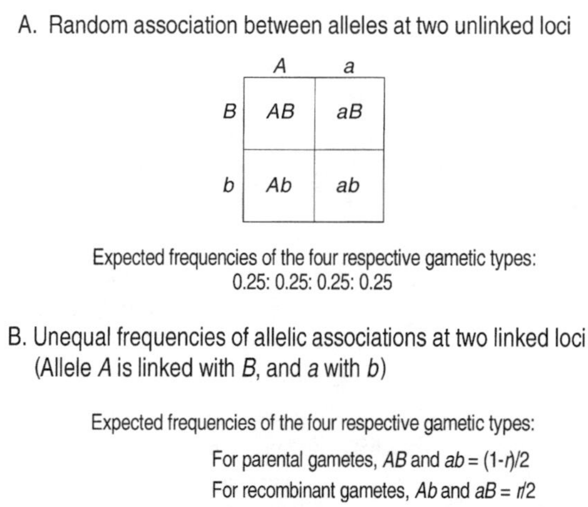

Figure 1.3 Expected gametic frequencies for (A) unlinked and (B) linked loci. The recombination frequency is denoted *r*.

The recombination fraction depends on the relative positions of two genes of interest on a chromosome. The largest possible recombination fraction is 0.50, which indicates that the genes reside on different chromosomes or quite far apart on the same chromosome. A recombination fraction less than 0.5 indicates that the genes are linked. The smallest possible recombination fraction is 0, which implies that two genes are so close together on the chromosome that recombination never occurs between them. In classical experimental designs for detecting linkage, an F_2 or backcross progeny group is screened for the frequencies of recombinant genotypes, and the results are used to establish the occurrence and closeness of any linkage relationships. The first such detection of genetic linkage was demonstrated by Bateson and Punnett in 1906 in sweet pea (King and Stansfield 1990:356).

Under conditions satisfying the assumptions underlying the Hardy–Weinberg model, linkage equilibrium between two linked genes eventually will be attained. The attainment of linkage equilibrium, however, is very gradual (Hartl 1988:40), particularly if two genes are nearby on a chromosome. This contrasts with the rate of attainment of Hardy–Weinberg proportions for alleles of a single gene, which requires just one to several generations.

Data on recombination frequencies among multiple marker loci can be used to calculate the distance between loci and to infer the order of genes on a chromosome. This is a useful tool for **genome mapping,** that is, for constructing an ordered listing of the genes or genetic markers occurring on the respective chromosomes of a species of interest. This is the primary means by which genomic maps for humans as well as for model agricultural and aquacultural species are constructed and by which quantitative trait loci are detected (Poompuang and Hallerman 1997). Detection of genes conferring fitness on their carriers also might be approached by linkage-based detection methods (Box 12.3).

1.2.6 Molecular Genetic Traits

Through the early decades of the twentieth century, methods for observing genetic variation were rather limited. In addition to observation of Mendelian traits described above, geneticists also could screen variation of chromosome number and morphology (Chapter 3) and of quantitative traits (Chapter 12). Population geneticists focused on study of Mendelian traits in laboratory model species, such as fruit flies, or on species for which fully inbred lines could be developed. Methods available were not well given for screening of populations in the wild and, hence, were not particularly amenable for practical applications in fisheries science and fisheries management.

The emergence of molecular biology at first entailed development of new methods for observing variation of proteins. Starch gel electrophoresis (Smithies 1955) and histochemical staining techniques (Chapter 2) were developed primarily for detecting variant forms of blood proteins and enzymes for study of their biochemistry. Broader application of these methods subsequently had a major impact on the field of population genetics. First of all, blood protein and allozyme screening methods contributed a dramatic expansion of the number of genetic markers available. Further, freed of the need for screening outwardly visible traits in inbred lines, screening of genetic variation in natural populations became a viable technical possibility. The advent of methods for screening variation at the protein level more than just improved the sensitivity and applicability of population genetic screenings, it led to recognition of unexpectedly high levels of genetic variation in a wide range of natural populations. Application of protein level methods revolutionized the population genetics worldview, led to new systematic inferences, and led to the selectionist–neutralist debate regarding the significance of the newly uncovered genetic variation (Chapter 9).

Methods for DNA cloning, sequencing, and hybridization (developed in the mid-1970s) and DNA amplification and automated sequencing (mid-1980s) led to the development of various classes of DNA level markers through the 1980s and 1990s (Chapters 4, 5, and 6). To understand the respective methodologies and their range of applications, the reader first must achieve some level of familiarity with genome structure and function, which is discussed in section 1.3.

1.3 Review of Molecular Genetics

1.3.1 *Genome Structure and Function*

During the early 1950s, Francis Crick, James Watson, and Maurice Wilkins established the currently accepted model of DNA structure. More importantly, they established how their physical model, the double helix, supported the concepts that coded information was, in fact, located within the DNA and that the DNA could self-replicate in a rather straightforward manner. These discoveries earned Crick, Watson, and Wilkins a Nobel Prize in Chemistry. Since then, we have learned much more about the structure and function of genes and DNA. In this section, we review some of the important ideas about nuclear genome organization in terms of DNA structure, genomic content, and gene function. We also will clarify the usage of several terms that may not be familiar to beginning students of fisheries genetics.

A number of subdisciplines in genetics and cell biology use the term gene as the functional unit of study. Introductory genetics textbooks often define a gene as the smallest portion of the genome that can transcribe a functional peptide (protein subunit). This "one gene, one peptide" concept is based primarily on the function of the gene product and became exceedingly important with the advent of protein detection methods discussed in Chapter 2. For example, in a simplified manner we often think of a gene for pigmentation or a specific enzyme. Multiple forms of a gene, or alleles, can be recognized at either the genotypic (or primary sequence) level or at the phenotypic (or gene product) level. The multiple allelic forms of some enzymes (allelic **isoenzymes** or **allozymes**) are due to differences in the encoding sequences within a gene. The location of a gene, or locus, is indirectly determined through laborious and time-consuming breeding studies. Prior to the advent of protein techniques, genetic mapping relied solely on the co-inheritance of many physical and behavioral traits. With the advent of molecular genetic techniques, the term gene has taken on a new meaning based on its structural identity. This sequence may or may not be transcribed or its specific location or function known. Clearly, the context in which the term gene is used is important and must be considered during investigations and when reporting genetic findings.

Genes do not randomly float within the nuclei of the cells of an organism but occur in a regular and organized manner. Nuclear DNA (nDNA) is organized into a hierarchy of structural architecture beginning with its primary structure and extending to its higher-order three-dimensional physical structure. **Primary structure** refers to the specific sequence of paired nucleotide bases (adenine [A], cytosine [C], guanine [G], and thymine [T]). Diversity at this level is due to **point mutations,** i.e., changes in composition or order of a single nucleotide caused by transcriptional errors usually made during the genomic replication process.

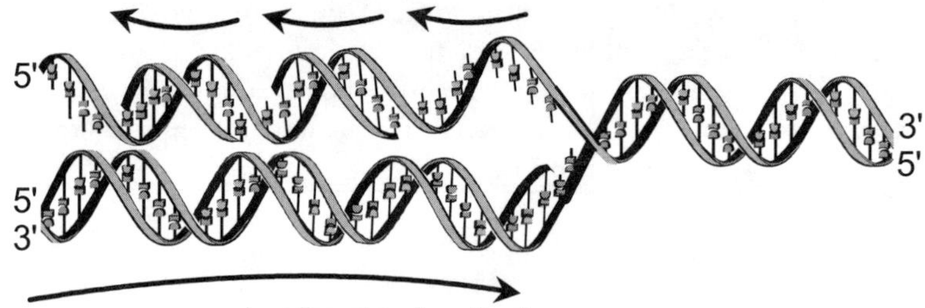

Figure 1.4 Continuous and discontinuous synthesis of DNA. Shown below, synthesis of DNA proceeds continuously in the 5′ to 3′ direction along the leading strand as the parental, or template, DNA is unwound. Shown above, synthesis occurs discontinuously in the 5′ to 3′ direction, the lagging strand replicated by creation of a series of short fragments. Each new fragment is synthesized near the point where the parental DNA double helix is being unwound. Later, the short fragments are joined together to create an intact DNA strand.

As outlined by Herskowitz (1967), higher-order DNA structure is often denoted as secondary, tertiary, and quaternary structure. **Second-order structure** includes the helical, or twisting, appearance of the molecule (Figure 1.4). The **double helix** is composed of two long strands of nucleotides spiraling or wrapping around each other. Nucleotides are linked side by side in a direction known as 5′ to 3′ along the length of a strand. The 5′ and 3′ refer to the numbered carbons of the pentose deoxyribose. The nucleotides in one strand also are bound to complementary nucleotides on the opposite DNA strand by hydrogen bonds. When uncoiled, double-stranded DNA looks like a ladder whose "rungs" are pairs of complementary nitrogenous bases (A pairs with T, and G pairs with C). New DNA can be synthesized from separated strands of an unwound helix (ladder) by enzymes known as **polymerases.** If a free 3′ end of DNA is available, a polymerase begins to add 5′ deoxynucleotides. As long as a single-stranded 3′ to 5′ DNA template and requisite deoxynucleotides are available, polymerases can help manufacture the complementary strand. Polymerases reading in a 3′ to 5′ direction also can manufacture complementary DNA to a 5′ to 3′ template but only in a discontinuous fashion and only with the assistance of several other enzymes. One of these enzymes, **DNA ligase,** frequently is used in genetic engineering procedures to chemically bond fragments of DNA from different sources. Both continuous and discontinuous synthesis are illustrated in Figure 1.4.

As you might expect, the cell's synthesis of complementary DNA is not perfect. Many polymerases have 3′ to 5′ **exonuclease** proofreading activity that allows them to recognize and excise mismatched nucleotides and then resume 5′ to 3′ synthesis. A related type of enzymes called **restriction endonucleases** are produced by bacteria and named for their ability to prevent viral infection by degrading invading viral nucleic acids. Each different restriction enzyme recognizes a specific nucleotide sequence on double-stranded DNA and breaks both strands within or near that sequence. For example, *Eco*R I is a restriction endonuclease from the bacterium *Escherichia coli* that recognizes the sequence GAATTC and cuts DNA each place that the sequence occurs. All of these enzymes, polymerases, ligases, exonucleases, and endonucleases are essential tools used by molecular geneticists to evaluate nDNA variation.

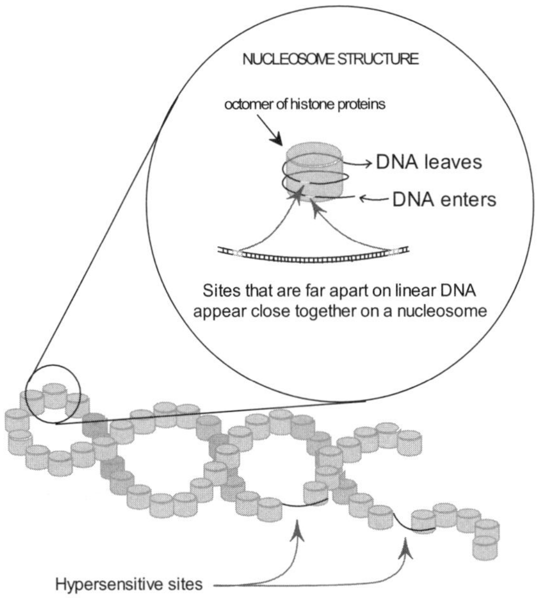

Figure 1.5 Chromosomes consist of thousands of nucleosomes. Each cylindrical nucleosome particle is composed of a histone octamer around which DNA stretches of approximately 150 base pairs wind up to two times. Strings of nucleosomes are further folded, forming chromatin. The net effect of this arrangement is that cleavage sites may be hidden or exposed depending both on their relative position around a nucleosome and on their absolute location along the DNA strand. Areas devoid of nucleosomes result in DNA stretches that are hypersensitive to nucleases and to certain chemicals. Hypersensitive sites are thought to be associated with replication, transcription, and gene regulation.

Third-order structure includes the three-dimensional conformation of nucleic acid chains such as folding and formation of intrastrand complementary base pairing. Higher-order structure includes the assembly of nDNA into chromosomes through the complex association of DNA with histones and other proteins into chromatin (Figure 1.5). Heat, salts, detergents, and the presence or absence of certain proteins are used to relax these higher-order structures, yielding linear nDNA that is receptive to molecular manipulation. Collection and accurate analysis of molecular data from nuclear genes depends on an understanding of the arrangement of DNA within chromosomes.

If one looks at the microscopic structure of condensed, chemically stained chromosomes during metaphase of the cell cycle, one sees alternating light and dark striations known as **bands,** as shown in Figure 1.6. Depending on the type of staining procedure, the bands of a chromosome indicate either the position of euchromatin, heterochromatin, centromeres, ribosomal DNA sequences, or dense arrangements of the DNA thread with proteins called **histones.** Histone proteins act as spools around which DNA wraps, resulting in structures known as **nucleosomes,** shown in Figure 1.5. Nucleosomes serve to compact DNA length by a factor of 10,000 as compared with the total length of a single nDNA thread. During **interphase**, the bulk of the cell cycle during which most of a chromosome is not condensed around histones (and does not display the conventional "X" shape), some of the DNA may remain highly coiled. These **heterochromatin** sections, located primarily in the centromeres and telomeres of chromosomes, are principally composed of repetitive DNA and are generally presumed to be genetically inactive.

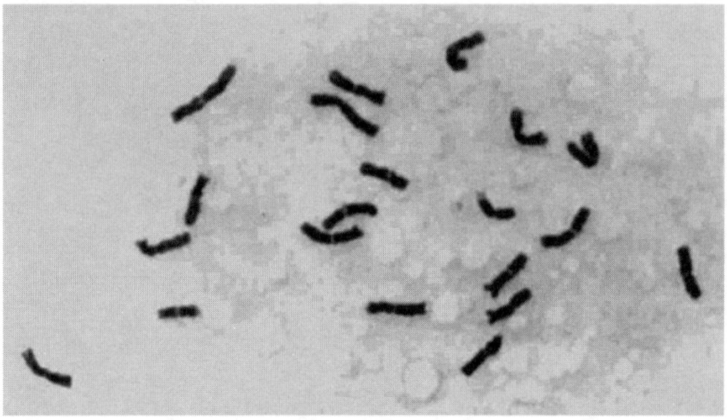

Figure 1.6 Diploid chromosome spread ($2n = 20$) of the Pacific oyster *Crassostrea gigas* prepared from two- to four-cell-stage embryos (Guo and Allen 1994). Early stage embryos yield condensed chromosomes, allowing observation of chromosome banding. Constitutive heterochromatin (C) bands and Giemsa-stained (G) bands of Pacific oyster chromosomes allow discrimination of chromosome pairs. Photograph provided courtesy of Stan Allen, Virginia Institute of Marine Science.

The packaging of DNA into distinct chromosomes is significant because it indicates that among the thousands of traits encoded by nuclear genes, many are located close together on a particular chromosome. Any complex of genes closely located together on a particular chromosome may exhibit physical, or classical, linkage. Alleles at genes that are linked often are transmitted as a unit from parent to offspring because they rarely if ever undergo crossing-over at meiosis. As discussed earlier (section 1.2.5), this association (linkage disequilibrium) must be accounted for during population-level analyses. Furthermore, linkage permits us to identify potential genetic markers (characters such as proteins or specific DNA sequences) that are closely associated with physical or behavioral traits of interest.

Substantial quantities of protein are associated with the DNA found in intact cells regardless of the cycle of cell division. The DNA–protein complexes inhibit the chemical treatments used to investigate DNA variation. Before genetically analyzing a specimen, we must first isolate the DNA from all other cellular components. When we "extract" DNA, the protein spools and other cellular materials are removed. The long threads and pieces of nDNA and the circular and linearized molecules of mitochondrial DNA remaining in aqueous solution can be used as the substrate for detecting DNA mutations at the molecular level.

1.3.2 Changes in DNA Structure

Mutation (Hallerman and Epifanio, Chapter 6) refers to a change in DNA sequence arising from substitution, insertion, deletion, or duplication of one or more nucleotides. These changes are the basis of all molecular genetic analyses in fishes and shellfishes. To a large extent, single nucleotide pair substitutions result from occasional mistakes that occur during the replication of nDNA. If a **purine** (adenine or guanine) is replaced by a different purine, or a **pyrimidine** (cytosine or thymine) is replaced by a different pyrimidine, the mutation is known as a **transition.** This is the most common type of mutation. A less common and energetically unfavorable type of substitution is a **transversion,** in which a purine is replaced by a pyrimidine, or vice versa. A phenomenon known as **slip-**

page during polymerization can lead to insertion or deletion of one or more base pairs. Because the genetic code is based on series of three nucleotides, if any number of nucleotides other than a multiple of three is affected, a **frame shift** will occur because all coding triplets downstream of the mutation will be altered. A final type of DNA mutation is **spontaneous lesion,** in which biochemically mediated changes in the condition of DNA (depurination and deamination) lead to pairing with normally noncomplementary bases. Despite the existence of repair enzymes mentioned earlier that correct such mistakes, redundancy in the genetic code and the fact that other base pairings can be stable (e.g., **tautomeric shifts** such as G-T) allow some mismatches to persist. Because of natural selection, mutations are more likely to persist in the noncoding, less functionally constrained segments of nDNA than in coding regions.

Variation in DNA structure can be observed in chromosomes by light microscopy when metaphase chromosome spreads are chemically stained (Chapter 3). Gross mutations at this level include substitutions, additions, insertions, deletions, inversions, or translocations of chromosomal sections. It is also possible for chromosomes to join via **Robertsonian fusion,** reducing the number of chromosomes without changing the total DNA content of a cell. There are basically two sources of nDNA variation that are pertinent to fish and shellfish genetic studies: chromosomal and mutational. The former category includes recombination, variation in nDNA content, and chromosome aberrations. Chromosomal rearrangements can be due to chance occurrence or may be due to exposure to toxic substances (termed mutagens) in the environment. Chapter 3 describes some of the significant events during fish evolution that appear to be related to alteration in DNA content. Sturgeon (Acipenseridae), armored catfishes (Callichthyidae and Loricariidae), suckers (Catastomidae), and salmonids (Salmonidae) are examples of entire families of fishes that evolved through either polyploidy or chromosomal rearrangements. **Chromosome aberrations** occur when chromosome segments are rearranged due to deletion, insertion, inversion, or duplication. The majority of deletions and insertions occur at repeat sequences (discussed in Chapters 3, 5).

1.4 References

Bondari, K. 1984. A study of abnormal characteristics of channel catfish and blue tilapia. Proceedings of the Southeastern Association of Fish and Wildlife Agencies 35(1981):547–553.

Bridges, W. R., and B. von Limbach. 1972. Inheritance of albinism in rainbow trout. Journal of Heredity 62:152–153.

Brown, B., P. Smouse, J. Epifanio, and C. Kobak. 1999. Mitochondrial DNA mixedstock analysis of American shad: coastal harvests are dynamic and variable. Transactions of the American Fisheries Society 128:977–994.

Brown, B., J. Epifanio, P. Smouse, and C. Kobak. 1996. Temporal stability of mtDNA haplotype frequencies in American shad stocks: to pool or not to pool across years? Canadian Journal of Fisheries and Aquatic Sciences 53:2274–2283.

Darwin, C., and A. R. Wallace. 1958. Evolution by natural selection. Cambridge University Press, New York.

Epifanio, J., P. Smouse, C. Kobak, and B. Brown. 1995. Measuring mitochondrial DNA divergence among populations of American shad *(Alosa sappidissima)*: how much variation is enough for mixed-stock analysis? Canadian Journal of Fisheries and Aquatic Sciences 52:1688–1702.

Falconer, D. S. 1989. Introduction to quantitative genetics, 2nd edition. Longman, Harlow, Essex, UK.

Grewe, P. M., C. C. Krueger, J. E. Marsden, C. F. Aquadro, and B. May. 1994. Hatchery origins of naturally produced lake trout fry captured in Lake Ontario: temporal and spatial variability based on allozyme and mitochondrial DNA data. Transactions of the American Fisheries Society 123:309–320.

Guo, X., and S. K. Allen. 1994. Viable tetraploids in the Pacific oyster (*Crassostrea gigas* Thunberg) produced by inhibiting polar body 1 in eggs from triploids. Molecular Marine Biology and Biotechnology 3:42–50.

Hardy, G. H. 1908. Mendelian proportions in a mixed population. Science 28:49–50.

Hartl, D. L. 1988. A primer of population genetics, 2nd edition. Sinauer Associates, Sunderland, Massachusetts.

Herskowitz, I. H. 1967. Basic principles of molecular genetics. Little, Brown and Co., Boston, Massachusetts.

Kincaid, H. L. 1975. Iridescent metallic blue color variant in rainbow trout. Journal of Heredity 66:100–102.

King, R. C., and W. D. Stansfield. 1990. A dictionary of genetics, 4th edition. Oxford University Press, New York.

Kirpichnikov, V. S. 1981. Genetic bases of fish selection. Springer-Verlag, New York.

Krueger, C. C., J. E. Marsden, H. L. Kincaid, and B. May. 1989. Genetic differentiation among lake trout strains stocked into Lake Ontario, Canada, USA. Transactions of the American Fisheries Society 118:317–330.

Lewontin, R. C. 1974. The genetic basis of evolutionary change. Columbia University Press, New York.

Matsui, Y. 1934. Genetical studies of goldfish in Japan. 2. On the Mendelian inheritance of telescope eyes of goldfish. Journal of the Imperial Fisheries Institute 30:37–46.

McAndrew, B. J., F. R. Roubal, R. J. Roberts, A. M. Bullock, and I. M. McEwen. 1988. The genetics and histology of red, blonde, and associated colour variants in *Oreochromis niloticus*. Genetica 76:127–137.

Moller, D. 1968. Genetic diversity in spawning cod along the Norwegian coast. Hereditas 60:1–32.

Pella, J. J., and G. B. Milner. 1987. Use of genetic marks in stock composition analysis. Pages 247–276 *in* N. Ryman and F. Utter, editors. Population genetics and fishery management. University of Washington Press, Seattle.

Poompuang, S., and E. M. Hallerman. 1997. Toward detection of quantitative trait loci and marker-assisted selection in fish. Reviews in Fisheries Science 5(3):253–277.

Smithies, O. 1955. Zone electrophoresis in starch gels: group variations in the serum proteins of normal human adults. Biochemical Journal 61:629–641.

Smouse, P. E., R. S. Waples, and J. A. Tworek. 1990. A genetic mixture analysis for use with incomplete source population data. Canadian Journal of Fisheries and Aquatic Sciences 47:620–634.

Tave, D., J. E. Bartels, and R. O. Smitherman. 1983. Saddleback: a dominant, lethal gene in *Sarotherodon aureus* (Steindachner) (= *Tilapia aurea*). Journal of Fish Diseases 6:59–73.

Waples, R. S., G. A. Winans, F. M. Utter, and C. Mahnken. 1990. Genetic approaches to the management of Pacific salmon. Fisheries 15:19–25.

Weinberg, W. 1908. Uber den Nachweis der Vererbung beim Menschen. Jahreshefte Verein, Naturk Wurtemberg 64:368–382.

Wohlfarth, G., M. Lahman, and R. Moav. 1963. Genetic improvement of carp. IV. Leather and line carp in fish ponds in Israel. Bamidgeh 15:3–8.

Yamamoto, T. 1969. Inheritance of albinism in the medaka, *Oryzias latipes*, with special attention to gene interaction. Genetics 62:797–809.

Measurement of Genetic Variation

2

Allozyme Variation

Bernie May

2.1 Introduction

Is this fish a hybrid between rainbow trout *Oncorhynchus mykiss* and cutthroat trout *O. clarki*? Do fish from these adjacent tributary streams interbreed with one another? Fisheries biologists have asked these types of questions for generations. Inherently, the basis of the answer to each question lies in the transmission of genetic information from parent to offspring, from one generation (or gene pool) to the next. Initially, only morphological or meristic characters were available (in addition to physical tagging studies) with which to answer these questions. The relationship between the morphological **phenotype** (how the organism appears to us) and the organism's **genotype** (the precise genetic information encoded by its DNA) is not easily discerned because of the major effects of environment in the expression of morphology. However, in the mid-1960s, a technique called allozyme electrophoresis (Harris 1966 and Lewontin and Hubby 1966) became available that increased the ability to infer genotype from an examination of phenotype (see Box 2.1). The ability to detect variation at single loci coding for allozymes revolutionized our ability to address the types of questions posed above. As with many new scientific advances, a fair amount of skepticism greeted the arrival of this new technique. However, after numerous inheritance studies demonstrated the direct relationship of allozyme phenotype (the banding pattern on a gel) to genotype, the general fisheries scientific community came to regard allozyme electrophoresis as a valuable technique for research and management.

Allozyme electrophoresis is an analytical technique that examines variation among that class of proteins that controls much of cell metabolism—bidirectional catalytic **enzymes,** which are molecules that convert one substrate to another without themselves being affected. The enzymes with which we are most familiar are those of the glycolytic pathway **(glycolysis),** in which glucose is broken down and the **Kreb's (or citric acid) cycle,** the pathway for energy generation in mitochondria. For example, the citric acid cycle enzyme malate dehydrogenase (MDH, enzyme number 1.1.1.37; IUBMBNC 1992) interconverts malate and nicotinamide adenine dinucleotide (NAD) to oxaloacetate and NADH, depending on in which direction the reaction is going.

■ Box 2.1
Phenotypes versus Genotypes

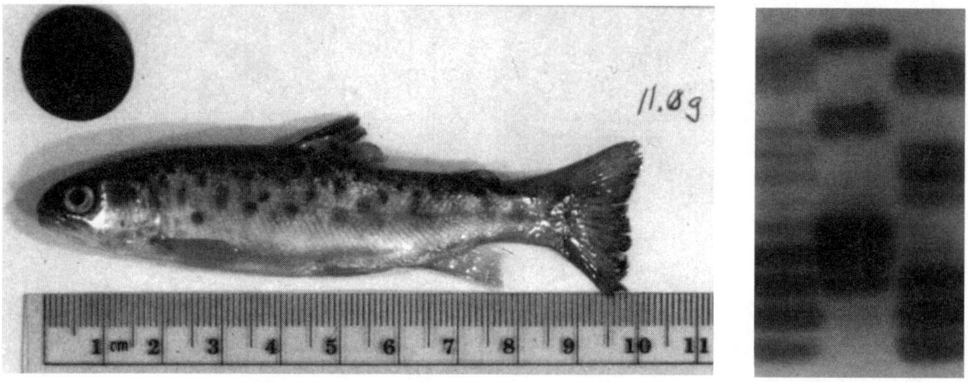

A B C

The left photograph is of a tiger trout (in this case a triploid hybrid with two copies of the female brown trout *Salmo trutta* genome and one copy of the male brook trout *Salvelinus fontinalis* genome); the right photograph is of the allozyme banding pattern for glucose-6-phosphate isomerase (GPI, 5.3.1.9) activity in this individual **(A)** and its parents, **(B)** paternal and **(C)** maternal. We recognize the body of a fish as a phenotype, a display of the genotype of the individual that has been affected to some degree by the environment to which this genotype has been exposed, from the environment of the egg to the current environment in which the fish resides. Likewise, an allozyme banding pattern (like the electropherogram on the right) is also just a phenotype. Some phenotypes are more influenced by the environment than are others. The power of allozyme and most DNA data are the direct relationship of phenotype to genotype, which is not true for morphological and meristic characters (e.g., length, weight, body condition, or number of fin rays) that can be dramatically affected by the environment. A difference in food abundance between two streams will affect the weight of the young of year in each stream, but the allozyme banding patterns will be unaffected.

The term **isozymes** refers to all of the different forms of an enzyme that perform the same catalytic function (e.g., all of the forms of MDH). These include forms encoded at the same locus as well as those encoded at multiple loci. Enzymes (isozymes) coded by multiple loci are often differentially expressed among different tissue types (e.g., of the five lactate dehydrogenase loci [*LDH**, 1.1.1.27] in salmonids, only *LDH-B2** is expressed in liver cells). The term **allozymes,** on the other hand, refers to only those genetically different forms of an enzyme that are encoded at the same locus. The majority of our interest is in allozymes because by studying allozyme variation we gain data on single-locus genetic variation. These single-locus genetic data permit us to answer many basic biological questions about fish and fish populations.

In the following sections of this chapter we will examine the general methods used to detect allozyme variation, the genetic basis of allozyme variation, interpretation of banding patterns on gels, and, finally, example applications of the data.

2.2 Methods of Isozyme Electrophoresis

The general method for detecting most isozymes includes the following steps (see Figure 2.1):

1. **extraction**—cell membranes are broken to release soluble enzymes into solution,
2. **electrophoresis**—the extract containing enzymes is inserted into a porous matrix (e.g., a gelatin-like block of hydrolyzed potato starch) and a voltage gradient is applied across the matrix to cause the charged molecules to migrate, and
3. **detection**—the enzymes are revealed by tying the specific chemical reaction of a given enzyme to a colorimetric or fluorescent detection system.

While I will discuss the general procedures used in my laboratory, more detailed information can be found for starch gels (Morizot and Schmidt 1990; Murphy et al. 1996; May 1998, and references therein), cellulose acetate (Hebert and Beaton 1989), and acrylamide (Hames 1990).

2.2.1 *Extraction*

Unlike DNA, which can be extracted from museum scale, fin, skin, or feather samples (or even many million-year-old samples in amber), isozymes are very temperature labile. Maximal enzymatic activity (and resolution) may be maintained for several months to years by holding tissue samples at ultracold temperatures (<–70°C in dry ice, liquid nitrogen, or ultracold freezers). Activity of some enzymes (e.g., LDH) can be maintained in a regular freezer (–20°C), but for most enzymes maximal enzymatic activity can be maintained only below –70°C.

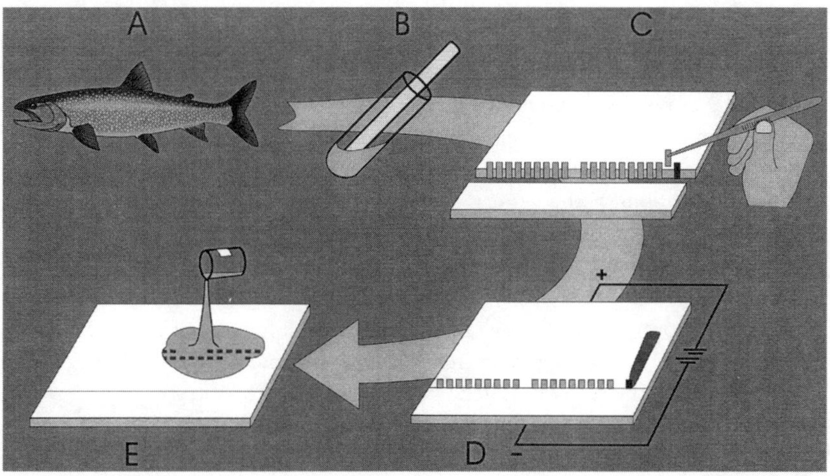

Figure 2.1 Starch gel electrophoresis of proteins. The five basics steps in allozyme electrophoresis are portrayed. **(A)** Tissue sample is taken from an organism. **(B)** The sample is homogenized in a dilute buffer solution thereby releasing the soluble proteins from the cells. **(C)** The supernatant containing the proteins is applied to the starch gel with a heavy weight filter paper wick. **(D)** An electric current causes proteins to migrate through the gel at varying rates depending on their size, shape, and, most importantly, charge. A wick containing dye tracks the migration of a known molecule. **(E)** A stain specific to the biochemical reaction catalyzed by a particular enzyme is poured over a slice of the gel, revealing bands of enzymatic activity.

Extraction of enzymes from the tissue of an individual is as simple as placing a pencil-eraser-sized piece of tissue in a small test tube, adding a lightly buffered solution to maintain enzyme stability (distilled water is often sufficient), and crushing the sample by hand with a solid glass rod. The rod breaks the cell membranes, releasing the soluble proteins into solution. The sample is centrifuged at low speed to separate the cellular debris from the protein-containing supernatant. Tissues, such as muscle, eye, liver, and kidney, of fish are often subjected to electrophoresis separately because of tissue specificity of expression of some multilocus enzymes. Different isozymes from different loci may co-migrate, causing scoring difficulties.

2.2.2 Electrophoresis

Electrophoresis of allozymes takes place in a matrix to separate the allozymes differentially, to preserve the separation after the electric field has been removed, and to serve as a base for subsequent visualization of the allozymes. The matrix for electrophoresis can be provided by a variety of media such as acrylamide, cellulose acetate, and hydrolyzed potato starch. Acrylamide typically has the best resolving power, especially when electrolytes are incorporated to permit separation based on **isoelectric point** (the pH at which the positive and negative charges on a protein are equal and the molecule no longer moves in the electric field); however, acrylamide is the most difficult to handle and is toxic. Cellulose acetate is the easiest to use; the cards or gels come pre-prepared and less extract and staining chemical are needed than with other media. However, cellulose acetate generally permits only a single enzyme to be stained per gel and typically only 12 to 20 individuals can be analyzed on a single gel. Potato starch has been the predominant medium of choice in applications requiring large amounts of data. Starch gels are easy to prepare and use, gels can be sliced for the staining of four to six different enzymes, and 40 to 50 individuals can be analyzed per gel. The explanations that follow are for horizontal starch gel electrophoresis, although both starch and acrylamide can be oriented vertically as well.

The samples are applied to the block of starch (the gel) by dipping thick filter paper wicks into the extracted supernatant that contains the soluble enzymes and applying the loaded wick to a cut in the gel (see Figure 2.1). Additionally, a wick is loaded with a dye marker to monitor the general movement of enzymes in the gel (the proteins are colorless with the exception of hemoglobin). Current is applied to the ends of the gel by means of thin sponges or cloths that connect the edge of the gel to a buffer tray that contains platinum electrodes connected to a DC power source. The current is applied, and the movement of the dye marker is followed for several hours. The gel must be covered with a coolant (e.g., blue-ice packs) or run in a cold chamber to prevent denaturation of the proteins by heat generated during electrophoresis.

At the conclusion of the run, the apparatus is disassembled, and the gel is cut into replicate gel slices (five slices for a 0.6-cm-thick gel) by sequentially placing 0.16-cm-thick Plexiglas™ strips along each side of the gel and drawing a taut 2–4-lb-test monofilament line through the gel. Each replicate slice is then placed in a separate tray and stained for a different enzyme.

2.2.3 Detection

A variety of methods are available for detecting specific enzymes, and coincidentally their isozymes, after electrophoresis. These methods include chromogenic reactions (soluble, often colorless component is converted into a colored product), fluorogenic reactions (fluorescence of substrate or product), autoradiography (labeling with radioisotopes), bioautography (detection with auxotrophic bacteria), two-dimensional gel spectroscopy (spectrophotometric detection of specific enzymes following two-dimensional electrophoresis), and immunoblotting (transfer of enzymes to a membrane and subsequent detection with specific antibodies) (see excellent review of enzymatic staining methods by Manchenko 1994). The most common method has been chromogenic, coupling a specific enzymatic reaction to the reduction of tetrazolium salts. Staining for a given enzyme is accomplished by adding only those components (substrates and cofactors) that are necessary for the enzyme in question to function. Dyes are added that can combine with one of the products of the enzymatic reaction and produce an insoluble product on the gel. The enzyme can repeat the reaction multiple times until a band is produced where each of the forms of the enzyme are found in the gel. For example, to stain for MDH one has to add only the salt malate and the cofactor NAD. Molecules of MDH will convert malate to oxaloacetate and NAD to NADH. The transient H^+ of NADH can be picked up by the intermediary phenazine methosulfate and transferred to thiazolyl blue or nitro-blue tetrazolium. Either of these latter two compounds will change upon reduction from a soluble yellow salt to an insoluble purple dye.

2.3 Genetic Basis of Isozyme Banding Pattern

The power of the isozyme methodology is the direct correlation of electrophoretic phenotype and the genotype of the individual at the Mendelian loci encoding the respective enzymes (see Box 2.2). Understanding the genetic basis behind the banding pattern observed on a gel permits us to assign a genotype to each individual for each enzyme system that we examine. The genetic basis for the variability observed for a single-enzyme banding pattern is as follows. The DNA sequence encoding an enzyme is first transcribed into an mRNA copy that is exported out of the nucleus. This mRNA copy is then translated into an amino acid sequence. Remember that triplets of the four nucleotides—guanine (G), thymine (T, which is replaced by uracyl [U] in RNA), adenine (A), and cytosine (C)—encode for specific amino acids (e.g., the triplet CUA codes for the amino acid leucine). In many cases, more than one of these amino acid sequences (also termed polypeptide chains or protein subunits) must combine to form the active enzyme. Enzymes can be monomers (each amino acid sequence is enzymatically active), dimers (two sequences or subunits must come together), trimers, tetramers, and so on. Most metabolic enzymes are dimers (e.g., MDH), monomers (e.g., phophoglucomutase, PGM, 5.4.2.2), or tetramers (e.g., LDH).

Variability for a single-locus-encoded enzyme is introduced by mutations (see Chapter 6). Mutations in the coding region of the nucleotide sequence of a gene encoding a particular metabolic enzyme will lead to a change in the triplet code. For example, replacing the U in CUA with a C changes the triplet code from one for leucine (CUA) to one for proline (CCA). Not all mutations lead to changes detectable on an allozyme gel. The change in code may not lead to the substitution of one amino acid for another because of the redundancy of the code. For example, many third-position base changes do not change the amino acid for which the codon codes; such changes are known as **silent substitutions.** Further, the change of one amino acid to another of like charge (from a positively charged amino acid to another positively charged one) will typically not lead to detectable changes in band mobility. Typically, only those mutations that lead

■ Box 2.2
DNA to Protein

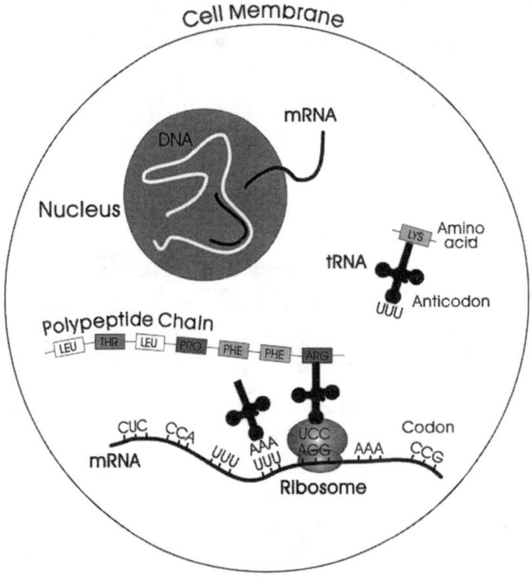

In the production of enzymatic proteins single genes are transcribed one gene at a time. A single stranded messenger RNA (mRNA) copy is transcribed from one strand of the double helix. This mRNA copy is exported out of the nucleus and subsequently translated into an amino acid sequence. Translation is mediated by ribosomes (protein and RNA complexes) in the presence of which amino acid specific transfer RNAs (tRNAs) bring in amino acids that correspond to the nucleotide sequence on the mRNA. Triplet nucleotides **(codons)** on the mRNA momentarily pair with their base pair equivalents **(anticodons)** on a specific tRNA. An amino acid sequence **(polypeptide chain)** is built as the specific tRNAs bring in their amino acid and the amino acid is bound by peptide bonds to the preceding amino acid. This nascent or growing chain coils up in a fashion determined by its sequence and enters a pool of these polypeptide chains. Multiple chains can be made from each mRNA molecule. These polypeptide chains are either immediately active (monomers) or combine into multimeric forms (e.g., dimers, trimers, and tetramers), depending on the specific enzyme (e.g., lactate dehydrogenase, LDH, is a tetramer in all vertebrates).

continued on next page

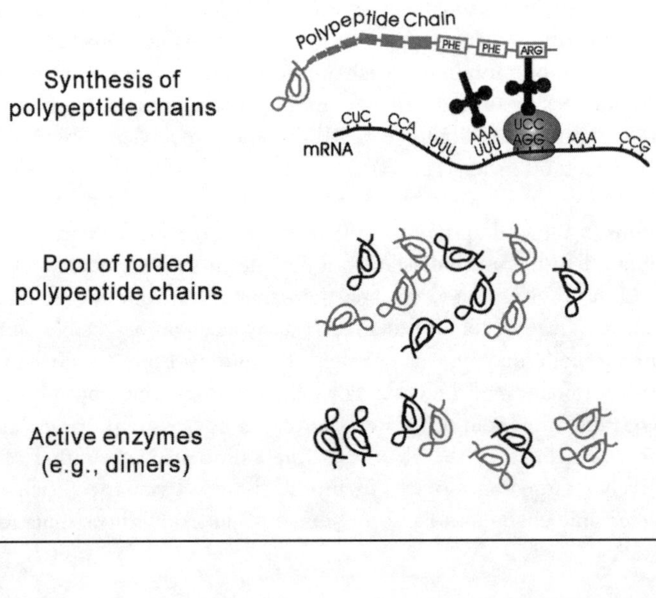

■ Box 2.2 *continued*

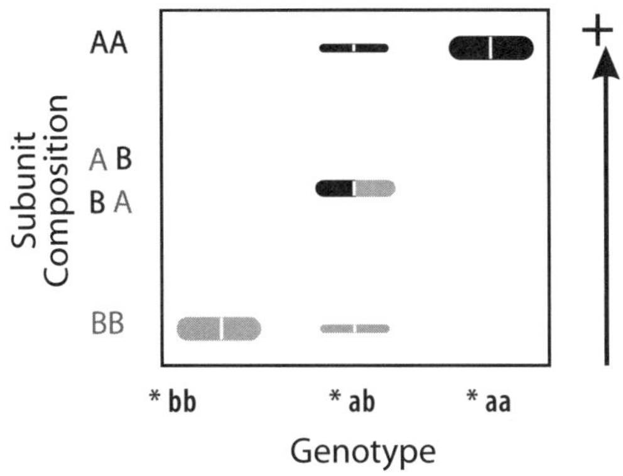

In the example given here, the enzyme is a dimer and the parental forms (maternal and paternal) differed in their DNA nucleotide sequence such that within this organism's cells two different forms of polypeptide chains are produced (light and dark). We can designate the alleles that code for these two polypeptide forms or protein subunits as **a* and **b*, and we can designate the dark amino acid subunit A and the light subunit as B. The two protein subunits can combine in four ways (AA, BB, AB, and BA). The first two are homodimers and the latter two are indistinguishable heterodimers. On a gel, this individual's phenotype would express three bands with the middle band twice as dark corresponding to the two heterodimers (AB and BA). We would designate this individual as genotype **ab* (a heterozygous individual). Phenotypes for homozygous individuals (genotypes **aa* and **bb*) would appear as single banded, with the bands in the same position as the two homodimers (AA and BB) in the heterozygous individual.

to the substitution of one amino acid for another of different charge will be detectable (i.e., about 30% of the changes). Finally, many mutations will not persist in the population because they lead to alterations in the amino acid sequence that consequently form an enzyme that is nonfunctional.

For most organisms there are two copies of each gene **(disomic),** one maternal and one paternal (two **homologous chromosomes** upon which the same sequence of genes reside). Historical gene duplication events have led to some metobolic enzymes being encoded by two or three different loci (e.g., MDH in vertebrates and invertebrates). In some cases, the evolution in the genes coding for these multiple loci have led to changes in the structure of the enzymes such that the products of one locus do not overlap with those of another on the stained gel.

The genomes of some fish taxa (e.g., salmonids [Salmonidae], sturgeons [Acipenseridae], and suckers [Catostomidae]) are the result of polyploidization events (total genomic duplication; see Chapter 3), and some of these species may possess even more single disomic loci that code for a given metabolic enzyme (Ferris and Whitt 1979). If a recent polyploidization event has occurred, there may be four gene copies at a single locus (a tetrasomic locus with four homologous chromosomes, two maternal and two paternal copies of each gene; Wright et al. 1983; Allendorf and Thorgaard 1984).

■ 2.4 Interpretation of Banding Patterns

Being able to look at a freshly stained gel for a particular enzyme and to interpret (read) a genetic basis for the observed variation is an exhilarating experience. The most important point in successfully reading gels is to be able to hold a mental picture of what a theoretical banding pattern would look like under the constraints of subunit composition, number of loci, and interaction of loci. We then must determine if what we actually see conforms to this perfect mental phenotype.

One starts to read a gel by assigning the observed phenotypes (banding patterns) to a discrete number of unambiguous phenotypic classes (see Box 2.3). Any ambiguity would suggest that some of the variation may not have a simple genetic basis but rather be due to **posttranslational modification** (the enzyme may have slightly different configurations as it folds following its formation) or to an inappropriate choice of gel buffer, which leads to poor resolution. A genotypic basis for the system is assigned based on the banding patterns of these discrete phenotypes. Likewise, each of these banding patterns for each of these classes must be predictable based on the assigned genotypes. One starts with the simplest explanation (one locus with two alleles) and works one's way up to more complexity (multiple alleles, multiple loci). The most common single-banded phenotype is classed as a homozygote and designated as genotype **aa*, having received one copy of allele **a* from each parent. (A systematic locus and allele nomenclature system is used in the literature that is based on relative mobilities of protein products [see Shaklee et al. 1990]). Other single-banded phenotypes are similarly classified as homozygous and given appropriate designations (e.g., **bb*). Multiple-banded individuals are classed as heterozygotes (**ab*), depending on with which homozygous individuals they share bands. Additional alleles are added to account for all observed phenotypes (e.g., **ac*, **bc*, and **cc*). It may be that not all possible phenotypes are observed because the frequencies of some alleles in the population and the number of individuals examined (sample size) lead to a low expectation of observing that phenotype. For example, the homozygous phenotypes for rare alleles often are not observed in small samples. Upon completion of reading a gel, one is left with a series of individual, locus-specific genotypes for groups of individuals. Although some applications of the data (e.g., mixed-stock fishery analysis [Chapter 13] and forensic [Chapter 17] or inheritance [Chapter 1] studies) make use of genotypic data for individuals, most applications use the results for groups of individuals from the same population. Allele frequencies at each locus can be calculated for a given population (see Box 2.3).

For further readings on enzymatic complications (artifact bands, multiple loci, and interactions of loci) see Utter et al. (1987), Buth (1990), and May (1998).

2.5 Applications

No other genetic tool has had a more profound effect on fisheries research and management than has allozyme analysis. Nearly every chapter in the practical applications section of this book (Chapters 13–18) will make reference to studies of allozyme variation. Below I simply introduce some of the varied types of possible applications.

A variety of applications of allozyme analysis are possible because allozymes are **codominant** Mendelian characters (both alleles are individually expressed in a heterozygous individual). Such characters (see other nuclear codominant characters such as polymerase-chain-reaction-amplified microsatellite loci or probes of single minisatellite loci, Chapter 5) are passed from parent to offspring in a predictable fashion (see section 2.3). Further, the loci encoding such characters and their respective alleles when combined across all members of a population constitute a **gene pool.** This gene pool is also passed from generation to generation in a predictable fashion. Barring mutation, selection, small

■ Box 2.3
Interpretation of an Allozyme Banding Pattern

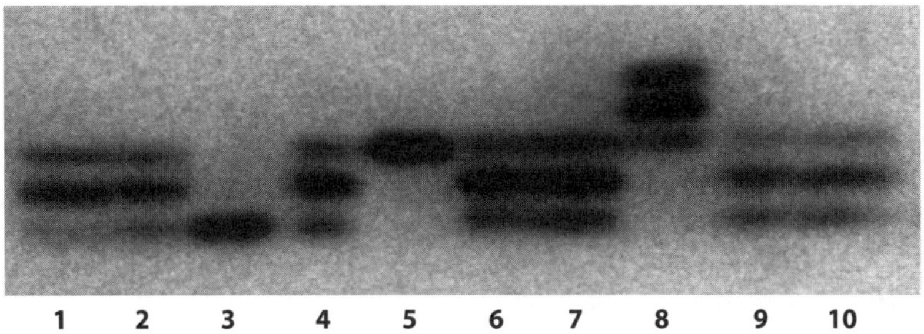

The displayed banding pattern was observed for glucose-6-phophate isomerase (GPI) activity in zebra mussel *Dreissena polymorpha*. Individual 3 is a homozygous ★*aa* and individual 1 is an ★*ab*. Assign genotypes for the other individuals. How many alleles must be segregating for this single-locus system? What additional phenotypes might be observable if more individuals were examined? What phenotypes (genotypes) would result from a cross of individuals 7 and 8?

From an allozyme banding pattern we can calculate allele frequencies. Assume these 10 individuals above come from a single population. What are the frequencies of the alleles in this sample? Remember each individual has two gene doses (one maternal and one paternal).

Answers

There must be three alleles segregating in this system. By convention, individual 8 must be a ★*bc*. We would also expect to see individuals ★*ac* (a wide three-banded pattern) and ★*cc* (an individual that has only the top band of individual 8). Crossing individuals 7 (★*ab*) and 8 (★*bc*) would lead to offspring ★*ab*, ★*ac*, ★*bb*, and ★*bc*, occurring in an approximately 1:1:1:1 ratio.

Each individual for a disomic locus has two gene doses, one maternal and one paternal. By counting the number of each allele possessed by each individual and dividing by two times the total number of individuals, we determine the frequency of the alleles in our collection, which is an estimate of the frequency in the population from which this collection came. Scoring across the gel from left to right, for the ★*a* allele we score one dose of the ★*a* allele in each of the first two individuals, two in the third (because it is homozygous), and so on for a total of nine ★*a* alleles among the 10 individuals. Dividing by 20 (total number of gene doses in the 10 individuals) gives us a frequency of 0.45 for the ★*a* allele. Likewise we count 10 ★*b* alleles and 1 ★*c* allele for frequencies of 0.50 and 0.05, respectively.

population sizes, assortative mating (nonrandom mating), and migration (successful matings due to immigrations or differential emigration), the gene pool (allele frequencies at all loci) will stay the same from generation to generation (Chapter 1).

2.5.1 Individual Identification

After early developmental stages, the allozyme phenotype does not change during an individual's lifetime, although there may be differential tissue expression of particular loci. Therefore, individual clones may be identified in clonally maintained populations (Monaco et al. 1984) depending on the level of allozyme variation in the population.

2.5.2 Inheritance and Mapping

Allozymes have proven to be informative nuclear genetic markers for monitoring the segregation (inheritance) of single genes. These loci can be used to monitor the gametic contribution of single parents, such as to document two contributions from the maternal parent in gynogenetic or triploid genome manipulation experiments. They can also be used to build linkage maps (May and Johnson 1993; Morizot et al. 1993) or to study evolution by genomic duplication exhibited by some fish families (e.g., Salmonidae, Wright et al. 1983 and Allendorf and Thorgaard 1984; Catostomidae, Ferris 1984).

2.5.3 Population Structure

The ability to calculate allele frequencies directly from stained allozyme gels (see Box 2.3) permits samples of putative reproductively isolated populations to be compared (see reviews in May and Krueger 1990; Shaklee and Bentzen 1998; Shaklee et al.1999; Grant et al. 1999; Chapter 13). Definitive answers can be obtained to such questions as, are these populations reproductively isolated; have hatchery spawning protocols led to losses of genetic variability; and how much variation exists within populations, among tributary populations, among river systems, and among major drainages? The genotype frequencies at each locus for each population can be tested for fit to expected Hardy–Weinberg proportions (Chapter 1). Hardy–Weinberg proportions is an expectation of how allelic frequencies should be exhibited based on the distribution of possible genotype frequencies in an idealized population where there is

—no differential selection among genotypes,
—no immigration or differential emigration,
—random mating among all genotypes,
—a large breeding population size, and
—no mutation.

Algorithms are available for determining whether the allele frequencies for two samples are significantly different (i.e., that the samples cannot be considered as being from a single population); partitioning the observed variation into within-individuals, within-populations, and among-populations components; and comparing which populations are the most similar. Additionally, there are computer programs available that use these algorithms (Weir 1996; http://www.mbg.cornell.edu/aquadro/software.html; http://www.cf.adfg.state.ak.us/geninfo/research/genetics/Software/anlink.htm).

2.5.4 *Population Marking*

A population of uniquely marked individuals or with a widely divergent suite of allele frequencies can be made readily (Seeb et al. 1986; Chapter 16). Members of a donor population are physically tagged and samples of muscle, fin, or blood are taken and typed for allozymes. Selected individuals are chosen based on allozyme type to be used as parents of a new, marked population. Alternatively, gametes can be taken and the fish sacrificed for tissues that cannot be sampled nondestructively (e.g., liver, heart, brain, or kidney). Subsequently, selected gametes can be combined to produce the marked families.

2.5.5 *Mixed-Stock Fishery Analysis*

One of the most practical of all applications of allozyme analysis has been the use of these data to measure the proportion of different stocks of salmonids harvested in a particular fishery **(mixed-stock analysis)**. Along the eastern Pacific rim (from Alaska to California), U.S., Canadian, and tribal agencies use estimates of component stocks to help regulate these mixed-stock fisheries, attempting to protect less abundant stocks (Shaklee and Phelps 1990, Chapter 13). Mixed-stock analysis has also been used to estimate the proportion of different strains contributing to the production of fry of natural lake trout *Salvelinus namaycush* (second-generation mixed-stock analysis; Marsden et al. 1989).

2.5.6 *Systematics*

As in population analysis (section 2.5.3), groups of species (usually within a genus) can be related to one another based on proportions of shared alleles (see Avise 1974 and Buth 1984). If two populations are reproductively isolated long enough, separate evolutionary histories of mutation, migration, selection, and genetic drift will cause the frequencies of alleles to diverge and ultimately to become fixed for alternate alleles at some allozyme loci (alternately homozygous). Such fixation is readily observed in mixed samples of two such reproductively isolated populations by the absence of heterozygous individuals at one or more loci (see also the Wahlund effect in Chapters 1, 7). In general, the more loci at which two populations are fixed for alternate alleles (often termed diagnostic loci), the greater is the length of evolutionary time that they have been reproductively isolated from one another (see Chapter 13).

2.5.7 *Hybrid Identification*

The fact that species will become alternately fixed at some allozyme-coding loci means that interspecific hybrids readily can be identified. First filial generation (F_1) hybrids will be heterozygous at all loci for which the two parental species are fixed. The F_2 and backcross individuals will be heterozygous or homozygous, depending on the nature of the cross. Such data can also be used to measure levels of introgression in hybridized populations (see Forbes and Allendorf 1991)

2.5.8 *Forensics*

Allozyme phenotypes make excellent forensic markers in fisheries law enforcement. The most common crime is fraudulently labeling one species as another, either for greater profit or to include a prohibited species (Chapter 17). Because fish and shellfish are either sold fresh or frozen, the enzymes are preserved and can be tested in the laboratory for species identity even though the shell or skin has been removed (Keenan and Shaklee 1985; Harvey 1990).

2.5.9 *Conservation Genetics*

Genetic information from the specific applications above can be combined with demographic data to address the conservation of threatened and endangered fish species (see Allendorf and Waples 1996; Nielsen 1996). Debates over the conservation of fish species will be one of the most politically active areas for fisheries biologists for the foreseeable future. Genetic data will form the foundation for much of the discussion.

■ 2.6 Summary

In this chapter we have discussed production of allozyme patterns from the tissue of the organism to the banding pattern on the gel; the genetic basis of allozyme variation; interpretation of banding patterns; analysis of the resultant data; and the varied applications of the data. Chapter 5 will introduce newer molecular procedures that examine nuclear DNA. These newer procedures overcome several of the following limitations of allozymes: only structural DNA is involved, representing less than 1% of the nuclear genome; each electromorph may involve multiple alleles, some of which may be more closely related to alleles in different electromorph classes; and average heterozygosity for polymorphic loci rarely exceeds 0.1. Although the data produced by these newer procedures offer several advantages over allozyme data, we nonetheless will continue to use allozyme data in fisheries research and management (see Ryman and Utter 1987; Whitmore 1990; Utter 1991). Allozymes remain a viable tool for analyzing genetic variation because they are coded by codominant alleles at single Mendelian loci, the procedures are relatively easy and inexpensive, large quantities of data can be produced quickly, and for many species we already have large baseline datasets.

■ 2.7 Acknowledgments

I am grateful to E. Hallerman, D. Perkins, A. Spidle, and two anonymous reviewers for critical reviews of earlier drafts of this manuscript. I, of course, owe a deep debt of thanks to F. Allendorf and F. Utter for sharing the magic with me so long ago.

■ 2.8 References

Allendorf, F. W., and G. H. Thorgaard. 1984. Polyploidy and the evolution of salmonid fishes. Pages 1–53 *in* B. J. Turner, editor. Evolutionary genetics of fishes. Plenum, New York.

Allendorf, F. W., and R. S. Waples. 1996. Conservation and genetics of salmonid fishes. Pages 238–280 *in* J. C. Avise and R. S. Waples, editors. Conservation genetics: case histories from nature. Chapman and Hall, New York.

Avise, J. C. 1974. Systematic value of electrophoretic data. Systematic Zoology 23:465–481.

Buth, D. G. 1984. The application of electrophoretic data in systematic studies. Annual Review of Ecology and Systematics 15:501–522.

Buth, D. G. 1990. Genetic principles and the interpretation of electrophoretic data. Pages 1–22 *in* D. H. Whitmore (1990).

Ferris, S. D. 1984. Tetraploidy and the evolution of the catostomid fishes. Pages 55–93 *in* B. J. Turner, editor. Evolutionary genetics of fishes. Plenum, New York.

Ferris, S. D., and G. S. Whitt. 1979. Evolution of the differential regulation of duplicate genes after polyploidization. Journal of Molecular Evolution 12:267–317.

Forbes, S. H., and F. W. Allendorf. 1991. Associations between mitochondrial and nuclear genotypes in cutthroat trout hybrid swarms. Evolution 45:1332–1349.

Grant, W. S., J. L. Garcia-Marin, and F. M. Utter. 1999. Defining population boundaries for fishery management. Pages 27–72 *in* S. Mustafa, editor. Genetics in sustainable fisheries management. Blackwell Scientific Publications, Oxford, UK.

Hames, B. D. 1990. One-dimensional polyacrylamide gel electrophoresis. Pages 1–147 *in* B. D. Hames and D. Rickwood, editors. Gel electrophoresis of proteins: a practical approach. IRL Press, Oxford, UK.

Harris, H. 1966. Enzyme polymorphisms in man. Proceedings of the Royal Society of London B 164:298–310.

Harvey, W. D. 1990. Electrophoretic techniques in forensics and law enforcement. Pages 157–172 *in* D. H. Whitmore (1990).

Hebert, P. D. N., and M. J. Beaton. 1989. Methodologies for allozyme analysis using cellulose acetate electrophoresis: a practical handbook. Helena Laboratories, Beaumont, Texas.

IUBMBNC (International Union of Biochemistry and Molecular Biology, Nomenclature Committee). 1992. Enzyme nomenclature, 1992. Academic Press, San Diego, California.

Keenan, C. P., and J. B. Shaklee. 1985. Electrophoretic identification of raw and cooked fish fillets and other marine products. Food Technology in Australia 37:117–128.

Lewontin, R. C., and J. L. Hubby. 1966. A molecular approach to the study of genetic heterozygosity in natural populations II: amount of variation and degrees of heterozygosity in natural populations of *Drosophila pseudoobscura*. Genetics. 54:595–609.

Manchenko, G. P. 1994. Handbook of detection of enzymes on electrophoretic gels. CRC Press, Boca Raton, Florida.

Marsden, J. E., C. C. Krueger, and B. May. 1989. Identification of parental origins of naturally produced lake trout in Lake Ontario: application of mixed-stock analysis to a second generation. North American Journal of Fisheries Management 9:257–268.

May, B. 1998. Starch gel electrophoresis of allozymes. Pages 1–28 and 371–378 *in* A. R. Hoelzel, editor. Molecular genetic analysis of populations: a practical approach, 2nd edition. Oxford University Press, New York.

May, B., and K. R. Johnson. 1993. Composite linkage map of salmonid fishes (*Salvelinus, Salmo,* and *Oncorhynchus*). Pages 151–159 *in* S. J. O'Brien, editor. Genetic maps 6. Volume 4, Nonhuman vertebrates. Cold Spring Harbor Laboratory Press, Cold Spring Harbor, New York.

May, B. P., and C. C. Krueger. 1990. Use of allozyme data for population analysis. Pages 157–172 *in* D. H. Whitmore (1990).

Monaco, P. J., E. M. Rasch, and J. S. Balsano. 1984. Apomictic reproduction in the Amazon molly, *Poecilia formosa,* and its triploid hybrids. Pages 311–328 *in* B. J. Turner, editor. Evolutionary genetics of fishes. Plenum, New York.

Morizot, D. C., J. Harless, R. S. Nairn, K. D. Kallman, and R. B. Walter. 1993. Linkage maps of non-salmonid fishes. Pages 151–159 *in* S. J. O'Brien, editor. Genetic maps 6. Volume 4, Nonhuman vertebrates. Cold Spring Harbor Laboratory Press, Cold Spring Harbor, New York.

Morizot, D. C., and M. E. Schmidt. 1990. Starch gel electrophoresis and histochemical visualization of proteins. Pages 23–80 *in* D. H. Whitmore (1990).

Murphy, R. W., J. W. Sites, D. G. Buth, and C. H. Haufler. 1996. Isozyme electrophoresis. Pages 51–120 *in* D. M. Hillis, C. Moritz, and B. K. Mable, editors. Molecular systematics, 2nd edition. Sinauer Associates, Sunderland, Massachusetts.

Nielsen, J. L. 1996. Molecular genetics and the conservation of salmonid biodiversity: *Oncorhynchus* at the edge of their range. Pages 383–398 *in* T. B. Smith and R. K. Wayne, editors. Molecular genetic approaches in conservation. Oxford University Press, New York.

Ryman, N., and F. Utter, editors. 1987. Population genetics and fishery management. University of Washington Press, Seattle and London.

Seeb, J. E., L. W. Seeb, and F. M. Utter. 1986. Use of genetic marks to assess stock dynamics and management programs for chum salmon. Transactions of the American Fisheries Society 115:448–454.

Shaklee, J. B., F. W. Allendorf, D. C. Morizot, and G. S. Whitt. 1990. Gene nomenclature for protein-coding loci in fish. Transactions of the American Fisheries Society 119:2–15.

Shaklee, J. B., T. D. Beacham, L. Seeb, and B. A. White. 1999. Managing fisheries using genetic data: case studies from four species of Pacific salmon. Fisheries Research 43:45–78.

Shaklee, J. B., and P. Bentzen. 1998. Genetic identification of stocks of marine fish and shellfish. Bulletin of Marine Science 62:589–621.

Shaklee, J. B., and S. R. Phelps. 1990. Operation of a large-scale, multiagency genetic stock identification program. Pages 817–830 *in* N. Parker, A. E. Giorgi, R. C. Heidinger, D. B. Jester, Jr., E. D. Prince, and G. A. Winans, editors. Fish-marking techniques. American Fisheries Society Symposium 7, Bethesda, Maryland.

Utter, F., P. Aebersold, and G. Winans. 1987. Pages 21–46 *in* N. Ryman and F. Utter, editors. Population genetics and fishery management. University of Washington Press, Seattle and London.

Utter, F. M. 1991. Biochemical genetics and fishery management: an historical perspective. Journal of Fish Biology 39 (Supplement A):1–20.

Weir, B. S. 1996. Genetic data analysis II: methods for discrete population genetic data. Sinauer Associates, Sunderland, Massachusetts.

Whitmore, D. H., editor. 1990. Electrophoretic and isoelectric focusing techniques in fisheries management. CRC Press, Boca Raton, Florida.

Wright, J. W., Jr., K. Johnson, A. Hollister, and B. May. 1983. Meiotic models to explain classical linkage, pseudolinkage, and chromosome pairing in tetraploid derivative salmonid genomes. Genetics and Evolution 10:239–260.

3

Chromosomal Variation

Ruth B. Phillips and Kent M. Reed

3.1 Introduction

3.1.1 *Structure of Chromosomes*

The genetic material of higher organisms is distributed among multiple chromosomes that are complexes of DNA, histone, and nonhistone chromosomal proteins. One linear, double-stranded DNA molecule runs throughout the length of each chromosome, and this DNA is wound onto histone complexes to form the basic **chromatin** fiber of 11 nm in width (Figure 1.5). This fiber is refolded many times to form the highly condensed structure of approximately 700 nm in width seen as the arms of the typical metaphase chromosome. Chromosomes of higher organisms are usually studied at mitotic metaphase because this is the stage at which they are most highly condensed. Also at this stage, the chromosomes are duplicated, so that two arms, known as sister **chromatids**, are clearly visible (Figure 3.1).

A typical metaphase chromosome seen with the aid of the light microscope consists of two chromatids attached at a primary constriction, or **centromere.** The centromere is of primary importance during cell division (mitosis and meiosis) because it is here that the spindle fibers attach for accurate segregation of the replicated chromosome to the daughter cells. The centromere is flanked by repetitive sequences that are typically species specific, chromosome specific, or both.

Secondary constrictions are often visible as unstained gaps along some of the chromosome arms. These usually are the sites of multiple copies of the ribosomal RNA (rDNA) cistron and form a nucleolar organizer region (NOR). Secondary constrictions are often found near chromosome ends where they may separate a small segment called a **satellite** from the main portion of the chromosome (Figure 3.1).

The morphology of a chromosome is determined by the placement of the centromere, which can be located anywhere along the length of the chromosome. Centromere position is usually assigned to one of four general classes. In a **metacentric** chromosome the centromere is approximately at the center of the chromosome so that the chromosome arms are similar in length. **Submetacentric** chromosomes have the centromere displaced toward one end, so the arms are unequal in size (termed long and

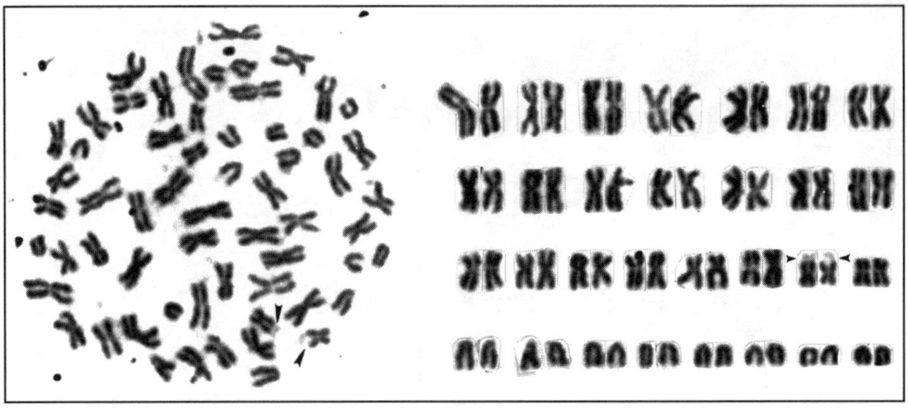

Figure 3.1 Giemsa-stained metaphase spread of rainbow trout *Oncorhynchus mykiss* chromosomes and karyotype showing metacentric, submetacentric, and acrocentric (or telocentric) chromosomes. The satellite regions on the chromosomes exhibiting nucleolar organizer regions (NOR) are indicated by arrowheads.

short arms). Chromosomes with very small short arms are called **acrocentric** or **subtelocentric,** and a **telocentric** chromosome has only one visible arm with the centromere at the end of the chromosome (Figure 3.1). Because most chromosomes probably have a small short arm even if it is difficult to see, some authors include telocentrics and subtelocentrics in the same group and identify them as uniarmed chromosomes, or acrocentrics. **Sex chromosomes (gonosomes)** are chromosomes that differ between males and females. Chromosomes other than sex chromosomes are called **autosomes.**

The regions of DNA at the ends of chromosomes are called **telomeres.** At the molecular level these regions are complex and often contain copies of tandemly repetitive DNA sequences. In addition to these subtelomeric repeats, the extreme ends of vertebrate chromosomes contain the common hexanucleotide repeat (thymine [T]-T-adenine [A]-guanine [G]-G-G)$_n$. Multiple copies of this repetitive sequence are required for chromosome stability and for complete replication of the chromosomal DNA.

The description of the complete set of chromosomes of an organism is called its **karyotype,** and it is species specific (Figure 3.1). In higher organisms the chromosomes in each somatic cell occur in pairs giving each species a characteristic **diploid** chromosome number, which is indicated by $2n$. The gametes contain one member of each pair, or one haploid set, and the **haploid** chromosome number is indicated as n. The chromosomes in the karyotype may differ from one another with respect to size, morphology, and banding pattern. They are grouped in homologous pairs and numbered from the largest to the smallest. A common descriptor of the karyotype is the total number of chromosome arms, the FN (fundamental number) or NF (numbre fundamental). Methods for calculating FN often vary between researchers. Some count all visible chromosomes arms whereas others do not count the minute arms of small submetacentrics or subtelocentric chromosomes.

When chromosomes condense, some regions are more tightly compacted than are others, giving rise to alternation of light and dark bands along the chromosomes. These may be visible in meiotic chromosomes without special staining techniques, but in mitotic chromosomes they must be enhanced by application of banding techniques in order to be seen (see section 3.2.2). Some regions of the chromosome remain highly condensed and darkly stained throughout the cell cycle. They are called **heterochro-**

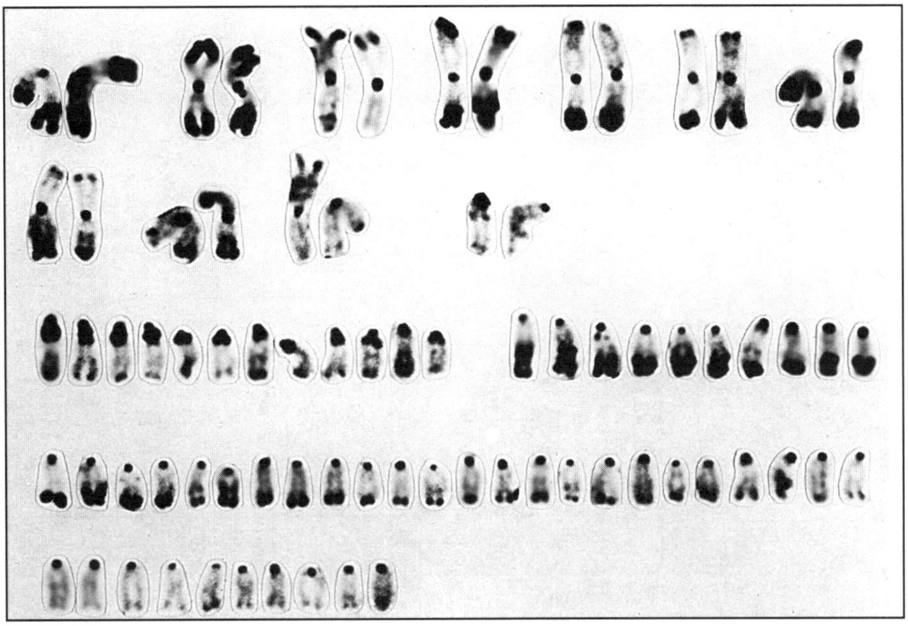

Figure 3.2 Karyotype of Arctic char *Salvelinus alpinus* displaying heterochromatin visualized by C banding. Photograph provided courtesy of Sheila Hartley.

matin, in contrast to the majority of chromosomal material, which is called **euchromatin.** The heterochromatin corresponds to highly repetitive DNA that does not usually code for proteins and is revealed by specific staining techniques (Figure 3.2).

3.1.2 *Chromosome Variation*

Karyotypes of individuals of closely related species or even individuals within a species may vary with respect to chromosome number, structure, or banding pattern.

3.1.2.1 **Chromosome Number.** Variation in the number of chromosomes in a cell may be due to **aneuploidy, monoploidy,** or **polyploidy.** Aneuploid cells have one, two, or several chromosomes more or less than the normal chromosome number for the species (e.g., $2n - 1$ is monosomy and $2n + 1$ is trisomy). Monoploid cells have only one set of chromosomes (n = the haploid number), and polyploid cells have additional sets of chromosomes (e.g., $3n$ is triploidy and $4n$ is tetraploidy). Aneuploidy is often the result of nondisjunction of particular chromosome pairs, whereas polyploidy can result if unreduced gametes fuse or if cytokinesis is blocked.

Another way in which cells may differ in chromosome number is by the presence of B chromosomes. The **B chromosomes** are "extra," or **supernumerary,** chromosomes that are not duplicates of the normal chromosomes as in aneuploidy. These chromosomes are usually heterochromatic and sometimes are so small that they can be classified as **microchromosomes.** Many are derived from hybridization events and often display irregular, non-Mendelian inheritance.

3.1.2.2 **Chromosome Structure.** Structural chromosomal changes may involve single chromosomes (intrachromosomal rearrangements) or interactions between chromosomes (inter-

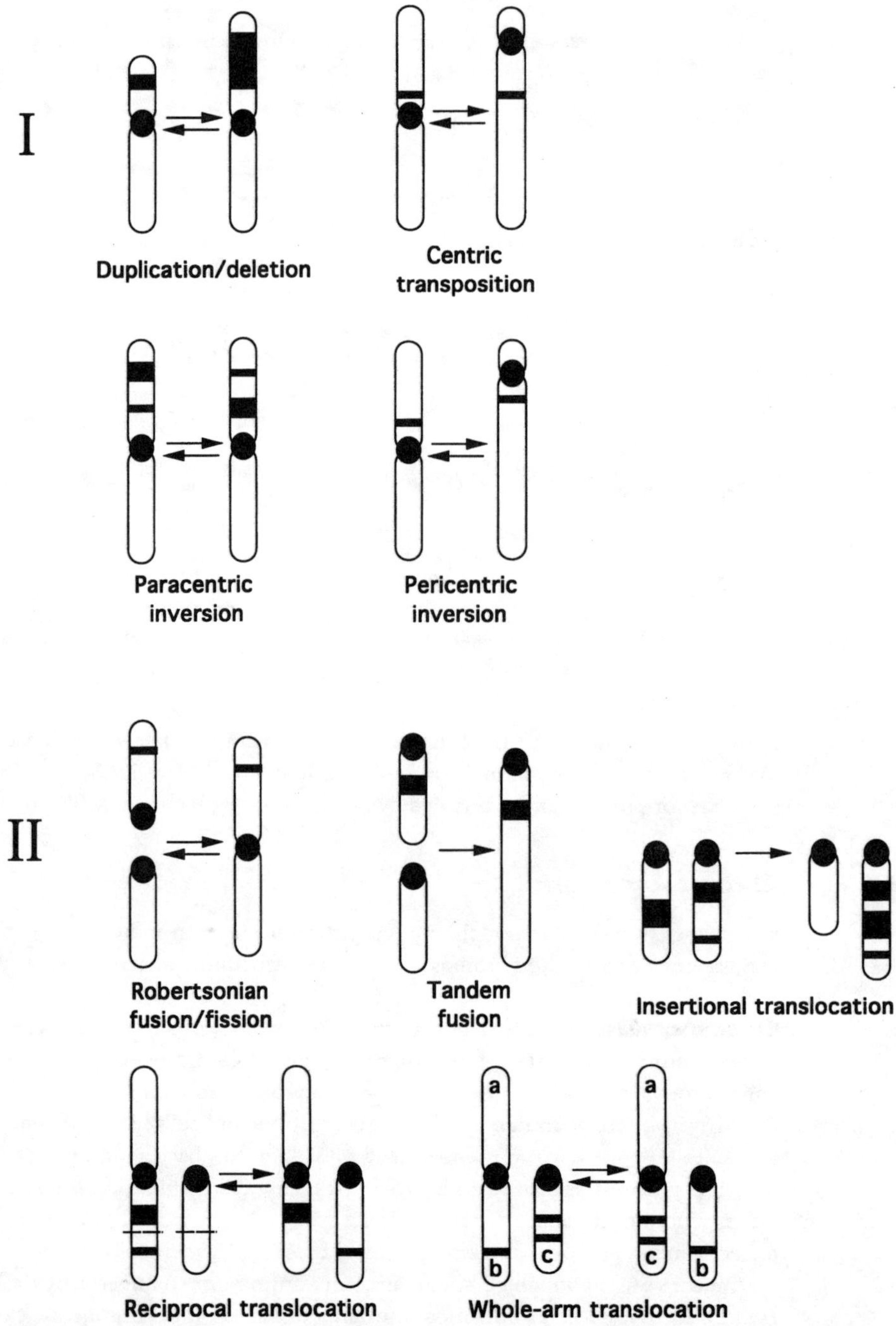

Figure 3.3 Idiograms of chromosome rearrangements. Group I includes intrachromosomal rearrangements, and group II includes interchromosomal rearrangements.

chromosomal rearrangements) (Figure 3.3). Intrachromosomal rearrangements include duplications, deletions, inversions, and centric transpositions. There are two major types of interchromosomal rearrangement: fusion–fissions (Robertsonian and tandem) and translocations.

Deletions (loss of a chromosomal segment) and **duplications** (doubling of a chromosomal segment) result in a change in chromosome length. Those involving euchromatin tend to be lethal unless only small segments are involved. When heterochromatin is involved, these types of chromosome change are typically selectively neutral although they may result in alteration of recombination patterns.

In the case of **inversions,** a chromosomal segment is excised, rotated 180°, and reinserted into the same chromosome. If the inverted segment does not include the centromere, it is a **paracentric inversion;** if it includes the centromere it is a **pericentric inversion.** The fitness effects of heterozygosity for inversions is dependent on the meiotic pairing configurations and frequency of crossing-over in heterozygotes. If meiotic pairing is exact, reverse (inversion) loops are formed. Crossing-over within homologously paired reverse loops can result in the production of chromatin bridges and acentric–dicentric fragments (paracentric inversions) or duplication–deletion chromatids (pericentric inversions). Meiotic data from several vertebrate taxa indicate that reduction in fertility for inversion heterozygotes cannot automatically be assumed. For example, in humans small inversions tend to have little or no measurable effects, whereas large inversions typically cause partial infertility in heterozygotes. As with heterochromatic additions, negative consequences may result not from meiotic pairing irregularities but from shifts in recombination. In natural populations, inversions result in the absence of recombinant progeny for the inverted portion of the chromosome. Thus inversions may fix coadapted gene complexes in the genome.

Centric transpositions represent a shift in the position of the centromere. These multiple-break rearrangements are likely to be rare and are probably the result of sequential rearrangements. They can manifest the same potential consequences as paracentric inversions, namely the production of chromatin bridges. Although poorly documented, meiotic problems may be more likely to occur in the case of this type of rearrangement because specific (reverse loop) meiotic configurations are not required for homologous pairing.

Robertsonian fusion or fission occurs when two uniarmed chromosomes fuse (fusion) to form a biarmed chromosome (metacentric or submetacentric; Figure 3.3) or when a biarmed chromosome splits (fission) at the centromere to produce two uniarmed chromosomes. **Tandem fusions** are the end-to-end fusion of two chromosomes into a single element and typically involve the loss or "silencing" of one centromere.

Translocations are rearrangements that change the position of a chromosomal segment within the genome. Insertion of a portion of chromatin from one chromosome into another is termed an **insertional translocation** (centric shifts are essentially intrachromosomal translocations). **Reciprocal translocations** are the exchange of chromosomal segments between two nonhomologous chromosomes. These may involve only a portion of a chromosome or may involve the swapping of entire chromosome arms (whole-arm translocations).

Interchromosomal rearrangements require complex pairing structures during meiosis in heterozygotes. In the heterozygous state, complete pairing during meiosis of Robertsonian fusions and fissions and tandem fusions results in trivalents (three chromosomes paired at meiosis). Reciprocal translocations require quadrivalents or higher-order multivalents. Meiotic problems in heterozygotes typically are the result of either failure of the elements to pair properly or nondisjunction or malsegregation, which cause aneuploidy. In the case of trivalents, if the metacentric chromosome segregates to one pole and the two acrocentrics to the other, normal or balanced gametes result. However, other segregation patterns yield duplication or deficient gametes. Heterozygosity for single rearrangements typically results in little or no reduction in fertility. However, heterozygosity for multiple rearrangements involving several chromosomes may produce massive germ cell death and sterility.

3.1.2.3 **Repetitive DNA and Chromosomal Banding Patterns.** The DNA of eukaryotic organisms is divided into highly repetitive DNA that does not code for proteins; moderately repetitive DNA that includes multiple-copy genes such as ribosomal genes, histone genes, and dispersed repetitive elements; and unique-sequence DNA that includes genes coding for specific proteins. The amount of repetitive DNA varies from 10% to more than 90% in different organisms. The multiple copies of repetitive DNA may be dispersed throughout the genome or tandemly arranged. The tandemly arranged DNA may be found in large blocks at centromeres, telomeres, or interstitial locations in the middle of chromosome arms. This highly repetitive DNA is typically condensed and seen under the light microscope as darkly staining heterochromatin, or C bands (see section 3.2.2.3). Often intraspecific or interspecific polymorphisms are present for the location or amount of heterochromatin. The B chromosomes or microchromosomes (see Figure 3.4) may consist almost entirely of repetitive DNA.

There is an alternation of AT-rich and GC (cytosine)-rich DNA along the chromosomes of warm-blooded animals. Several chemical treatments or banding techniques are available that reveal chromosome-specific banding patterns. Application of these techniques has enabled accurate chromosome identification and characterization of chromo-

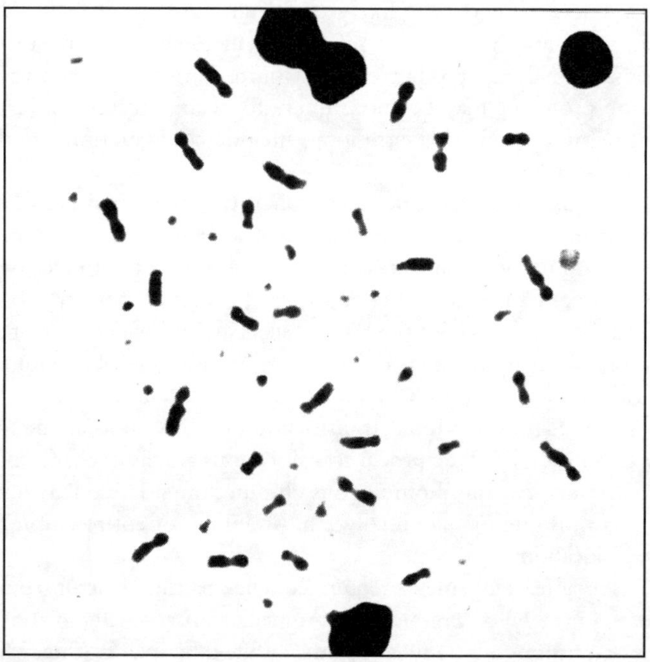

Figure 3.4 Metaphase of the longnose gar *Lepisosteus osseus* showing macrochromosomes and microchromosomes.

somal rearrangements. Fish chromosomes are much more homogeneous with regard to base composition, but some banding patterns are inducible (see section 3.2.2).

3.2 Methods and Applications of Fish Chromosome Analysis

3.2.1 *Preparation of Chromosome Slides*

Methods used in analysis of fish chromosomes have been reviewed recently (Gold et al. 1990; Thorgaard and Disney 1990; F. Foresti, L. Foresti, and A. Toledo, Institute of Biosciences, University of Sao Paulo, and C. Ozouf-Costaz, National Museum of Natural History-Paris, unpublished manuscript), and a visual summary is presented in Figure 3.5. In order to study chromosomes, dividing tissue must be used. Noninvasive methods that do not require sacrifice of the animal include blood culture and culture of regenerating tissue (e.g., fin clips). Fins or blood can be taken from the animal in the field and brought back to the laboratory for culture. Invasive methods include preparations from kidneys and of whole embryos; these methods usually require that living material be brought back to the laboratory.

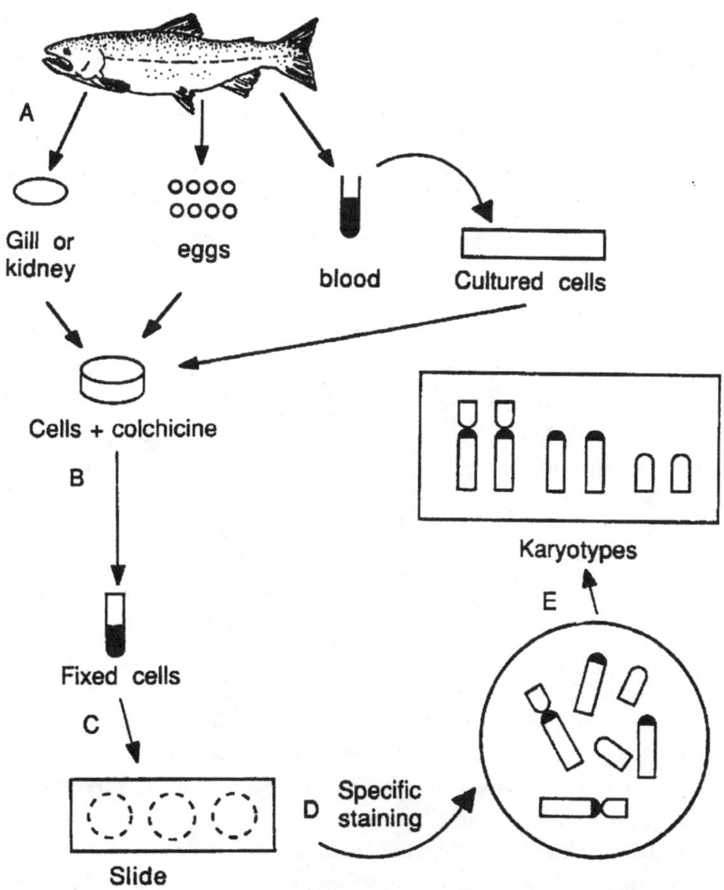

Figure 3.5 Summary of methodologies used to prepare chromosomes of fishes.

Kidney and embryo preparations are usually made after incubation of the cells in colchicine, a chemical that delays the formation of the spindle. In the case of kidney tissue, colchicine can be injected into the animal before removal of the kidney. For some species the number of dividing cells (mitotic index) is low, and better preparations are obtained if mitosis in the fish is stimulated by injection of the mitogen phytohemmagutinin or cobalt chloride a few days prior to sacrifice. In embryos, mitosis may be stimulated by raising the incubation temperature (heat shock).

Once dividing tissue is obtained, the procedure follows the steps shown in Figure 3.5. Dividing cells are treated with colchicine, exposed to a hypotonic solution that swells the cells to enhance chromosome spreading, and then fixed in Carnoy's fixative (3:1 methanol to acetic acid). The suspension of fixed cells is dropped onto slides, and the slides are stained. Finally, the chromosomes are examined under the microscope, and photographs are made from which karyotypes can be prepared. The conventional method of preparing karyotypes is to photograph the cells, make prints, and cut out and arrange the chromosomes. An alternative method is to capture digital images directly from the microscope or from negatives for computer-based image analysis and karyotyping.

3.2.2 *Chromosome Banding Methods*

3.2.2.1 **Giemsa (G), Fluorochrome, and Reverse (R) Bands.** In warm-blooded animals, there is an alternation in the base content and relative condensation of chromatin along the chromosome that can be enhanced so that each chromosome can be identified by its specific banding pattern after staining with Giemsa **(G bands).** These G bands are typically induced by digestion with trypsin in warm-blooded animals, but this technique has generally not been effective with fish chromosomes. In mammals and birds, the darkly staining regions are AT-rich and replicate late in the cell cycle, and the lightly staining regions are GC-rich and replicate early in the cell cycle. Therefore, the G-banding pattern is also obtained by staining with fluorochromes that are specific for AT-rich DNA, such as quinacrine (or DAPI, 4′,6-diamidino-2 phenyl-indole), resulting in **Q bands.** A reverse pattern (**R bands,** section 3.2.2.2) is obtained with chromomycin A_3 or mithromycin, which stain GC-rich DNA. Cold-blooded animals do not have much GC-rich DNA, and their chromosomes are not compartmentalized into GC-rich and AT-rich domains. Thus, these fluorochromes usually stain only the highly repetitive DNAs (see section 3.2.2.3) and do not produce a G-banding pattern. There are a few exceptions among fishes: the chromosomes of the European eel *Anguilla anguilla* show a banding pattern after staining with quinacrine, and trypsin has induced G banding in some fishes (reviewed in Gold et al. 1990).

3.2.2.2 **Replication Banding.** The R-banding pattern (reverse of G-banding in warm-blooded animals) can be obtained in the chromosomes of most vertebrates by means of the replication-banding technique. In this method, cultured cells are synchronized by addition of methyltrexate, and 5 bromo-2′ dexoyuridine (BrdU) is added at the appropriate time of the cell cycle so that it is incorporated into the late-replicating regions. When stained with Giemsa or acridine orange, these late-replicating regions (bands) appear more lightly stained. Because replication occurs discontinuously along most eukaryotic chromosomes, theoretically it should be possible to produce bands on fish chromosomes by means of replication banding. In several North American cyprinids and neotropical

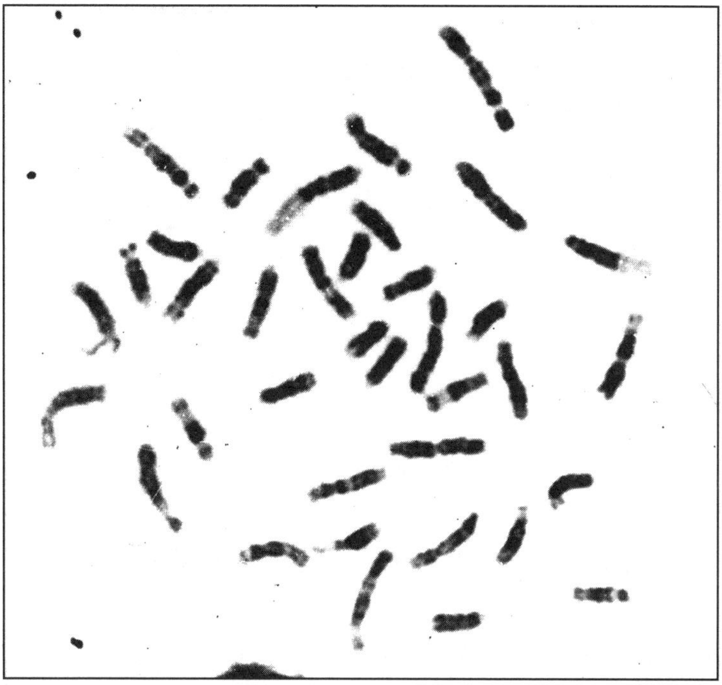

Figure 3.6 Replication-banding technique applied to chromosomes of a knifefish *Eigenmannia* sp. Photograph provided courtesy of L. Almeida-Toledo.

species, this technique has been successfully applied (Amores and Postlethwait 1999) (Figure 3.6). In most fish species, however, banding patterns have not been adequate for identification of each chromosome pair, probably due to problems with synchronizing the cells and the small size of many fish chromosomes. A new approach that may produce banding patterns detailed enough for accurate identification of fish chromosomes is fluorescence in situ hybridization (FISH) by use of probes to dispersed repetitive DNAs (see section 3.2.2.6).

3.2.2.3 **Constitutive Heterochromatin (C) Bands.** Constitutive heterochromatin is visualized by C-banding techniques that involve selective removal of the less-compacted DNA with alkali reagents such as barium hydroxide, treatment with salt solutions, or both, followed by staining with Giemsa (Figure 3.2). Intraspecific variation in the size and location of C bands has been observed in natural populations and can serve as population markers. The **C bands** represent tandemly arranged highly repetitive DNA with the typical repeat unit being 100–500 base pairs in length. Most genomes contain several subclasses of these sequences, which vary in the base sequence of the repeating unit. Subclasses of C bands that have a different base composition can be identified using AT-specific and GC-specific fluorochromes as described above. Minor changes in the amount and distribution of the AT-rich sequences in closely related species may result in bright staining of the heterochromatin with quinacrine in one species but not in another species. For example, most of the telomeric heterochromatin on the metacentric chromosomes of lake trout *Salvelinus namaycush* stains quinacrine positive, producing Q bands, but similar sites in the closely related species brook trout *S. fontinalis* generally do not. Thus, the parental origin of most of these chromosomes in the hybrid splake *S. fonti-*

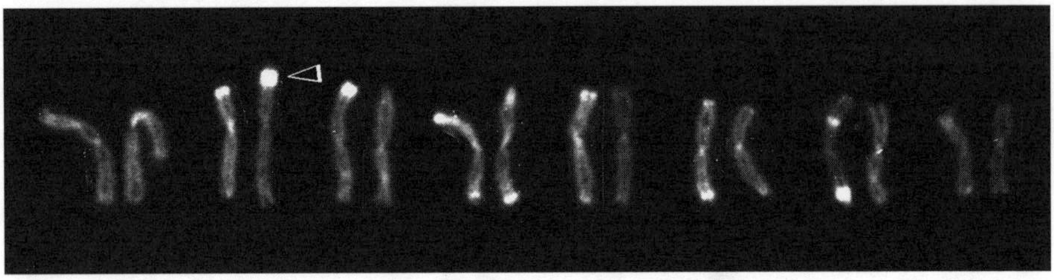

Figure 3.7 Partial karyotype showing quinacrine-positive bands (Q bands) obtained for metaphase chromosomes of a male F_1 hybrid splake (female brook trout × male lake trout). Lake trout chromosomes are on the left of each pair. Note the large Q band on the X chromosome of brook trout (arrowhead).

nalis × *S. namaycush* can be identified using quinacrine (Figure 3.7). Accurate identification of the subclasses of C bands that contain repeat units with different base sequences is now possible using the FISH technique (see section 3.2.2.6).

3.2.2.4 **Nucleolar Organizer Region (NOR) Banding.** **Nucleolar organizer regions (NOR)** are the sites of the rDNA genes, some of which are actively transcribed during the cell cycle. These regions are often decondensed (unwound) and may be visualized as secondary constrictions or gaps in the chromosome, particularly when located at interstitial locations. Active NORs can be visualized by means of the silver staining technique (Goodpasteur and Bloom 1975), which stains the ribosomal proteins (Figure 3.8). In fishes, the rDNA is GC-rich, and NORs can be visualized regardless of activity in most species by using chromomycin A_3 (CMA_3) and mithromycin (Figure 3.8). A new method (denaturation–propidium iodide staining) based on the thermal characteristics of rDNA has recently been described for the staining of NORs in fishes (Rab et al. 1996). Bands produced with this technique directly correspond with those produced by staining with CMA_3. However, like CMA_3 staining, this technique also stains GC-rich heterochromatin in addition to NORs. Thus, for unambiguous identification of NORs, it is necessary to use in situ hybridization (FISH) with probes specific to rDNA.

3.2.2.5 **Restriction Enzyme Banding.** **Restriction enzymes** digest DNA by making cuts at specific restriction sites: for example, the enzyme *Alu* I cuts DNA at the sequence AG↓CT. When a particular restriction enzyme is applied to a chromosome slide, it will digest the tandemly repetitive DNA (C bands) if the repeat monomers contain a restriction site for that enzyme. If the repetitive DNA is digested, the C band at that location will appear faintly stained as compared with repetitive DNA that lacks the proper site. Thus, different restriction enzymes can be used to identify subclasses of heterochromatin (C bands). It is possible that C bands with the same restriction enzyme patterns may have completely different sequences, although they must have at least one restriction site in common. A more direct approach to classifying repetitive DNAs is to digest genomic DNA with a battery of restriction enzymes, identifying those enzymes that produce large numbers of monomers (visualized as discrete bands on gels). These monomers can then be cloned and sequenced (e.g., Reed and Phillips 1995). Individual clones can then be labeled as probes to identify the location of these repeats in the genome with FISH (Figure 3.9).

3.2.2.6 **Fluorescence In Situ Hybridization (FISH).** Sequences of DNA can be localized directly on chromosomes by means of the technique of **fluorescence in situ hybridiza-**

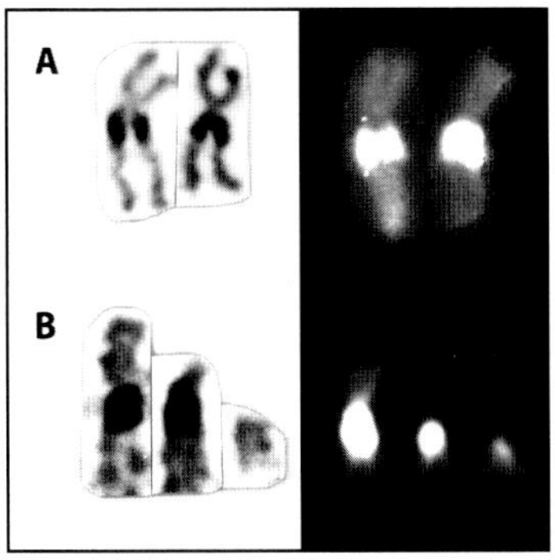

Figure 3.8 Partial karyotypes showing the chromosome pair with NORs in pink salmon *Oncorhynchus gorbuscha*. (A) Chromosomes from a fish with a $2n$ of 52. (B) Chromosomes from a fish heterozygous for a Robertsonian translocation with a $2n$ of 53. Chromosomes on the left are stained with silver nitrate; those on the right are stained with chromomycin A_3.

tion (FISH). Briefly, the technique involves denaturation of DNA probes and chromosomes, hybridization of probes to chromosomal DNA, posthybridization washes, and detection and visualization of the signal.

This technique has been used for many applications in human genetics including identification of specific chromosomes, detection and characterization of specific chromosome abnormalities, and mapping of genes to specific chromosomal locations. Specific chromosome regions, entire chromosomes, or genomes of one species in interspecific hybrids can be "painted" using the appropriate probes. "Paint probes" for specific chromosomes are produced by isolation of the chromosomes by flow sorting (separation of chromosomes by means of flow cytometry) or microdissection followed by amplification of the DNA by use of degenerate primers in the polymerase chain reaction (PCR). Although repetitive sequences are easier to visualize, new methodology including primed in situ hybridization (PRINS) and use of large insert clones (cosmids, yeast artificial chromosomes, and P_1-derived and F-factor-derived bacterial artificial chromosomes) as probes has allowed visualization of single-copy sequences on chromosomes.

Applications of FISH techniques to fish genetics have been recently reviewed (Phillips and Reed 1996; Phillips, in press). Different repetitive and moderately repetitive DNAs have been localized to centromeres, telomeres, and sex chromosomes of several fishes (Figure 3.9). Paint probes have been produced by microdissection followed by amplification of fragments using PCR (Figure 3.10A, B). Single-copy genes have been localized to specific chromosomes of several fish species including zebrafish *Danio rerio* by means of large insert clones (Figure 3.11). Major advances in the construction of physical maps for selected fish genomes are likely to occur in the near future.

3.2.3 *Meiotic Studies*

Meiotic studies are important for the detection and characterization of chromosomal rearrangements and for confirming the designation of sex chromosomes. Studies of meiotic chromosomes in hybrids can provide information on interspecific chromosomal homologies and mechanisms of hybrid sterility. Examination of fish meiotic chromosomes has been limited partly by the difficulty in obtaining material at the right meiotic stage. However, good methods are now available for preparation and analysis of the

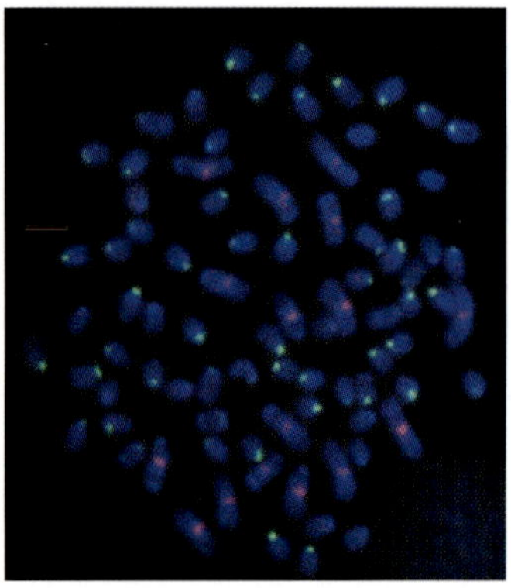

Figure 3.9 Localization of two repetitive sequences (*SnAlu* I-5 in red and *SnAlu* I-10 in green) isolated from lake trout by means of fluorescence in situ hybridization (FISH) to centromeres of different lake trout chromosomes. The repetitive sequence *SnAlu* I-5 is found at centromeres of the metacentric pairs, whereas *SnAlu* I-10 is found at centromeres of many acrocentric pairs.

A **B**

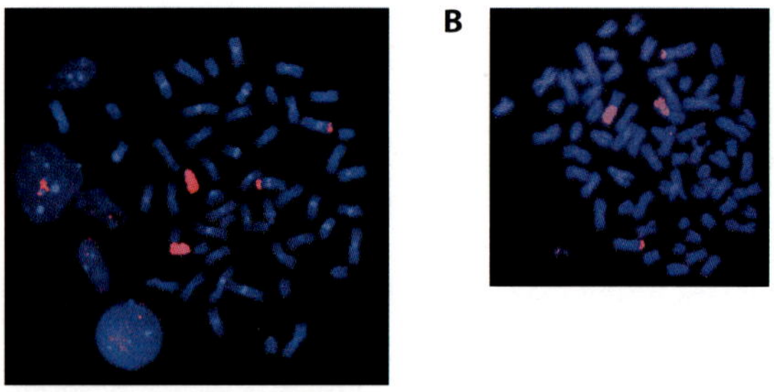

Figure 3.10 Hybridization of a paint probe obtained by microdissection of the largest acrocentric chromosome pair in rainbow trout to **(A)** rainbow trout chromosomes and **(B)** chinook salmon *Oncorhynchus tshawytscha* chromosomes. The probe also highlights one end of a second, presumably homeologous chromosome pair.

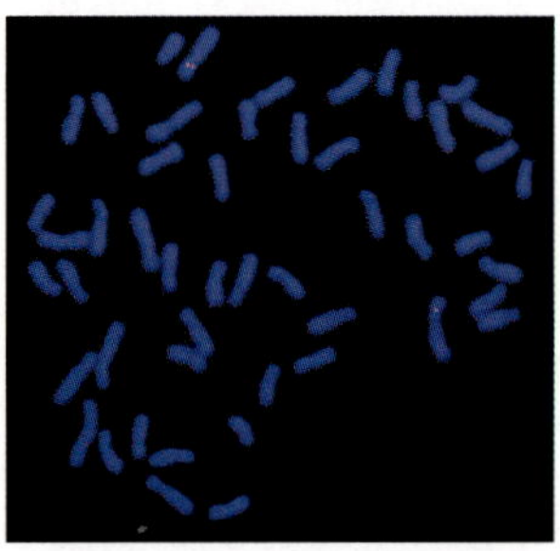

Figure 3.11 Localization of the trka (neurotrophic tyrosine kinase) gene on the short arm of the largest chromosome pair in zebrafish by means of FISH.

synaptonemal complexes in fishes (Figure 3.12). The **synaptonemal complex** is a linear proteinaceous structure that forms between homologous chromosomes during meiotic prophase I. Synaptonemal complexes provide increased resolution for examining both meiotic pairing behavior and chromosomal morphology. With a few exceptions, this technique has not been used in studies of fish chromosomes. The technique has proven useful in analysis of chromosome polymorphisms in other vertebrates including inversions, deletions–additions, and translocations.

3.2.4 Cytogenetic Methods for Toxicity Testing

Cytogenetic methods used in toxicity testing include scoring of chromosome aberrations, sister chromatid exchanges (SCE), anaphase aberrations, and micronuclei induction following exposure of cells to toxic substances (Al-Sabti 1991). The SCEs result from exchange between sister chromatids during the repair process after chromosome breakage, and scoring SCEs is more sensitive than scoring chromosome breaks. The frequency of SCEs is well correlated with doses of environmental chemicals and radiation. Studies on SCEs have been done *in vivo*, on cultured lymphocytes from experimental fish, and cells exposed to toxic substances *in vitro*. Chromosome aberrations such as lagging chromosomes at anaphase and production of micronuclei are easy to score and are especially useful for embryonic material and in fishes with high chromosome numbers.

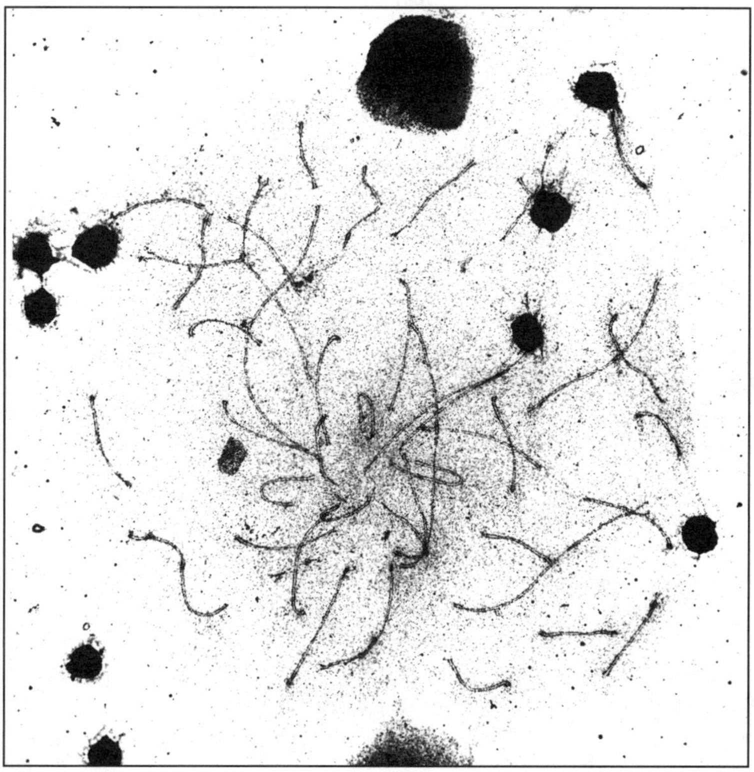

Figure 3.12 Electron micrograph of silver-stained pachytene nucleus of lake trout showing the synaptonemal complexes.

3.3 Chromosomes, Speciation, and Phylogenetic Analysis

Evolution is based on changes in the genetic material, and these changes may include chromosomal rearrangements. The role that structural changes in chromosomes play in the speciation process continues to be a point of debate. Any discussion of the importance of chromosomal variation in speciation must consider the types and meiotic consequences of individual chromosomal mutations because structural rearrangements alter genetic linkage relationships and recombination patterns in addition to changing gross chromosomal morphology.

Reduced fertility of heterozygotes is an assumption common to models of chromosomal speciation (King 1994; Sites and Reed 1994). Most models assume that chromosomal rearrangements are initially underdominant because of an increased frequency of aneuploidy in heterozygotes. The more underdominant the rearrangement is, the lower the probability of fixation. However, the probability of fixation is enhanced in small populations through random drift (Chapter 9), inbreeding (Chapter 10), or some form of **meiotic drive** (preferential recovery of a specific allele in gametes of a heterozygote). The subsequent fixation of rearrangements in isolated populations functions to reduce hybrid fitness and restrict gene flow between populations, thereby promoting speciation. As discussed previously (section 3.1.2.2), the type of rearrangement and the number of chromosomes involved often determine the effects these rearrangements have on the fertility of heterozygous carriers. Some chromosomal changes may show negative effects, whereas others are meiotically neutral. In mammals, there are a number of examples, such as the monkey owl and the black rhino, in which chromosomal variation among subspecies leads to reduced fertility in hybrids (see Section 11.3.2). The important point here is that negative effects (reduced fertility) cannot be assumed *a priori* for any chromosomal heterozygote (or fixed difference between closely related taxa) but should be tested directly (for example, in hybrids) before being incorporated into speciation models or hypotheses.

3.3.1 Polyploidy and Speciation

One chromosomal change that can result in "instant" speciation is polyploidy. **Polyploidy** results when an entire chromosome set is duplicated. **Autopolyploidy** results when one set of diploid chromosomes in a given individual is duplicated, whereas **allopolyploidy** results from hybridization between two species followed by duplication. In a new autotetraploid, multivalent pairing and tetrasomic inheritance will be common, whereas in an allotetraploid, bivalent pairing and disomic inheritance will predominate. Bivalent pairing and disomic inheritance is restored in autotetraploids through the process of **diploidization** (reversion to normal pairing). Proper segregation of chromosomes is impossible in odd-numbered polyploids such as triploids, which usually can reproduce only asexually through parthenogenesis. Some fish genera contain closely related diploid, tetraploid, and higher-level ploidy species (e.g., sturgeons and paddlefishes [Acipenseriformes] and *Barbus* spp. [Cyprinidae]). Other cases are known in which an entire family is the result of radiation following a polyploidy event (e.g., trouts [Salmonidae] and suckers [Catostomidae]).

3.3.2 Phylogenetic Analysis Based on Cytogenetic Characters

Any discrete character data can be used in a phylogenetic analysis based on maximum parsimony or cladistic methods (Chapter 13). Nucleolar organizer region phenotypes were useful in phylogenetic analysis of North American cyprinids (Amemiya and Gold 1990), which have relatively conserved diploid chromosome numbers ($2n$ = 48–50). In fishes, the use of chromosome characters has been limited by the apparent lack of chromosome variation in some groups and, in other groups, the inability to identify homology of specific chromosomal arms. It is very important to determine chromosome arm homologies when using karyotypic data for phylogenetic analysis. In mammals, where chromosome homologies have been determined using G banding, a number of cases are known in which chromosome fissions have occurred in one lineage and chromosome fusions in another. With the advent of FISH, it should be possible to better identify homologies. Then discrete karyological events such as recognizable fusions, fissions, or inversions involving specific arms and chromosomes can be identified. Using related taxa for comparison, these events can then be analyzed using cladistic methods and compared with other data sets.

3.4 Variation in Fish Chromosomes

Although approximately 40% of the 4,200 species of mammals have been karyotyped, karyological information is available for only 13% of the 23,000 described species of fishes (Gold 1979; Gold et al. 1980; Ojima 1980; Sola et al. 1981; Oliveria et al. 1988; Phillips and Ihssen 1990; Klinkhardt et al. 1995). For many of these fishes, the data consist of only the diploid chromosome number. Analysis of C-banding patterns has been done on approximately 6% of the karyotyped fishes and staining of NORs on 12%. Heteromorphic sex chromosomes have been identified in roughly 6% of the species examined. A subset of these data has been compiled for approximately 2,400 species, and this subset constitutes a computerized cytogenetic database (Klinkhardt et al. 1995).

Diploid chromosome numbers reported in fishes vary between a $2n$ of 16 in the gourami *Sphaerichtys espromenarides* (Belontiidae) and a $2n$ of 446 in *Ptychobarbus (Diptychus) dipogon* (Cyprinidae). However, the most common chromosome number in fishes is a $2n$ of 48, and this has been considered as ancestral for teleosts (Figure 3.13). Greater variation in chromosome number seems to be the case in freshwater fishes versus marine fishes, which tend to have larger population sizes and a more constant environment. High diploid numbers and the presence of microchromosomes are found primarily in the more primitive fishes (such as lampreys [Petromyzontidae], cartilaginous fishes [Elasmobranchiomorphi], gars [Lepisosteidae], and sturgeons [Acipenseridae]).

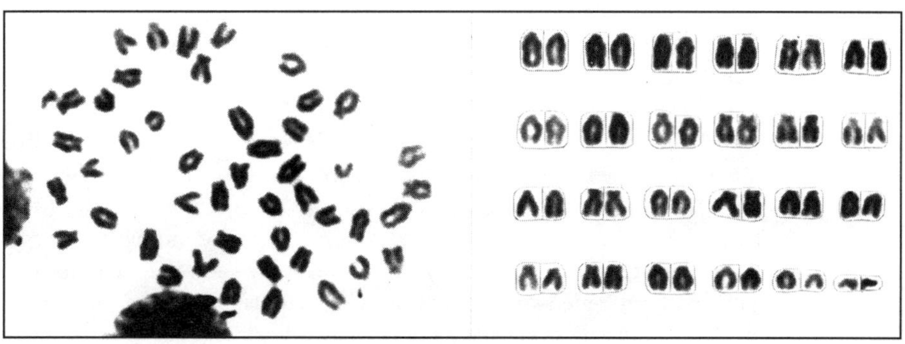

Figure 3.13 Geimsa-stained metaphase nucleus of Eurasian perch *Perca fluviatilis* and karyotype ($2n$ = 48) showing all acrocentric chromosomes.

The sizes of fish genomes are considerably more variable than are mammalian genomes (the human haploid genome is 3,000 million base pairs [mbp]). For example, the haploid genome is approximately 400 mbp in the Japanese pufferfish *Fugu rubripes*; 1,200 mbp in rainbow smelt *Osmerus mordax;* 2,500 mbp in rainbow trout; and 7,000 mbp in the green sturgeon *Acipenser medirostris*. Chondricthyans have even larger genome sizes with an estimated 15,000 mbp in the Brazilian electric ray *Torpedo ocellata,* and 34,000 mbp in the angular rough shark *Oxynotus centrina*. Each of these has high amounts of AT-rich repetitive DNAs. Among chondricthyans the more advanced species have lower chromosome numbers and smaller genome sizes and lack microchromosomes. The presence of more metacentric chromosomes in the advanced species suggests that most of the changes in chromosome number have occurred via Robertsonian fusion rearrangements.

3.4.1 Variation in Chromosome Number: Polyploidy

There are two well-studied cases in which an entire family of fishes has diverged following a polyploidy event: the Salmonidae, thought to be autotetraploids, and the Castotomidae, which appear to be allotetraploids. In many other cases, both diploid and tetraploid species are found within the same genus or family. Examples of this include the sturgeons and paddlefishes (Acipenseriformes), in which one group of species has approximately a $2n$ of 120–140 chromosomes (including microchromosomes) and another group has a $2n$ of 240–280 chromosomes. A second example includes minnows and carp (Cyprinidae), loaches (Cobitidae), plated catfishes (Callichthyidae), and suckermouth catfishes (Loricariidae), in which some species have a $2n$ of 50 and others a $2n$ of 100. In the cyprinid genus *Barbus,* a polyploid series exists in which some species are diploid ($2n$ = 50), some tetraploid ($2n$ = 100), and some hexaploid ($2n$ = 150) (see Figure 3.14). Even higher chromosome numbers are found in a few related *Barbus* spp. in northern India. Another cyprinid, *Ptychobarbus dipogon,* has a $2n$ of 446, suggesting that it may be a triploid of one of the hexaploid species.

After polyploidization, multivalent pairing is often common, but diploidization eventually occurs. In the case of salmonids, there is good evidence from cytogenetics and observation of tetrasomic inheritance at some loci inherited from the male parent that diploidization is incomplete. Duplicate gene copies may be silenced, remain as isoloci, or evolve new functions. In catastomids, it is estimated that 50% of the duplicate copies have been silenced. In salmonids, it is estimated that approximately 30% of enzyme loci have been silenced, 30% remain as isoloci, and 30% have diverged in function (Allendorf and Thorgaard 1984). Some of the diverged enzyme loci are now expressed in different tissues or at different times in development. It has been hypothesized that the availability of duplicate gene copies that are free to evolve rapidly to fulfill new functions has been important in the adaptive radiation of salmonids.

Polyploid species with even numbers of chromosomes ($4n$, $6n$, and so on) are fertile, but odd-numbered polyploids, such as triploids ($3n$), usually cannot form viable gametes. Triploids are not found in natural populations of higher vertebrates (birds and mammals), but in fishes there are many documented cases of unisexual triploids that reproduce parthenogenetically, often by different mechanisms. For example, unisexual fishes occur in several genera of the livebearers (Poeciliidae) from the southern United States and Mexico. Among cyprinids there is a diploid–triploid complex in the genus *Rivulus* in the southeastern United States and in the genus *Rutilus* on the Iberian Peninsula. In

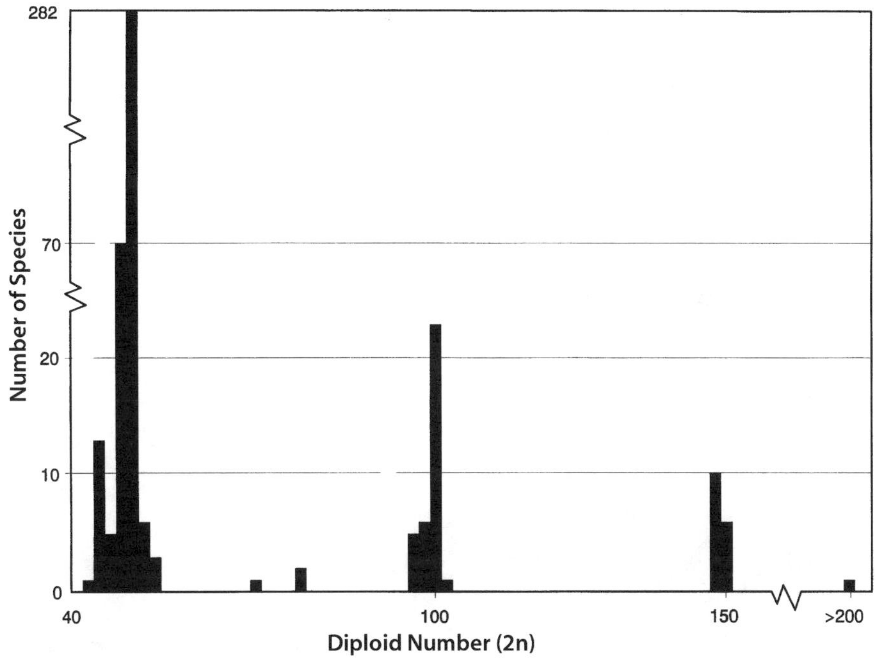

Figure 3.14 Frequency histogram of the diploid numbers of 315 cyprinid fishes. Data adapted from Buth, Dowling, and Gold 1991.

many cases, including the Amazon molly *Poecilia formosa* (Figure 3.15), the unisexuals are hybrids between two species that must coexist with a bisexual sister species, which provides the sperm needed to activate the eggs. The triploids are often found in the harsher environments, such as small pools, that show extreme fluctuations in temperature and dissolved oxygen. New triploid lineages appear to be constantly generated, especially in disrupted or harsh environments.

3.4.2 Variation in Chromosome Number: Robertsonian Rearrangements

Robertsonian rearrangements in which acrocentric chromosomes fuse to form metacentric chromosomes or metacentrics split to form acrocentrics are very common in natural populations. There are many examples in fishes. Perhaps the best known cases are the extensive Robertsonian rearrangements that have occurred in salmonid fishes following polyploidization (reviewed in Phillips and Rab 2001). The chromosome arm number (FN) is 100 in most salmonid species, but the diploid number of chromosomes varies between 52 and 102. There is an evolutionary trend toward lower chromosome number in the salmonids, with most of the species in the most advanced genus having 52 to 68. Intraspecific Robertsonian polymorphisms have been found in many species of salmonid fishes including rainbow trout, Atlantic salmon *Salmo salar,* brown trout *Salmo trutta,* pink salmon, lake herring *Coregonus artedi,* and Arctic char. These intraspecific polymorphisms have been used for stock identification, identification of subspecies, and analysis of hybrid zones (Phillips and Kapuscinski 1988; Ostberg and Thorgaard 1999).

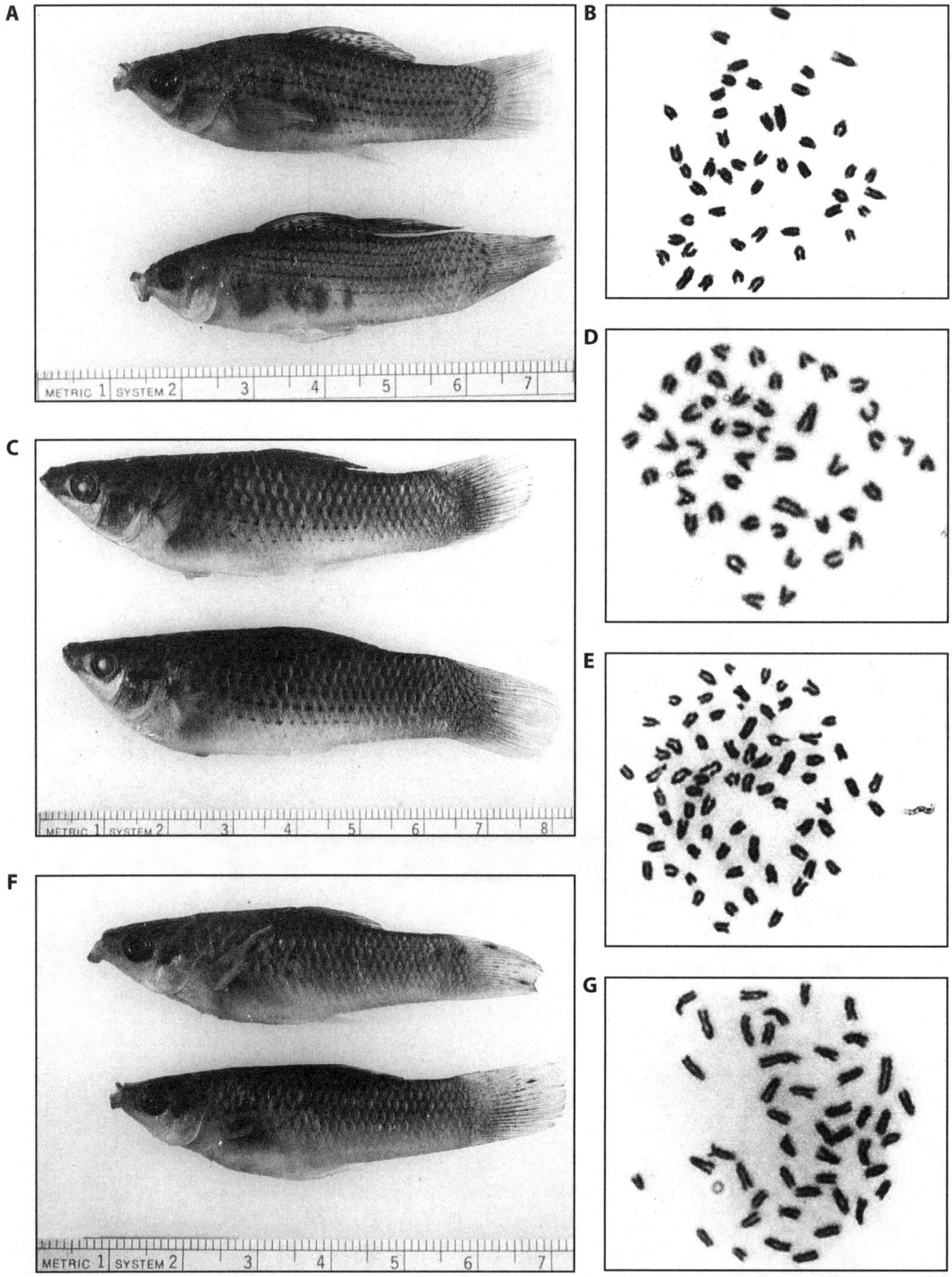

Figure 3.15 Bisexual parental sailfin molly *Poecilia latipinna* and shortfin molly *P. mexicana* and dipolid and triploid (unisexual) Amazon molly. **(A)** Sailfin molly, female above and male below. **(B)** Karyotype of female sailfin molly. **(C)** Amazon molly, diploid above and triploid below. **(D, E)** Karyotypes of Amazon molly, diploid (D) and triploid (E). **(F)** Shortfin molly, female above and male below. **(G)** Karyotype of female shortfin molly. Photographs provided courtesy of E. M. Rasch; karyotypes provided courtesy of L. Sola.

3.4.3 Variation in Chromosome Arm Number: Tandem Translocations

Tandem translocations in which small acrocentric chromosomes fuse "head to tail" to form large acrocentric chromosomes are also found in fishes but are not as common as are Robertsonian translocations. Sometimes a block of repetitive DNA (a C band) is retained at the site of fusion in the middle of the chromosome arm. This type of translocation results in a reduction in FN, or the number of chromosome arms. An example is Atlantic salmon. Unlike other salmonid species, which have undergone Robertsonian translocations and retain an FN close to 100, there have been extensive tandem translocations in Atlantic salmon, and the FN in this species is 74. All of the large acrocentric chromosomes have a large interstitial C band in the long arm (reviewed in Phillips and Rab, in press).

3.4.4 Variation in Chromosome Arm Size: Inversions

Documented cases of inversions are few in fishes because of the absence of reliable methods for chromosome identification. In the absence of specific banding patterns, only pericentric inversions, which change the arm ratio, are detected. Staining of NORs has allowed identification of numerous inversions in chromosomes, such as the inversion polymorphism of pink salmon. It is likely that improved techniques will reveal many hidden inversions in fish chromosomes. Intraspecific variation in the number of small submetacentric chromosomes with visible short arms has been found in many species. Many of these are likely the result of heterochromatic additions (see section 3.4.4), but several certainly represent small inversions.

3.4.5 Variation in Heterochromatin: C Bands

Data from C-band staining are important when first comparing the karyotypes of fish species in order to distinguish between short arms that are entirely heterochromatic (representing heterochromatin additions) and those that are euchromatic (the result of small inversions). As specific repetitive probes become available, it will be possible to compare the location of specific repetitive DNA in related species and stocks. Intraspecific variation in the size of C bands or the size or number of heterochromatic short arms is very common in fishes. Stock-specific differences in the number of heterochromatic bands that stain positive with quinacrine have been found in Atlantic salmon, Arctic char, and lake trout. Supernumerary (B) chromosomes composed primarily of heterochromatin have been found in a variety of fish groups including sturgeon and whitefishes *Coregonus* spp. Usually these are small chromosomes, and the number per individual varies between 0 and 6. However, very large B chromosomes showing population-level variation have been found in characin fishes of the genus *Astyanax* and catfish genera *Microlepidogaster* (Loricariidae) and *Rhamdia* (Heptapturidae) in Brazil.

3.4.6 Variation in Number and Position of NORs

Variation in number and position of NORs is very common among closely related fish species and has been used in phylogenetic analysis of salmonids (Phillips, Pleyte, and Ihssen 1989) and cyprinids (Gold, Zoch, and Amemiya 1988). Intraspecific variation in the number of NORs and their chromosomal location has been reported for a number of species, including the salmonids brook trout, lake trout, and Arctic char of the genus

Salvelinus, ciscoes and whitefishes of the genus *Coregonus*, and brown trout of the genus *Salmo* as well as the Amazon molly. Stock-specific differences in the number and location of NORs in hatchery and natural populations have been documented in brown trout, lake trout, Arctic char, and the sailfin molly. It is not known why the number and position of NORs is very stable in some species and variable in others. Nucleolar organizer regions are usually found at telomeric sites, which are more likely to be involved in chromosome rearrangements and transposition events. There is evidence that transposition followed by crossing-over may be important in the generation of NORs at new chromosomal sites in the *Salvelinus* spp. Because rDNA may be a hotspot for crossing-over, it is not surprising that many intraspecific rearrangements of the NOR-bearing chromosomes are found.

3.4.7 *Presence of Sex Chromosomes*

In some species of fishes, males are the **heterogametic** sex; that is, the males produce two types of sperm, one giving rise to males, the other to females. In other species, females are the heterogametic sex. Both systems are found in different populations of a few fish species. In male heterogametic systems, males may be XY or XO and females are XX. More complex differences (e.g., XY_1Y_2/XX and $X_1X_2Y/X_1X_1X_2X_2$) also may be present. In female heterogametic systems, females are ZW and males ZZ. Morphologically distinct sex chromosomes have been reported in only 5.8% of fish species examined. However, in many cases the sex chromosomes differ by only small heterochromatin additions to the Y or W chromosome, an addition that would not be seen without C-band staining. As mentioned earlier, this technique was not used in many previously reported studies. In several cases, molecular analysis of repetitive DNA has revealed the presence of sex-specific sequences. It is important to confirm the presence of sex chromosomes by doing crosses and correlating cytotypes with sex because spurious associations of heterochromatic polymorphisms with a given sex are possible in small samples.

3.5 Perspective: Applications and Limitations of Cytogenetic Methods

Cytogenetic methods have been used extensively in fish systematics for species identification and for phylogenetic analysis, but relatively few studies have been done on intraspecific chromosome variation in natural populations. In the past, cytogenetic analysis of fishes has been limited by the need for live material and the lack of good methods for chromosome identification. However, good chromosome preparations from a variety of fish species can be obtained from blood or fin clips taken from the fish at remote locations and brought to the laboratory within 24 h for culturing. Accurate chromosome identification for many fish species by means of dispersed repetitive sequences, whole chromosome paint probes, or probes to single-copy genes should become available in the near future. Because FISH can be applied to interphase cells, it is possible to identify marker chromosomes in nondividing cells. Chromosome counts may even be attainable on historical material using probes to centromeres and telomeres!

Cytogenetic methods have been used to a limited extent in analysis of triploid fish that have been produced using chromosome manipulation techniques. With development of species-specific paint probes, one can anticipate increased applications in production and evaluation of interspecific diploid and triploid hybrids. Several of these hybrids are now being tested for aquaculture, and it is important to estimate their relative fertility. With paint probes it should be possible to determine which chromosomes are missing in aneuploid cells and whether tripolar mitoses have occurred. Sectioned tissues could be examined with computer-assisted image analysis in order to estimate the relative amount of aneuploidy in different tissue types.

Tremendous advances in human and mammalian genetics have occurred within the past few years as a result of the human and mouse genome projects. Mapping of informative polymorphic microsatellite loci has provided markers that have greatly facilitated positional cloning and sequencing of genes important to human health and development. Genome projects are underway for most domesticated animals, and several important quantitative trait loci (see Box 12.3) have already been mapped. Because of the conserved nature of vertebrate genes, we can anticipate that many genes isolated in connection with the mammalian genome projects will yield probes that can be used to isolate and characterize the homologous fish genes.

The number of microsatellite loci (see Box 5.3) that have been isolated from fishes has been increasing steadily in the last couple of years. Linkage maps based on microsatellite loci, various other DNA polymorphisms, and allozyme loci have been prepared for a few fishes, including the zebrafish and some livebearers *Xiphophorus* spp., tilapia *Tilapia* spp., catfishes, medaka *Oryzias* spp., and salmonids. Currently, single-copy genes are being added to these genetic linkage maps, with the most detailed map available for the zebrafish. Comparison of the zebrafish map with mammalian maps has shown that blocks of conserved genes have been retained for over more than 450 million years, although intrachromosomal rearrangements have been common. The complete genome sequence of the Japanese pufferfish is expected to be available before the end of 2001 and that of the zebrafish by 2002–2003. The availability of these two complete teleost genome sequences coupled with much more detailed linkage maps of single-copy genes for several other species will set the stage for a detailed analysis of the evolution of teleost chromosomes.

With genetic maps based on polymorphic microsatellite markers, it will be possible to map the genes that determine traits important in adaptation to specific environments and in aquaculture. With the FISH technique, it should be possible to determine the chromosomal location of these genes for microdissection-based cloning. This would greatly enhance the analysis of the molecular basis of phenotypic variation in fishes.

3.6 References

Allendorf, F. W., and G. H. Thorgaard. 1984. Tetraploidy and the evolution of salmonid fishes. Pages 1–53 *in* B. J. Turner, editor. Evolutionary genetics of fishes. Plenum, New York.

Al-Sabti, K. 1991. Handbook of genotoxic effects and fish chromosomes. J. Stefan Institute, Ljubljana, Yugoslavia.

Amemiya, C. T., and J. R. Gold 1990. Chromosomal NOR phenotypes of seven species of North American Cyprinidae, with comments on cytosystematic relationships of the *Notropis volucellus* species-group. Copeia 1990:68-78.

Amores, A., and J. H. Postlethwait. 1999. Banding chromosomes and the zebrafish kayotype. Chapter 19 *in* Methods in cell biology, volume 60. Academic Press, New York.

Buth, D. G., T. E. Dowling and J. R. Gold 1991. Molecular and cytological investigations Chapter 4 *in* I. J. Winfield and J. S. Nelson, editors. Cyprinid fishes. Chapman and Hall, New York.

Gold, J. R. 1979. Fish cytogenetics. Fish Physiology 8:353–404.

Gold, J. R., W. J. Karel, and M. R. Strand. 1980. Chromosome formulae of North American fishes. Progressive Fish-Culturist 42:10–23.

Gold, J. R., Y. C. Li, N. S. Shipley, and P. K. Powers. 1990. Improved methods for working with fish chromosomes with a review of metaphase chromosome banding. Journal of Fish Biology 37:563–575.

Gold, J.R., P. K. Zoch, and C. T. Amemiya. 1988. Cytogenetic studies in North American minnows (Cyprinidae). XIV. Chromosomal NOR phenotypes of eight species from the genus *Notropis.* Cytobios 54:137–147.

Goodpasteur, C., and S. E. Bloom. 1975. Visualization of nucleolar organizer regions in mammalian chromosomes using silver staining. Chromosoma 53:37–50.

King, M. 1994. Species evolution: the role of chromosome change. Cambridge University Press, Cambridge.

Klinkhardt, M., M. Tesche, and H. J. Greven. 1995. Database of fish chromosomes. Westarp Wissenschaften, Magdeburg, Germany.

Ojima, Y. 1983. Fish cytogenetics. Pages 111–145 *in* A. K. Sharma and A. Sharma, editors. Chromosomes in evolution of eukaryotic groups, volume 1. CRC Press, Boca Raton, Florida.

Oliveria, C., L. R. Almeida Toledo, F. Foresti, H. A. Britski, and S. A. Toledo Filho. 1988. Chromosome formulae of neotropical freshwater fishes. Revista Brasileira de Genetica 121:577–624.

Ostberg, C. O., and G. H. Thorgaard. 1999. Geographic distribution of chromosome and microsatellite DNA polymorphisms in *Oncorhynchus mykiss* native to western Washington. Copeia 1999:287–289.

Phillips, R. B. 2001. Application of fluorescence in situ hybridization (FISH) to fish genetics and genome mapping. Marine Biotechnology 3(Supplement 1):5145–152.

Phillips, R. B., and P. E. Ihssen. 1990. Genetic marking of fish using variation in chromosomes and nuclear DNA. Pages 499–513 *in* N. C. Parker, A. E. Giogri, R. C. Heidinger, D. B. Jerter, Jr., E. D. Prince, and G. A. Winans, editors. Fish-marking techniques. American Fisheries Society Symposium 7, Bethesda, Maryland.

Phillips, R. B., and A. R. Kapuscinski. 1988. High frequency of translocation heterozygotes in odd year populations of pink salmon *(Oncorhynchus gorbuscha).* Cytogenetics and Cell Genetics 48:178–182.

Phillips, R.B., K. A. Pleyte, and P. E. Ihssen. 1989. Patterns of chromosomal nucleolar organizer regions (NOR) variation in fishes of the genus *Salvelinus.* Copeia 1989:47–53.

Phillips, R. B., and P. Rab. 2001. Chromosome evolution in the Salmonidae (Pisces): an update. Biological Reviews 76:1–25.

Phillips, R. B., and K. M. Reed. 1996. Application of fluorescence in situ hybridization (FISH) to fish genetics. Aquaculture 140:197–216.

Rab, P., K. M. Reed, A. Ponce De Leon, and R. Phillips. 1996. A new method for detecting nucleolus organizer regions in fish chromosomes using denaturation and propidium iodide staining. Biotechnic & Histochemistry 71:157–162.

Reed, K. M., and R. B. Phillips. 1995. Molecular characterization and cytogenetic analysis of highly repetitive DNAs of lake trout, *Salvelinus namaycush.* Chromosoma 104:242–251.

Sites, J. W., and K. M. Reed. 1994. Chromosomal evolution, speciation, and systematics: some relevant issues. Herpetologica 50:237–249.

Sola, L., S. Cataudella, and E. Capanna. 1981. New developments in vertebrate cytotaxonomy. III. Karyology of bony fishes: a review. Genetica 54:285–328.

Thorgaard, G. H., and J. E. Disney. 1990. Chromosome preparation and analysis. Chapter 6 *in* C. B. Schreck and P. B. Moyle, editors. Methods for fish biology. American Fisheries Society, Bethesda, Maryland.

4

Mitochondrial DNA

Neil Billington

4.1 Introduction

In this chapter, I describe basic features of animal mitochondrial DNA (mtDNA) that make it a useful genetic marker for studies of population genetics and molecular systematics of fishes. I also outline various strategies and methods for the analysis of fish mtDNA. Finally, I present a number of examples drawn from the fisheries science literature of the use of mtDNA analysis to determine relationships among populations, to identify the female parent of a hybrid, and to examine molecular systematic relationships.

4.2 Features of Animal Mitochondrial DNA

A small proportion (<1%) of the DNA of eukaryotic cells is nonnuclear; it is located within organelles in the cytoplasm called mitochondria and appears to have been endosymbiotic in early eukaryotic cells (Gray 1992). **Mitochondrial DNA** differs from nuclear DNA in that (1) the entire genome is transcribed as a unit, using a genetic code that is distinct from the "universal" genetic code; (2) replication is continuous, unidirectional, and asymmetrical without any apparent editing or repair mechanism; and (3) there is an economy of size, with no introns present; that is, genes are tightly packed with some sequence overlap and there are few intergenic sequences (Brown 1983).

Animal mtDNA is a double-stranded, circular molecule usually ranging in size from 14,000 to 26,000 base pairs (bp), although a mitochondrial genome of up to 42 kilobase pairs (kbp) has been observed in the scallop *Placopecten megellanicus* (Snyder et al. 1987). Generally, fish mtDNA is **homoplasmic;** that is, all the molecules in an organism are identical. Therefore, any tissue can be used as a DNA source. Mitochondrial DNA is present in multiple copies per cell, thus making it relatively easy to isolate and purify (Brown 1983). Animal mitochondrial genomes carry genes encoding 13 proteins (seven NADH dehydrogenase subunits, three cytochrome oxidase subunits, ATPase 6, ATPase 8, and cytochrome *b*), 22 transfer RNAs (tRNAs), and 2 ribosomal RNAs (rRNA; 12S and 16S, S being a measure of relative size in Svedberg units) molecules involved in electron transport and ATP synthesis (Figure 4.1). In addition, a control region is present

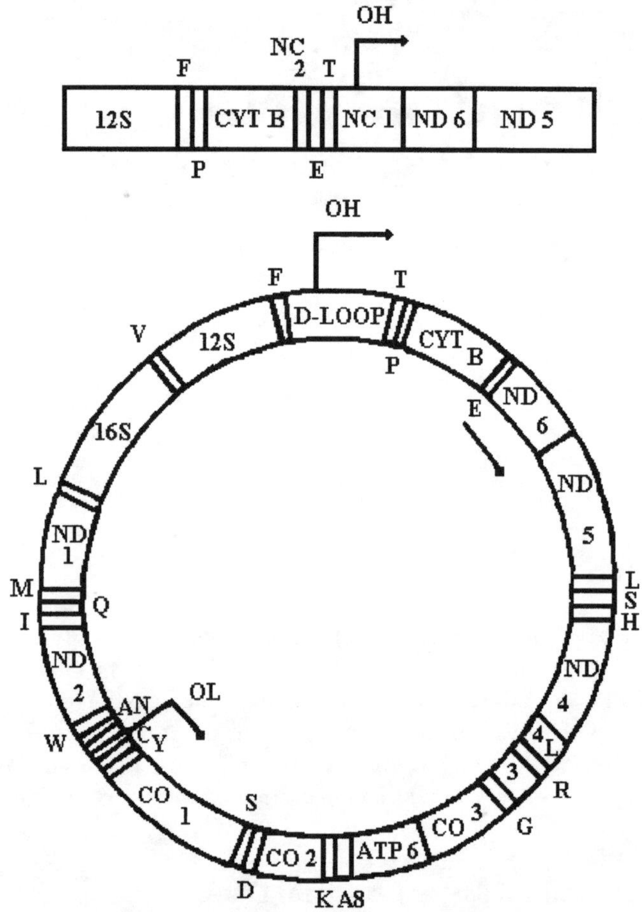

Figure 4.1 Organization of the fish mitochondrial genome (circle); differences in genome organization in the sea lamprey *Petromyzon marinus* between the 12S ribosomal RNA (rRNA) and NADH dehydrogenase (ND) 5 genes are shown above (linear map). The outer circle depicts the heavy (H) strand and the inner circle depicts the light (L) strand. The origins of H-strand replication (OH) and L-strand replication (OL) are shown, and the directions of replication indicated by arrows. The genes are as follows (clockwise): cytochrome *b* (CYT B), NADH subunit 6 (ND 6), NADH subunit 5 (ND 5), NADH subunit 4 (ND 4), NADH subunit 4L (4L), NADH subunit 3 (3), cytochrome oxidase I (CO 3), ATPase subunit 6 (ATP 6), ATPase subunit 8 (A8), cytochrome oxidase subunit 2 (CO 2), cytochrome oxidase subunit 1 (CO 1), NADH subunit 2 (ND 2), NADH subunit 1 (ND 1), large (16S) rRNA (16S), and small (12S) rRNA (12S). The D-loop is a nontranscribed region that contains the origin of H-strand replication. The coding (template) sequence of all protein and rRNA genes except ND 6 are on the H-strand. The direction of transcription for all of the genes on the H-stand is anticlockwise, and transcription of all genes encoded on the L-strand is clockwise (shown by arrow for ND 6). The transfer RNA (tRNA) genes are identified by the International Union of Biochemistry and Molecular Biology letters for the various amino acids and are (clockwise) threonine (T), proline (P), glutamic acid (E), leucine (L), serine (S), histidine (H), arginine (R), glycine (G), lysine (K), aspartic acid (D), tyrosine (Y), cystine (C), asparagine (N), alanine (A), tryptophan (W), isoleucine (I), glutamine (Q), methionine (M), valine (V), and phenylalanine (F); the tRNA genes are labeled on the strand on which their coding sequences occur. The sea lamprey has two noncoding regions (NC 1 and NC 2), one of which (NC 1) contains the origin of H-strand replication.

that contains the origin of heavy-strand replication (known as the D-loop region in vertebrates).

There is some variation in content, however; certain nematodes appear to lack the ATPase 8 gene (Okimoto et al. 1992), and the blue mussel *Mytilus edulis* mitochondrial genome contains an additional tRNA gene (Hoffmann et al. 1992). Although slight differences have been observed in birds and marsupials (Avise 1994), gene order is highly conserved among most vertebrate taxa (Moritz et al. 1987). Studies based on **DNA sequencing** (a collection of procedures used to determine the exact nucleotide sequence of a target fragment of DNA) of the complete mitochondrial genome of Atlantic cod *Gadus morhua* (Johansen et al. 1990; Johansen and Johansen 1994), tasseled-mouth loach *Crossostoma lacustre* (Tzeng et al. 1992), common carp *Cyprinus carpio* (Chang et al. 1994), rainbow trout *Oncorhynchus mykiss* (Zardoya et al. 1995), and Atlantic salmon *Salmo salar* (Hurst et al. 1999) have shown that advanced teleosts exhibit the typical vertebrate gene order (Figure 4.1). However, differences in the number and position of the noncoding regions relative to tRNA genes were observed in the sea lamprey when its whole mitochondrial genome was sequenced (Lee and Kocher 1995; Figure 4.1). The same genomic organization was also observed in the European river lamprey *Lampetra fluviatilis* (Delarbre et al. 2000). It is still not known whether the lamprey gene order represents the vertebrate lineage's ancestral state or a uniquely derived state.

Three other features of animal mtDNA make it a useful genetic marker for studies of molecular systematics and, in particular, for population genetic studies and examination of phylogenetic relationships among closely related species. First, mtDNA is maternally transmitted (Hutchinson et al. 1974; Avise and Vrijenhoek 1987; May and Grewe 1993); that is, all offspring, regardless of sex, inherit their mother's mtDNA haploid genotype (**haplotype**). However, rare cases of paternal leakage have been reported, including one involving a fish, the European anchovy *Engraulis encrasicolus* (Magoulas and Zouros 1993). Second, mtDNA is not subject to recombination (Brown 1983) and therefore provides a set of completely linked, homologous markers that permit clear definition of maternal genealogies. This allows excellent discrimination between common ancestry and convergence (Avise 1994). Third, the average rate of nucleotide substitution in the mtDNA of higher vertebrates is approximately 5 to 10 times that of the nuclear genome, enhancing its resolution capacity for population level studies (Brown 1983; Moritz et al. 1987).

In philopatric fish species, these mtDNA characteristics have the effect of generating more genetic divergence in the mitochondrial genome than exists in most nuclear genes. Assuming a fixed mutation rate, the extent of genetic divergence between both segments of the genome is determined by the interaction of processes such as natural selection (Chapter 8) and genetic drift (Chapter 9), which promote divergence, and gene exchange (migration, Chapter 7), which reduces divergence. When populations are reasonably large and selection pressures weak, gene frequency divergence in the nuclear genome can be suppressed by modest gene flow arising from the movement of individuals of either sex. Thus, in the case in which females are philopatric and males disperse during spawning, no divergence in the nuclear genome will occur. However, this same pattern of gene exchange permits divergence of the mitochondrial genome because of its maternal inheritance.

Maternally inherited mtDNA is also more likely to show differences among populations than is single-copy nuclear DNA due to its smaller effective population size (one quarter that of the bisexually inherited diploid nuclear genome if the sex ratio is 1:1), meaning that mtDNA is more responsive to reductions in genetic variability due to population bottlenecks (Birky et al. 1983; Birky 1991). Moreover, assuming equal migration ability of both sexes, four times as much interchange among populations is required to prevent divergence in the mitochondrial, as opposed to the nuclear, genome in populations that are initially genetically similar. Conversely, four times as much migration is required to erode existing divergence in mtDNA haplotype frequencies as is required to erode gene frequency differences at nuclear loci.

The fact that mtDNA can have its own distribution of variability largely independent of that of the nuclear genome has important implications. For example, mtDNA can exhibit a limited migration across a hybrid zone without concomitant nuclear gene flow, a feature termed invasiveness by Ferris and Berg (1987). Under the assumption of selective neutrality for mtDNA, invasiveness is largely a function of the maternal immigration rate and not selection operating in the nuclear genome (Takahata and Slatkin 1984). Moreover, if migrating females distribute mtDNA between populations, and nuclear genes are carried by both sexes, a population may be subdivided for mitochondrial genes and yet appear panmictic for nuclear genes under differential rates of male and female migration (Birky et al. 1983). For example, if a single population becomes subdivided into two isolates, gene frequencies will diverge in both the nuclear and mitochondrial genomes as a result of drift and mutation. In fact, populations may become rapidly fixed for different mtDNA types due to the effects of drift (Birky 1991) and stochastic lineage extinction (Avise 1994). If these populations come into secondary contact, gene frequencies at variable loci in the nuclear genome initially develop into a steep cline, which gradually erodes until there is no evidence of past isolation. The erosion of mtDNA divergence follows a similar pattern but occurs more slowly. Furthermore, at the end of the process, the populations contain two distinct mitochondrial lineages, which provide a record of the historic isolation event (Billington and Hebert 1991b). For example, Ward et al. (1989) found evidence of secondary contact in their comparison of mtDNA and allozyme variation in walleye *Stizostedion vitreum* in Canada. Walleye mtDNA haplotypes separate into three main groups: an A group that predominates in the eastern Great Lakes drainage, a B group predominant in the central region, and a C group in the west. This distribution appears to reflect the recolonization of Canadian lakes from three separate glacial refugia. By contrast, allozyme data revealed only a western group, whose distribution corresponded with that of group C, and a second group whose range encompassed the whole of the Great Lakes drainage (Ward et al. 1989). Thus, mtDNA retained a history of the isolation of walleye in an Atlantic refugium, whereas nuclear DNA did not.

4.2.1 Genome Size, Length Variation, and Heteroplasmy

Mitochondrial genomes of fish vary little in size, with a range of 15.2 to 19.8 kbp reported among the 47 species reviewed by Billington and Hebert (1991b) that were studied by whole-molecule restriction fragment length polymorphism analysis. A **restriction fragment length polymorphism (RFLP)** is a polymorphism in an individual defined by restriction fragment sizes of distinctive lengths produced by a specific restriction endonuclease. The polymorphism is usually caused by the gain or loss of a restriction site but may also result from an insertion or deletion of a fragment of DNA between two conserved restriction sites. However, some of the smaller size values reported in the review by Billington and Hebert (1991b) are probably underestimates, as they were obtained in studies that employed techniques that did not permit detection of restriction fragments of less than 0.5 kbp. Based on this size stability, one might expect that length variation in mitochondrial genomes among closely related fish species would be rare, but this is not the case. Length variation has been noted among fishes of different genera, among conspecific individuals, and even among the mitochondria of single individuals (**length heteroplasmy** is the presence in an individual of mtDNA molecules of more than one size; see Table 4.1). In all cases in which such length inserts have been mapped or sequenced, they were located in or adjacent to the D-loop region (e.g., Broughton and Dowling 1994; Faber and Stepien 1997; Bentzen et al. 1998). Length variation in mtDNA of other animal species also has been mapped to the same region, suggesting that it is less affected by insertion and deletion effects than are those areas of the molecule that encode functional genes (Brown 1983; Moritz et al. 1987).

Possible mechanisms for the production of length variation include replication slippage, duplication, and deletion (Moritz et al. 1987). However, most inserts in fish mtDNA appear to be due to **tandem duplications** (adjacent repeated units of short DNA sequences) (Buroker et al. 1990; Broughton and Dowling 1994; Miracle and Campton 1995; Faber and Stepien 1997). **Restriction site heteroplasmy,** the presence in an individual of more than one type of mtDNA molecule, each type having different restriction sites for an endonuclease, has also been observed in fishes but usually at low frequency (Bentzen et al. 1988; Billington 1993). Extreme restriction site heteroplasmy observed in three individual European anchovies was attributed to paternal leakage (Magoulas and Zouros 1993).

Chapman (1987, 1989) and Wirgin et al. (1989) noted that length variation was more common than was restriction site variation in striped bass and that the incidence of length variants differed among stocks. However, the usefulness of length variants for stock discrimination remains uncertain until their transmission genetics are more completely understood. For example, it is uncertain if heteroplasmic individuals transmit variants to their offspring (Mulligan and Chapman 1989). Moreover, Stellwag et al. (1994) found a lack of congruence between individual striped bass length variants and specific genotypes identified by restriction site polymorphisms. The geographically and genotypically independent distributions of the length variants suggested that the mtDNA length variants had converged in striped bass, and Stellwag et al. (1994) cautioned against conducting phylogeographic analyses based solely on mtDNA length variants. Therefore, because of the concerns that have been raised about the heritability of length variants (Stellwag et al. 1994), they should be used only as genetic markers for the analysis of population subdivision when there is good evidence that they are stably inherited.

4.2.2 Molecular Clock Calibration

The average rate of mtDNA sequence divergence as determined by whole-molecule RFLP analysis for several mammalian species is approximately 2% per million years and remains linear for 8 to10 million years (Brown 1983). This rate has been generally confirmed for other homeotherms (Brown 1983), including birds (Shields and Wilson 1987). This molecular clock calibration has been employed to estimate the divergence times among several families of fishes (Salmonidae: Bernatchez et al. 1989; Grewe et al. 1990; Centrarchidae: Bermingham and Avise 1986; Percidae: Billington et al. 1990, 1991; and Cyprinodontidae: Brown and Chapman 1991).

Although Billington et al. (1990, 1991) found good agreement in divergence times among *Stizostedion* spp. estimated from allozymes, mtDNA data, and related geological events, it is clear that caution should be employed in assuming rate uniformity of

Table 4.1 Some examples of length variation (insertion [I]; tandem repeat [TR]; and deletion [D]) and heteroplasmy (length heteroplasmy [LH] and site heteroplasmy [SH]) in fish mitochondrial DNA (mtDNA), with frequency of occurrence in each survey. All lengths are in base pairs (bp).

Species	Frequency (%)	Variant size (bp)	Type	Reference
Gulf of Mexico Atlantic sturgeon *Acipenser oxyrhynchus desotoi*	18.5	81 1–4 copy repeat	TR LH	Miracle and Campton (1995)
White sturgeon *Acipenser transmontanus*	41	82 3–5 copy repeat	TR LH	Buroker et al. (1990)
	42	82 1–6 copy repeat	TR LH	Brown et al. (1992)
American shad *Alosa sapidissima*	12 4	1,500	LH SH	Bentzen et al. (1988)
Bowfin *Amia calva*	8	450–900	LH	Bermingham et al. (1986)
Spotfin shiner *Cyprinella spiloptera*	100	260 2 or 3 copy repeat	TR	Broughton and Dowling (1994)
Atlantic cod	100	40 2–6 copy repeat	TR LH	Arnason and Rand (1992)
Cape hake *Merluccius capensis*	12 25	200 400	D I	Becker et al. (1988)
White perch *Morone americana*	100	110 3–10 copy repeat	LH	Mulligan and Chapman (1989)

Striped bass[a]	B 45	100	TR	Chapman (1987, 1989)
Morone saxatilis	C 38	200	TR	
	D/E 3	150/250	TR LH	
	F 6	300	TR	
	B 7	200	TR	Wirgin et al. (1989)
	C 58	300	TR	
	D 28	400	TR	
	20		LH	
Yellow perch	4		SH	Billington (1993)
Perca flavescens				
Lake trout	1	20	LH	Grewe and Hebert (1988)
Salvelinus namaycush				
Sebastes spp.: Deepwater redfish		275	TR LH	Bentzen et al. (1998)
S. mentella, Acadian redfish	33	2 repeats		
S. fasciatus, and redfish *S. marinus*	9	3 repeats		
Walleye	2	1,700	I	Billington and Hebert (1988)
	2	130	I	Ward et al. (1989)

[a] Letters identifying size-classes for striped bass are different in each of the two studies.

mtDNA evolution. For example, the rate of mtDNA evolution in the herring genus *Alosa* was only one-tenth of that found in higher vertebrates (Bentzen et al. 1993). Dowling and Brown (1989) suggested that mtDNA divergence rate heterogeneity might be responsible for incongruencies between allozyme and mtDNA phylogenies for four *Notropis* spp. Lower rates of amino acid substitution have been observed in the mtDNA-encoded proteins of fishes and other cold-blooded vertebrates than in warm-blooded vertebrates (Kocher et al. 1989; Thomas and Beckenbach 1989). Sharks (Martin et al. 1992b) and antarctic fishes (Bargelloni et al. 1994) also appear to have much lower rates of mtDNA protein evolution. There is increasing evidence for an association between the thermal habit of organisms and their rate of mtDNA evolution (Rand 1994).

Clearly, the "universal" mtDNA clock rate should not be assumed to apply in mtDNA studies of fishes. Each case should be calibrated against a fossil record or other biogeological events, such as the rise of the Isthmus of Panama (3.5 million years ago) as a barrier to gene flow between Atlantic and Pacific Ocean fishes (Bermingham et al. 1997). Ultimately, resolution of the apparent rate heterogeneity in fish mtDNA will be decided by more detailed analysis of fish taxa that have diverged within the last 10 million years and for which there are comprehensive fossil records. Even so, the calculated confidence intervals around the calibrated clock regression line can be so large that the clock is likely to be imprecise (Hillis et al. 1996).

4.3 Methods of Obtaining Data

A number of different strategies can be used to examine mtDNA variation in fishes, depending on the level of resolution desired and the type and condition of tissues available (Figure 4.2). There are two basic approaches: (1) RFLP analysis of whole-molecule mtDNA by digesting it with restriction endonucleases, or (2) RFLP analysis or DNA sequencing of small segments of the mtDNA molecule obtained by means of amplification with the polymerase chain reaction (PCR).

4.3.1 *Whole-Molecule Analyses*

Whole-molecule mtDNA analyses are more traditional and will be considered first. They can be divided into two strategies: (1) in which purified mtDNA is obtained, and (2) in which mtDNA bands are resolved from total DNA digests by the **Southern blot** procedure. For this procedure, DNA fragments are denatured, transferred to a membrane from an electrophoresis gel by capillary action, and visualized by hybridization with a probe.

Purified mtDNA usually is obtained from fresh liver or gonadal tissues because of their high mitochondrial densities; however, essentially any tissue may be used (Chapman and Brown 1990; White and Densmore 1992). Normally, nuclei are removed from the tissue homogenate by low-speed centrifugation, mitochondria are then pelleted, and mtDNA is released by membrane lysis. In rapid preparation methods, protein contaminants are removed by organic extraction with phenol and chloroform, and the DNA is collected by ethanol precipitation. Alternatively, mtDNA may be highly purified, if needed, by cesium chloride density-gradient centrifugation (White and Densmore 1992).

The preservation history of the sample is important because it affects the choice of isolation method. Highly purified mtDNA can be obtained from only fresh tissue because the ultracentrifugation method relies on the mtDNA circles being intact. They

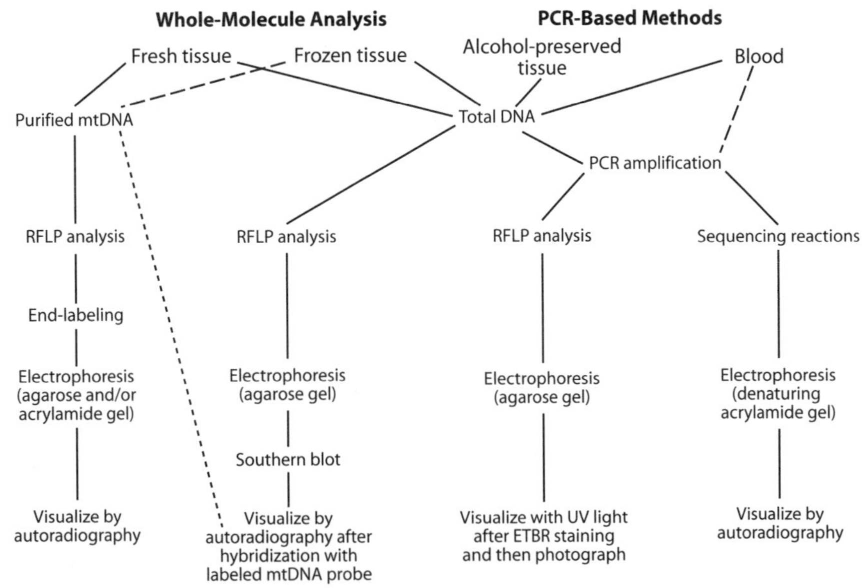

Advantages

Small fragments (10–400 bp) can be visualized	Frozen samples easier to transport and store	Analysis is rapid	Variation at the nucleotide level can be studied
4-base RE can be used efficiently	Nonlethal sampling for muscle can be used	Nonlethal sampling can be used	Analysis good for systematic studies
High-resolution restriction site maps easier to contrast	Blots can be re-probed to score nuclear DNA variation	Highly variable regions of the molecule can be selected	

Disadvantages

Extraction is time-consuming	Fragments <400 bp cannot be resolved	PCR primers needed	PCR primers needed
Ultracentrifuge is needed	Analyses often limited to 6-base RE	Variable regions must be found	Analysis costly, especially for population-level studies
mtDNA circles must be intact	High-resolution restriction site maps hard to construct	Amplfied fragments often small and RFLP patterns have many bands <400 bp	Too much variation sometimes for frequency-based analyses
Radioactive label required	Radioactive label often required		Radioactive label usually required
	Pure mtDNA probe needed		

Figure 4.2 Different strategies for the examination of mitochondrial DNA (mtDNA) variation in fishes. Solid lines indicate main pathways, dashed lines indicate possible but suboptimal pathways, and dotted lines indicate that purified mtDNA is needed to make the hybridization probe used to visualize the mtDNA bands in total DNA digests by the Southern blot procedure. Procedures include polymerase chain reaction (PCR), restriction fragment length polymorphism (RFLP) analysis, and ethidium bromide (ETBR) staining; other abbreviations given are base pairs (bp) and restriction enzymes (RE).

become broken by the freeze–thaw process associated with frozen material and with ethanol preservation. Thus, total DNA extractions are performed on frozen and ethanol-preserved tissue, and fragments are resolved by Southern blot analysis. The PCR amplification may be performed on DNA extracted from fresh, frozen, or ethanol-preserved tissue (Figure 4.2). It has even been possible to obtain mtDNA data from formalin-preserved museum collections by using PCR analysis of very short (<200 bp) segments (Wirgin et al. 1997).

The study of mtDNA variation is based upon digestion of the whole molecule by means of **restriction endonucleases,** commercially available enzymes that cleave DNA at specific recognition sequences (Roberts 1984). The resulting fragments are separated according to their molecular weight by gel electrophoresis, with the smaller fragments migrating faster than larger ones. The endonucleases most commonly used in RFLP surveys typically have a recognition sequence of 4, 5, or 6 bases and cut a typical mtDNA molecule 16.7 kbp long approximately 64, 16, and 4 times, respectively (assuming a random distribution of nucleotides). If a nucleotide substitution occurs that coincides with the recognition sequence of the restriction endonuclease, then a gain or loss of a restriction site will occur. A gain or loss results in a change in the number and lengths of the observed fragments; a restriction site gain in a large fragment will result in that fragment's replacement by two smaller fragments whose total lengths will be equal to the length of the original fragment. As mtDNA is a circular molecule, the number of fragments observed equals the number of restriction sites for a particular endonuclease. Fragments may be visualized either by ethidium bromide staining or by **autoradiography** (the making of an image on an X-ray film by exposing it to a gel or Southern blot containing radioactively labeled DNA fragments in a light-proof container) following **end-labeling** (the molecular genetic process in which radioactively labeled nucleotides are attached to the end of a DNA fragment using a polymerase enzyme) with radioactively (Figure 4.3) or chemically labeled nucleotides. Variability in fragment patterns is interpreted as genetic variation at the nucleotide level. Thus, a nucleotide substitution in the molecule causing the loss or gain of a nucleotide sequence recognized by a specific restriction enzyme will cause a change in the observed fragment pattern (Figure 4.3). Length variants are identified when larger or smaller bands than usual are observed.

An alternative method to obtaining purified mtDNA is to visualize the mtDNA bands in total DNA digests with a labeled mtDNA **probe,** a radioactively or chemically labeled DNA fragment that hybridizes to an homologous target DNA sequence, permitting its visualization. Total DNA is extracted from tissue samples (e.g., muscle or liver) (Grewe et al. 1993) or blood (Billington and Hebert 1990) and then digested with endonucleases. The restricted DNA fragments are separated by electrophoresis, the DNA is denatured to make it single stranded, and it is transferred and bound to a nylon membrane (Chapman and Brown 1990). A probe is made from purified mtDNA, usually from the species of interest, but mtDNA from a more distantly related species often will work (González-Villaseñor et al. 1986). This probe mtDNA is labeled with radioactively labeled nucleotides, denatured to make it single stranded, and exposed to the nylon membrane. There it hybridizes to the mtDNA fragments in the total DNA digest. Then the mtDNA fragments are visualized by autoradiography after excess probe has been washed away. Alternatively, it is possible to label the mtDNA probe with digoxigenin-labeled nucleotides and visualize the results by chemiluminescence (Holtke et al. 1992), thus avoiding the use of radioisotopes.

Detailed accounts of whole-molecule mtDNA analytical protocols can be found elsewhere (e.g., Chapman and Brown 1990; Tegelstrom 1992; White and Densmore

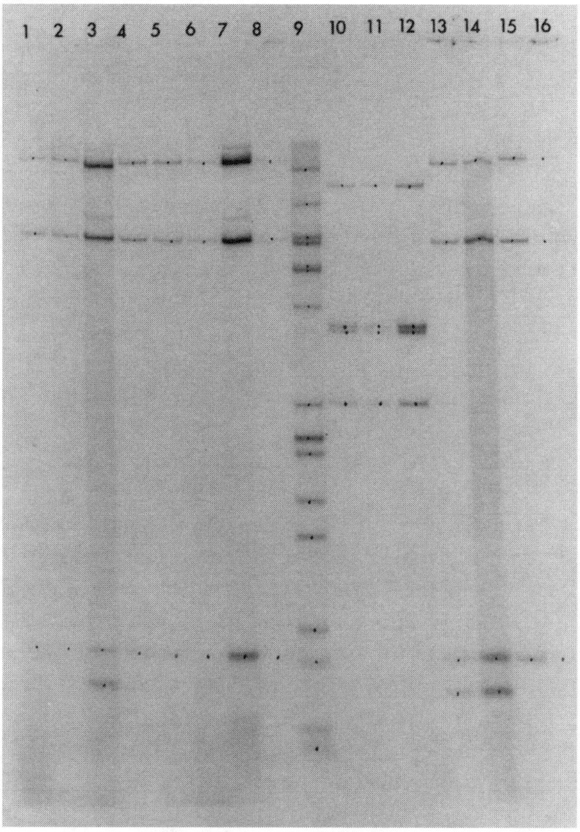

Figure 4.3 Autoradiograph showing fragment patterns between walleye and zander *Stizostedion lucioperca*. Purified mtDNA from walleye (lanes 1–8 and 13–16) and zander (lanes 10–12) were digested with the restriction endonuclease *Nco* I, end-labeled with ^{32}P-labeled nucleotides, and subjected to electrophoresis in a 1% agarose gel along with (in lane 9) an EHH λ DNA size standard (consisting of λ phage DNA digested with *Hind* III mixed with λ phage DNA double-digested with *Eco*R I and *Hind* III to make a DNA size standard). Note that the fragment patterns between walleye and zander are quite different. This endonuclease reveals an intraspecific polymorphism in walleye mtDNA. There are two fragment patterns observed: a site loss occurs between pattern A (lanes 3, 13, and 14), with four restriction sites, and pattern B (lanes 1, 2, 4–8, 15, and 16), with three restriction sites.

1992; Grewe et al. 1993; Dowling et al. 1996). Chapman and Brown (1990) review the various methods for examining whole-molecule mtDNA variation in fishes and discuss the criteria for choosing a particular method. However, they stress that techniques that may work well for some fish species may not work for others.

4.3.2 Polymerase Chain Reaction Fragment Analyses

Use of PCR to amplify DNA obtained from extremely small amounts of tissue is now common. The **polymerase chain reaction** technique allows short segments (300–5,000 bp) of DNA to be amplified, thereby producing a large number of copies, so that variation in these DNA segments can be examined either by RFLP analysis or directly by sequencing. The PCR-amplification technique is important in studies of rare and endangered species because small amounts of tissue can be collected by nonlethal sampling methods. In fact, DNA can be isolated from preserved or even partly degraded material. Thus, the PCR method provides an opportunity for comparison of DNA sequence divergence in recently extinct species for which only museum specimens are available (Arnheim et al. 1990). Moreover, the epithelial cells on fish scales can be used as a source of DNA that can be amplified by PCR (Whitmore et al. 1992) so that mtDNA variation can be examined by either RFLP or sequence analysis. In fact, it is possible to look at historical levels of genetic variation in fish populations and to follow any changes over time by determining mtDNA variation following PCR amplification of DNA obtained from archived scale samples that originally were collected by fisheries management agencies to obtain age and growth data (Gatt et al. 2002). Such an approach has also been undertaken on fish species with microsatellite DNA markers (e.g., Nielsen et al. 1999).

The use of PCR amplification to study both intraspecific and interspecific variation in fish mtDNA is described in detail by Beckenbach (1991). **Primers** (small pieces of single-stranded DNA usually about 20 to 30 nucleotides in length and complimentary in sequence to regions of the molecule that flank the area or gene of interest) need to be obtained or synthesized. Usually these primers are targeted to regions of the molecule where nucleotide sequences are conserved among many organisms, such as in tRNA genes or highly conserved parts of the gene of interest. Some primers (referred to as universal primers) have been identified that allow the amplification of mitochondrial genes, or parts of genes, from many species of vertebrates and even from some invertebrates (e.g., Kocher et al. 1989; Hoelzel and Green 1992). Often, however, primers that are more specific to the species or genera of interest need to be developed for optimum PCR amplification (Palumbi 1996). The PCR primers that allow the amplification of specific regions and genes from fish mtDNA also have been described (e.g., Cronin et al. 1993; Meyer 1994).

The template DNA is mixed with the two primers that flank the region to be amplified, a thermally stable polymerase and its buffer, and the four nucleotides (adenine [A], cytosine [C], guanine [G], and thymine [T]). The PCR amplification process involves heating the template DNA (usually to 94°C) to denature it. Then, the mixture is cooled to allow the primers to anneal to their complimentary sequences (the temperature for this depends on the primers being used, but is often 50°C for many universal primers). Finally, the reaction is heated (usually to 72°C) to allow the polymerase to incorporate the nucleotides and synthesize the complimentary DNA strand to the template. Usually, the process is repeated for approximately 30 cycles, after which 2^n times the original quantity of the desired sequence will have been amplified (where n is the number of cycles). It is essential to run a control reaction, containing all of the components used in the PCR reaction except the template DNA, to confirm that these components have not become contaminated with nontarget DNA. The PCR reaction products, along with the control and a size standard, are electrophoresed in an agarose mini-gel to confirm that no contamination has occurred and that the PCR product is of the expected size.

Once the region of interest has been amplified, the amplified DNA can be cut with restriction endonucleases, and the resulting fragments screened on an agarose gel to identify RFLPs. Alternatively, the amplified DNA can be sequenced. Usually, RFLP analysis of amplified products is used in population level studies, whereas sequencing is most commonly used for systematic studies. Occasionally, sequencing is used in population level studies if levels of variation are low in the species being studied. However, sequencing in population level studies is often an inefficient approach because it is more expensive in materials and time than are RFLP surveys, and it may generate too much variation (i.e., each individual examined may have a unique sequence) for frequency-based analyses.

4.4 Data Analysis

4.4.1 *Haplotype Diversity*

Fish that have the same mtDNA sequence, or that have the same composite restriction fragment pattern, are classed as having the same mtDNA haplotype. Once the different haplotypes in a sample have been identified, their relative frequencies in the sample can be calculated. In most fish species examined to date, a few common haplotypes have been identified along with a substantially larger number of rare haplotypes, which are the common haplotype mutational derivatives (Billington and Hebert 1991b). Comparisons of mtDNA diversity among species are complicated, however, by the fact that the number of haplotypes detected, particularly in RFLP surveys, is a function of both the number of fish examined and of the intensity of sampling the genome (i.e., the number of fragments surveyed, which itself is a function of the number and type of endonucleases used). Thus, there is ordinarily a positive association between haplotype diversity and both the number of fish examined and the number of endonucleases employed in an analysis. The effect of sample size on the number of haplotypes detected can be removed through the use of diversity indices (Box 4.1), but standardization of sampling effort on the genome is more difficult to achieve.

Apart from the desirability of comparing levels of haplotype diversity among different taxa, it is also important to ascertain whether sufficient effort has been expended in a particular study to recognize essentially all of the mtDNA diversity present. This problem is analogous to the measurement of clonal richness in asexual organisms, where a combinatorial analytical technique (Hebert et al. 1988) has been employed. This same procedure, involving incremental random resampling of individuals and endonucleases, also has been used to estimate the likelihood of detecting mtDNA diversity in fishes (Bernatchez et al. 1989; Bernatchez and Dodson 1990; Brown et al. 1992).

4.4.2 *Genetic Divergence among Haplotypes*

Once mtDNA haplotype diversity has been assessed, it becomes important to examine the extent of genetic divergence among haplotypes. High genetic similarity among haplotypes suggests that they were derived recently from a common ancestral haplotype, whereas greater divergence suggests a longer interval since their origin. Sequence divergence estimates can be derived from the comparison of aligned nucleotide sequences. In RFLP surveys, sequence divergence can be estimated from a comparison of congruence in fragment patterns or from a direct comparison of shared restriction sites. The fragment method suffers from the disadvantage that taxa sharing restriction sites may share no fragments. Moreover, site characters are preferred to fragment characters because fragment data provide nonindependent information concerning a single evolutionary event (the gain or loss of a restriction site). When the haplotypes are closely related, site gains and losses can be inferred from fragment data. However, analysis of shared restriction sites among more distantly related haplotypes requires construction of detailed

■ Box 4.1
Haplotype Diversity

The diversity of mitochondrial DNA (mtDNA) lineages within a population can be evaluated by estimating the haplotype (nucleon) diversity (h) by the method of Nei and Tajima (1981).

$$\hat{h} = \frac{n\left(1 - \sum_{i=1}^{l} x_i^2\right)}{n - 1},$$

where x_i is the frequency of the ith type of mtDNA in a sample of n individuals, and l is the number of haplotypes observed in the sample. The haplotype diversity index (h) estimates the probability that two randomly chosen individuals from the population will have different haplotypes. Calculation of h permits comparisons among published mtDNA diversity estimates, with values of h ranging from 0.0 (when there is only one haplotype observed in a population) to 1.0 (when every individual exhibits a different haplotype). For haploid genomes, that is, for mtDNA, haplotype diversity is analogous to expected heterozygosity (Nei and Tajima 1981).

An estimate of nucleotide diversity (π) can be calculated as

$$\hat{\pi} = \frac{n}{n-1} \sum_{ij} \hat{x}_i \hat{x}_j \pi_{ij},$$

where x_i and x_j are the frequencies of the ith and jth type of mtDNA in a sample, and π_{ij} is the fraction of mismatched sites between the two types (Nei and Tajima 1981). This index considers both the frequency of occurrence of haplotypes in a population and the divergence among them.

restriction-site maps (Figure 4.4), which can be time-consuming. Direct sequence analysis of PCR-amplified segments of the mitochondrial genome avoids the difficulties inherent in map construction and offers the best possibility of rapid assessment of sequence divergence. Even so, insertions and deletions (indels) of nucleotides can cause gaps in some sequences that may result in problems in sequence alignment.

4.4.3 *Evolutionary Relationships among Haplotypes*

The evolutionary relationships among haplotypes can be examined by using three general approaches: distance-based analyses, parsimony, or maximum likelihood. Distance, parsimony, and maximum likelihood approaches also are used to study phylogenetic relationships among fishes at and above the species level based on mtDNA sequence divergence data (e.g., Sloss 2001). Page and Holmes (1998) provide a detailed discussion of each of these methods along with their advantages and disadvantages.

The genetic distance between haplotypes can be calculated and the evolutionary relationships among them can be reconstructed in a phylogenetic tree. Tree-building algorithms such as unweighted pair-group method with arithmetic means (UPGMA) clus-

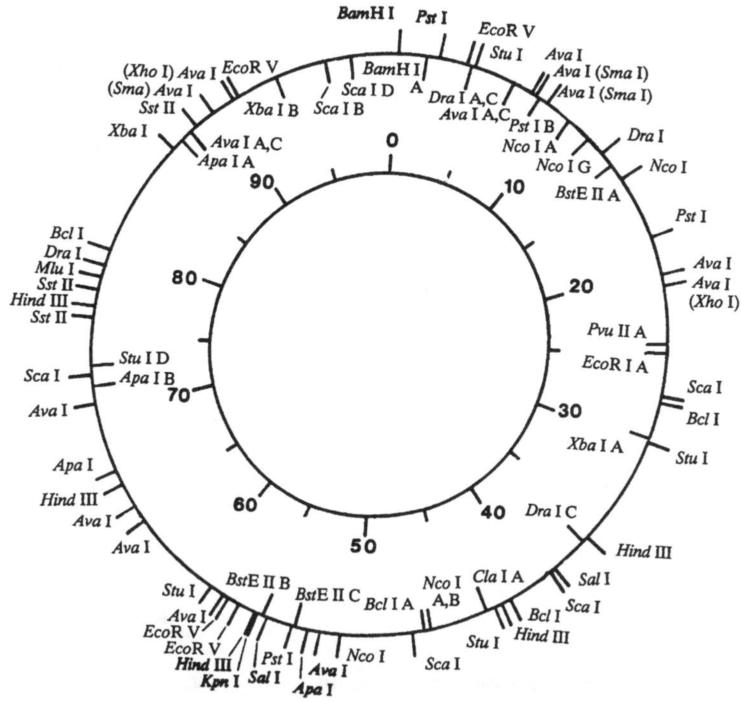

Figure 4.4 Restriction map of the walleye mtDNA molecule for all restriction sites found in haplotypes 1–10 and 34 (data from Billington and Strange 1995), with abbreviations for restriction endonucleases following Roberts (1984). Invariant sites are shown on the exterior of the circle, and variable sites are shown on the interior. Capital letters designate restriction fragment patterns that the variable sites (interior circle) represent. Semi-isoschizomers of *Ava* I are shown in parentheses. The scale is in 0–100 map units.

tering and neighbor joining (Saitou and Nei 1987) are commonly used, but they tend to rely on the rate of evolution being approximately equal among lineages.

Alternatively, a parsimony network can be constructed showing the restriction-site changes among the haplotypes so that they can be phylogenetically ordered within a species. The nonrecombining, maternal mode of mtDNA inheritance means that such a network reflects the maternal gene genealogy of the organism. Moreover, this parsimony network can be superimposed on a map of sampling locations, and often this juxtaposition provides a valuable insight into the historical processes governing the geographical distributions of the genealogical lineages of the haplotypes. This area of study has been called phylogeography (Avise 1994, 2001).

Maximum likelihood methods involve finding the likelihood of obtaining the observed data set given a particular phylogenetic tree; the tree that makes a given data set the most probable evolutionary outcome is the maximum likelihood estimate of the phylogeny. Maximum likelihood methods of inference can incorporate explicit models of sequence evolution and permit statistical tests of evolutionary hypotheses.

It is now common practice to calculate confidence estimates on the branches of phylogenetic trees by bootstrapping. The original data set is resampled with replacement many times to provide a series of pseudoreplicate data sets. A phylogeny is inferred for each of these new data sets, and the number of times that a corresponding monophyletic group is supported is expressed as a percentage that enables the strength of a phylogenetic hypothesis to be assessed.

There is one complication with inferring phylogenies based on mtDNA sequence variation, however, because mtDNA is inherited as a single linkage group. Thus, any phylogeny will reflect that of the gene (gene tree), but it may not reflect the phylogeny of the species (species tree). Even in the absence of hybridization, the phylogeny of the gene may differ in topology from that of the species (or population) due to the process of stochastic lineage sorting of ancestral mtDNA polymorphisms (Avise 1994). In contrast, if phylogenies are inferred from sequence analysis of different sets of nuclear genes selected from different chromosomes, each nuclear gene tree will provide an independent estimate of the species tree. Neutral polymorphisms will be pruned out of mtDNA gene trees in $4N_f$ generations, where N_f is the number of breeding females (Avise 1994). Therefore, if more than $4N_f$ generations have elapsed since a species last split from its nearest relatives, it will be monomorphic in its matriarchal ancestry. At intermediate times since the speciation of two sister taxa, their gene trees are likely to be poly- or paraphyletic, so haplotypes in one taxon may be more closely related to haplotypes in the other taxon. Furthermore, if the branch length between the first and second bifurcation in a three-species phylogeny is short, the probability that the gene tree and the species tree will be incongruent is high. However, the probability that a gene tree will coincide with the species tree in this case also depends on the effective population sizes of the taxa during the first and second bifurcation. Thus, Moore (1995) has shown that because the effective population size of a mitochondrial gene is less (¼) than that of an autosomal nuclear gene, a mitochondrial gene tree has a better chance of tracking the species tree than any single nuclear gene tree.

4.4.4 *Population Level Analyses*

In population surveys, homogeneity chi-square tests usually are performed to determine whether the haplotype frequencies differ significantly among populations. As most haplotypes observed in population level mtDNA studies on fish occur only rarely, these low sample numbers per cell can invalidate the assumptions of the chi-square analysis. This problem can be overcome with a Monte-Carlo approach in which the magnitude of the chi-square value obtained is compared with the magnitude of many pseudo-chi-squared values obtained from random rearrangements of the data table, while maintaining the row and column totals (Roff and Bentzen 1989). If significant heterogeneity is found in the data set, further tests can be performed to determine the partitioning of this heterogeneity. It is important to remember that if multiple post hoc chi-square tests are to be performed on subsets of the data, the alpha level for these tests should be adjusted by α/n (where n is the number of post hoc tests) (Cooper 1968).

The presence of population subdivision can be evaluated by conducting gene diversity analyses, either by using restriction sites as alternative alleles (Takahata and Palumbi 1985) or by treating mtDNA haplotypes as alternative alleles (Birky et al. 1989; Ward et al. 1989). Then, gene flow can be evaluated by estimating the effective migration rate ($N_e m$) from the genetic variation among populations, G_{ST} (e.g., Brown and Chapman 1991; Grewe et al. 1994a; Chapter 7) or by other approaches (i.e., from private alleles, via phylogenetic analyses, or from F_{ST}, the fixation index) (e.g., Brown and Chapman 1991). It is also possible to partition mtDNA haplotypic diversity at different hierarchical levels of subdivision by performing an analysis of molecular variance (AMOVA; Excoffier et al. 1992). This method was used by Danzmann and Ihssen (1995) to partition mtDNA variation in brook trout from Algonquin Park in Ontario among populations (47%) and among drainages (14%).

Finally, likelihood-based methods have been developed to estimate the relative proportions of different fish stocks in a mixed population (also known as genetic stock identification or mixed-stock analysis; see Chapter 13) from mtDNA haplotype frequency data (Xu et al. 1994; Brown et al. 1999). Many recent studies have found that a combination of both mtDNA and allozyme analyses is most useful for mixed-stock analysis (e.g., Grewe et al. 1994a; McParland et al.1999; Seeb and Crane 1999).

A number of computer programs are available for conducting the above analyses, and Swofford et al. (1996) list the sources of many of them. Some of these programs can be downloaded directly to your computer by electronic mail, whereas others are available either commercially or from their authors.

4.5 Applications

Mitochondrial DNA analysis lends itself to a number of applications in fisheries biology and management. These include the determination of variation among individuals, through the examination of variation within and among populations, to the phylogenetic examination of variation among species and other higher taxonomic categories (e.g., genera). Within these categories, mtDNA studies can contribute to the identification of fish stocks and the determination of the contribution of these stocks in a mixed fishery (Chapter 13), provide information relevant to the study of hybridization and introgression between fishes (Chapter 15), serve as a genetic marker in forensic analysis (Chapter 17), and provide critical information for use in the conservation and rehabilitation of fish species (Chapters 14, 15).

4.5.1 *Examination of Genetic Variation among Individuals within Populations*

It is often possible to discriminate among individual fish within a population by mtDNA analysis. For example, using a combination of 4-, 5-, and 6-base restriction endonucleases, Billington (1998) was able to discriminate 32 of 33 walleyes in the central Great Lakes region. Similarly, Avise et al. (1989) observed that menhaden *Brevoortia* spp. exhibited sufficient mtDNA size and restriction site polymorphism to permit the unambiguous discrimination of a high proportion (32/33) of the conspecific individuals assayed. Such extreme mtDNA variation also occurs in other fish species. Individual walleyes and individuals of other fish species can be distinguished by their unique DNA sequences (Stepien 1995). The ability to use such mtDNA fingerprints, or to identify individuals by DNA sequencing following PCR amplification of DNA obtained from tissue collected by a nonlethal sampling procedure (e.g., Whitmore et al. 1992), should prove important in maternity analyses. Thus, it should be possible to discriminate among the progeny of individual females and to compare the relative contribution of individual females to juvenile recruitment.

4.5.2 *Population Level Mitochondrial DNA Diversity*

Mitochondrial DNA analysis has been used in numerous population level studies of genetic diversity in fishes. In many cases, the aim of these studies was to determine the presence of stock structure in exploited fishes; in other cases, it was the identification of historical biogeographic partitioning of mtDNA variation. Practical applications of such work include the development of genetic markers for use in the management of fish stocks (Chapter 16). Population level mtDNA studies of exploited fishes are numerous and will be discussed below. However, there are relatively few studies of mtDNA variation among populations of nongame fish species. Brown and Chapman (1991) found that there was no evidence of geographic population structure in the mummichog based on whole-molecule mtDNA RFLP analysis, with most variation being found within, rather than among, sampling locations. Because estimated gene flow was much greater than one migrant per generation (N_em ranged from 11.4 to 49.4, depending on the method of estimation used), gene flow was sufficient to prevent population differentiation in this species due to genetic drift. Sequence analysis of the mtDNA D-loop region of the Trinidad guppy *Poecilia reticulata* revealed variation among natural populations that existed primarily among drainages (Fajen and Breden 1992). Whole-molecule RFLP analysis of freshwater populations of the threespine stickleback *Gasterosteus aculeatus* revealed two divergent groups of mtDNA haplotypes, which suggested the presence of a Pleistocene glacial refugium for one of these lineages on the Queen Charlotte Islands (O'Reilly et al. 1993). It is surprising that more studies of population level processes in fishes have not been conducted on nongame species considering fisheries management agencies tend to enhance gene flow by the stock transfer of game fishes within the natural range of a species (Ferguson 1990; Billington and Hebert 1991a; Weiss et al. 2001). Such studies may be particularly useful in identifying zoogeographic distribution patterns that may be impossible to reconstruct from game fish species for which stock transfers have confused mtDNA geographic distributions.

4.5.3 Stock Discrimination

Allozyme analyses have been successful in stock discrimination of many fish species (Allendorf et al. 1987; Utter 1991), but in some species comprehensive allozyme studies have failed to reveal significant genetic variation (Todd 1981; Leary and Booke 1982). It is now apparent that there is substantial mtDNA variation in many fish species and that this variation can serve as a basis for stock discrimination (Billington and Hebert 1991b). There are a number of instances in which mtDNA analyses have revealed considerably more genetic variation than have allozyme analyses. For example, Billington (1993) found seven mtDNA haplotypes in 22 yellow perch in western Lake Erie, where a previous allozyme study of 54 loci had failed to reveal any genetic variation (mean heterozygosity, $\bar{H}$, was 0). Ward et al. (1989) showed that, consistent with theoretical expectations (Birky et al. 1983), mtDNA data were more effective for the discrimination of walleye stocks than were allozyme data; 37% of the total variation among walleye mtDNA haplotype frequencies could be attributed to interpopulational variation, whereas only 10% of allele frequency differentiation was linked to interpopulational variation.

There have been some occasions, however, in which mtDNA analysis has revealed equally informative, or even less useful, genetic data than has allozymes. For example, Ferguson et al. (1991) found substantial allozyme variation among brook trout sampled from five rivers on Cape Race, Newfoundland (42 loci, $\bar{H}$ ranged from 0.02 to 0.06, Nei's D ranged from 0.02 to 0.48), whereas fish from three of these rivers exhibited only a single mtDNA haplotype (haplotype diversity, $\hat{h}$, was 0). Fish from a fourth river had only two haplotypes (0.31% sequence divergence between them, $\hat{h}$ = 0.534), one of which was the same as in the other three populations. Thus, in this case, mtDNA analysis did not provide greater resolution of population structure than did allozymes. O'Connell et al. (1995) found that allozyme and mtDNA RFLP data were equally informative for distinguishing populations of Atlantic salmon. They suggested that the increased sensitivity of mtDNA to fluctuations in population size can lead to depressed levels of variability, so that mtDNA may not prove to be more informative than allozymes in salmonids. Sampling families rather than populations may also result in biased estimates of mtDNA haplotype frequencies, as was documented by Hansen et al. (1997) in a study involving both mtDNA and microsatellite analyses of brown trout. In two populations surveyed, mtDNA haplotype frequencies differed significantly between age-classes, and 17 of 18 individuals shared the same mtDNA haplotype in one sample due to them likely representing only three full-sib families. This problem appeared to be more severe for juveniles than for older brown trout.

Initially, it was hoped that mtDNA analysis would be an exceptionally useful tool for stock discrimination, but the early optimism that fixed haplotype differences would be found among stocks was not realized. Instead, the ability to discriminate stocks usually depends on the statistical significance of haplotype frequency differences, in a similar fashion to the analysis of allele frequencies for allozyme data. Other recent methods of detecting nuclear DNA variation, especially microsatellites (see Chapter 5), often have revealed variation in cases in which mtDNA analyses have not. For example, Ruzzante et al. (1996) showed that microsatellite DNA analysis permitted discrimination between Atlantic cod overwintering inshore in Newfoundland and those overwintering offshore, whereas mtDNA analysis of these same Atlantic cod populations did not (Carr et al. 1995).

Most of the early attempts to use mtDNA analysis to examine genetic variation in fishes involved whole-molecule RFLP analysis (reviewed by Billington and Hebert 1991b, with additional references herein). More recently, it has been found that resolution of population level genetic variation can be improved and data collection achieved in a much more timely fashion by PCR–RFLP analysis of regions of the mtDNA molecule that evolve rapidly (Martin et al. 1992a; Cronin et al. 1993; Merker and Woodruff 1996). Merker and Woodruff (1996) were able to discriminate two of three walleye populations from the western basin of Lake Erie using PCR–RFLP analysis, whereas whole-molecule RFLP analysis failed to separate these populations (Billington et al. 1992). Interestingly, while analysis of the noncoding D-loop region by PCR–RFLP analysis has proved informative on occasion (e.g., Martin et al. 1992a; Tabata and Mizuta 1997; Gatt et al. 2000; Szalanski et al. 2000), it often has failed to reveal significant variation in some species (Hall and Nawrocki 1995; Hansen and Loeschcke 1996; Merker and Woodruff 1996). In addition, considerable variation has been found in other regions of the molecule, especially the NADH dehydrogenase (ND) complex genes: *ND1* (Hall and Nawrocki 1995; Hansen and Loeschcke 1996; Williams et al. 1997a; Nielsen et al. 1998; Paragamian et al. 1999; McCusker et al. 2000: Nilsson et al. 2001), *ND2* (Williams et al. 1997a; Paragamian et al. 1999), *ND3/4* (Merker and Woodruff 1996; Nielsen et al. 1998), and *ND5/6* (Hall and Nawrocki 1995; Hansen and Loeschcke 1996; Williams et al. 1997a; Nielsen et al. 1998; Pigeon et al. 1998; Paragamian et al. 1999). Occasionally, some problems in detecting genetic variation in early studies were related to using universal primers (e.g., Cronin et al. 1993) rather than more specific primers developed for the species in question (or another closely related species), and variation was revealed in further studies when more specific primers were developed (e.g., Nielsen et al. 1998). Sufficient variation for population level analyses has been found in the cytochrome-*b* gene (e.g., Williams et al. 1997a; Wood and Raley 2000) and even in the more conserved *12S* and *16S* rRNA genes in some species (e.g., Martin et al. 1992a; Kreiser et al. 2001).

Direct sequencing of mtDNA regions following PCR amplification also has uncovered genetic variation (e.g., Carr and Marshall 1991; Finnerty and Block 1992; Stepien 1995; Dougherty et al. 1996; Stepien et al. 1998; Stepien and Faber 1998; Arnason et al. 2000; Wirgin et al. 2000; Muss et al. 2001) and has revealed additional variation compared with whole-molecule RFLP analysis (Stepien 1995; Stepien and Faber 1998). However, one problem with this approach is that occasionally it can reveal too much variation for population level analyses—in extreme cases almost every individual has a unique mtDNA sequence (e.g., Stepien 1995). Sequencing is also relatively expensive and time-consuming, although the more widespread availability of automatic sequencers may help reduce these problems. Therefore, although sequencing will be useful in some circumstances, particularly in conservation genetic studies in which sample sizes may be small, it would appear that the PCR–RFLP approach offers a more practical method of rapid population level mtDNA analysis.

Programs to ensure the conservation of genetically divergent stocks of a fish species require as a first step the identification of such stocks. The value of frequency shifts in mtDNA haplotypes for stock discrimination has clearly been shown in a number of studies (reviewed by Billington and Hebert 1991b). More recent examples include work on barramundi perch *Lates calcarifer* (Chenoweth et al. 1998), brown bullhead *Ameiurus nebulosus* (Murdoch and Hebert 1997), capelin (Dodson et al. 1991), Atlantic cod (Dahle 1991), common snook *Centropomus undecimalis* (Tringali and Bert 1996), black mudfish *Neochanna diversus* (Gleeson et al. 1999), striped mullet *Mugil cephalus* (Crosetti et al. 1994), haddock (Zwanenburg et al. 1992), perches (Percidae) (Refseth et al. 1998; Stepien and Faber 1998; McParland et al. 1999; Nesbø et al. 1999; Gatt et al. 2000), red drum (Gold and Richardson 1991), trouts (Salmonidae) (Birt et al. 1991; King et al. 1993; Hartley 1995; Brykov et al. 1996; Hynes et al. 1996; Vitic and Strobeck 1996; Danzmann et al. 1998; Redenbach and Taylor 1999; King et al. 2000; Bernatchez 2001; Brunner et al. 2001; Nilsson et al. 2001), sharks (Heist et al. 1996a), and sheatfishes (Siluridae) (Triantafyllidis et al. 1999). Fortunately, it appears that mtDNA haplotype frequencies in populations appear to be temporally stable on a time scale appropriate for fisheries management (Gold et al. 1993; Grewe et al. 1994b; Brykov et al. 1996; Stepien and Faber 1998), although the issue of haplotype temporal stability is addressed only rarely.

It is important to consider the limitations of mtDNA-based stock recognition. Assuming an average rate of mtDNA evolution of 2% per million years (Brown 1983), then 330 bp substitutions would be expected per million years in a 16,500-bp-long mtDNA molecule. If the entire molecule is surveyed, this corresponds to 3.3 substitutions per 10,000 years. Therefore, even with complete isolation, populations that have colonized habitats since the end of the Pleistocene will show little divergence. Moreover, as most whole-molecule mtDNA RFLP studies are based on surveys of 3 to 5% of the genome, the chance of identifying divergence is further attenuated. Hence, it is unrealistic to expect to be able to differentiate stocks in nearby river or lake systems on the basis of divergence, which has arisen since their relatively recent colonization. Fortunately, the ability to discriminate stocks does not rest solely on divergence, which postdates isolation, but instead may also be based on preexisting mtDNA polymorphisms (Billington and Hebert 1991b). Such divergence arises as a consequence of sampling effects associated with population founding. Where individual populations originate from a few females, then founder effects, genetic drift (Birky 1991), and stochastic lineage extinction (Avise 1994) may lead to fixation for different haplotypes, thus permitting unambiguous discrimination of the stocks.

In a number of managed fish species, the identification of genetically distinct stocks has been complicated by recent stock transfers, including those of hatchery origins (Bermingham and Avise 1986; Ovenden et al. 1988; Billington and Hebert 1991b; Billington et al. 1992; Bernardi et al. 1995; Jones et al. 1997; Hansen et al. 2000; Weiss et al. 2001). In some cases, the stocking of fish will complicate stock identification based on mtDNA analysis, but in other cases, useful information about the source of the stocked fish may become apparent. For example, walleye populations stocked with fish of known origin show similar proportions of mtDNA haplotypes to their source populations. Populations that were supplemented by stocking now possess haplotypes that serve as genetic markers for the populations used to stock them (Billington et al. 1992).

4.5.4 Geographic Patterning in Mitochondrial DNA Diversity

Peak levels of mtDNA divergence based on whole-molecule RFLP analyses in freshwater fish species (10.0%) are greater than those in anadromous (5.0%), catadromous (2.0%), brackish (2.0%), or marine (4.4%) fishes (Billington and Hebert 1991b). However, this low level of divergence in marine fishes is in contrast to the large number of haplotypes (haplotype diversity) in marine systems (Table 4.2). Ovenden (1990) also documented generally low mtDNA divergence values in marine organisms and suggested a number of possible explanations. These included a possible slower rate of mtDNA evolution in marine organisms (this may be true in some [e.g., Bentzen et al. 1993; Bargelloni et al. 1994] but not all cases), a lower diversity due to bottleneck effects (not true in terms of the number of haplotypes, or haplotype diversity), and family-specific mortality in species in which the offspring of individual females are grouped into egg-masses or schools (but, again, most marine species have high haplotype diversity). Although significant population structure has been observed in some marine species (Mulligan et al. 1992; Zwanenburg et al. 1992; Crosetti et al. 1994), in many cases there seems to be significant gene flow occurring in many marine systems (Kornfield and Bogdanowicz 1987; Brown and Chapman 1991; Martin et al. 1992a; Scoles and Graves 1993; Grewe et al. 1994a; Heist et al. 1995, 1996a, 1996b; McGauley and Mulligan 1995; Graves 1996), which would reduce mtDNA divergence among populations. By contrast, in freshwater systems populations generally are more isolated (so there is likely less gene flow) and population sizes tend to be smaller, so that different haplotypes are more likely to become fixed as a result of genetic drift and mtDNA lineage extinction processes. Furthermore, in freshwater systems that were affected by glaciation, there is evidence that divergent haplotypes are likely the result of recolonization of fish from separate glacial refugia (see below). Finally, other historical factors, such as geology and climate, have been implicated for the regional biogeographic distribution of highly divergent haplotypes in freshwater fishes (Bermingham and Avise 1986; Billington and Hebert 1991b; Avise 1992, 1994, 1996; Bernatchez and Wilson 1998; Bernatchez 2001; Brunner et al. 2001). Some of these events also appear to have affected some coastal marine systems (Avise 1992, 1994, 1996).

The extent of mtDNA variation in freshwater fishes at northern latitudes also appears related to geological events, with the least amount of sequence divergence observed in northern species and the greatest divergence occurring in species from unglaciated regions (Billington and Hebert 1991b; Bernatchez and Wilson 1998). Some of the lowest values of sequence divergence occur in species whose current ranges were largely glaciated during the Pleistocene. The reduced variation in these species is most probably due to the effects of reduced population size in glacial refugia causing bottlenecks (e.g.,

Table 4.2 Documented mtDNA haplotype diversity values in fishes from studies of whole-molecule restriction fragment length polymorphisms.

Species	Number of populations (number with 0 values)	Haplotype diversity mean (SD)	Reference
		Freshwater	
Northern pike *Esox lucius*	2 (1)	0.080 (0.08)	Brzuzan et al. (1998)
Sauger *Stizostedion canadense*	16 (9)	0.224 (0.285)	Billington (1996a)
Yellow perch	11 (4)	0.227 (0.239)	Billington (1996a)
Brook trout *Salvelinus fontinalis*	58 (27)	0.262 (0.284)	Danzmann and Ihssen (1995)
Powan *Coregonus lavaretus*	7 (1)	0.344 (0.294)	Hartley (1995)
Rainbow trout	6 (2)	0.391 (0.375)	Danzmann et al. (1993)
Threespine stickleback *Gasterosteus aculeatus*	10 (2)	0.398 (0.235)	O'Reilly et al. (1993)
Cisco *Coregonus artedi*	8 (2)	0.416 (0.376)	Snyder et al. (1992)
Walleye	15 (3)	0.429 (0.265)	Ward et al. (1989)
White sturgeon	5	0.434 (0.194)	Brown et al. (1992)
		Anadromous	
Lake whitefish *Coregonus clupeaformis*	6 (1)	0.252 (0.260)	Bernatchez et al. (1989)
Brown trout *Salmo trutta*	7 (1)	0.567 (0.271)	Hansen et al. (1995)
Atlantic salmon	22 (4)	0.578 (0.336)	O'Connell et al. (1995)
Cisco	8	0.780 (0.166)	Bernatchez and Dodson (1990)
Pink salmon *Oncorhynchus gorbuscha*	12	0.851 (0.049)	Brykov et al. (1996)
Powan	7	0.895 (0.060)	Bernatchez et al. (1989)
		Marine	
Mummichog *Fundulus heteroclitus*	5	0.305 (0.096)	Brown and Chapman (1991)
Atlantic cod	16	0.325 (0.158)	Carr et al. (1995)
Threespine stickleback	2	0.550 (0.020)	O'Reilly et al. (1993)
Tarakihi (jackass morwong) *Cheilodactylus macropterus*	9	0.634 (0.085)	Grewe et al. (1994a)
Shortfin mako *Isurus oxyrinchus*	5	0.747 (0.165)	Heist et al. (1996)
Haddock *Melanogrammus aeglefinus*	5	0.800 (0.045)	Zwanenburg et al. (1992)
Capelin *Mallotus villosus*	9	0.843 (0.098)	Dodson et al. (1991)
Yellowfin tuna *Thunnus albacares*	6	0.852 (0.021)	Scoles and Graves (1993)
Atlantic herring *Clupea harengus*	4	0.912 (0.061)	Kornfield and Bogdanowicz (1987)
Walleye pollock *Theragra chalcogramma*	4	0.916 (0.054)	Mulligan et al. (1992)
Red drum *Sciaenops ocellatus*	3	0.945 (0.037)	Gold and Richardson (1991)

Bernatchez et al. 1989). Population bottlenecks associated with the Pleistocene glaciation appear to have had severe effects on the extent of both haplotype diversity and divergence in many temperate zone fishes. For example, Bernatchez et al. (1989) compared the levels of mtDNA variation in the lake whitefish *Coregonus clupeaformis* from the Hudson–James Bay region in North America and powan from the Bothnian Gulf in Europe. These species have similar abundances, life histories, and potential for intra- and interpopulational gene flow and may represent allopatric populations of the same taxon. Lake whitefish survived the Pleistocene in three refugia, only one of which colonized Hudson–James Bay, whereas much of the range of powan was unglaciated. As expected, lake whitefish exhibited far lower levels of mtDNA diversity than did powan. Similarly, Bernatchez and Dodson (1990) found higher haplotype diversity and divergence in cisco than in lake whitefish populations from Hudson–James Bay. They attributed this difference to cisco having recolonized the area from two separate refugia compared with lake whitefish, which had a single refugium. Reduced mtDNA variation in fishes affected by glaciation has also been observed in the Southern Hemisphere (Ovenden and White 1990).

4.5.5 Genetic Markers

Many questions concerning the effects of stocking and other management practices on fish populations can be examined by the use of genetically marked fish (Chapter 16). Potential applications of genetic markers in fisheries management include monitoring levels of genetic variation in hatcheries (Grewe and Hebert 1988; Hynes et al. 1989; Danzmann et al. 1991, 1993; Morán et al. 1996), discriminating between wild- and hatchery-reared fish (Danzmann et al. 1991; Hansen et al. 1995; Morán et al. 1996; Williams et al. 1996, 1997b; Jones et al. 1997; Hansen et al. 2000), monitoring introduced or transplanted stocks (Bernardi et al. 1995; Nilsson et al. 2001; Weiss et al. 2001), monitoring escapes of cage-cultured fishes (Skaala et al. 1990; Knox and Verspoor 1991; Clifford et al. 1998), estimating the relative contribution of stocks to a fishery (mixed-stock analysis) (Wirgin et al. 1993; Epifanio et al. 1995; Waldman et al. 1996; Brown et al. 1999; McParland et al. 1999; Seeb and Crane 1999), discriminating species (sections 4.5.6 and 4.5.7), and conducting forensic analyses (section 4.5.8 and Chapter 17). Mitochondrial DNA markers also could be used as a method of identification or marker of propriety in broodstocks that have been selectively developed for aquaculture (Billington 1996b).

Several studies of mtDNA variation have shown that many stocks are discriminated easily and that hatchery broodstocks often differ from wild populations. In those fisheries that rely solely upon stocking programs, this variation is the only information required for assessment of differential growth and survival of stocked fish. In those cases in which stocking programs act to augment resident fish stocks, discrimination of stocked and resident fish first requires an assessment of natural mtDNA diversity so that stocked fish can be identified. On occasion, the donor population will be fixed for a different mtDNA haplotype than possessed by the residents. However, many stocks share common haplotypes, making discrimination of stocked and resident fish impossible (Billington and Hebert 1991b). There is, however, the opportunity to establish new broodstocks marked with unique mtDNA haplotypes, which would permit discrimination of stocked and resident fish. The establishment of nonlethal sampling techniques permits mtDNA variation to be assessed in broodstocks, leading to the selection of certain females for the establishment of lineages with known mtDNA haplotypes (Billington and Hebert 1990). Such mitochondrially marked fish could be used for conducting experimental stocking studies. The use of PCR-based mtDNA techniques will greatly improve the speed of these analyses and the sample sizes that can be processed.

4.5.6 Systematic Analysis and Molecular Phylogeny

Mitochondrial DNA analysis has been used to construct molecular phylogenies of fish species in order to make inferences about their systematic relationships. Early studies were based on whole-molecule RFLP analysis, with the better studies including comprehensive restriction-site mapping. For example, Grewe et al. (1990) mapped restriction sites in seven members of the salmonid genus *Salvelinus* and constructed a molecular phylogeny of them using the salmonid *Hucho* as an outgroup. The phylogenetic and zoogeographic relationships among members of the percid genus *Stizostedion* also were examined by mtDNA restriction-site analysis (Billington et al. 1990, 1991). A RFLP analysis of mtDNA was used by Bernatchez et al. (1991) to examine the phylogenetic relationships among four *Coregonus* species, by Seyoum and Kornfield (1992) to examine the evolutionary relationships among the Nile tilapia *Tilapia nilotica* subspecies complex, by Dowling et al. (1992) to study the evolutionary relationships among shiners in the cyprinid genus *Luxilus*, by Echelle and Dowling (1992) to examine the phylogenetic relationships among Death Valley pupfishes of the genus *Cyprinodon*, and by Bentzen et al. (1993) to examine genetic relationships among the shads in the genus *Alosa*.

More recently, the trend has been to use PCR to amplify and sequence portions of specific genes, most commonly the cytochrome-*b* gene because it has a rate of base substitution that makes it particularly suitable for comparisons of closely related species and genera (Whitmore et al. 1994). An early example of this approach was the identification of differences in the cytochrome-*b* sequences of four different tuna *Thunnus* spp. (Bartlett and Davidson 1991). Numerous additional molecular systematic studies of fishes have been published using the cytochrome-*b* gene, including studies of salmonid phylogenies (McVeigh and Davidson 1991; Patarnello et al. 1994; Park et al. 2000), relationships within the smelt genus *Osmerus* (Taylor and Dodson 1994), genetic identification of sharks (Heist and Gold 1998), and phylogenetic relationships among African cichlid fishes (Meyer et al. 1990; Sturmbauer et al. 1994), carps (Cyprinidae) (Gilles et al. 1998; Tsigenopoulos and Berrebi 2000), cods (Gadidae) (Carr et al. 1999), and pangasiid catfishes (Pangasiidae) (Pouyard et al. 2000). Other regions of the mtDNA molecule also have been used for molecular systematic analyses, including the D-loop or control region (Meyer et al. 1990; Shedlock et al. 1992; Brown et al. 1996; Faber and Stepien 1997; Stepien et al. 1998), *ND2* and *ND4 L* genes (Broughton and Gold 2000), the ribosomal RNA (*12S* and *16S*) genes (Bargelloni et al. 1994; Tringali et al. 1999; Sloss 2001), and cytochrome oxidase subunit I (Yokobori et al. 1994; Carr et al. 1999). More recent studies involve examination of two or more mitochondrial genes (Gilles et al. 1998; Birstein and DeSalle 1998; Holder et al. 1999; Krieger et al. 2000; Sloss 2001).

Some workers have used a combination of PCR amplification of specific genes and subsequent RFLP analysis to determine variation among fish species. For example, Chow et al. (1993) used PCR–RFLP analysis of cytochome *b* and *12S* rRNA to examine relationships among western Atlantic snappers of the subfamily Lutjaninae. Tagliavini et al. (1995) were able to find a single RFLP with the endonucluease *Hinf* I in a PCR-amplified portion of the cytochrome-*b* gene that permitted discrimination between European eel *Anguilla anguilla* and American eel *A. rostrata*.

The use of mtDNA sequences to examine phylogenetic relationships among fishes is currently an important area of research as evidenced by its use in 15 of 17 chapters in a recent volume on the molecular systematics of fishes (Kocher and Stepien 1997). The choice of sequence used depends upon the phylogenetic level of the hypothesis that is being tested. These levels range from rapidly evolving sequences (e.g., D-loop, *ND1*, *ND3/4*, and *ND5/6*) for examining intraspecific relationships and interspecific relationships between closely related organisms through moderately evolving genes (e.g., cytochrome *b*) for examination of relationships among genera to the slowly evolving *12S* and *16S* rRNA and cytochrome oxidase I genes for family level comparisons. Apart from this use in constructing molecular phylogenies, DNA sequence data can provide important information about the tempo and mode of nucleotide substitution and the dynamics of sequence evolution (reviewed by Meyer 1993, 1994). One such example is that transitions (A↔G and C↔T) are more common than are transversions (all other substitutions) in recently derived species. In more divergent species, transitions become saturated and transversions are used as characters for systematic comparisons. Relationships among highly divergent forms may be determined by inferring amino acid substitutions from protein coding sequence data and using these data to infer the molecular phylogeny of the organisms.

4.5.7 *Hybridization, Introgression, and Unisexual Species*

As a group, fishes show high levels of **interspecific hybridization** (hybridization between two different fish species) primarily resulting from their employment of external fertilization. Hybridization is considerably more common among freshwater than among marine fishes, perhaps because most freshwater habitats are of relatively recent origin compared with oceans (Mayr 1963; Campton 1987). Thus, many freshwater species in the Northern Hemisphere evolved during the Pleistocene, and isolating mechanisms may not have had time to develop completely. In other cases, hybridization may have been enhanced by the secondary contact of fish species with partially developed isolating mechanisms as a result of introductions, the invasion of nonnative fishes, or interdrainage contacts due to water diversion projects.

Although study of genetic variation in the nuclear genome (e.g., allozyme or microsatellite DNA analyses) is essential to ascertain the incidence of hybrids, mtDNA analysis can provide supplemental information on both the direction of hybridization and the incidence of **introgression,** the repeated backcrossing of hybrid descendants with a parental species resulting in the incorporation of genes from one species into the gene pool of another (Campton 1987, 1990). There are a growing number of examples of mtDNA exchange between closely related fish species (Billington and Hebert 1991b). In some cases, hybridization was unidirectional. For example, five F_1 hybrids between northern pike and chain pickerel *E. niger*, identified by allozyme analysis, all possessed chain pickerel mtDNA (Herke et al. 1990). Based on morphological and, more reliably, allozyme analysis, hybridization between brown trout and Atlantic salmon has been confirmed on both sides of the Atlantic (Crozier 1984; McGowan and Davidson 1992; Youngson et al. 1992). However, mtDNA analysis revealed that, in Scotland, the female parent of all hybrids was Atlantic salmon (Youngson et al. 1992), whereas in Newfoundland the maternal parent of all hybrids was the brown trout (McGowan and Davidson 1992). In the latter case, it was suggested that the presence of an abundance of precocious male Atlantic salmon parr was responsible for the frequency and direction of hybridization observed.

Asymmetrical hybridization between bluegill *Lepomis macrochirus* and pumpkinseed *L. gibbosus* in Lake Opinicon, Ontario, was confirmed by mtDNA analysis. All 44 F_1

hybrids identified by protein electrophoresis possessed pumpkinseed mtDNA (Konkle and Philipp 1992). Moreover, Konkle and Philipp (1992) suggested that precociously maturing male bluegills that attempted to sneak matings from larger nest-building males were a possible cause of this particular case of unidirectional hybridization.

In other cases, bidirectional transmission or introgression of mtDNA (or both processes) has been noted (Billington and Hebert 1991b). For example, the introgression of sauger mtDNA into walleye appeared to be caused by the inadvertent stocking of a few female sauger, or sauger × walleye hybrids, into a walleye population (Billington et al. 1988); bidirectional transmission of walleye and sauger mtDNA also has been documented (Billington et al. 1992; Billington 1996a). Bidirectional hybridization was also reported between black crappie *Pomoxis nigromaculatus* and white crappie *P. annularis* in Weiss Reservoir, Alabama, based on allozyme and mitochondrial cytochrome-*b* RFLP analysis (Smith et al. 1994; Travnichek et al. 1997). Hybridization and introgression between the pallid sturgeon *Scaphirhynchus albus* and shovelnose sturgeon *S. platorynchus* has complicated the taxonomic relationships between these species and has serious implications for the management of and conservation plans for the federally endangered pallid sturgeon (Campton et al. 2000; Tranah et al. 2001).

Bidirectional hybridization between Atlantic salmon and brown trout was confirmed by mtDNA analysis in a stream in the English Lake District, with 18.2% of juvenile Atlantic salmon being identified as hybrids based on hypervariable single-locus minisatellite profiling (Hartley 1996). Based on allozyme and mtDNA markers, Wilson and Hebert (1993) detected hybridization between Arctic char *Salvelinus alpinus* and lake trout occurring in the Canadian Arctic. Hybridization was largely unidirectional between lake trout females and Arctic char males, although one hybrid was found that possessed Arctic char mtDNA. An ancient introgression of Arctic char mtDNA into brook trout in a watershed lacking present-day Arctic char populations.was proposed by Bernatchez et al. (1995) to account for a population of brook trout that possessed Arctic char mtDNA but had morphological and diagnostic alleles of brook trout. Further studies revealed that seven populations in a single river subdrainage of the Portneuf basin consisted of introgressed brook trout, all of which had Artic char mtDNA but allozymes typical of brook trout (Glémet et al. 1998). Glémet et al. (1998) suggested that a combination of biogeographical conditions and positive selection for mtDNA introgression led to the presence of these introgressed brook trout in northeastern North America.

Hybridization has been observed commonly in the cyprinid genus *Luxilus* (formerly *Notropis*; Meagher and Dowling 1991). In the James River drainage of southwestern Virginia, hybrids between common shiner *L. cornutus* and crescent shiner *L. cerasinus* all possessed crescent shiner mtDNA; in addition, an individual with morphological and genetic characteristics of white shiner *L. albeolus* was collected (Meagher and Dowling 1991). However, the question of the hybrid origin of white shiner could not be resolved clearly due to difficulty in discriminating among ancestral polymorphism, convergence, and past hybridization among *Luxilus* spp. Striped shiner *L. chrysocephalus* also is known to hybridize with common shiner (Dowling and Hoeh 1991). Extensive asymmetrical introgression of common shiner mtDNA into striped shiner has been observed south of their hybrid zone, with common shiner mtDNA being found in 75 to 100% of individuals that, based on morphology and allozymes, are striped shiner (Dowling and Hoeh 1991; Duvernell and Aspinwall 1995).

In general, hybridization events have the potential to complicate assessment of systematic relationships if introgressed mtDNA genomes rise to high frequencies in the recipient taxon but are lost from the donor taxon. However, it is important to note that highly divergent lineages found within a taxon may be the result of retained ancestral polymorphisms rather than introgression (Avise et al. 1990). Mitochondrial DNA analysis also has been used to determine the maternal parentage of unisexual fishes, whose origin has been linked to interspecific hybridization (e.g., Avise and Vrijenhoek 1987; Quattro et al. 1991).

4.5.8 Forensic Analysis

It is often important for government agencies to be able to identify filleted or processed fish samples in order to prevent fraudulent marketing and the illegal harvest of protected fish populations or species. Although it often is possible to identify certain fish species based on the shape, color, and rib numbers of their fillets or by their scale characteristics (Oates et al. 1993), it is not possible to use these methods to determine if protected populations of a species have been harvested illegally. Genetic markers can prove useful for forensic analysis in fisheries management. For example, an illegal harvest of a closed population of the red king crab *Paralithodes camtschaticus* was confirmed by forensic analysis based upon the geographical distribution of genetic variation in this species at two polymorphic protein-coding loci (Seeb et al. 1990; Chapter 17). However, because protein electrophoretic methods work on only fresh or frozen tissue, and the population- or species-diagnostic loci may not be found in the muscle tissue of the fillet, it may be preferable to use DNA analysis for forensic work. The DNA is often more stable than are proteins, and it may be extracted from any tissue (Bartlett and Davidson 1992). Due to its high copy number per cell and relative ease of analysis, mtDNA offers to provide a useful genetic marker for forensic analysis of filleted and processed fish. For example, Shields et al. (1992) reported that diagnostic differences between the RFLP patterns of mtDNA from cisco and lake whitefish were useful in identifying fresh fillets and fresh or frozen carcasses of these species. Moreover, Bartlett and Davidson (1992) have extended the forensic use of mtDNA analysis to include processed fish (including canned, partially cooked, pickled, salted, and smoked fish). They have done so by optimizing DNA extraction protocols and identifying the fish samples by a phylogenetic analysis following PCR amplification and sequencing of a portion of the mtDNA cytochrome-*b* gene. Given the considerable amount of population structure that can be resolved by mtDNA analysis in many fish species (reviewed above and by Billington and Hebert 1991b), mtDNA analysis would appear to be a promising technique for conducting forensic analyses of suspected illegal harvests in closed fisheries, provided sufficient haplotype frequency differences exist between the open and closed fish populations.

4.5.9 Conservation Genetics and Rehabilitation Programs

In a review of the applications of mtDNA analysis in conservation, Moritz (1994) suggested that the uses of mtDNA analysis in conservation fall into two areas: gene conservation, which involves the identification and management of genetic diversity, and molecular ecology, which involves the use of mtDNA variation to guide and assist demographic studies of populations. Within the category of gene conservation, which often is seen as a long-term goal of maintaining genetic diversity and therefore evolutionary potential, Moritz (1994) suggests that mtDNA analyses have three main uses: (1) to measure genetic variation within populations, especially ones thought to have recently declined, (2) to identify evolutionarily divergent sets of populations (i.e., evolutionarily significant units; Chapters 15, 18), and (3) to assess the conservation value of populations or areas from an evolutionary or phylogenetic perspective. Within the category of molecular ecology, in which the applications are often simpler and more relevant to short-term management issues, Moritz (1994) suggests that there are three areas in which genetics can be used as a tool to provide ecologists with information. These are (1) to define an appropriate geographic scale for monitoring and management (e.g., defining management units as routinely is done in fisheries), (2) to provide a means to identify the origin of individuals in migratory species, and (3) to test for dramatic changes in population size and connectedness.

In a fisheries context, I already have covered the use of mtDNA analyses to define management units and identify the stock origin of migratory species. Moritz (1994) points out that while using mtDNA analysis to make inferences about the timing and sizes of population bottlenecks may be interesting intellectual exercises, they are unlikely to provide estimates that are sufficiently accurate for effective management due to the large stochastic variances involved in the responses of individual loci to population bottlenecks. For example, Miracle and Campton (1995) suggested that the lack of DNA sequence variation among individual Gulf of Mexico Atlantic sturgeon, except for length heteroplasmy in the D-loop region commonly seen in other *Acipenser* species, was consistent with the hypothesis that this subspecies had undergone a population or evolutionary bottleneck. An additional criticism contends that mtDNA is just one locus and, thus, is information poor as compared with nuclear DNA analyses. Therefore, I will focus on providing examples from the fisheries literature of gene conservation.

One particularly useful application of mtDNA analysis to conservation genetics has been the identification or confirmation of the unique genetic status of rare, endangered, or threatened species, subspecies, or distinct population segments. For example, Bernatchez et al. (1991) confirmed by mtDNA analysis that the endangered Atlantic whitefish *Coregonus huntsmani* (also known as Acadian whitefish) was genetically distinct from lake whitefish, Arctic cisco *C. autumnalis*, and cisco (also known as lake cisco or lake herring). Kornfield and Kircheis (1994) found a unique mtDNA banding pattern (haplotype) that distinguishes the silver char *Salvelinus alpinus oquassa* of Floods Pond in Maine from other eastern North American populations of Arctic char. They suggest that these Floods Pond silver char are appropriate for preservation because the silver char represent a genetically identified lineage in an early stage of divergence from other Arctic char lineages. Based on mtDNA analysis, southern populations of walleye in the Mobile Basin of Mississippi and Alabama were shown to be highly distinct (2.34% sequence divergence) from Mississippi River and other northern walleye populations (Billington and Strange 1995; Billington et al. 1997; Billington and Maceina 1997). Confirmation that these southern walleye represented a genetically distinct population segment, whose numbers were low and whose habitat was under significant threat, led to the filing of a petition (U.S. Office of the Federal Register 60 [13 March 1995]:13397) to list the southern walleye under the U.S. Endangered Species Act of 1973 (16 U.S.C. §§ 1531 to 1544). Mitochondrial DNA data were critical to the genetic discrimination of the federally endangered Alabama sturgeon *Scaphirhynchus suttkusi* from the shovelnose sturgeon and the federally endangered pallid sturgeon (Campton et al. 2000). However, discrimination of the pallid sturgeon from the shovelnose sturgeon appeared to be compromised in some areas where they are sympatric because of hybridization and introgression (Campton et al. 2000; Tranah et al. 2001). Mitochondrial DNA analyses also provided critical resolution of Atlantic salmon stocks; in particular it permitted some unique genetic markers to be identified in the last remaining U.S. stocks of Atlantic salmon in Downeast Maine that are candidates for listing under the U.S.

Endangered Species Act (King et al. 2000). Mitochondrial DNA analyses also have great potential for the identification of evolutionarily significant units in fishes (Bernatchez 1995; Chapters 15, 18). These analyses have been particularly important in resolving areas that had different biogeographical histories and in providing an assessment of the conservation value of these areas from an evolutionary or phylogenetic perspective (Bermingham and Avise 1986; Avise 1992, 1994, 1996).

Another area in which mtDNA analysis has provided useful genetic data has been in the design and operation of recovery programs. One particularly good example is the recovery program for the lake trout in the Great Lakes. The lake trout had disappeared from all of the Great Lakes with the exception of Lake Superior by the 1960s due to a combination of predation from the sea lamprey, overfishing, and habitat degradation. Efforts were made to re-establish self-sustaining lake trout populations from hatchery stocks following the control of sea lamprey populations (Eshenroder et al. 1984). Grewe and Hebert (1988) surveyed mtDNA variation in lake trout populations and in broodstocks that were used in lake trout rehabilitation programs across the Great Lakes. They found that there were dramatic shifts in the relative proportions of mtDNA haplotypes among the broodstocks. This variation facilitated the identification of stocked lake trout strains that were reproducing successfully in the Great Lakes so that stocking effort could focus on these successful strains. Grewe et al. (1993) then were able to perfect a rapid method of lake trout mtDNA analysis that made it feasible to obtain the large sample sizes necessary for the collection of baseline data to be used for mixed-stock analysis of the stocked lake trout strains. With a combination of allozymes and mtDNA, Grewe et al. (1994b) were able to use mixed-stock analysis to determine that the Seneca strain contributed 67 to 90% of naturally produced lake trout fry captured at the Stoney Island Reef in Lake Ontario. They suggested that the consistent success of this strain relative to its low stocking level indicated its suitability for restoration of lake trout at shallow-water reef locations in Lake Ontario. Kircheis et al. (1995) recommended against using Arctic char that had been stocked into lakes in Idaho but had originated from Sunapee Lake in New Hampshire (where they are now extirpated) to restock New Hampshire lakes including Sunapee Lake. They determined the fish from Idaho were not genetically unique and might have been affected by hybridization and introgression with other chars.

Mitochondrial DNA studies are likely to remain important in the field of conservation genetics. In particular, because of the use of PCR to amplify DNA from small pieces of tissue, it will be much easier to obtain mtDNA data from endangered and threatened species, subspecies, and populations. Earlier mtDNA studies often required the sacrifice of the fish, a practice that is obviously undesirable for an endangered or threatened species. Now it is possible to use nonlethal sampling techniques to obtain blood, muscle clips from the base of fins, barbel clips, muscle biopsy, or removal of scales to obtain small amounts of tissue from which DNA can be extracted. Then, specific regions of the mtDNA molecule, suitable to address the particular conservation question that is being addressed, can be amplified by PCR, and the DNA either sequenced or subjected to RFLP analysis.

4.6 Acknowledgments

I thank Eric Hallerman, Carey Krajewski, Gregory Moyer, Brian Sloss, Richard Williams, and three anonymous reviewers for their comments on earlier versions of this chapter.

4.7 References

Allendorf, F. W., N. Ryman, and F. M. Utter. 1987. Genetics and fishery management, past, present and future. Pages 1–19 *in* N. Ryman and F. Utter, editors. Population genetics and fisheries management. University of Washington Press, Seattle.

Arnason, E., P. H. Petersen, K. Krisinsson, H. Sigurgislason, and S. Palsson. 2000. Mitochondrial cytochrome *b* DNA sequence variation of Atlantic cod from Iceland and Greenland. Journal of Fish Biology 56:409–430.

Arnason, E., and D. M. Rand. 1992. Heteroplasmy of short tandem repeats in mitochondrial DNA of Atlantic cod, *Gadus morhua*. Genetics 132:211–220.

Arnheim, N., T. White, and W. E. Rainey. 1990. Application of PCR: organismal and population biology. Bioscience 40:174–181.

Avise, J. C. 1992. Molecular population structure and the biogeographic history of a regional fauna: a case history with lessons for conservation biology. Oikos 63:62–76.

Avise, J. C. 1994. Molecular markers, natural history and evolution. Chapman and Hall, New York.

Avise, J. C. 1996. Toward a regional conservation genetics perspective: phylogeography of faunas in the southeastern United States. Pages 431–470 *in* J. C. Avise and J. L. Hamrick, editors. Conservation genetics. Chapman and Hall, New York.

Avise, J. C. 2001. Phylogeography: the history and formation of species. Harvard University Press, Cambridge, Massachusetts.

Avise, J. C., C. D. Ankney, and W. S. Nelson. 1990. Mitochondrial gene trees and the evolutionary relationship of mallard and black ducks. Evolution 44:1109–1119.

Avise, J. C., B. W. Bowen, and T. Lamb. 1989. DNA fingerprints from hypervariable mitochondrial genotypes. Molecular Biology and Evolution 6:258–269.

Avise, J. C., and R. C. Vrijenhoek. 1987. Mode of inheritance and variation of mitochondrial DNA in hybridogenetic fishes of the genus *Poeciliopsis*. Molecular Biology and Evolution. 4:514–525.

Bargelloni, L., P. A. Ritchie, T. Patarnello, B. Battaglia, D. M. Lambert, and A. Meyer. 1994. Molecular evolution at subzero temperatures: mitochondrial and nuclear phylogenies of fishes from Antarctica (Suborder Notothenioidei), and the evolution of antifreeze glycopeptides. Molecular Biology and Evolution 11:854–863.

Bartlett, S. E., and W. S. Davidson. 1991. Identification of *Thunnus* tuna species by the polymerase chain reaction and direct sequence analysis of their mitochondrial cytochrome *b* genes. Canadian Journal of Fisheries and Aquatic Sciences 48:309–317.

Bartlett, S. E., and W. S. Davidson. 1992. FINS (Forensically Informative Nucleotide Sequencing): a procedure for identifying the animal origin of biological specimens. BioTechniques 12:408–411.

Beckenbach, A. T. 1991. Rapid mtDNA sequence analysis of fish populations using the polymerase chain reaction (PCR). Canadian Journal of Fisheries and Aquatic Sciences 48 (Supplement 1):95–98.

Becker, I. I., W. S. Grant, R. Kirby, and F. T. Robb. 1988. Evolutionary divergence between sympatric species of southern African hakes, *Merluccius capensis* and *M. paradoxus*. II. Restriction enzyme analysis of mitochondrial DNA. Heredity 61:21–30.

Bentzen, P., W. C. Leggett, and G. G. Brown. 1988. Length and restriction site heteroplasmy in the mitochondrial DNA of American shad *(Alosa sapidissima)*. Genetics 118:509–518.

Bentzen, P., W. C. Leggett, and G. G. Brown. 1993. Genetic relationships among the shads *(Alosa)* revealed by mitochondrial DNA analysis. Journal of Fish Biology 43:909–917.

Bentzen, P., J. M. Wright, L. T. Bryden, M. Sargent, and K. C. T. Zwanenberg. 1998. Tandem repeat polymorphism and heteroplasmy in the mitochondrial control region of redfishes (*Sebastes*: Scorpaenidae). Journal of Heredity 89:1–7.

Bermingham, E., and J. C. Avise. 1986. Molecular zoogeography of freshwater fishes in the southeastern United States. Genetics 113:939–965.

Berminghan, E., T. Lamb, and J. C. Avise. 1986. Size polymorphism and heteroplasmy in the mitochondrial DNA of lower vertebrates. Journal of Heredity 77:249–252.

Bermingham, E., S. S. McCafferty, and A. P. Martin. 1997. Fish biogeography and molecular clocks: perspectives from the Panamanian Isthmus. Pages 113–128 *in* T. D. Kocher and C. A. Stepien, editors. Molecular systematics of fishes. Academic Press, San Diego, California.

Bernardi, G., C. Fernandez-Delgado, M. Gomez-Chiarri, and D. A. Powers. 1995. Origin of a Spanish population of *Fundulus heteroclitus* inferred by cytochrome *b* sequence analysis. Journal of Fish Biology 47:737–740.

Bernatchez, L. 1995. A role for molecular systematics in defining evolutionary significant units in fishes. Pages 114–132 *in* J. L. Neilsen, editor. Evolution and the aquatic ecosystem: defining unique units in population conservation. American Fisheries Society Symposium 17, Bethesda, Maryland.

Bernatchez, L. 2001. The evolutionary history of brown trout (*Salmo trutta* L.) inferred from phylogeographic, nested clade, and mismatch analysis of mitochondrial DNA variation. Evolution 55:351–379.

Bernatchez, L., and J. J. Dodson. 1990. Mitochondrial DNA variation among anadromous populations of cisco *(Coregonus artedii)* as revealed by restriction analysis. Canadian Journal of Fisheries and Aquatic Sciences 47:533–543.

Bernatchez, L., J. J. Dodson, and S. Boivin. 1989. Population bottlenecks: influence on mitochondrial DNA diversity and its effect in coregonine stock discrimination. Journal of Fish Biology 35 (Supplement A):233–244.

Bernatchez, L., T. A. Edge, J. J. Dodson, and S. U. Qadri. 1991. Mitochondrial DNA and isozyme electrophoretic analyses of the endangered Acadian whitefish, *Coregonus huntsmani* Scott, 1987. Canadian Journal of Zoology 69:311–316.

Bernatchez, L., H. Glmet, C. C. Wilson, and R. G. Danzmann. 1995. Introgression and fixation of Arctic char *(Salvelinus alpinus)* mitochondrial genome in an allopatric population of brook trout *(Salvelinus fontinalis)*. Canadian Journal of Fisheries and Aquatic Sciences 52:179–185.

Bernatchez, L., and C. C. Wilson. 1998. Comparative phylogeography of Nearctic and Palearctic fishes. Molecular Ecology 7:431–452.

Billington, N. 1993. Genetic variation in Lake Erie yellow perch *(Perca flavescens)* demonstrated by mitochondrial DNA analysis. Journal of Fish Biology 43:941–943.

Billington, N. 1996a. Geographical distribution of mitochondrial DNA (mtDNA) variation in walleye, sauger, and yellow perch. Annales Zoologici Fennici 33:699–706.

Billington, N. 1996b. Genetic markers and stock identification. Pages 323–330 *in* R. C. Summerfelt, editor. The walleye culture manual. NCRAC Culture Series 101. North Central Regional Aquaculture Center Publications Office, Iowa State University, Ames.

Billington, N. 1998. Genetic variation in percids determined by mitochondrial DNA analysis. Italian Journal of Zoology 65 (Supplement):35–40.

Billington, N., R. J. Barrette, and P. D. N. Hebert. 1992. Management implications of mitochondrial DNA variation in walleye stocks. North American Journal of Fisheries Management 12:276–284.

Billington, N., R. G. Danzmann, P. D. N. Hebert, and R. D. Ward. 1991. Phylogenetic relationships among four members of *Stizostedion* (Percidae) determined by mitochondrial DNA and allozyme analysis. Journal of Fish Biology 39 (Supplement A):251–258.

Billington, N., and P. D. N. Hebert. 1988. Mitochondrial DNA variation in Great Lakes walleye *(Stizostedion vitreum)* populations. Canadian Journal of Fisheries and Aquatic Sciences 45:643–654.

Billington, N., and P. D. N. Hebert. 1990. Technique for determining mitochondrial DNA markers in blood samples from walleyes. Pages 492–498 *in* N. C. Parker, A. E. Giorgi, R. C. Heidinger, D. B. Jester Jr., E. D. Prince, and G. A. Winans, editors. Fish-marking techniques. American Fisheries Society Symposium 7, Bethesda, Maryland.

Billington, N., and P. D. N. Hebert, editors. 1991a. The ecological and genetic implications of fish introduction (FIN). Canadian Journal of Fisheries and Aquatic Sciences 48 (Supplement 1):1–181.

Billington, N., and P. D. N. Hebert. 1991b. Mitochondrial DNA diversity in fishes and its implications for introductions. Canadian Journal of Fisheries and Aquatic Sciences 48 (Supplement 1):80–94.

Billington, N., P. D. N. Hebert, and R. D. Ward. 1988. Evidence of introgressive hybridization in the genus *Stizostedion*: interspecific transfer of mitochondrial DNA between sauger and walleye. Canadian Journal of Fisheries and Aquatic Sciences 45:2035–2041.

Billington, N., P. D. N. Hebert, and R. D. Ward. 1990. Allozyme and mitochondrial DNA variation among three species of *Stizostedion* (Percidae): phylogenetic and zoogeographical implications. Canadian Journal of Fisheries and Aquatic Sciences 47:1093–1102.

Billington, N., and M. J. Maceina. 1997. Genetic and population characteristics of walleyes in the Mobile Drainage of Alabama. Transactions of the American Fisheries Society 126:804–814.

Billington, N., and R. M. Strange. 1995. Mitochondrial DNA analysis confirms the existence of a genetically divergent walleye population in northeastern Mississippi. Transactions of the American Fisheries Society 124:770–776.

Billington, N., R. M. Strange, and M. J. Maceina. 1997. Mitochondrial-DNA confirmation of southern walleye in the Mobile Basin, Alabama. Proceedings of the Annual Conference Southeastern Association of Fish and Wildlife Agencies 49(1995):112–117.

Birky, C. W. 1991. Evolution and population genetics of organelle genes: mechanisms and models. Pages 112–134 *in* R. K. Selander, A. G. Clark, and T. S. Whittam, editors. Evolution at the molecular level. Sinauer, Sunderland, Massachusetts.

Birky, C. W., P. Fuerst, and T. Maruyama. 1989. Organelle gene diversity under migration, and drift: equilibrium expectations, approach to equilibrium, effects of heteroplasmic cells, and comparison to nuclear genes. Genetics 121:613–627.

Birky, C. W., T. Maruyama, and P. Fuerst. 1983. An approach to population and evolutionary genetic theory for genes in mitochondria and chloroplasts, and some results. Genetics 103:513–527.

Birstein, V. J., and R. DeSalle. 1998. Molecular phylogeny of Acipenserinae. Molecular Phylogenetics and Evolution 9:141–155.

Birt, T. P., J. M. Green, and W. S. Davidson. 1991. Mitochondrial DNA variation reveals genetically distinct sympatric populations of anadromous and nonanadromous Atlantic salmon, *Salmo salar*. Canadian Journal of Fisheries and Aquatic Sciences 48:577–582.

Broughton, R. E., and T. E. Dowling. 1994. Length variation in mitochondrial DNA of the minnow *Cyprinella spiloptera*. Genetics 138:179–190.

Broughton, R. E., and J. R. Gold. 2000. Phylogenetic relationships in the North American cyprinid genus *Cyprinella* (Actinopterygii: Cyprinidae) based on sequences of the mitochondrial ND2 and ND4L genes. Copeia 2000:1–10.

Brown, B. L., and R. W. Chapman. 1991. Gene flow and mitochondrial DNA variation in the killifish, *Fundulus heteroclitus*. Evolution 45:1147–1161.

Brown, B. L., P. E. Smouse, J. M. Epifanio, and C. J. Kobak. 1999. Mitochondrial DNA mixed-stock analysis of American shad: coastal harvests are dynamic and variable. Transactions of the American Fisheries Society 128:977–994.

Brown, J. R., A. T. Beckenbach, and M. J. Smith. 1992. Influence of Pleistocene glaciations and human intervention upon mitochondrial DNA diversity in white sturgeon *(Acipenser transmontanus)* populations. Canadian Journal of Fisheries and Aquatic Sciences 49:358–367.

Brown, J. R., K. Beckenbach, A. T. Beckenbach, and M. J. Smith. 1996. Length variation, heteroplasmy and sequence divergence in the mitochondrial DNA of four species of sturgeon *(Acipenser)*. Genetics 142:525–535.

Brown, W. M. 1983. Evolution of animal mitochondrial DNAs. Pages 62–88 *in* M. Nei and R. K. Koehn, editors. Evolution of genes and proteins. Sinauer, Sunderland, Massachusetts.

Brunner, P. C., M. R. Douglas, A. Osinov, C. C. Wilson, and L. Bernatchez. 2001. Holacrtic phylogeogeography of Arctic char (*Salvelinus alpinus* L.) inferred from mitochondrial DNA sequences. Evolution 55:573–586.

Brykov, V. A., N. Polyakova, L. A. Skurikhina, and A. D. Kukhlevsky. 1996. Geographical and temporal mitochondrial DNA variability in populations of pink salmon. Journal of Fish Biology 48:899–909.

Brzuzan, P., M. Luczynski, and P. A. Kuznair. 1998. Mitochondrial DNA variation in two samples of northern pike, *Esox lucius* L. Aquaculture Research 29:521–526.

Buroker, N. E., and six coauthors. 1990. Length heteroplasmy of sturgeon mitochondrial DNA: an illegitimate elongation model. Genetics 124:157–163.

Campton, D. E. 1987. Natural hybridization and introgression in fishes: methods of detection and genetic interpretations. Pages 161–192 *in* N. Ryman and F. Utter, editors. Population genetics and fishery management. University of Washington Press, Seattle.

Campton, D. E. 1990. Application of biochemical and molecular genetic markers to analysis of hybridization. Pages 241–263 *in* D. H. Whitmore, editor. Electrophoretic and isoelectric focusing techniques in fisheries management. CRC Press, Boca Raton, Florida.

Campton, D. E., A. L. Bass, F. A. Chapman, and B. W. Bowen. 2000. Genetic distinction of pallid, shovelnose, and Alabama sturgeon: emerging species and the U.S. Endangered Species Act. Conservation Genetics 1:17–32.

Carr, S. M., D. S. Kivlichan, P. Pepin, and D. C. Crutcher. 1999. Molecular systematics of gadid fishes: implications for the biogeographic origins of Pacific species. Canadian Journal of Zoology 77:19–26.

Carr, S. M., and H. D. Marshall. 1991. Detection of intraspecific DNA sequence variation in the mitochondrial cytochrome *b* gene of Atlantic cod *(Gadus morhua)* by the polymerase chain reaction. Canadian Journal of Fisheries and Aquatic Sciences 48:48–52.

Carr, S. M., A. J. Snellen, K. A. Howse, and J. S. Wroblewski. 1995. Mitochondrial DNA sequence variation and genetic stock structure of Atlantic cod *(Gadus morhua)* from bay and offshore locations on the Newfoundland continental shelf. Molecular Ecology 4:79–88.

Chang, Y-S., F-L. Huang, and T-B. Lo. 1994. The complete nucleotide sequence and gene organization of carp *(Cyprinus carpio)* mitochondrial genome. Journal of Molecular Evolution 38:138–155.

Chapman, R. W. 1987. Changes in the population structure of male striped bass, *Morone saxatilis,* spawning in the three areas of the Chesapeake Bay from 1984 to 1986. U.S. National Marine Fisheries Service Fishery Bulletin 85:167–170.

Chapman, R. W. 1989. Spatial and temporal variation of mitochondrial DNA haplotype frequencies in the striped bass *(Morone saxatilis)* 1982 year class. Copeia 1989:344–348.

Chapman, R. W., and B. L. Brown. 1990. Mitochondrial DNA isolation methods. Pages 107–129 *in* D. H. Whitmore, editor. Electrophoretic and isoelectric focusing techniques in fisheries management. CRC Press, Boca Raton, Florida.

Chenoweth, S. F., J. M. Hughes, C. P. Keenan, and S. Lavery. 1998. Concordance between dispersal and mitochondrial gene flow: isolation by distance in a tropical teleost, *Lates calcarifer* (Australian barramundi). Heredity 80:187–197.

Chow, S., M. E. Clarke, and P. J. Walsh. 1993. PCR–RFLP analysis on thirteen western Atlantic snappers (subfamily Lutjaninae): a simple method for species and stock identification. U.S. National Marine Fisheries Service Fishery Bulletin 91:619–627.

Clifford, S. L., P. McGinnity, and A. Ferguson. 1998. Genetic changes in an Atlantic salmon population resulting from escaped juvenile farm salmon. Journal of Fish Biology 52:118–127.

Cooper, D. W. 1968. The significance level in multiple tests made simultaneously. Heredity 23:614–617.

Cronin, M. A., W. J. Spearman, R. J. Wilmot, J. C. Patton, and J. W. Bickham. 1993. Mitochondrial DNA variation in chinook *(Oncorhynchus tshawytscha)* and chum salmon *(O. keta)* detected by restriction enzyme analysis of polymerase chain reaction (PCR) products. Canadian Journal of Fisheries and Aquatic Sciences 50:708–715.

Crosetti, D., W. S. Nelson, and J. C. Avise. 1994. Pronounced genetic structure of mitochondrial DNA among populations of the circumglobally distributed grey mullet *(Mugil cephalus)*. Journal of Fish Biology 44:47–58.

Crozier, W. W. 1984. Electrophoretic identification and comparative examination of naturally occurring F_1 hybrids between brown trout (*Salmo trutta* L.) and Atlantic salmon (*S. salar* L.). Comparative Biochemistry and Physiology 78B:785–790.

Dahle, G. 1991. Cod, *Gadus morhua* L., populations identified by mitochondrial DNA. Journal of Fish Biology 38:295–303.

Danzmann, R. G., M. M. Ferguson, and S. K. A. Arndt. 1993. Mitochondrial DNA variability in

Ontario and New York rainbow trout *(Oncorhynchus mykiss)*. Canadian Journal of Zoology 71:1923–1933.

Danzmann, R. G., and P. E. Ihssen. 1995. A phylogeographic survey of brook char *(Salvelinus fontinalis)* in Algonquin Park, Ontario, based upon mitochondrial DNA variation. Molecular Ecology 4:681–697.

Danzmann, R. G., P. E. Ihssen, and P. D. N. Hebert. 1991. Genetic discrimination of wild and hatchery populations of brook char, *Salvelinus fontinalis* (Mitchill), in Ontario using mitochondrial DNA analysis. Journal of Fish Biology 39 (Supplement A):69–77.

Danzmann, R. G., R. P. Morgan II, M. W. Jones, L. Bernatchez, and P. E. Ihssen. 1998. A major sextet of mitochondrial DNA phylogenetic assemblages extant in eastern North American brook trout *(Salvelinus fontinalis)*: distribution and postglacial dispersal patterns. Canadian Journal of Zoology 76:1300–1318.

Delarbre, C., and seven coauthors. 2000. The complete nucleotide sequence of the mitochondrial DNA of the Agnathan *Lampetra fluviatilis*: bearings on the phylogeny of Cyclostomes. Molecular Biology and Evolution 17:519–529.

Dodson, J. J., J. E. Carscadden, L. Bernatchez, and F. Colombani. 1991. Relationship between spawning mode and phylogeographic structure in mitochondrial DNA of North Atlantic capelin *Mallotus villosus*. Marine Ecology Progress Series 76:103–113.

Dougherty, J. D., W. S. Moore, and J. L. Ram. 1996. Mitochondrial DNA analysis of round goby *(Neogobius melanostomus)* and tubenose goby *(Proterorhinus marmoratus)* in the Great Lakes basin. Canadian Journal of Fisheries and Aquatic Sciences 53:474–480.

Dowling, T. E., and W. M. Brown. 1989. Allozymes, mitochondrial DNA, and levels of phylogenetic resolution among four minnow species (*Notropis*: Cyprinidae). Systematic Zoology 38:126–143.

Dowling, T. E., and W. R. Hoeh. 1991. The extent of introgression outside the contact zone between *Notropis cornutus* and *Notropis chrysocephalus* (Teleostei: Cyprinidae). Evolution 45:944–956.

Dowling, T. E., W. R. Hoeh, G. R. Smith, and W. M. Brown. 1992. Evolutionary relationships of shiners in the genus *Luxilus* (Cyprinidae) as determined by analysis of mitochondrial DNA. Copeia 1992:306–322.

Dowling, T. E., C. Moritz, J. D. Palmer, and L. H. Rieseberg. 1996. Nucleic acids III: analysis of fragments and restriction sites. Pages 249–320 *in* D. M. Hillis, C. Moritz, and B. K. Mable, editors. Molecular systematics. Sinauer, Sunderland, Massachusetts.

Duvernell, D. D., and M. Aspinwall. 1995. Introgression of *Luxilus cornutus* mtDNA into allopatric populations of *Luxilus chrysocephalus* (Teleostei: Cyprinidae). Molecular Ecology 4:173–181.

Echelle, A. A., and T. E. Dowling. 1992. Mitochondrial DNA variation and evolution of the Death Valley pupfishes (*Cyprinodon*, Cyprinodontidae). Evolution 46:193–206.

Epifanio, J. M., P. E. Smouse, C. J. Kobak, and B. L. Brown. 1995. Mitochondrial DNA divergence among populations of American shad *(Alosa sapidissima)*: how much variation is enough for mixed-stock analysis? Canadian Journal of Fisheries and Aquatic Sciences 52:1688–1702.

Eshenroder, R. L., T. P. Poe, and C. H. Oliver, editors. 1984. Strategies for the rehabilitation of lake trout in the Great Lakes: proceedings of a conference on lake trout research, August 1993. Great Lakes Fishery Commission Technical Report 40, Ann Arbor, Michigan.

Excoffier, L., P. E. Smouse, and J. M. Quattro. 1992. Analysis of molecular variance inferred from metric distances among DNA haplotypes: applications to human mitochondrial DNA restriction data. Genetics 131:479–491.

Faber, J. E., and C. A. Stepien. 1997. The utility of mitochondrial DNA control region sequences for analyzing phylogenetic relationships among populations, species, and genera of the Percidae. Pages 129–143 *in* T. D. Kocher and C. A. Stepien, editors. Molecular systematics of fishes. Academic Press, San Diego, California.

Fajen, A., and F. Breden. 1992. Mitochondrial DNA sequence variation among natural populations of the Trinidad guppy, *Poecilia reticulata*. Evolution 46:1457–1465.

Ferguson, M. M. 1990. The genetic impact of introduced fishes on native species. Canadian Journal of Zoology 68:1053–1057.

Ferguson, M. M., R. G. Danzmann, and J. A. Hutchings. 1991. Incongruent estimates of population differentiation among brook char, *Salvelinus fontinalis*, from Cape Race, Newfoundland, Canada, based upon allozyme and mitochondrial DNA variation. Journal of Fish Biology 39 (Supplement A):79–85.

Ferris, S. D., and W. J. Berg. 1987. The utility of mitochondrial DNA in fish genetics and fishery management. Pages 277–299 *in* N. Ryman and F. Utter, editors. Population genetics and fishery management. University of Washington Press, Seattle.

Finnerty, J. R., and B. A. Block. 1992. Direct sequencing of mitochondrial DNA detects highly divergent haplotypes in blue marlin *(Makaira nigricans)*. Molecular Marine Biology and Biotechnology 1:206–214.

Gatt, M. H., M. M. Ferguson, and A. P. Liskauskas. 2000. Comparison of control region sequencing and fragment RFLP analysis for resolving mitochondrial DNA variation, and phylogenetic relationships among Great Lakes walleyes. Transaction of the American Fisheries Society 129:1288–1299.

Gatt, M. H., D. J. Fraser, A. P. Liskauskas, and M. M. Ferguson. 2002. Mitochondrial DNA variation and stock structure of walleyes from eastern Lake Huron: an analysis of contemporary and historical samples. Transactons of the American Fisheries Society 131:99–108.

Gilles, A., G. Lecointre, E. Faure, R. Chappaz, and G. Brun. 1998. Mitochondrial phylogeny of the European cyprinids: implications for their systematics, reticulate evolution, and colonization time. Molecular Phylogeny and Evolution 10:132–143.

Gleeson, D. M., R. L. J. Howitt, and N. Ling. 1999. Genetic variation, population structure and cryptic species within the black mudfish, *Neochanna diversus,* an endemic galaxiid from New Zealand. Molecular Ecology 8:47–57.

Glémet, H., P. Blier, and L. Bernatchez. 1998. Geographical extent of Arctic char *(Salvelinus alpinus)* mtDNA introgression in brook char populations *(S. fontinalis)* from eastern Québec, Canada. Molecular Ecology 7:1655–1662.

Gold, J. R., and L. R. Richardson. 1991. Genetic studies in marine fishes. IV. An analysis of population structure in the red drum *(Sciaenops ocellatus)*. Fisheries Research 12:213–241.

Gold, J. R., L. R. Richardson, T. L. King, and G. C. Matlock. 1993. Temporal stability of nuclear gene (allozyme) and mitochondrial DNA genotypes among red drums from the Gulf of Mexico. Transaction of the American Fisheries Society 122:659–668.

González-Villaseñor, L. I., A. M. Burkhoff, V. Corces, and D. A. Powers. 1986. Characterization of cloned mitochondrial DNA from the teleost *Fundulus heteroclitus* and its usefulness as an interspecies hybridization probe. Canadian Journal of Fisheries and Aquatic Sciences 43:1866–1872.

Graves, J. E. 1996. Conservation genetics of fishes in the pelagic marine realm. Pages 335–366 *in* J. C. Avise and J. L. Hamrick, editors. Conservation genetics. Chapman and Hall, New York.

Gray, M. W. 1992. The endosymbiont hypothesis revisited. International Review of Cytology 141:233–357.

Grewe, P. M., N. Billington, and P. D. N. Hebert. 1990. Phylogenetic relationships among members of *Salvelinus* inferred from mitochondrial DNA divergence. Canadian Journal of Fisheries and Aquatic Sciences 47:984–991.

Grewe, P. M., and P. D. N. Hebert. 1988. Mitochondrial DNA diversity among broodstocks of the lake trout, *Salvelinus namaycush*. Canadian Journal of Fisheries and Aquatic Sciences 45:2114–2122.

Grewe, P. M., C. C. Krueger, C. F. Aquadro, E. Bermingham, H. L. Kincaid, and B. May. 1993. Mitochondrial DNA variation among lake trout *(Salvelinus namaycush)* strains stocked into Lake Ontario. Canadian Journal of Fisheries and Aquatic Sciences 50:2397–2403.

Grewe, P. M., C. C. Krueger, J. E. Marsden, C. F. Aquadro, and B. May. 1994a. Hatchery origins of naturally produced lake trout fry captured in Lake Ontario: temporal and spatial variability based on allozyme and mitochondrial DNA data. Transactions of the American Fisheries Society 123:309–320.

Grewe, P. M., A. J. Smolenski, and R. D. Ward. 1994.b. Mitochondrial DNA diversity in jackass morwong (*Nemadactylus macropterus*: Teleostei) from Australian and New Zealand waters. Canadian Journal of Fisheries and Aquatic Sciences 51:1101–1109.

Hall, H. J., and L. W. Nawrocki. 1995. A rapid method for detecting mitochondrial DNA variation in brown trout, *Salmo trutta*. Journal of Fish Biology 46:360–364.

Hansen, M. M., R. A. Hynes, V. Loeschcke, and G. Rasmussen. 1995. Assessment of the stocked or wild origin of anadromous brown trout (*Salmo trutta* L.) in a Danish river system, using mitochondrial DNA analysis. Molecular Ecology 4:189–198.

Hansen, M. M., and V. Loeschcke. 1996. Genetic differentiation among Danish brown trout populations, as detected by RFLP analysis of PCR amplified mitochondrial DNA segments. Journal of Fish Biology 48:422–436.

Hansen, M. M., E. E. Nielsen, and K.-L. D. Mensberg. 1997. The problem of sampling families rather than populations: relatedness among individuals in samples of juvenile brown trout *Salmo trutta* L. Molecular Ecology 6:469–474.

Hansen, M. M., D. E. Ruzzante, E. E. Nielsen, and K-L. D. Mensberg. 2000. Microsatellite and mitochondrial DNA polymorphism reveals life-history dependent interbreeding between hatchery and wild brown trout (*Salmo trutta* L.). Molecular Ecology 9:583–594.

Hartley, S. E. 1995. Mitochondrial DNA analysis distinguishes between British populations of whitefish. Journal of Fish Biology 47 (Supplement A):145–155.

Hartley, S. E. 1996. High incidence of Atlantic salmon × brown trout hybrids in a Lake District stream. Journal of Fish Biology 48:151–154.

Hebert, P. D. N., R. D. Ward, and L. J. Weider. 1988. Clonal-diversity patterns and breeding-system variation in *Daphnia pulex*, an asexual–sexual complex. Evolution 42:147–159.

Heist, E. J., and J. R. Gold. 1998. Genetic identification of sharks in the U.S. Atlantic large coastal shark fishery. U.S. National Marine Fisheries Service Fishery Bulletin 97:53–61.

Heist, E. J., J. E. Graves, and J. A. Musick. 1995. Population genetics of the sandbar shark *(Carcharhinus plumbeus)* in the Gulf of Mexico and Mid-Atlantic Bight. Copeia 1995:555–562.

Heist, E. J., J. A. Musick, and J. E. Graves. 1996a. Genetic population structure of the shortfin mako *(Isurus oxyrinchus)* inferred from restriction fragment length polymorphism analysis of mitochondrial DNA. Canadian Journal of Fisheries and Aquatic Sciences 53:583–588.

Heist, E. J., J. A. Musick, and J. E. Graves. 1996b. Mitochondrial DNA diversity and divergence among sharpnose sharks, *Rhizoprionodon terrraenovae,* from the Gulf of Mexico and Mid-Atlantic Bight. U.S. National Marine Fisheries Service Fishery Bulletin 94:664–668.

Herke, S. W., I. Kornfield, P. Moran, and J. R. Moring. 1990. Molecular confirmation of hybridization between northern pike *(Esox lucius)* and chain pickerel *(E. niger)*. Copeia 1990:846–850.

Hillis, D. M., B. K. Mable, and C. Moritz. 1996. Applications of molecular systematics. Pages 515–543 *in* D. M. Hillis, C. Moritz, and B. K. Mable, editors. Molecular Systematics, 2nd edition. Sinauer, Sunderland, Massachusetts.

Hoelzel, A. R., and A. Green. 1992. Analysis of population-level variation by sequencing PCR-amplified DNA. Pages 159–187 *in* A. R. Hoelzel, editor. Molecular genetic analysis of populations: a practical approach. IRL Press/Oxford University Press, Oxford, UK.

Hoffmann, R. J., J. L. Boore, and W. M. Brown. 1992. A novel mitochondrial genome organization for the blue mussel, *Mytilus edulis*. Genetics 131:397–412.

Holder, M. T., M. V. Erdmann, T. P. Wilcox, R. L. Caldwell, and D. M. Hillis. 1999. Two living species of coelacanths? Proceedings of the National Academy of Sciences of the USA 96:12616–12620.

Holtke, H. J., G. Sanger, C. Kessler, and E. Schmitz. 1992. Sensitive chemiluminescent detection of digoxigenin-labeled nucleic acids: a fast and simple protocol and its applications. BioTechniques 12:104–112.

Hurst, C. D., S. E. Bartlett, W. S. Davidson, and I. J. Bruce. 1999. The complete mitochondrial DNA sequence of the Atlantic salmon, *Salmo salar*. Gene 239:237–242.

Hutchinson, C. A., J. E. Newbold, S. S. Potter, and M. H. Edgell. 1974. Maternal inheritance of mammalian mitochondrial DNA. Nature (London) 251:536–538.

Hynes, R. A., E. J. Duke, and P. Joyce. 1989. Mitochondrial DNA as a genetic marker for brown trout, *Salmo trutta* L., populations. Journal of Fish Biology 35:687–701.

Hynes, R. A., A. Ferguson, and M. A. McCann. 1996. Variation in mitochondrial DNA and post-glacial colonization of north western Europe by brown trout. Journal of Fish Biology 48:54–67.

Johansen, S., P. H. Guddal, and T. Johansen. 1990. Organization of the mitochondrial genome of Atlantic cod, *Gadus morhua*. Nucleic Acids Research 18:411–419.

Johansen, S., and T. Johansen. 1994. Sequence analysis of 12 structural genes and a novel non-coding region from mitochondrial DNA of Atlantic cod, *Gadus morhua*. Biochimica et Biophysica Acta 1218:213–217.

Jones, M.W., R. G. Danzmann, and D. Clay. 1997. Genetic relationships among populations of wild resident, and wild and hatchery anadromous brook char. Journal of Fish Biology 51:29–40.

King, D. P. F., S. J. Hovey, D. Thompson, and A. Scott. 1993. Mitochondrial DNA variation in Atlantic salmon, *Salmo salar* L., populations. Journal of Fish Biology 42:25–33.

King, T. L., A. P. Spidle, M. S. Eackles, B. A. Lubinski, and W. B. Schill. 2000. Mitochondrial DNA diversity in North American and European Atlantic salmon with emphasis on the Downeast rivers of Maine. Journal of Fish Biology 57:614–630.

Kircheis, F. W., I. Kornfield, and S. Seyoum. 1995. Genetic identity of transplanted Arctic char: implications for restocking programs. North American Journal of Fisheries Management 15:54–59.

Knox, D., and E. Verspoor. 1991. A mitochondrial DNA restriction fragment length polymorphism of potential use for discrimination of farmed Norwegian and wild Atlantic salmon populations in Scotland. Aquaculture 98:249–257.

Kocher, T. D., and six coauthors. 1989. Dynamics of mitochondrial DNA evolution in animals: amplification and sequencing with conserved primers. Proceedings of the National Academy of Sciences of the USA 86:6196–6200.

Kocher, T. D., and C. A. Stepien, editors. 1997. Molecular systematics of fishes. Academic Press, San Diego, California.

Konkle, B. R., and D. P. Philipp. 1992. Asymmetric hybridization between two species of sunfishes (*Lepomis*: Centrarchidae). Molecular Ecology 1:215–222.

Kornfield, I., and S. Bogdanowicz. 1987. Differentiation of mitochondrial DNA in Atlantic herring, *Clupea harengus*. U.S. National Marine Fisheries Service Fishery Bulletin 85:561–568.

Kornfield, I., and F. W. Kircheis. 1994. Mitochondrial DNA and conservation of an aboriginal Arctic char *(Salvelinus alpinus oquassa)* from Floods Pond, Maine. Canadian Journal of Fisheries and Aquatic Sciences 51:62–67.

Kreiser, B. R., J. B. Mitton, and J. D. Woodling. 2001. Phylogeography of the plains killifish, *Fundulus zebrinus*. Evolution 55:339–350.

Krieger, J., P. A. Fuerst, and T. M. Cavender. 2000. Phylogenetic relationships of the North American sturgeons (Order Acipenseriformes) based on mitochondrial DNA sequences. Molecular Phylogenetics and Evolution 16:64–72.

Leary, R., and H. E. Booke. 1982. Genetic stock analysis of yellow perch from Green Bay and Lake Michigan. Transactions of the American Fisheries Society 111:52–57.

Lee, W-J., and T. D. Kocher. 1995. Complete sequence of a sea lamprey *(Petromyzon marinus)* mitochondrial genome: early establishment of the vertebrate genome organization. Genetics 139:873–887.

Magoulas, A., and E. Zouros. 1993. Restriction-site heteroplasmy in anchovy *(Engraulis encrasicolus)* indicates incidental biparental inheritance of mitochondrial DNA. Molecular Biology and Evolution 10:319–325.

Martin, A. P., R. Humphreys, and S. R. Palumbi. 1992a. Population genetic structure of the armorhead, *Pseudopentaceros wheeleri,* in the North Pacific Ocean: application of the polymerase chain reaction to fisheries problems. Canadian Journal of Fisheries and Aquatic Sciences 49:2386–2391.

Martin, A. P., G. J. P. Naylor, and S. R. Palumbi. 1992b. Rates of mitochondrial DNA evolution in sharks are slow compared with mammals. Nature (London) 357:153–155.

May, B., and P. M. Grewe. 1993. Fate of maternal mtDNA following ^{60}Co inactivation of maternal nuclear DNA in unfertilized salmonid eggs. Genome 36:725–730.

Mayr, E. 1963. Animal species and evolution. Belknap Press, Cambridge, Massachusetts.

McCusker, M. R., E. Parkinson, and E. B. Taylor. 2000. Mitochondrial DNA variation in rainbow trout *(Oncorhynchus mykiss)* across its native range: testing biogeographical hypotheses and their relevance to conservation. Molecular Ecology 9:2089–2108.

McGauley, K., and T. J. Mulligan. 1995. Polymerase chain reaction–restriction fragment length polymorphism analysis of mitochondrial rRNA genes from yellow rockfish. Journal of Fish Biology 47:744–747.

McGowan, C., and W. S. Davidson. 1992. Unidirectional natural hybridization between brown trout *(Salmo trutta)* and Atlantic salmon *(Salmo salar)* in Newfoundland. Canadian Journal of Fisheries and Aquatic Sciences 49:1953–1958.

McParland, T. L., M. M. Ferguson, and A. P. Liskauskas. 1999. Genetic population structure and mixed-stock analysis of walleyes in the Lake Erie–Lake Huron corridor using allozyme and mitochondrial DNA markers. Transactions of the American Fisheries Society 128:1055–1067.

McVeigh, H. P., and W. S. Davidson. 1991. A salmonid phylogeny inferred from mitochondrial cytochrome *b* gene sequences. Journal of Fish Biology 39 (Supplement A):277–282.

Meagher, S., and T. E. Dowling. 1991. Hybridization between the cyprinid fishes *Luxilus albeolus, L. cornutus,* and *L. cerasinus* with comments on the proposed hybrid origin of *L. albeolus*. Copeia 1991:979–991.

Merker, R. J., and R. C. Woodruff. 1996. Molecular evidence for divergent breeding groups of walleye *(Stizostedion vitreum)* in tributaries to western Lake Erie. Journal of Great Lakes Research 22:280–288.

Meyer, A. 1993. Evolution of mitochondrial DNA in fishes. Pages 1–38 *in* P. W. Hochachka and T. P. Mommsen, editors. Biochemistry and molecular biology of fishes, volume 2. Elsevier Publishers, Amsterdam.

Meyer, A. 1994. DNA technology and phylogeny of fish. Pages 389–415 *in* A. R. Beaumont, editor. Genetics and evolution of aquatic organisms. Chapman and Hall, London.

Meyer, A., T. D. Kocher, P. Basasibwaki, and A. C. Wilson. 1990. Monophyletic origin of Lake Victoria cichlid fishes suggested by mitochondrial DNA sequences. Nature (London) 347:550–553.

Miracle, A. L., and D. E. Campton. 1995. Tandem repeat sequence variation and length heteroplasmy in the mitochondrial DNA D-loop of the threatened Gulf of Mexico sturgeon, *Acipenser oxyrhnchus desotoi*. Journal of Heredity 86:22–27.

Moore, W. S. 1995. Inferring phylogenies from mtDNA variation: mitochondrial gene trees versus nuclear gene trees. Evolution 49:718–726.

Morán, P., A. M. Pendás, and E. García-Vázquez. 1996. Mitochondrial DNA variation in wild and hatchery brown trout (*Salmo trutta* L.) populations from Spain. Aquaculture 141:59–65.

Moritz, C. 1994. Applications of mitochondrial DNA analysis in conservation: a critical review. Molecular Ecology 3:401–411.

Moritz, C., T. E. Dowling, and W. M. Brown. 1987. Evolution of animal mitochondrial DNA: relevance for population biology and systematics. Annual Review of Ecology and Systematics 18:269–292.

Mulligan, T. J., and R. W. Chapman. 1989. Mitochondrial DNA analysis of Chesapeake Bay white perch, *Morone americana*. Copeia 1989:679–688.

Mulligan, T. J., R. W. Chapman, and B. L. Brown. 1992. Mitochondrial DNA analysis of walleye pollock, *Theragra chalcogramma,* from the eastern Bering Sea and Shelikof Strait, Gulf of Alaska. Canadian Journal of Fisheries and Aquatic Sciences 49:319–326.

Murdoch, M. H., and P. D. N. Hebert. 1997. Mitochondrial DNA evidence of distinct glacial refugia for brown bullhead *(Ameiurus nebulosus)* in the Great Lakes. Canadian Journal of Fisheries and Aquatic Sciences 54:1450–1460.

Muss, A., D. R. Robertson, C. A. Stepien, P. Wirtz, and B. W. Bowen. 2001. Phylogeography of *Ophioblennius*: the role of ocean currents and geography in reef fish evolution. Evolution 55:561–572.

Nei, M., and F. Tajima. 1981. DNA polymorphism detectable by endonucleases. Genetics 97:145–163.

Nesbø, C. L., T. Fossheim, L. A. Vøllestad, and K. S. Jakobsen. 1999. Genetic divergence and phylogeographic relationships among European perch *(Perca fluviatilis)* populations reflect glacial refugia and postglacial colonization. Molecular Ecology 8:1387–1404.

Nielsen, E. E., M. Hansen, and V. Loeschcke. 1999. Analysis of DNA from old scale samples: technical aspects, applications and perspectives for conservation. Hereditas 130:265–276.

Nielsen, E. E., M. M. Hansen, and K-L. D. Mensberg. 1998. Improved primer sequences for the mitochondrial ND1, ND3/4 and ND5/6 segments in salmonid fishes: application of RFLP analysis of Atlantic salmon. Journal of Fish Biology 53:216–220.

Nilsson, J., and 15 coauthors. 2001. Matrilinear phylogeography of Atlantic salmon (*Salmo salar* L.) in Europe, and postglacial colonization of the Baltic Sea area. Molecular Ecology 10:89–102.

Oates, D. W., L. M. Krings, and K. Ditz. 1993. Field manual for the identification of selected North American freshwater fish by fillets and scales. Nebraska Game and Parks Commission, Nebraska Technical Series 19, Lincoln.

O'Connell, M., D. O. F. Skibinski, and J. A. Beardmore. 1995. Mitochondrial DNA and allozyme variation in Atlantic salmon *(Salmo salar)* populations in Wales. Canadian Journal of Fisheries and Aquatic Sciences 52:171–178.

Okimoto, R., J. L. Macfarlane, D. O. Clary, and D. R. Wolstenholme. 1992. The mitochondrial genomes of two nematodes, *Caenorhabditis elegans* and *Ascaris suum*. Genetics 130:471–498.

O'Reilly, P., T. E. Reimchen, R. Beech, and C. Strobeck. 1993. Mitochondrial DNA in *Gasterosteus* and Pleistocene glacial refugium on the Queen Charlotte Islands, British Columbia. Evolution 47:678–684.

Ovenden, J. R. 1990. Mitochondrial DNA and marine stock assessment: a review. Australian Journal of Marine and Freshwater Research 41:835–853.

Ovenden, J. R., and R. G. White. 1990. Mitochondrial and allozyme genetics of incipient speciation in a landlocked population of *Galaxias truttaceus* (Pisces: Galaxidae). Genetics 124:701–716.

Ovenden, J. R., R. G. White, and A. C. Sanger. 1988. Evolutionary relationships of *Gadopsis* spp. inferred from restriction enzyme analysis of their mitochondrial DNA. Journal of Fish Biology 32:137–148.

Page, R. D. M., and E. C. Holmes. 1998. Molecular evolution: a phylogenetic approach. Blackwell Scientific Publications, Oxford, UK.

Palumbi, S. R. 1996. Nucleic acids II: the polymerase chain reaction. Pages 205–247 *in* D. M. Hillis, C. Moritz, and B. K. Mable, editors. Molecular systematics, 2nd edition. Sinauer, Sunderland, Massachusetts.

Paragamian, V. L., M. S. Powell, and J. C. Faler. 1999. Mitochondrial DNA analysis of burbot stocks in the Kootenai River basin of British Columbia, Montana, and Idaho. Transactions of the American Fisheries Society 128:868–874.

Park, J. Y., H. J. Lee, W. J. Kim, J. H. Lee, and K. S. Min. 2000. Mitochondrial cytochrome *b* sequence variation in Korean salmonids. Journal of Fish Biology 56:1029–1293.

Patarnello, T., L. Bargelloni, F. Caldara, and L. Columbo. 1994. Cytochrome *b* and 16S rRNA sequence variation in the *Salmo trutta* (Salmonidae, Teleostei) species complex. Molecular Phylogenetics and Evolution 3:69–74.

Pigeon, D., J. J. Dodson, and L. Bernatchez. 1998. A mtDNA analysis of spatiotemporal distribution of two sympatric larval populations of rainbow smelt *(Oserus mordax)* in the St. Lawrence River estuary, Quebec, Canada. Canadian Journal of Fisheries and Aquatic Sciences 55:1739–1747.

Pouyard, L., G. G. Teugels, R. Gustiano, and M. Legendre. 2000. Contribution to the phylogeny of pangasiid catfishes based on allozymes and mitochondrial DNA. Journal of Fish Biology 56:1509–1538.

Quattro, J. M., J. C. Avise, and R. C. Vrijenhoek. 1991. Molecular evidence for multiple origins of hybridogenetic fish clones (Poeciliidae: *Poeciliopsis*). Genetics 127:391–398.

Rand, D. M. 1994. Thermal habitat, metabolic rate and the evolution of mitochondrial DNA. Trends in Ecology and Evolution 9:125–131.

Redenbach, Z., and E. B. Taylor. 1999. Zoogeographical implications of variation in mitochondrial DNA of Arctic grayling *(Thymallus arcticus)*. Molecular Ecology 8:23–35.

Refseth, U. H., C. L. Nesbø, J. E. Stacy, L. A. Vøllestad, E. Fjeld, and K. S. Jakobsen. 1998. Genetic evidence for different migration routes of freshwater fish into Norway revealed by analysis of current perch *(Perca fluviatilis)* populations in Scandinavia. Molecular Ecology 7:1015–1027.

Roberts, R. J. 1984. Restriction and modifying enzymes and their recognition sequences. Nucleic Acids Research 12:r167–r204.

Roff, D. A., and P. Bentzen. 1989. The statistical analysis of mitochondrial DNA polymorphisms: χ^2 and the problem of small samples. Molecular Biology and Evolution 6:539–545.

Ruzzante, D. E., C. T. Taggart, D. Cook, and S. Goddard. 1996. Genetic differentiation between inshore and offshore Atlantic cod *(Gadus morhua)* off Newfoundland: microsatellite DNA variation and antifreeze level. Canadian Journal of Fisheries and Aquatic Sciences 53:634–645.

Saitou, N., and M. Nei. 1987. The neighbor-joining method: a new method for reconstructing phylogenetic trees. Molecular Biology and Evolution 4:406–425.

Scoles, D. R., and J. E. Graves. 1993. Genetic analysis of the population structure of yellowfin tuna, *Thunnus albacares,* from the Pacific Ocean. U.S. National Marine Fisheries Service Fishery Bulletin 91:690–698.

Seeb, J. E., G. H. Kruse, L. W. Seeb, and R. G. Weck. 1990. Genetic structure of red king crab populations in Alaska facilitates enforcement of fishing regulations. Pages 491–502 *in* Proceedings of the international symposium on king and tanner crabs. Alaska Sea Grant Report 90-4, Fairbanks, Alaska.

Seeb, L. W., and P. A. Crane. 1999. Allozymes and mitochondrial DNA discriminate Asian and North American populations of chum salmon in mixed-stock fisheries along the south coast of the Alaska Peninsula. Transactions of the American Fisheries Society 128:88–103.

Seyoum, S., and I. Kornfield. 1992. Taxonomic notes on the *Oreochromis niloticus* subspecies-complex (Pisces: Cichlidae), with a description of a new subspecies. Canadian Journal of Zoology 70:2161–2165.

Shedlock, A. M., J. D. Parker, D. A. Crispin, T. W. Pietsch, and G. C. Burmer. 1992. Evolution of the salmonid mitochondrial control region. Molecular Phylogenetics and Evolution 1:179–192.

Shields, B. A., A. R. Kapuscinski, and K. S. Guise. 1992. Mitochondrial DNA variation in four Minnesota populations of lake whitefish: utility as species and population markers. Transactions of the American Fisheries Society 121:21–25.

Shields, G. F., and A. C. Wilson. 1987. Calibration of mitochondrial DNA evolution in geese. Journal of Molecular Evolution 24:212–217.

Skaala, Ø., G. Dahle, K. E. Jørstad, and G. Naevdal. 1990. Interactions between natural and farmed fish populations; information from genetic markers. Journal of Fish Biology 36:449–460.

Sloss, B. L. 2001. A molecular phylogeny of the Percidae. Doctoral dissertation. Southern Illinois University, Carbondale.

Smith, S. M., M. J. Maceina, and R. A. Durham. 1994. Natural hybridizaton between black crappie and white crappie in Weiss Lake, Alabama. Transactions of the American Fisheries Society 123:71–79.

Snyder, M., A. R. Fraser, J. Laroche, K. E. Gartner-Kepkay, and E. Zouros. 1987. Atypical mitochondrial DNA from the deep-sea scallop *Placopecten magellanicus.* Proceedings of the National Academy of Sciences of the USA 84:7595–7599.

Snyder, T. P., R. D. Larsden, and S. H. Bowen. 1992. Mitochondrial DNA diversity among Lake Superior and inland ciscoes (*Coregonus artedi* and *C. hoyi*). Canadian Journal of Fisheries and Aquatic Sciences 49:1902–1907.

Stellwag, E. J., E. S. Payne, and R. A. Rulifson. 1994. Mitochondrial DNA diversity of Roanoke River striped bass. Transactions of the American Fisheries Society 123:321–334.

Stepien, C. A. 1995. Population genetic divergence and geographic patterns from DNA sequences: examples from marine and freshwater fishes. Pages 263–287 *in* J. L. Nielsen, editor. Evolution and the aquatic ecosystem: defining unique units in population conservation. American Fisheries Society Symposium 17, Bethesda, Maryland.

Stepien, C. A., A. K. Dillon, and M. D. Chandler. 1998. Genetic identity, phylogeography, and systematics of ruffe *Gymnocephalus* in the North American Great Lakes and Eurasia. Journal of Great Lakes Research 24:361–378.

Stepien, C. A., and J. E. Faber. 1998. Population genetic structure, phylogeography and spawning philopatry in walleye *(Stizostedion vitreum)* from mitochondrial DNA control region sequences. Molecular Ecology 7:1757–1769.

Sturmbauer, C., E. Verheyen, and A. Meyer. 1994. Mitochondrial phylogeny of the Lamprologini, the major substrate spawning lineage of cichlid fishes from Lake Tanganyika in eastern Africa. Molecular Biology and Evolution 11:691–703.

Swofford, D. L., G. J. Olsen, P. J. Waddell, and D. M. Hillis. 1996. Phylogenetic inference. Pages 407–514 *in* D. M. Hillis, C. Moritz, and B. K. Mable, editors. Molecular systematics, 2nd edition. Sinauer, Sunderland, Massachusetts.

Szalanski, A. L., R. Bischoff, and G. Mestl. 2000. Population genetic structure of Nebraska paddlefish based on mitochondrial DNA variation. Transactions of the American Fisheries Society 129:1060–1065.

Tabata, K., and A. Mizuta. 1997. RFLP analysis of the mtDNA D-loop region in Red Sea bream *Pagrus major* population from four locations of western Japan. Fisheries Science 63:211–217.

Tagliavini, J., I. J. Harrison, and G. Gandolfi. 1995. Discrimination between *Anguilla anguilla* and *A. rostrata* by polymerase chain reaction–restriction fragment length polymorphism analysis. Journal of Fish Biology 47:741–743.

Takahata, N., and S. R. Palumbi. 1985. Extranuclear differentiation and gene flow in the finite island model. Genetics 109:441–457.

Takahata, N., and M. Slatkin. 1984. Mitochondrial gene flow. Proceedings of the National Academy of Sciences of the USA 81:1764–1767.

Taylor, E. B., and J. J. Dodson. 1994. A molecular analysis of relationships and biogeography within a species complex of Holarctic fish (genus *Osmerus*). Molecular Ecology 3:235–248.

Tegelstrom, H. 1992. Detection of mitochondrial DNA fragments. Pages 89–113 *in* A. R. Hoelzel, editor. Molecular genetic analysis of populations: a practical approach. IRL Press/ Oxford University Press, Oxford, UK.

Thomas, W. K., and A. T. Beckenbach. 1989. Variation in salmonid mitochondrial DNA: evolutionary constraints and mechanisms of substitutions. Journal of Molecular Evolution 29:233–245.

Todd, T. N. 1981. Allelic variability in species and stocks of lake Superior ciscoes (Coregoninae). Canadian Journal of Fisheries and Aquatic Sciences 38:1808–1813.

Tranah, G. J., H. L. Kincaid, C. C. Kreuger, D. E. Campton, and B. May. 2001. Reproductive isolation in sympatric populations of pallid and shovelnose sturgeon. North American Journal of Fisheries Management 21:367–373.

Travnichek, V. H., M. J. Maceina, M. C. Wooten, and R. A. Dunham. 1997. Symmetrical hybridization between black crappie and white crappie in an Alabama reservoir based on analysis of the cytochrome-*b* gene. Transactions of the American Fisheries Society 126:127–132.

Triantafyllidis, A., T. J. Abatzopoulos, and P. S. Economidis. 1999. Genetic differentiation and phylogenetic relationships among Greek *Silurus glanis* and *Silurus aristotelis* (Pices, Siluridae) populations, assessed by PCR–RFLP analysis of mitochondrial DNA segments. Heredity 82:503–509.

Tringali, M. D., and T. M. Bert. 1996. The genetic stock structure of common snook *(Centropomus undecimalis)*. Canadian Journal of Fisheries and Aquatic Sciences 53:974–984.

Tringali, M. D., T. M. Bert, and S. Seyoum. 1999. Genetic identification of Centropomine fishes. Transactions of the American Fisheries Society 128:446–458.

Tsigenopoulos, C. S., and P. Berrebi. 2000. Molecular phylogeny of north Mediterranean freshwater barbs (genus *Barbus*: Cyprinidae) inferred from cytochrome *b* sequences: biogeographic and systematic implications. Molecular Phylogenetics and Evolution 14:165–179.

Tzeng, C-S., C-F Hui, S-C Shen, and P. C. Huang. 1992. The complete nucleotide sequence of the *Crossostoma lacustre* mitochondrial genome: conservation and variations among vertebrates. Nucleic Acids Research 20:4853–4858.

Utter, F. M. 1991. Biochemical genetics and fishery management: an historical perspective. Journal of Fish Biology 39 (Supplement A):1–20.

Vitic, R., and C. Strobeck. 1996. Mitochondrial DNA analysis of populations of lake trout *(Salvelinus namaycush)* in west-central Canada: implications for stock identification. Canadian Journal of Fisheries and Aquatic Sciences 53:1038–1047.

Waldman, J. R., J. T. Hart, and I. I. Wirgin. 1996. Stock composition of the New York Bight Atlantic sturgeon fishery based on analysis of mitochondrial DNA. Transactions of the American Fisheries Society 125:364–371.

Ward, R. D., N. Billington, and P. D. N. Hebert. 1989. Comparison of allozyme and mitochondrial DNA variation in populations of walleye, *Stizostedion vitreum*. Canadian Journal of Fisheries and Aquatic Sciences 46:2074–2084.

Weiss, S., C. Schlotterer, H. Waidbacher, and M. Jungwirth. 2001. Haplotype (mtDNA) diversity of brown trout *Salmo trutta* in tributaries of the Austrian Danube: massive introgression of Atlantic basin fish—by man or nature? Molecular Ecology 10:1241–1246.

White, P. S., and L. D. Densmore III. 1992. Mitochondrial DNA isolation. Pages 29–58 *in* A. R. Hoelzel, editor. Molecular genetic analysis of populations: a practical approach. IRL Press/Oxford University Press, Oxford, UK.

Whitmore, D. H., T. H. Thai, and C. M. Craft. 1992. Gene amplification permits minimally invasive analysis of fish mitochondrial DNA. Transactions of the American Fisheries Society 121:170–177.

Whitmore, D. H., T. H. Thai, and C. M. Craft. 1994. The largemouth bass cytochrome *b* gene. Journal of Fish Biology 44:637–645.

Williams, R. N., R. P. Evans, and D. K. Shiozawa. 1997a. Mitochondrial DNA diversity patterns of bull trout in the upper Columbia River basin. Pages 283–297 *in* W. C. Mackay, M. K. Brewin, and M. Monita, editors. Friends of the bull trout conference proceedings. The Bull Trout Task Force, Calgary, Alberta.

Williams, R. N., R. F. Leary, and K. P. Currens. 1997b. Localized genetic effects of a long-term hatchery stocking program on resident rainbow trout in the Metolius River, Oregon. North American Journal of Fisheries Management 17:1079–1093.

Williams, R. N., D. K. Shiozawa, J. E. Carter, and R. F. Leary. 1996. Genetic detection of putative hybridization between native and introduced rainbow trout populations of the Upper Snake River. Transactions of the American Fisheries Society 125:387–401.

Wilson, C. C., and P. D. N. Hebert. 1993. Natural hybridization between Arctic char *(Salvelinus alpinus)* and lake trout *(S. namaycush)* in the Canadian Arctic. Canadian Journal of Fisheries and Aquatic Sciences 50:2652–2658.

Wirgin, I., L. Maceda, J. Stabile, and C. Messing. 1997. An evaluation of introgression of Atlantic coast striped bass mitochondrial DNA in a Gulf of Mexico population using formalin-preserved museum collections. Molecular Ecology 6:907–916.

Wirgin, I., L. Maceda, J. R. Waldman, and R. N. Crittenden. 1993. Use of mitochondrial DNA polymorphisms to estimate the relative contributions of the Hudson River and Chesapeake Bay stripped bass stocks to the mixed fishery on the Atlantic coast. Transactions of the American Fisheries Society 122:669–684.

Wirgin, I. I., R. Proenca, and J. Grossfield. 1989. Mitochondrial DNA diversity among populations of striped bass in the southeastern United States. Canadian Journal of Zoology 67:891–907.

Wirgin, I., and six coauthors. 2000. Genetic structure of Atlantic sturgeon populations based on mitochondrial DNA control region sequences. Transactions of the American Fisheries Society 129:476–486.

Wood, R. M., and M. E. Raley. 2000. Cytochrome *b* sequence variation in the crystal darter *Crystallaria asprella* (Actinopterygii: Percidae). Copeia 2000:20–26.

Xu, S., C. J. Kobak, and P. E. Smouse. 1994. Constrained least squares estimation of mixed population stock composition from mtDNA haplotype frequency data. Canadian Journal of Fisheries and Aquatic Sciences 51:417–425.

Yokobori, S., M. Hasegawa, T. Ueda, N. Okada, K. Nishikawa, and K. Watanabe. 1994. Relationships among coelacanths, lungfishes, and tetrapods: a phylogenetic analysis based on mitochondrial cytochrome oxidase I gene sequences. Journal of Molecular Evolution 38:602–609.

Youngson, A. F., D. Knox, and R. Johnstone. 1992. Wild adult hybrids of *Salmo salar* L. and *Salmo trutta* L. Journal of Fish Biology 40:817–820.

Zardoya, R., A. Garrido-Pertierra, and J. M. Bautista. 1995. The complete nucleotide sequence of the mitochondrial DNA genome of the rainbow trout, *Oncorhynchus mykiss*. Journal of Molecular Evolution 41:942–951.

Zwanenburg, K. C. T., P. Bentzen, and J. M. Wright. 1992. Mitochondrial DNA differentiation in western North Atlantic populations of haddock *(Melanogrammus aeglefinus)*. Canadian Journal of Fisheries and Aquatic Sciences 49:2527–2537.

5

Nuclear DNA

Bonnie Brown and John Epifanio

5.1 Introduction

As fisheries students or professionals, we often desire to detect the amount and pattern of genetic divergence within and among taxonomic units of interest. Not surprisingly, genetic differences within and among groups fall along a continuum ranging from the near identity within populations of clonal species (e.g., some livebearers [Poeciliidae]) to great divergence among phylogenetic classes (e.g., Cyclostomata versus Osteichthyes). We often ask the rather vague question, are these fish genetically different? Except for identical twins or nonrecombinant clones, the answer to this question must be, "yes."

Perhaps a more salient question should be, are the genetic differences among these populations important enough to concern us in a management or conservation context? To answer this question, we conceivably could determine the exact nucleotide sequence of all genes in individuals and directly compare the amount and pattern of differences within or among populations. Generally, the degree of sequence divergence between populations, in at least some parts of the genome, reflects the amount of time since the populations were isolated. We also would want to sequence samples from related outgroups to provide context for the genetic differences observed. Given the size of fish and shellfish genomes, however, the sequencing approach would be a cumbersome and time-consuming proposition and likely would prove impractical in light of current funding limitations. Extensive efforts directed toward sequencing and mapping complex genomes, such as the Human Genome Project and the National Animal Genome Research Project, highlight the difficulties and expense associated with such undertakings. A more reasonable approach, particularly from a population genetic perspective, is to survey (or sample) classes of genes that represent the diversity present in populations and other related groupings. In fact, we can use a number of indirect techniques to sample genetic diversity. These methods and the characters they uncover can be used as indices of genetic diversity if these characters are structured in meaningful geographical, temporal, or life history specific patterns that reflect their evolutionary history. The patterns also may be useful for uncovering the consequences of management practices, recent catastrophic events, or ongoing processes associated with habitat changes.

The most recent and perhaps promising approaches to collecting data relevant to fisheries science come from direct assessments of nuclear DNA (nDNA) sequence variation. (DNA, or deoxyribonucleic acids, are complex molecules that contain the inherited blueprints for all living organisms.) A wide spectrum of nDNA screening technologies have application to important fisheries issues, including conservation, domestication, forensics, phylogeography, reproductive success, stock identification, mixed-stock analysis, and taxonomy. In **eukaryotic** cells (i.e., cells having a structurally defined or discrete nucleus), most of the functional, protein-encoding DNA, as well as the noncoding, or "junk," DNA, resides within the nucleus. There are, of course, other protein-encoding DNAs in mitochondria, which encode compounds important in the energy-producing cytochrome pathway, and in plant chloroplasts, which encode photosynthetic enzymes. This chapter is concerned primarily with the organization and variation of nuclear DNA. Although some comparisons and contrasts will be made to such extra-nuclear DNA, readers interested in that topic are referred to Chapter 4. Three broad categories of nDNA investigation have evolved during the past decade: single-copy nuclear DNA (Southern 1975), hypervariable DNA (Davies 1985), and primer-amplified DNA (Saiki et al. 1988). Utilizing spin-off technologies from molecular biology, such as conventional cloning, genomic libraries, and the polymerase chain reaction (PCR; see Box 5.1), geneticists currently are contributing to fisheries management and conservation by genetic characterization of biological units that previously were intractable, such as individual fish larvae, small populations, and metapopulations.

Because PCR amplification primers generally are at least 20 nucleotides in length, most single-nucleotide mutations occurring within priming sites have little effect on the annealing ability of amplification primers. However, variations in the DNA sequence annealing to the 3′ end of a primer will result in nonamplification of the expected segment. Such a genomic sequence is referred to as a **null allele.** Furthermore, insertions and deletions of multiple base pairs that occur within the stretch between amplification primers are detected easily as increases or decreases in size of the resulting amplification products. Another type of analysis that actually is based on detection of DNA insertions or deletions such as these is variable number of tandem repeats, or VNTR, analysis (section 5.2.2). Like restriction fragment length polymorphism (RFLP; section 5.3.1) analysis, VNTR techniques may involve Southern blotting and probing (Figure 5.1).

■ 5.2 Variation in Nuclear DNA

5.2.1 Genomic Content

How much nDNA exists, and how many protein-encoding genes are there in fishes? The amount of genetic material in cells varies both among fish species and populations. In fact, the amount of DNA available in cells also can vary within individuals and among cell lines. The most obvious source of this latter kind of variation occurs between somatic and germ lines. A single sperm or egg, for example, has approximately one-half the nuclear DNA of a **somatic** (nonreproductive) cell. Diversity in genomic content among individuals or groups of individuals is due to small chromosomal mutations that alter total DNA content subtly. Changes in ploidy (Chapters 2, 6) impart more global changes to the genome. In animals, spontaneous ploidy changes are rare events and most often are associated with the evolution of novel species.

Genomic content is measured as a concentration compared with a human-based **lymphocyte** (white blood cell) standard resulting in a relative measure called the **C-value** (the characteristic amount of DNA in the unreplicated haploid nucleus of an

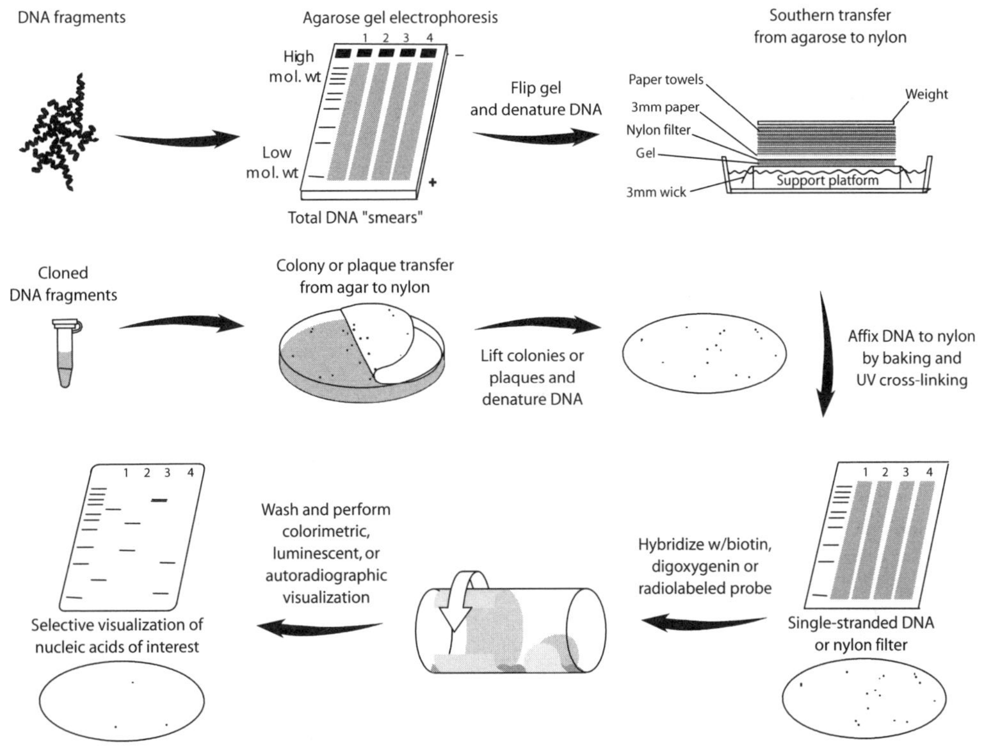

Figure 5.1 Schematic representation of a Southern (1975) transfer and heterologous (derived from a different species) hybridization technique for visualizing restriction fragment length polymorphisms (RFLPs), repetitive DNAs, and other nucleic acids of interest.

organism). In humans, C-values typically approximate 7 picograms (pg) of DNA per cell. Estimates of the amount of nDNA in typical piscine cells are in the range of 1 to 3 pg DNA as opposed to 6 pg for mammalian cells, although some fish species (e.g., lungfishes [Ceratodidae, Lepidosirenidae, and Protopteridae]) have as much as 100 pg DNA per genome. Determination of C-values is a useful tool complementing karyotyping in the study of diploid–tetraploid sibling species evolution in fishes. Based primarily on an anthropocentric view of organismal complexity, we might expect that advanced, or complex, organisms would require greater amounts of DNA to encode that complexity. However, a strict relationship between organismal complexity and DNA content has not been established. The failure to observe such a relationship often is called the **C-value paradox** (i.e., organisms with similar levels of evolutionary complexity often differ dramatically in genome size). In a review of information on genome evolution, Cavalier-Smith (1985) describes the C-value of a number of vertebrates. Of particular interest to students of fish biology is the range of nuclear genome sizes of the lungfishes. A size range of 80 to 142 pg has been observed, which is considerably more DNA than estimated for most other fishes and the so-called higher vertebrates. Maintenance of such a large genome is considered by some to be a costly genetic load; therefore, the benefits of an ultra-large nuclear genome are unclear. In fish families such as the trouts (Salmonidae), suckers (Catostomidae), and sturgeons (Acipenseridae), large genome size is a function of polyploidy. The origin of these groups and a discussion of ancestral polyploidy can be found in Chapter 3.

■ Box 5.1
Amplifying DNA Segments by Means of the Polymerase Chain Reaction

Prior to the late 1980s, the sole method of obtaining multiple copies of a DNA strand of interest was by the process of **cloning.** This essential molecular tool involves inserting foreign DNA into a vector such as a plasmid or phage that then carries the foreign DNA into bacteria in which the chimeric DNAs are replicated. Although any DNA can be cloned, the process of ligating the segment of interest into a plasmid requires that fairly large amounts are initially available (e.g., 0.1 μg of purified foreign DNA is used in a typical ligation). Thus, nDNA sequences obtained from only very small samples were not amenable to cloning.

Because of the sensitivity of more recent molecular genetic techniques, small samples, small life stages, and previously refractory tissues can be analyzed for genetic variation. Polymerase chain reaction (PCR)-based studies theoretically need only a single molecule of template DNA for successful amplification; in practice, we use several thousand copies for a single amplification reaction. The PCR is a chemical amplification process comprising three cycles: melting, annealing, and extension. A typical PCR reaction includes a small amount of target DNA (usually less than 50 ng), one or more oligonucleotide primers, free nucleotides, and thermally stable DNA polymerase in a suitable reaction buffer. **Primers** are short, single-stranded oligonucleotides (ranging from 10–50 base pairs [bp] in length) that flank both ends of the DNA segment under investigation. Thermostable DNA polymerases have been isolated from numerous bacteria that thrive in thermal springs or deep-sea hydrothermal vents and are stable at higher temperatures.

The amplification mixture first is heated to separate the double-stranded DNA. Melting breaks the hydrogen bonds between DNA strands, thereby removing any secondary structure of the molecule. However, melting does not break down the covalent phosphodiester bonds between the nucleotides, leaving primary structure intact. The melted DNA mixture is then cooled to permit single-stranded primers to anneal, or hybridize, to complementary sequences of the single-stranded genomic DNA. These DNA–primer hybrids serve as initiation templates for replication by the DNA polymerase. The mixture is then heated again but not as hot as the melting temperature. The DNA polymerase catalyzes the sequential addition of free nucleotides beginning at the 3′ end of the annealed primers (see Chapter 1). The mixture is reheated to the melting temperature, and the cycle repeats (see figure). At the end of each cycle, the mixture contains twice the former concentration of the targeted sequence and, therefore, each cycle effectively doubles the concentration of the desired DNA segment (see figure). Usually after 30 to 40 cycles, the reaction is stopped and the resulting fragments are separated by size, typically by means of gel electrophoresis. Because the concentration of amplified fragments is very high, they can be visualized with ethidium bromide or SYBRGreen®, alleviating the need for costly and time-consuming Southern hybridizations. There are many variants of PCR: producing double-stranded product, single-stranded product, or extra-long strands, among other techniques. See White (1993) for a thorough coverage of this essential molecular genetic tool.

continued on next page

■ Box 5.1 *continued*

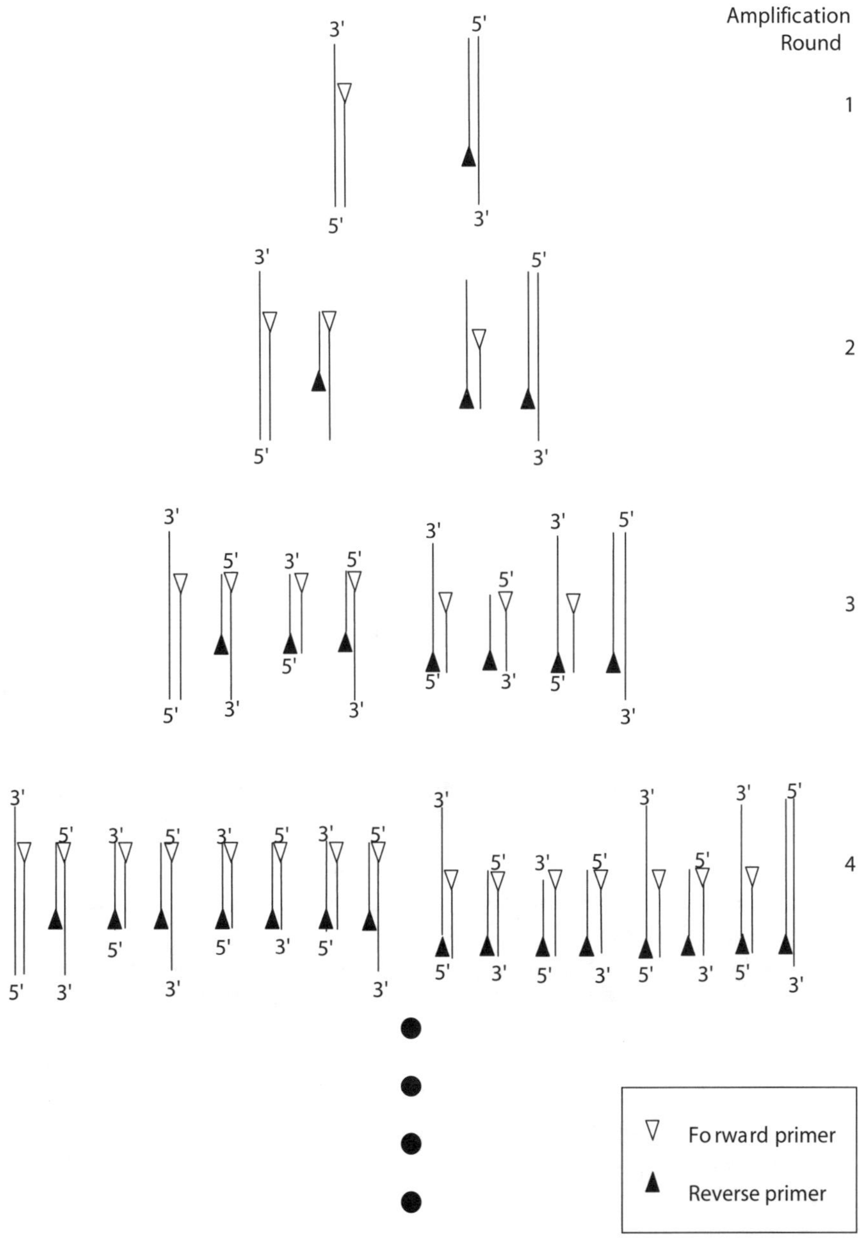

An individual's C-value can be measured with a spectrophotometer or a flow cytometer. Procedures require treating tissue or blood homogenates (fish erythrocytes are nucleated) with stains such as propidium iodide or Feulgen, which bind specifically to DNA and fluoresce under an ultraviolet light source. The stained samples are sorted through a particle counter and total fluorescence is partitioned into per-cell equivalents. Most studies of teleost C-values have employed a technique known as **cytophotometry,** by which an instrument combines microscopy with spectrophotometry to analyze Feulgen-stained tissue homogenates. A more recent and sensitive technique for measuring cellular DNA content is flow cytometry combined with propidium iodide staining of tissue homogenates. This latter method is becoming more widely used in studies of fish and shellfish DNA content.

Investigation of the genomic size or DNA content of a variety of teleost species indicates considerable variation in C-value within populations. However, no trend has been detected in the available data. For example, some minnow (Cyprinidae) species average 7% difference in genome content among conspecific individuals. The C-value within subfamilies of killifishes (Cyprinodontidae) in China is highly variable, and three minnows of the subfamily Schizothoracinae range from approximately 3 pg per nucleus to greater than 19 pg per nucleus (Jianxun et al. 1991). As a result, no clear relationship has been uncovered connecting fish or shellfish phylogenetic relatedness and variation in genome content.

5.2.2 *Repetitive DNA*

Another aspect of nDNA variation is the amount of repetitive DNA and its degree of repetition at particular loci. The nuclear genome of fishes, like all eukaryotes, contains segments of DNA that are repeated tens or even hundreds of thousands of times. These repeated sequences are known as **repetitive DNAs** and are classified and defined as follows.

One class of repetitive DNA includes **nonmobile repetitive DNA sequences,** that is, repeats not associated with a mechanism to move to another site in the genome.

- —**Dispersed gene families** are homologous sequences for genes, including those that encode actins, myosin, some histones, and other proteins that are repeated and spread throughout the genome.
- —**Tandem gene families** are genes whose products are needed in large quantities, for example, ribosomal RNA (rRNA), transfer RNA (tRNA), and some histones. They occur in arrays of repeated, adjacent copies.
- —**Noncoding telomeric sequences** at the tips of chromosomes are short, simple sequences that fold back on themselves in a hairpin structure so that lagging 5′ ends can have 3′ ends for priming of polymerization.
- —**Satellite DNA** refers to highly repetitive DNA, sometimes found in the heterochromatin flanking centromeres. This class of repetitive DNA includes repetitive monomers and megasatellites.
- —**Variable number of tandem repeats (VNTRs)** are DNA sequence units repeated in tandem and dispersed throughout the genome. They vary in number at different loci and in different individuals. The VNTRs include minisatellites and microsatellites.

Another class of repetitive DNA includes the **mobile repetitive DNA sequences,** that is, repeats that are associated with a mechanism to move to another site in the genome.

Transposed sequences include the following repetitive DNAs.

— **Transposons** are coding DNA sequences (also known as DNA intermediate transposons) flanked by short terminal inverted repeats. Transposons have the ability to move from one chromosomal position to another.
— **Retrotransposons** are structurally related to retroviruses and are flanked by long terminal direct repeats. They multiply by reverse transcription of RNA into DNA, and the resulting copies insert themselves into the chromosome at numerous locations..
— **Retroposons** are structurally related to retroviruses but are not flanked by repeats. They multiply in the same way as retrotransposons. Two types of retroposon are
  — **short interspersed elements (SINEs),** whose units are hundreds to thousands of base pairs (bp), are homologous to DNA coding sequences (e.g., tRNA), and
  — **long interspersed elements (LINEs),** whose units are thousands to tens of thousands of bp, and whose sequences are homologous to retroviruses.

Repetitive DNAs are of significant importance in studies of fish and shellfish genetics (Figure 5.2). One category commonly referred to in the literature is satellite DNA (stDNA) which consists of short to long sequences (called motifs and gene families) that are tandemly repeated a varying number of times. Satellite DNAs are so named because they are rich in either adenine (A) and thymine (T) (making them lighter than the majority of chromosomal DNA) or rich in guanine (G) and cytosine (C) (making them heavier), leading to separation from other nuclear DNA in a cesium chloride gradient following ultracentrifugation. Investigators categorize satellites roughly by the size and location of a repeat. **Repetitive monomers** and **megasatellites** are adjacent repeats hundreds to thousands of nucleotide base pairs in length and sometimes are found clustered in the centromeres; **minisatellites** are repeats of motifs 9 to 65 bp and are dispersed throughout the genome; and **microsatellites** are tandem arrays of very short repeating motifs 2 to 8 bp and are dispersed throughout the genome.

Mini- and microsatellites have become important population genetics tools, and increasingly they are applied to the study of fish and shellfish genetics. Both exhibit their diversity as variation in the number of tandem repeats. The VNTRs are thought to originate in a two-step process in which saltatory replication, or slippage during DNA polymerase reaction, first results in multiple copies of a DNA sequence in the region between two nucleosomes followed by **unequal crossing-over.** Recall that crossing-over normally results in an equal exchange of base pairs between two arms of a chromosome. In some crossovers, however, alignment of the two chromosomes is staggered, forming one longer and one shorter recombinant molecule. In the longer, a tandem duplication occurs, while a deletion of the same sequence occurs in the shorter recombinant molecule. Most duplications are removed from a population by genetic drift (Chapter 9), but some duplications increase in frequency and spread through the population. It is thought that certain regions of the genome may continue to undergo unequal crossing-over, resulting in a chain of repeats as illustrated in Figure 5.2. Notice that if crossing-over occurs at the end of the chain of repeating units, a geometric expansion of repeats occurs from two to four to eight and so on. Conversely, if crossing over occurs within the chain of repeating units, an arithmetic expansion occurs, incrementing the chain of repeats by one unit with each round of replication.

5.2.3 *Single-Copy DNA: Structural and Regulatory Genes*

Nonrepetitive DNA sequences also provide information about genetic variation. Each cell in an individual contains the same suite of chromosomes as all other cells in that

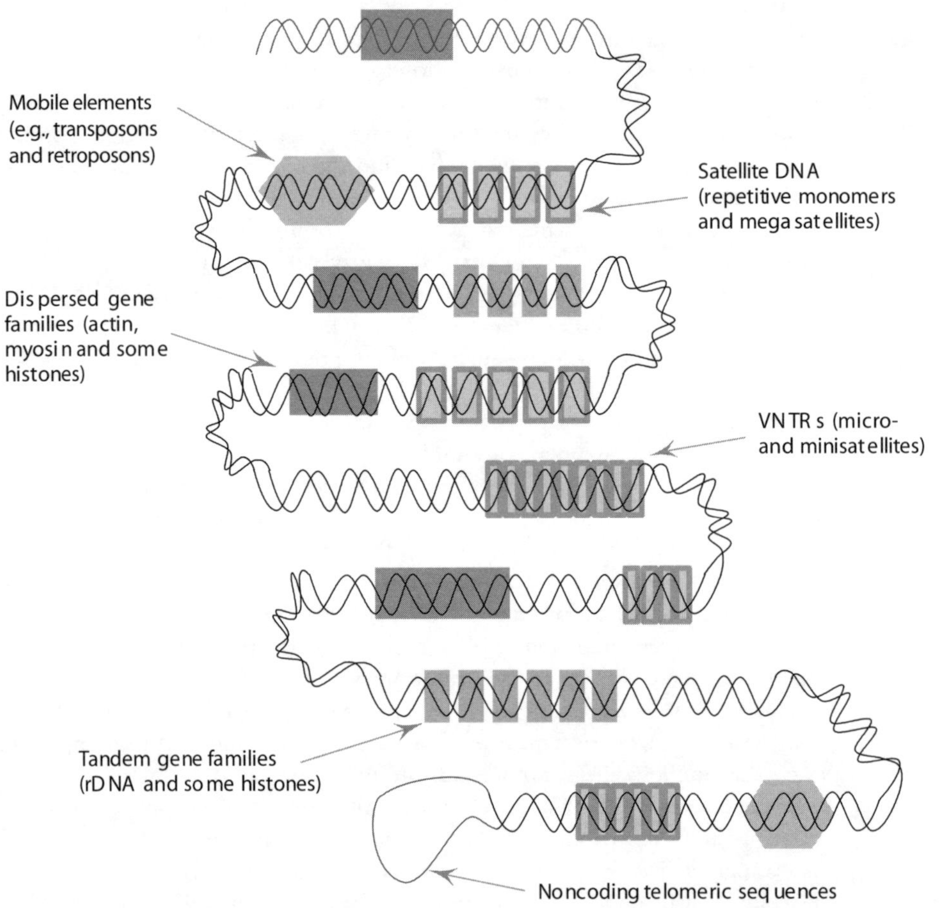

Figure 5.2 Schematic representation of different types of repetitive DNA, including ribosomal DNA (rDNA) and variable number of tandem repeats (VNTRs).

individual. Genes on the chromosomes, essentially the chemically encoded programs for all biological processes in that organism, are hereditary units that can undergo replication, mutation, and expression. Genes that encode allozymes (Chapter 2) and all other proteins are located on chromosomes as short sequences (generally hundreds to thousands of nucleotides in length) called **exons** that are interrupted by noncoding **introns** (Figure 5.2). In the laboratory, once the nDNA is separated from DNA-binding proteins, coding sequences (i.e., exons) are indistinguishable from noncoding introns and from the repetitive DNA. Thus, a total DNA preparation from an individual (Box 5.2) provides the material necessary to conduct literally dozens of different types of screenings.

There are functional distinctions between several classes of genes. Most of the examples we have discussed in this text represent the class of genes called structural genes. **Structural genes** encode proteins important as building blocks and as enzyme subunits for catalyzing metabolic and synthetic biochemical pathways required within all cells. Mutations that alter codons may affect amino acid sequence and thereby the size, shape, or charge of an enzyme. A mutation that drastically alters an enzyme's active site may render the enzyme inactive in the organism's current environment. If carried into a population, this mutation becomes a deleterious allele. Many genetic diseases are caused by

■ Box 5.2 Extracting Nuclear DNA

To analyze nuclear DNA (nDNA) of fishes or shellfishes, only very small samples of tissue from each individual are necessary. Virtually any tissue source may be used: blood, hemolymph, fins, barbels, flesh, mantle, gill, or gonad. Minute quantities of fresh, alcohol preserved, or frozen tissue (1–100 mg) will yield sufficient quantities of nDNA to perform most analyses. Hillis et al. (1996) describe numerous techniques that have been examined for isolating and purifying DNA from animal tissues. The methods described below work for most tissues from both fishes and shellfishes (see Grewe et al. 1994 for hexadecyl trimethyl ammonium bromide [CTAB] protocol and Shure et al. 1983 and modifications by Chen et al. 1992 for liquid N_2 protocol). The CTAB protocol yields high quantities of moderate- to small-size DNA fragments, and the liquid N_2 protocol yields moderate quantities of relatively intact, long DNA strands. Commercially available kits also are available that allow total genomic DNA extraction without the use of toxic organic solvents; however, integrity of DNA in such preparations may decline dramatically over time if stored at greater than –80°C.

Summary of protocol for DNA isolation by use of CTAB buffer

1. Place 100 to 200 μg of tissue into an Eppendorf-type tube.
2. Dispense 700 μL stock CTAB buffer (50 mM Tris base, pH 8.0; 0.7 M NaCl; 10 mM EDTA; 1% CTAB; and 0.1% β-mercaptoethanol) along with 100 μg of proteinase K (an enzyme that degrades proteins including enzymes) into the sample tube.
3. Gently hand-grind tissue with a micropestle that fits snugly into the base of the tube.
4. Heat the buffered mixture to 60°C for at least 1 h. This step breaks down membranes and permits DNA to elute.

continued on next page

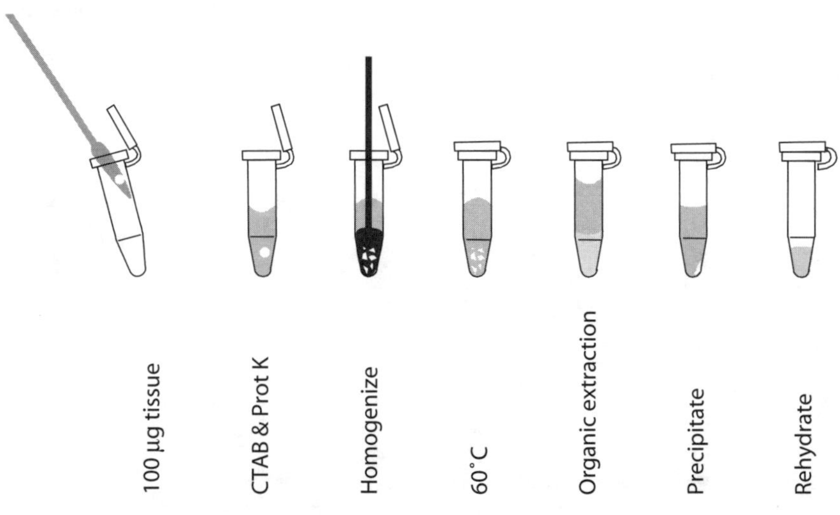

■ Box 5.2 *continued*

5. After incubation, spin tubes in a microcentrifuge at room temperature for 1 min to form pellets of cellular debris.
6. Extract proteins from the supernatant with phenol:chloroform:isoamyl alcohol (25:24:1), spin to separate phases, and recover upper aqueous phase.
7. Estimate volume of the resulting supernatant and add 0.1 volume 4.4 M ammonium acetate and an equal volume of isopropanol, mix, and hold –20°C to precipitate DNA. Spin, decant alcohol supernatant, and wash pellet with 70% ethanol. Dry and then dissolve DNA pellet in 100 µL TE buffer (10 mM Tris base, pH 8.0, 1 mM EDTA).
8. Quantify DNA by electrophoresing an aliquot of the sample alongside samples of known DNA concentration ranges, by employing DNA detection strips, or by using a more sensitive fluorimetric method

Summary of protocol for liquid N_2 isolation of DNA

1. Dispense 1 mL stock Fedoroff buffer (50 mM Tris base, pH 8.0; 0.35 M NaCl; 20 mM EDTA; 7.0 M urea; and 2% Sarkosyl) into a microfuge tube.
2. Fill a small sterilized ceramic mortar with liquid N_2. When one-half of the N_2 has evaporated, add 100 to 200 mg of tissue. Tissue will freeze to a white blob.
3. Just as the last of the liquid N_2 evaporates, begin grinding tissue. Grind very hard and very rapidly for 30–40 strokes to yield a fine powder.
4. Immediately scrape powder into microfuge tube with buffer by means of a flame-sterilized, metal, round-ended spatula (scoopula) and shake well to mix (don't form vortex). This step elutes DNA.
5. Immediately add 1 mL phenol:chloroform:isoamyl alcohol (25:24:1) and shake hard (don't form vortex). Spin in a microcentrifuge at room temperature for 2–5 min to form pellet of debris.
6. To precipitate DNA, add 200 µL of 7.5 M ammonium acetate, mix gently, split into two tubes, add an equal volume of isopropanol to each, and mix gently.

continued on next page

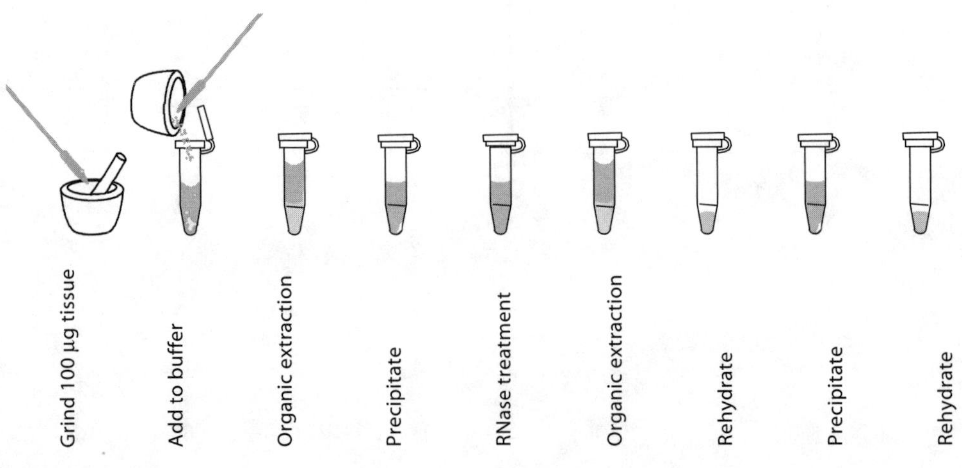

■ Box 5.2 *continued*

Strings of DNA should become apparent immediately. Continue mixing until a compact clot forms. Spin 30 s to form pellet of DNA.

7. Wash pellet and sides of tube with 70% ethanol. Dry pellet and rehydrate in 500 µL of TE buffer plus 2 µl of DNase-free RNase A. Hold at 37°C for 30 min.
8. Add 250 µL of 7.5 M ammonium acetate and mix gently. Repeat organic extraction and precipitation (steps 5 and 6 above), then rehydrate DNA in 100 µL TE buffer.
9. Quantify DNA by electrophoresing an aliquot alongside samples of known DNA concentration ranges, by employing DNA detection strips, or by using a more sensitive fluorimetric method.

mutations that alter enzyme activity by changing the active site or producing nonfunctional protein subunits.

Genes that control the expression of structural genes are referred to as **regulatory genes.** Regulatory genes often are thought of as on-and-off switches for structural genes. Structural genes do not transcribe products at constant rates throughout the life of a cellular lineage. Nor do cell lines in every tissue type transcribe all structural genes. Rather, structural genes are turned on at programmed times during development. Furthermore, differentiated cell lines are selective as to the genes that are transcribed; this is a response to organ and tissue specialization. In fishes, important developmental periods are accompanied by, and perhaps are a response to, increased transcription of important structural genes. For example, the physiological changes associated with salmon smoltification are likely to be under the partial control of regulatory genes. Other developmental milestones occurring during embryogenesis and reproductive maturation are associated with regulatory gene-programmed activation. The emerging field of developmental genetics to a great extent focuses on regulatory gene control of structural gene expression and differentiation (Epifanio and Philipp 1994; Cardwell et al 1998; Marcus and McCune 1999).

Genetic variation due to regulatory gene expression has received considerable attention in trouts, sunfishes (Centrarchidae), killifishes *Fundulus* spp., zebra danio *Danio rerio*, tilapias *Tilapia* spp., and bonefish *Albula vulpes*. The laboratory methodologies for RNA isolation, complementary DNA (cDNA) cloning, and sequencing now are easily applicable to the study of developmental and tissue-specific gene expression in aquatic species. As a component of the "genomics revolution," we likely will see renewed interest in gene regulation and development from a more molecular perspective. In particular, temporal differences in gene expression **(heterochrony)** among fish stocks may become an important area of investigation in determining the effects of stock mixing.

A final class of structural genes includes **ribosomal DNA,** DNA regions that code for ribosomal RNA components and therefore are transcribed but not translated into protein products. Coding sequences for **ribosomal RNAs** (rRNAs; the transcripts of ribosomal genes) have been evolutionarily conserved across taxa and therefore have been important to our understanding of macroevolutionary processes and broad phylogenetic analyses. This family of genes occurs as a single unit encoding 5.8 S, 18 S, and 28 S rRNA molecules (S being a measure of size), which assist in the ribosome's role in protein synthesis. As in other eukaryotes, hundreds of copies of these units are dispersed throughout the genome. Arrays of ribosomal genes reside as tandem repeats in the nucleolar organizer region (NOR) of chromosomes, and each is separated from the next by a nontranscribed spacer region. The rRNA gene clusters (along with spacer regions) occupy a substantial portion of fish DNA and constitute a considerable portion of their repetitive DNA sequences. For both fishes and shellfishes, studies of rRNAs and NORs have broad utility, including phyletic resolution of genera and species, ploidy verification, metabolic studies, identification of different hatchery lines, and verification of interspecific hybridization (Chapter 3).

5.3 Techniques Used to Determine Nuclear DNA Variation

In all nDNA studies, biochemical and mechanical techniques are used to manipulate the small quantities of DNA isolated from single individuals. The methods are described in this chapter, but to truly understand nDNA techniques and the results generated by each of these methods, we must first understand the basic structure and function of the DNA molecule as outlined in Chapter 1.

There are essentially two approaches one can use to compare individuals for genetic divergence across taxa. The first is to acquire directly the entire nucleotide sequences among individuals representing each taxon. The second is an indirect, or indexing, approach in which homologous reaches of the genome are compared using either sequence information or gene product information. Whereas detailed gene sequence information is essential for some phylogenetic analyses, it is not generally an appropriate genetic character to use to examine population level phenomena. Furthermore, comparison of direct nucleotide sequence data for a statistically significant sample size from a fishery species and related outgroups poses some profound analytical, data reduction, and statistical challenges. Thus, indirect assessment of nuclear sequence differences is generally the optimal means for obtaining population data.

5.3.1 Restriction Fragment Length Polymorphism Analysis

Indirect assessment of DNA sequence variation often relies on restriction enzyme digestion and detection of RFLPs. However, because of its great complexity, nDNA variation cannot be detected and quantified in the same manner as described for mitochondrial DNA (mtDNA) variation (Chapter 4) (i.e., by simple gel electrophoresis of restriction digest fragments). Detection of nuclear RFLPs typically is accomplished by either probing restriction digests of total nDNA with specific single-stranded DNA sequences called **probes** or by first amplifying specific nDNA segments and then performing restriction digests on the amplified material (Figure 5.3). The former can be accomplished by using a colorimetrically, fluorimetrically, or radioactively labeled probe, whereas the latter is accomplished by using cloning vectors or PCR to amplify portions of the original DNA for further study.

A method related to RFLP analysis is an analysis of restriction sites and concurrent site mapping. One of the proven difficulties with RFLPs is that electrophoretic comigration of fragments among individuals does not necessarily indicate homology. Erroneous conclusions can, therefore, be drawn based on the presence or absence of a particular fragment. Restriction site analyses extend the information value of the data by identifying the relative position of sites on a chromosome in relation to other sites recognized by the same restriction endonuclease as well as those recognized by other

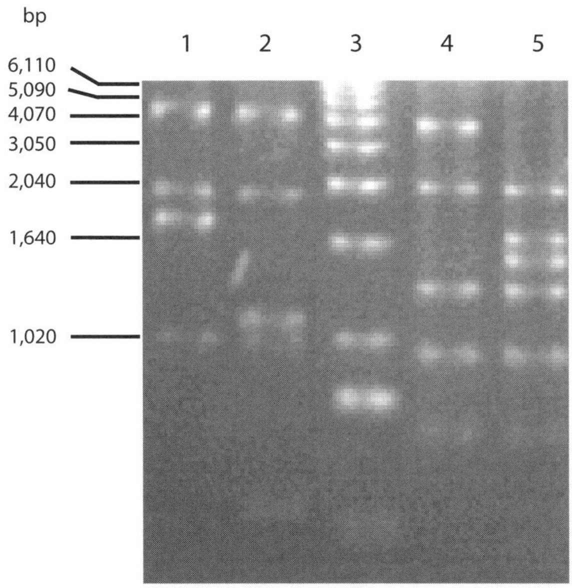

Figure 5.3 Composite gel photo showing polymerase chain reaction (PCR)–RFLP haplotypes for two individual blackbanded sunfish *Enneacanthus chaetodon*. Bands were produced by first amplifying three large nonoverlapping segments of each individual's mitochondrial DNA followed by digesting the amplified products with the restriction enzymes *Hind* III (lanes 1 and 2) and *Dra* I (lanes 4 and 5). Lanes 1 and 4 represent the first individual and lanes 2 and 5 represent the second. Fragment size in number of base pairs (bp) is shown for the molecular weight standard (lane 3).

enzymes. Documenting the presence or absence of restriction sites is a more technically demanding and time-consuming approach because a series of double enzyme digestions are required; however, the information content tends to be more accurate than RFLP screening without site mapping.

5.3.2 *Repetitive DNA Analysis*

The various methods used to examine repetitive DNA yield data over a broad range of sensitivity to genetic processes, data that can be used to address many fisheries and aquaculture issues. Repetitive DNA can be particularly useful for investigating population structure (Chapter 13). The sequences generally are not located within functional gene coding sequencess and are therefore considered to be selectively neutral, allowing for the assumption that the estimated amount of sequence divergence between units of interest is directly proportional to the length of time since separation. To estimate genetic relatedness among populations, banding patterns on Southern blots or actual sequences may be employed to calculate nucleotide sequence divergence based on equations outlined by Nei (1987a, 1987b). Because there are virtually an unlimited number of loci, the use of multiple, independent repetitive DNA markers has been proposed as a means of increasing the reliability of estimates of heterozygosity for the total genome. Improved estimates of heterozygosity are useful for both the study of population genetics and genetic improvement of fishery species.

The nuclear genome of pollock *Pollachius virens* contains at least 13% repetitive DNA (Denovan and Wright 1990), whereas that of rainbow trout *Oncorhynchus mykiss* contains approximately 84% repetitive DNA (Schmidtke et al. 1979). This excess constitutes a sizable reservoir of potentially useful genetic information. Several of the major repetitive DNA classes observed in the nuclear genome of eukaryotes can be used to assess nDNA sequence variation in fishes. First, because rRNAs evolve at different rates (Muslimov 1993), studies of the functional rRNA tandem gene families can be used to address issues ranging from phylogenetic analysis to assessment of population structure simply by selecting the appropriate coding or spacer regions and preparing restriction maps of the genes. Second, nonfunctional repetitive sequences, in particular centromeric and VNTR satellite DNAs, are useful for population level and microevolutionary investigations including selective breeding and forensics. Satellite monomers, minisatellites, and microsatellites have been investigated in molluscs, crustaceans, echinoderms, and in trouts, cods (Gadidae), killifishes, ayus (Plecoglossidae), sunfishes, temperate basses (Moronidae = Percichthyidae), and several other fish taxa (satellites in several aquatic species are reviewed by Beridze 1986). For example, mini- and microsatellites have been used to identify species within several teleost genera, multilocus VNTR sequences have been used to discriminate microgeographic population structure in liverbearers *Poecilia* spp. (Laughlin and Turner 1994), satellite monomers have shown dichotomous differences between populations of sheepshead minnow *Cyprinodon variegatus* (Turner et al. 1991), and dispersed repetitive sequences have been used to evaluate ancestry of striped bass *Morone saxatilis* broodstocks (Turner et al. 1991).

Minisatellites also have been applied in fisheries for forensics, to estimate mating success, and for confirming gynogenesis. Using a suite of individual minisatellite loci to investigate the breeding structure of populations of wild smallmouth bass *Micropterus dolomieu* in northern Ontario, Gross et al. (1994) demonstrated that nest-specific "fingerprints" could be identified and that the loci displayed both temporal and spatial patterns of philopatry. Similar results have been obtained using single-locus probes (e.g., derived from phage M13) as a population marker in rainbow trout (Fields et al. 1989). Several examples of the methods and techniques used to generate repetitive DNA information for fishes and shellfishes are outlined in Box 5.3.

5.3.3 Single-Copy DNA Analysis

The use of cloned **single-copy nuclear DNA** (scnDNA; nonrepetitive nuclear sequences that occur with a frequency of one per haploid genome) as described by Avise (1994) provides fisheries geneticists with a method that avoids many of the concerns associated with allozymes and some other genetic characters. Single-copy nuclear DNA loci are especially useful because they originate from multiple unlinked genes and allow the rapid survey of many individuals in a population for genetic variation in a short period of time. In contrast to genetic analysis for mtDNA haplotypes, which is hampered by a loss of a Hardy–Weinberg reference frame (see Chapter 1), scnDNA segregates as a Mendelian trait (biparentally transmitted and expressed), which permits testing for a Wahlund effect (Chapter 7) and other consequences of disequilibrium. Single-copy nuclear DNA loci are more likely to be physically unlinked than are mtDNA restriction sites, which can increase statistical power because of the ability to perform repeatable and independent testing. Compared with metabolic allozyme variants, some of which may not be selectively neutral, scnDNA variants generally are considered selectively neutral. Lastly, because scnDNA variants occur throughout the nuclear genome and are highly abundant, they afford a much greater coverage of the genome than do the limited number of allozymes available for genetic screening.

■ Box 5.3
Examining Repetitive DNA

Methods for examining variability differ for the various classes of repetitive DNAs.

Satellite monomers

The procedure for analyzing satellite monomers involves restriction digestion of nDNA to find discrete monomer repeats that can be seen as bands that overlay the smear of random-sized DNA fragments separated on an agarose gel (see figure below). Excision and purification of a band allows one to clone the monomer and construct primers to amplify satellite sequences from individuals in several populations, sequence each individual's amplification product, and ultimately assess sequence differences within and among populations.

Minisatellites

Techniques for minisatellite analysis involve probing complete restriction digests of nDNA with commercially available repetitive DNA sequences such as Jeffreys et al. (1985) probes (33.15 and 33.6) or the M13 repeat (e.g., Fields et al. 1989) to generate individual-specific multilocus fingerprints. Alternatively, one can clone random

continued on next page

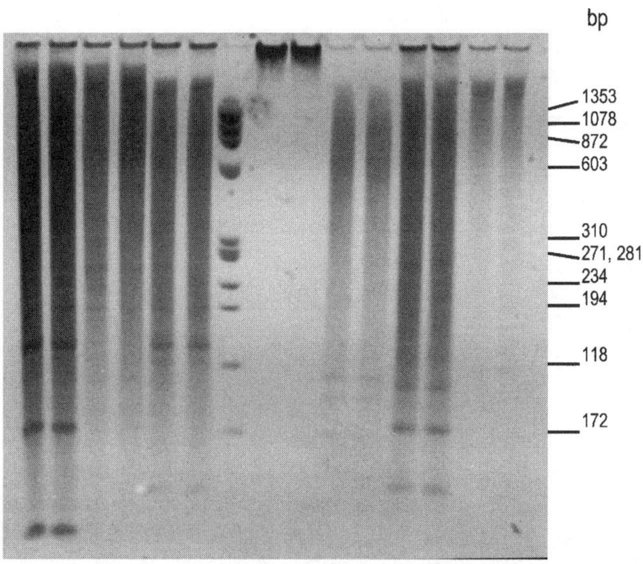

This polyacrylamide gel has been stained with ethidium bromide and shows monomer bands overlaying smears of total genomic DNA. Each lane contains approximately 10 μg of genomic DNA digested with *Alu* I. Duplicate samples from left to right are coho salmon *Oncorhynchus kisutch,* rainbow trout, cutthroat trout *O. clarki,* molecular weight size marker, uncut coho salmon DNA, brook trout *Salvelinus fontinalis,* lake trout *S. namaycush,* and a Lake Malawi cichlid *Pseudotropheus zebra.* Photograph provided courtesy of Bruce Turner, Virginia Polytechnic Institute and State University.

■ Box 5.3 *continued*

nDNA bands from partial restriction digests, select those cloned segments that hybridize strongly to Jeffreys probes, and use the cloned inserts to probe complete restriction digests of total DNA from many individuals to assess differences among families and among individuals within families. This method also is useful across taxa to assess inter-individual variation in related genera.

Microsatellites

The technique for microsatellite analysis requires probing complete digests of nDNA with simple sequence repeats (i.e., di-, tri-, or tetra-nucleotide repeats) to generate individual-specific fingerprints. Alternatively, primers that begin PCR amplification from unique (nonrepeat) sequences that flank a microsatellite repeat region can be created from repeat sequences detected in a DNA library. The PCR primers are used to amplify microsatellites to generate individual-specific genotypes or to evaluate familial relationships. Shown on the following page is an example of microsatellite analysis.

continued on next page

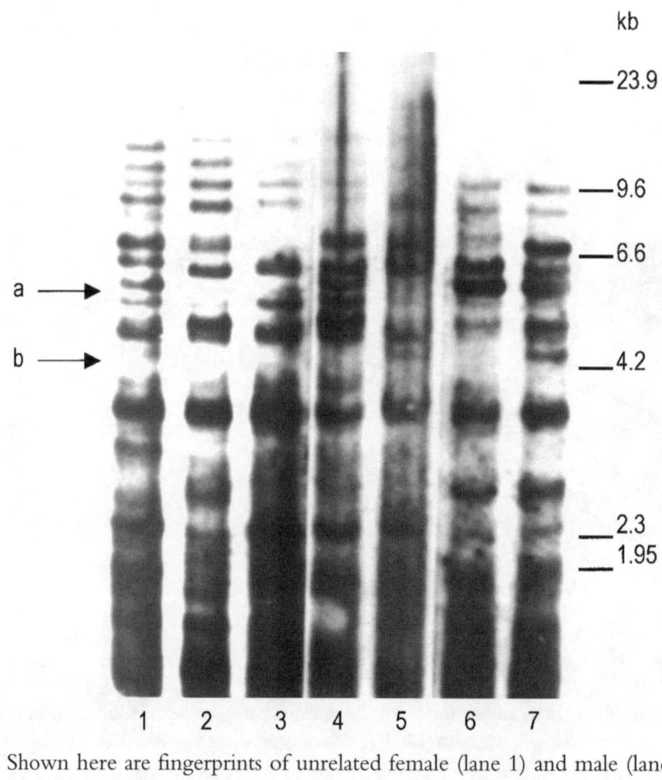

Shown here are fingerprints of unrelated female (lane 1) and male (lane 2) rainbow trout and five offspring (lanes 3–7), revealed by hybridization with M13 DNA. Scale of band molecular weights, measured in kilobases (kb), is shown to the right of the gel. Alternate segregation in offspring is shown by two female-specific bands (a and b). Photograph provided courtesy of Bob Fields, Illinois Natural History Survey.

■ Box 5.3 *continued*

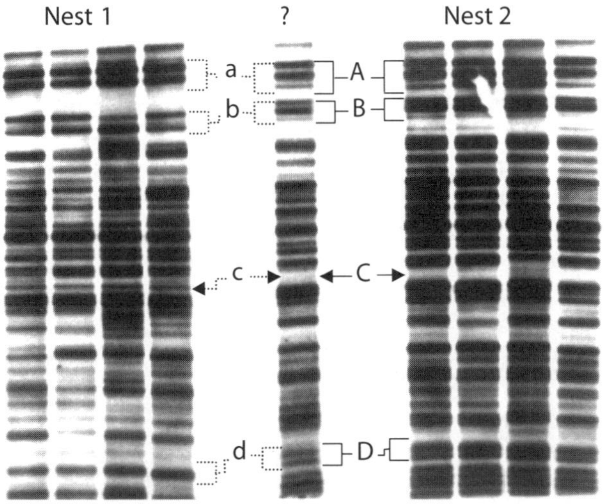

DNA fingerprints used in a blind test to determine the nest of origin of a smallmouth bass fry (Gross and Kapuscinski 1994). Dotted brackets and lowercase letters show regions where bands from the unknown fry do not match with fry known to originate from nest 1. Solid brackets and uppercase letters show regions where bands from the unknown fry match bands seen in other fry known to be from nest 2. Photograph provided courtesy of Mark Gross, University of Minnesota.

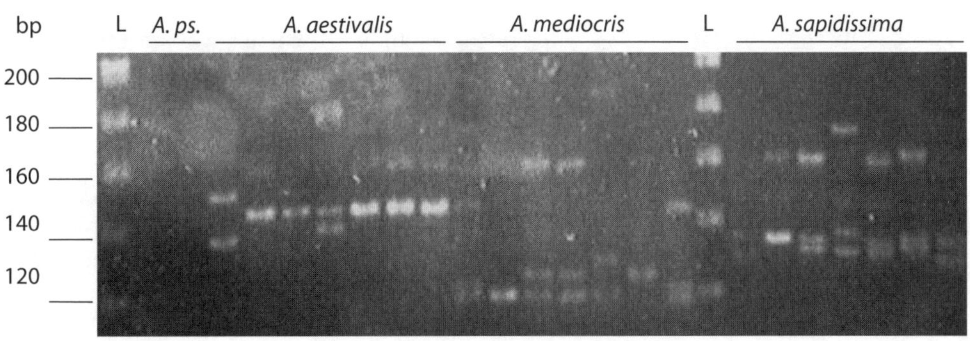

Population genetic variation for the *Asa4* microsatellite for individuals from four herring species, alewife *Alosa pseudoharengus (A. ps.)*, blueback herring *A. aestivalis*, hickory shad *A. mediocris,* and American shad *A. sapidissima*. Molecular weight standard was loaded in lanes marked L. Photo provided courtesy of David Taylor, Virginia Commonwealth University.

Applications of scnDNA in fisheries science have focused on population structure and the extent of gene flow. The utility and selective neutrality of a PCR-based scnDNA assay has been demonstrated by verifying pronounced population subdivision in the Eastern oyster *Crassostrea virginica* (Karl and Avise 1992). Several types of scnDNA studies have been useful in the study of striped bass population genetics, illustrating geographically partitioned genetic variation (Wirgin and Maceda 1991). Investigators also have identified hybridization among congeneric *Morone* taxa by means of oncogene probes and have used similar probes to determine ancestral history of individual fish (Wirgin et al. 1992). Oncogene probes are of interest in fisheries genetics studies because they have been shown to reveal a broad range of population level variation, from no intraspecific polymorphism (i.e., fixed differences between taxa) to population-specific variation. In Box 5.4, three specific examples of the use and application of scnDNA are outlined.

5.3.4 Single Nucleotide Polymorphisms

A variant of scnDNA polymorphism is based on detection of individual nucleotide substitutions termed SNP (pronounced snip) analysis. Some SNP analyses detect substitutions in the recognition sequence for a restriction endonuclease, many employ PCR, some employ microchip arrays and fluorescence technology, and some merely employ dot blot hybridization using a pool of mutation-specific oligonucleotides. To date, the main applications for SNPs have been in genomic studies and as markers diagnostic for disease; however, because they are a major component of many gene chip technologies, SNPs may soon become a useful tool in fisheries.

5.3.5 Multiple Arbitrary Amplicon Profiling

The specific locations and functions of many nDNA sequences we can survey are unknown. A general class of PCR-based techniques used to detect such anonymous, or arbitrary, sequences is called **multiple arbitrary amplicon profiling (MAAP).** Variation detected by MAAP is useful for several kinds of genetic inquiries in fishes and shellfishes, including chromosomal mapping, stock identification, and breeding analyses. Of the various types of MAAP methods commonly used to detect genetic variation to date, the procedure known as randomly amplified polymorphic DNA (RAPD, pronounced rapid) has been most widely employed in studies of fishes. Other MAAP methods currently in use include DNA amplification fingerprinting (DAF) and sequence characterized amplified region (SCAR).

In theory, detection of variation by means of RAPD is relatively straightforward, but a moderate level of laboratory competence is required to conduct this kind of work. Randomly amplified polymorphic DNA is generated by PCR (Box 5.1, Figure 5.4). A typical RAPD reaction includes a small amount of genomic DNA (usually less than 50 ng), one or more oligonucleotide primer, free nucleotides, and *Taq* polymerase in a suitable reaction buffer. Primer sequences are short, usually about 10 bp in length, although other sizes can work. The specific amount and concentration of template DNA, as well as buffer and salt conditions, must be empirically determined for each new species of fish or shellfish before beginning a large-scale investigation. The resulting amplified fragments may vary in length (tens to thousands of base pairs) and are presumed to be scattered throughout the genome. A mutation within the sequence complementary to the 3′ end of a primer (i.e., substitution, addition, or deletion) will prevent primer annealing

■ Box 5.4
Examining Single-Copy Nuclear DNA Variation

Several methods for observing variability of single-copy nuclear DNA (scnDNA) have been employed in fisheries and aquaculture.

Single-copy nuclear restriction fragment length polymorphism (scnRFLP) analysis

A genomic **DNA library** is produced by cloning partial-digest DNA fragments. Inserts are assayed by dot blot analysis (Sambrook et al. 1989) to determine if they contain single-copy sequences. Individual samples are digested with restriction enzymes, electrophoresed to separate fragments, and undergo Southern hybridization with the cloned single-copy sequences to identify scnRFLP bands. This technique can be used to verify the extent of hybridization between species in wild populations or to determine geographic and population level variation. A variation of this technique employs commercially available oncogene probes instead of cloned species-specific probes (cf., Wirgin and Maceda 1991).

Polymerase chain reaction and restriction fragment length polymorphism (PCR RFLP) analysis

Both ends of random single-copy clone inserts are sequenced to provide the information necessary to design PCR primers for anonymous regions of the nDNA. Genetic polymorphisms are identified by amplifying individual samples from a population with these primers and then using the amplified DNA as substrate for several restriction enzyme analyses. The PCR-based assays can be used to estimate among-population variance and gene flow (cf., Karl and Avise 1993).

Exon-primed, intron-crossing (EPIC) analysis

This method requires the design of PCR primers that anneal to the exons of highly conserved genes (such as actin, aldolase, globin, and major histocompatibility complexes) flanking rapidly evolving noncoding intron regions. Exon-primed, intron-crossing PCR amplification products can be digested with various restriction enzymes and visualized following electrophoresis through agarose gels *or* they can be cloned and sequenced to separate the alleles contributed by both parents. Exon-primed, intron-crossing products can reveal evidence of parental lineage or can be used to show species and population level variation, depending on selection of the gene to be amplified (cf., Palumbi and Baker 1994).

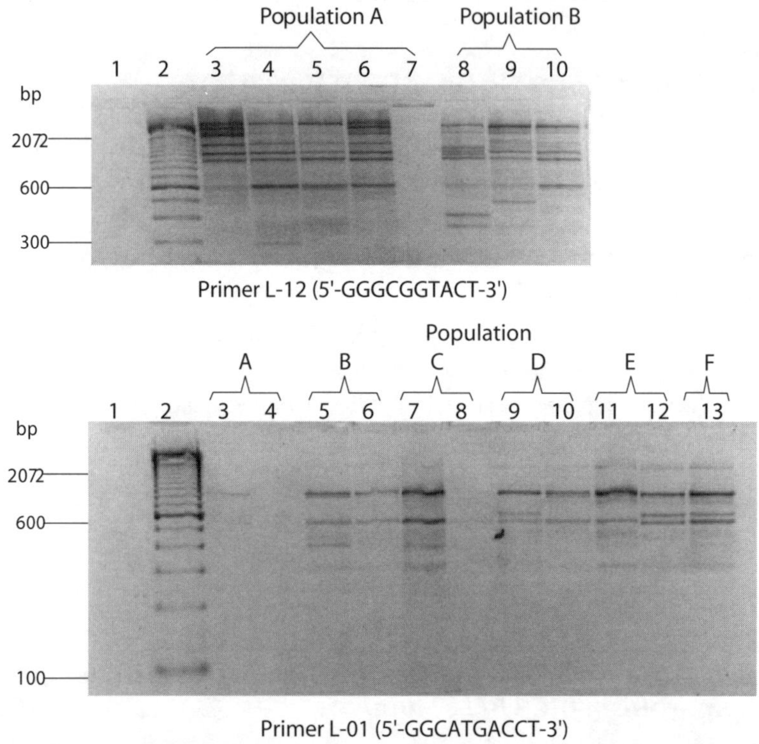

Figure 5.4 Within and among population genetic variation in bluegill *Lepomis macrochirus*. Populations indicated are from the headwaters of different watersheds in the upper Midwest. Randomly amplified polymorphic DNA (RAPD) analysis was used in this study as a stock identification tool for fishery management. Lane 1 in both gels is the negative control. Lane 2 in both gels contains a 100 bp DNA ladder as a molecular weight standard. Photographs provided courtesy of Vince Tranquilli and David Philipp, Illinois Natural History Survey.

and amplification. Upon electrophoresis, such a mutation will be noted as the absence of a fragment. Therefore, the presence or absence of a targeted sequence can be used as a character state, or allele, of a locus and potentially serves as a genetic marker.

Characters assessed through RAPD are useful for genetic studies because they provide population-specific data. The technique is rapid and inexpensive compared with other molecular-based methods that require blotting and visualizing with a labeled probe. Unlike PCR amplification of scnDNA, mtDNA, and specific coding genes, no prior sequence information is needed to screen and develop markers for a new species of interest because a single primer is used as both the points of amplification initiation and termination. A preliminary search for variable markers with a series of commercially produced random primers can indicate likely candidates for more intensive screening. The use of RAPD, however; requires some specific precautions to avoid interpreting procedural artifacts as data. Generally accepted protocols require that both negative (no DNA template) and positive (amplicons previously scored) controls be included with each PCR assay to account for false-positive and false-negative results. Repeatability of results among laboratories sometimes has proven problematic. Such artifactual results are due mostly to nuances of the PCR procedure or to contamination of the polymerase with extraneous DNA. The potential for generating artifacts in all techniques that employ

PCR should be critically considered. Therefore, confirmation and reproduction of the results through independent runs is a generally accepted procedure. Another difficulty concerns the handling and analysis of data. Although a number of analytical approaches have been explored, few have received exhaustive empirical challenge and validation.

The hereditary nature of RAPD characters has received limited attention and is largely unknown. Until the mode of transmission is firmly established for each RAPD locus (i.e., band), the use of RAPDs as a genetic marker must be viewed with caution. Randomly amplified polymorphic DNAs are analogous to a dominant character; in a typical analysis, the heterozygous condition is indistinguishable from the homozygous genotype expressing two copies of a given aplified sequence. Therefore, RAPD characters are not as informative individually as codominantly expressed ones. For dominant or recessive characters, allele and genotype frequencies are estimated by assuming Hardy–Weinberg equilibrium conditions. The frequency of the homozygous recessive genotype (fragment absence) is used to back-calculate the frequencies of the heterozygote, dominant homozygote, and dominant allele.

Finally, RAPDs are primarily sensitive to nucleotide substitutions within the *priming* regions and are not sensitive to any but large-scale length mutations that occur within *amplified* regions. Therefore, variation is underestimated and large suites of characters must be assayed to achieve statistical confidence. Individuals can be scored for presence or absence of fragments from a suite of "loci," the number of loci varying among primers for each new species. One of the most versatile methods for analyzing RAPD population data is **analysis of molecular variance** (AMOVA, Excoffier et al. 1992). This statistical procedure partitions molecular variation in a manner analogous to analysis of variance (ANOVA) and allows complex experimental designs for breeding studies as well as for stock identification studies.

Randomly amplified polymorphic DNAs have gained considerable attention for their use in population-level studies, linkage group identification, genome mapping, and as a potential source for single-locus genetic fingerprints. Historically, RAPDs have received attention especially from plant biologists, presumably because plants often exhibit low levels of among-population allozyme or extranuclear variation. Randomly amplified polymorphic DNAs provide access to genetic variation in both coding and noncoding sequences. Application to fish and shellfish biology is growing. Once technical difficulties associated with reproducibility of results are overcome, modes of inheritance are firmly established, and mathematical procedures for analyzing data are further developed, RAPD and other MAAP markers should provide economical tools of broad applicability for fisheries biologists.

5.4 Perspective

Because nDNA exhibits the greatest variability of all genetic systems that a fisheries scientist can study, examination of the patterns of nDNA variation will be a highly productive avenue for research on both cultured and wild populations. Nuclear DNA variation is particularly suited to studies of species that exhibit low levels of population variation in their mtDNA or isozymes. Furthermore, because of the wide variety of methods available, nDNA variation spans every level of genetic variation from comparison across upper taxonomic levels to assessment of variation between genotypes of individuals. Applications in fishes and shellfishes for which nDNA variation has been or would be useful include the following.

— *Domestication* Documenting the effects of selection for traits that allow strains to perform better in captive versus wild environments (Chapters 8, 14).
— *Reproductive success* Studying the effects of selection and fitness of alleles or genotypes relative to one another (Chapter 8).
— *Hybridization* Investigating gene flow among groups, focusing on the fitness and performance consequences of interbreeding, particularly when there is nonrandom mating with members of another population, subspecies, or species (Chapters 7, 11, 15).
— *Phylogenetics and phylogeography* Examining the temporal and spatial distribution of genetic diversity (Chapters 8, 13).
— *Population diversity and heterozygosity* Estimating the relationship between within- and among-group variation (Chapter 13).
— *Conservation of biodiversity* Exploring relationships of diversity loss to management, restoration, or natural events and processes (Chapter 18).

Of particular concern to fisheries restoration and aquaculture is the fact that a population's history determines how much genetic variation is currently within the gene pool. Circumstances in the past such as habitat modification, overharvesting, or selective breeding may have caused extreme selective pressures for traits, alleles, or both. Subsequently, these characters may have been driven to fixation by genetic drift, denying current and future populations the variation normally provided by the process of recombination. In extreme cases, a population's history can result in strong founder effects and near genetic homogeneity. Using molecular markers and knowledge of genome expression, aquaculturists can accelerate domestication, more fully study genotype–environment interactions, or assess the impact of native and cultured stocks by means of single and multilocus microsatellites. Fisheries scientists may enhance genetic variability of depleted populations through replenishment with appropriate stocks by the use of carefully planned programs based upon knowledge of life history, quantitative genetics, and nDNA variation.

In recent years, a host of new molecular methods have been explored and reported for fishes and shellfishes. The dizzying pace of these advancements precludes an exhaustive review of all recently reported classes of nuclear genetic characters. Characters revealed with these methods undoubtedly will be similar to other, time-tested characters. In fact, before conclusions confidently can be drawn from or management decisions based on these new characters, certain information should be demonstrated. First, the mode of inheritance should be understood. For example, characters inherited from only one parent will behave differently than will characters inherited from both parents. Second, the rate of mutation must be considered and appropriately applied. Questions concerning ancient divergences are not suitably investigated with characters evolving at a rapid rate. Conversely, recent divergences may go undetected if a highly conserved marker is employed. Third, the extent to which natural selection constrains variation must be considered. In summary, one must be alert to the potential pitfalls of choosing a technique because it is the newest tool. The most important consideration is the biological question being asked. One must select the nDNA technique that best addresses the fisheries biology question at hand rather than selecting a question that neatly fits a method.

5.5 References

Avise, J. C. 1994. Molecular markers, natural history and evolution. Chapman and Hall, New York.
Beridze, T. 1986. Satellite DNA. Springer-Verlag, New York.

Cardwell, T. N., R. J. Sheffer, and P. W. Hedrick. 1998. Genetic differences in male development time among populations of the endangered Gila topminnow. Journal of Heredity 89:353–355.

Cavalier-Smith, T. 1985. The evolution of genome size. John Wiley and Sons, New York.

Davies, K. E. 1985. New hypervariable DNA markers for mapping human genetic disease. Trends in Genetics 1:97.

Denovan, E. M., and J. M. Wright. 1990. A satellite DNA family from pollock *(Pollachius virens)*. Gene 87:279–283.

Epifanio, J. M., and D. P. Philipp. 1994. Ontogeny of gene expression in *Pomoxis* (Pisces: Centrarchidae). Developmental Genetics 15:119–128.

Excoffier, L., P. E. Smouse, and J. M. Quattro. 1992. Analysis of molecular variance inferred from metric distances among DNA haplotypes: application to human mitochondrial DNA restriction data. Genetics 131:479–491.

Fields, R. W., K. R. Johnson, and G. H. Thorgaard. 1989. DNA fingerprints in rainbow trout detected by hybridization with DNA of bacteriophage M13. Transactions of the American Fisheries Society 118:78–81.

Grewe, P. M., C. C. Krueger, J. E. Marsden, C. F. Aquadro, and B. May. 1994. Hatchery origins of naturally produced lake trout fry captured in Lake Ontario: temporal and spatial variability based on allozyme and mitochondrial DNA data. Transactions of the American Fisheries Society 123:309–320.

Gross, M. L., A. R. Kapuscinski, and A. J. Faras. 1994. Nest-specific DNA fingerprints of smallmouth bass in Lake Opeongo, Ontario. Transactions of the American Fisheries Society 123:449–459.

Jeffreys, A. J., V. Wilson, and S.-L. Thein. 1985. Individual-specific "fingerprints" of human DNA. Nature 316:76–79.

Jianxun, C., R. Xiuhai, and Y. Qixing. 1991. Nuclear DNA content variation in fishes. Cytologia 56:425–429.

Karl, S. A., and J. C. Avise. 1992. Balancing selection at allozyme loci in oysters: implications from nuclear RFLPs. Science 256:100–102.

Laughlin, T. F., and B. J. Turner. 1994. Multilocus DNA fingerprinting detects population differentiation in the outbred and abundant fish species *Poecilia latipinna*. Molecular Ecology 3:263–266.

Marcus, J. M., and A. R. McCune. 1999. Ontogeny and phylogeny in the northern swordtail clade of *Xiphophorus*. Systematic Biology 48:491–522.

Muslimov, I. A. 1993. The ribosome as a source of genome hypervariability? Medical Hypotheses 41(3):201–204.

Nei, M. 1987a. Genetic distance and molecular phylogeny. Pages 193–224 *in* N. Ryman and F. Utter, editors. Population genetics and fishery management. University of Washington Press, Seattle, Washington.

Nei, M. 1987b. Molecular evolutionary genetics. Columbia University Press, New York.

Saiki, R. K., and seven coauthors. 1988. Primer-directed enzymatic amplification of DNA with a thermostable DNA polymerase. Science 239:487–491.

Schmidtke, J., E. Schmitt, E. Matzke, and W. Engel. 1979. Repetitive DNA sequences in the rainbow trout genome. Chromosoma 75:185–198.

Southern, E. M. 1975. Detection of specific sequences among DNA fragments separated by gel electrophoresis. Journal of Molecular Biology 98:503–517.

Turner, B. J., J. F. Elder, and T. F. Laughlin. 1991. Repetitive DNA sequences and the divergence of fish populations: some hopeful beginnings. Journal of Fish Biology 39 (Supplement A):131–142.

Wirgin, I. I., and L. Maceda. 1991. Development and use of striped bass-specific RFLP probes. Journal of Fish Biology 39(Supplement A):159–167.

Wirgin, I. I., L. Maceda, and C. Mesing. 1992. Use of cellular oncogene probes to identify *Morone* hybrids. Journal of Heredity 83:375–382.

Population Genetic Processes

6

Mutation

Eric Hallerman and John Epifanio

6.1 Introduction

Mutation is a process through which heritable genetic changes are acquired by an organism. Mutation is the ultimate source of all new genetic variation (Wallace 1981:185). When we consider two allelic forms of a gene at a locus, A and a, we assume that one arose via mutation from the other or that each arose via mutation from an earlier, common form. Different allelic states of a gene arise only as a result of mutational changes. Within a given population, migration (or gene flow) (Chapter 7) can act as an alternative source of variation; however, without mutation having occurred somewhere in the past, natives and immigrants would have identical alleles. Similarly, recombination (Chapter 1) can give rise to new combinations of variation, *once* allelic variation arises at a number of gene loci. Without mutation at particular loci, there would be nothing different to recombine. Migration and recombination, therefore, are secondary generators of genetic variation, but mutation remains the only original source.

The process of mutation supplies genetic variation, the building blocks from which evolutionary changes are mediated by natural selection. Mutation, however, is a random process with regard to the adaptive needs of a species. Mutation most frequently is the product of mistakes by the replication machinery for DNA (see section 6.2). Further, the chances for mutation are probabilistic (or random) rather than deterministic. As a result of these two features, the vast majority of mutations are detrimental or neutral rather than beneficial for their carriers. Hence, mutation alone, uncontrolled by natural selection, would result in the breakdown of adaptation, not in adaptive evolution (Dobzhansky 1970:65).

6.2 Types of Mutation

Mutations change all types of genetically determined characters (or traits), including those that are morphological, physiological, and behavioral. Classical mutants of the fruit fly *Drosophila melanogaster* have provided a useful model system for study of unambiguous, visible traits such as eye and body color; size, shape, and venation of wings; number of abdominal bristles; and other morphological traits. In fish, comparable examples have been provided by the many morphological, coloration, and scale pattern traits of the goldfish *Carassius auratus* (Figure 6.1). Recessive lethal genes, both sex-linked and autosomal, soon proved to be considerably more frequent than are classical, visible traits (Dobzhansky 1970). The kinds of mutations studied in a given organism depend in large part on the methods used for detection. Mutations in microbes, such as yeast *Saccharomyces cerevesiae,* change mostly biochemical pathways, such as the ability to metabolize particular substances. Studies of mutation in higher organisms most often focus on changes in DNA sequences or expressed proteins.

Mutations can be conceived as occurring on any of three genetic levels: DNA sequence, chromosome, or whole genome. Key differences among mutations at these levels are the frequencies at which they occur and the consequences of their occurrence. Mutations at the DNA sequence level are relatively common. Within **functional genes** (i.e., DNA sequences that code for a product or regulate expression of the coding), the spontaneous mutation rate typically ranges from 10^{-6} to 10^{-4} mutations per gene per generation. The rate is somewhat higher for nonexpressed, repeated DNA sequences, reaching as high as 10^{-3} mutations per locus. Mutations at the DNA sequence level include substitutions of one nucleotide for another, deletions of one or more nucleotides, inversions of DNA sequences, duplications of one or more nucleotides or of a motif of repeated nucleotides, and transpositions of DNA sequences from one locus to another, including transposition of a mobile genetic element from one locus to another (Shapiro 1983; Lambert et al. 1988).

6.2.1 Mutations at the DNA Sequence Level

The effects of DNA sequence level mutations will differ depending on where they occur relative to functional elements of the genome (Wallace 1981:187–193). Mutations occurring in **structural genes,** that is, those encoding the amino acid sequence of a protein, may be classified according to their effect on the expressed protein. Mutations resulting from substitution of one base pair for another are termed **missense mutations.** **Transitional mutations** are those in which a purine (adenosine or guanine) replaces a purine, or a pyramidine (cytosine or thymidine) replaces a pyramidine. Because of the redundancy of the genetic code, about a quarter of transitional mutations (0.23–0.24; Nei 1975:24–25) do not change the amino acid sequence of the expressed protein and hence are termed **synonymous mutations.** Because synonymous mutations do not confer a change to proteins, the mutation often is unrecognized and generally will have little or no effect on the organism. The replacement of a purine by a pyramidine results in a **transversion,** few of which prove synonymous mutations. Mutations that result in production of a termination codon for translation of a protein are termed **nonsense mutations;** because 3 of 64 codons are termination signals, about 5% of single-base mutations fall in this category. The effect of a nonsense mutation is that the protein is either shortened or not produced. Therefore, nonsense mutations can have an effect on the organism, the magnitude of which depends on the protein's function. Many of the congenital diseases caused by single genes are the result of nonsense mutations, that is, important metabolic proteins are incorrectly produced. **Frameshift mutations** arise from the gain or loss of one or two base pairs from the DNA sequence encoding a protein. Because the coding for a protein is organized into three-base codons, all amino acids downstream from a frameshift mutation will be changed relative to the original protein. Again, the extent of the change to the protein and its function in the organism will shape the net effect of these mutations on the organism.

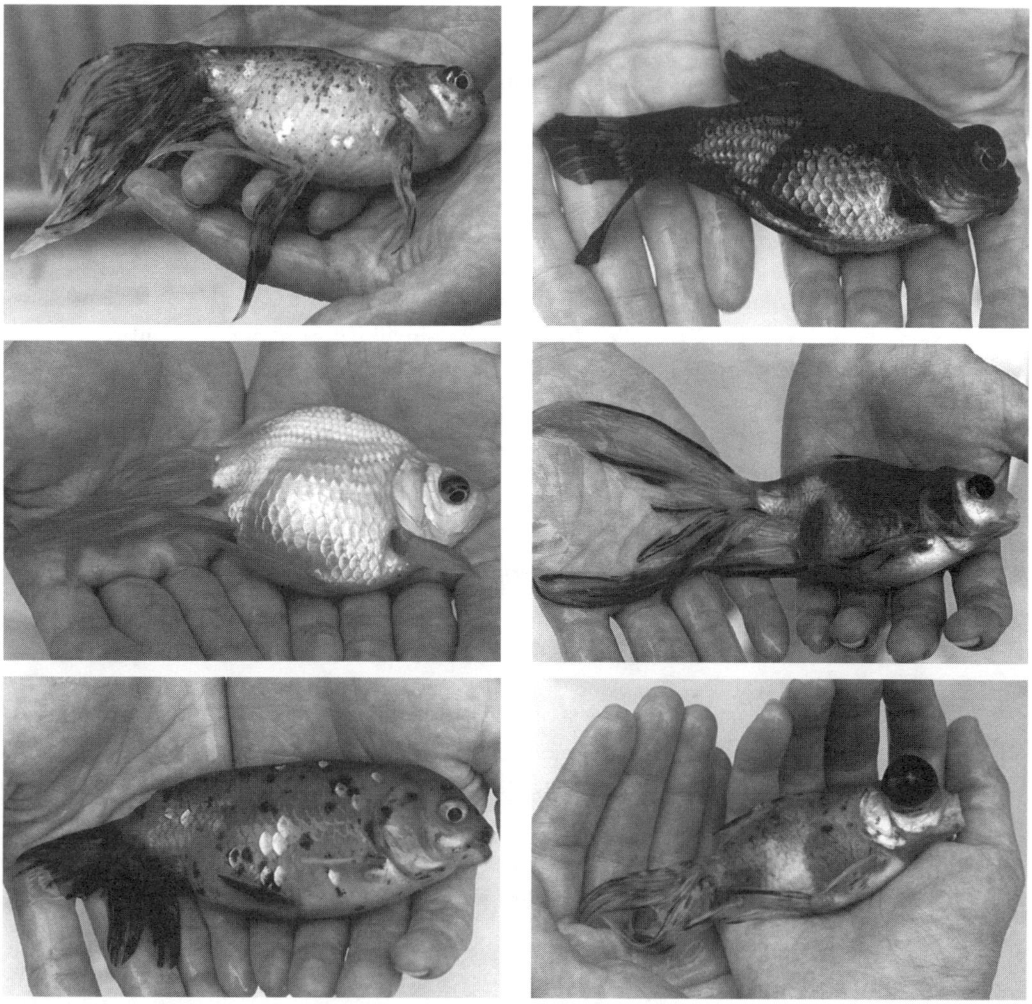

Figure 6.1 Many of the valued morphological, coloration, and scale pattern phenotypes of goldfish are the consequence of mutations of single genes.

The **expression** of genes (i.e., the timing, cellular or tissue location, and level of production of a gene product) is controlled by **regulatory elements**, DNA sequences that most frequently are located just upstream of structural genes. Mutations of regulatory elements might affect the level of production or even the presence or absence of expression of the protein. For example, a mutation might affect a binding site and hence the affinity for a key protein or RNA molecule that stimulates or represses expression of the gene. Mutations in the promoter element might affect the attachment of RNA polymerase needed for transcription of the gene. The presence or absence of activity of a particular protein may result from a regulatory mutation rather than a structural mutation. Mutation in regulatory elements is believed to play an important role in evolution.

Transposable elements are DNA sequences that have the ability to excise themselves from one point in the genome and insert themselves elsewhere in the genome. The mechanisms of transposition are similar to those used by a class of viruses called **retroviruses**. Retroviruses use the replication machinery of their host's genome to reproduce and spread. Because of this similarity, the origin of transposable elements is hypothesized to have come from an evolutionary event in which a eukaryotic progenitor incorporated sequences from an ancient retrovirus. The action of transposable elements can result in the insertional mutagenesis of the receiving DNA sequence, generally silencing any functional gene. In some cases, excision of the transposable element will restore expression of the gene, in others not. McClintock (1948) first uncovered the action of transposable elements that effectively turned on and off coloration genes for kernels of Indian corn *Zea mays* (see also Federoff 1983). Transposable elements have not been studied extensively in fishes, though Duvernell (1998) noted their presence and genetic variation among pupfishes *Cyprinodon* spp.

Not all of an individual's DNA encodes functional structural or regulatory genes. These **noncoding regions** also are subject to mutations. Mutations in noncoding regions frequently have no impact on the phenotype of the organism and are not subject to selection. Nonetheless, mutations in noncoding regions are important to population geneticists because they give rise to molecular genetic variation useful for population genetics studies (Chapter 5).

6.2.2 *Mutations at the Chromosome Level*

Mutations at the chromosome level (Chapter 3) include inversions of chromosomal segments, translocations of segments from one chromosome to another, fusion or fission of segments or whole chromosomes, and deletions or duplications of chromosomal segments or entire chromosomes. Such mutations are relatively uncommon because they can have a large effect on the carrier or its offspring. Because chromosomes can carry thousands to many thousands of genes, loss of part or all of a chromosome can affect many functional genes.

6.2.3 *Mutations at the Genome Level*

Mutations at the whole genome level (Chapter 3) are quite rare. **Polyploidization** is the process whereby the entire chromosomal complement of a species is duplicated. Polyploidization has occurred numerous times in the evolutionary history of plants, proving an important process of genetic differentiation (Stebbins 1976), for example, among different lineages in wheat. Polyploidization is a very rare event among animals but has proven important in certain evolutionary lines of aquatic organisms, including fish—carps and minnows (Cyprinidae), salmon and trout (Salmonidae), suckers (Catosotomidae), and probably sturgeons (Ascipenseridae) and paddlefishes (Polyodontidae), as well as certain shrimps and salamanders. Polyploidization generally is associated with a major cladogenic event resulting in the emergence of new phylogenetic lines.

6.2.4 *Rates of Mutation*

From the descriptions above, it should be clear that mutation rates may differ by orders of magnitude among different regions of the genome. Mutation frequencies per generation also differ over several orders of magnitude (from 10^{-10} to 10^{-4}) among species (Dobzhansky 1970:71). However, these differences are much less pronounced when mutation rates are calculated per unit time. Of related note, functionally less important molecules, or portions of molecules, evolve faster than do more important ones (Kimura 1992). Mutation rates also differ among varieties of a single organism and are themselves under genetic control (Dobzhansky 1970). For example, some strains of *Drosophila melanogaster* have higher mutation rates than do others, differences that are caused by mutability enhancer genes that can be mapped to linkage groups (Ives 1950; Thompson 1962).

The approximate constancy of the rate for amino acid or nucleotide substitutions among protein or gene regions provides the basis for the **molecular evolutionary clock hypothesis.** That mutations are accumulated at relatively constant rates has proved useful for estimating the time since divergence of related evolutionary lines. For example, if we estimate the number of mutations between two lines and have some idea of the rate at which mutations accumulate, we can estimate the time since the lines shared a common ancestor. There is no universal molecular clock. Functionally important genes are less likely to accumulate mutations than are noncoding sequences, hence the rate of ticking for each clock may vary greatly. As a result, a prerequisite for applying a clock is its calibration.

6.3 Allele Frequency Changes Due to Mutation Pressure

Because mutation rates are small, **mutation pressure,** the tendency for allele frequencies to change as a result of recurrent mutation, is negligible over the course of a few generations (Hartl 1988). However, the cumulative effects of mutation over long periods of time can become appreciable. Consider a locus with two alleles *A* and *a*, with frequencies of p_t and q_t ($= 1 - p_t$) in generation *t*. For this example, assume that a simple mutation can convert *A* to *a*, and vice versa. The mutation rate per generation from *A* to *a* is μ, and from *a* to *A* is ν. When μ is greater than ν, we generally refer to μ as the rate of the **forward mutation** and ν as the rate of the **back mutation.** In generation *t*, an *A* allele could have been either an *A* allele that did not undergo mutation in generation $t - 1$ (probability $= 1 - \mu$), or it could have been an *a* allele that mutated to *A* (probability $= \nu$). Mathematically, this can be expressed as

$$p_t = p_{t-1}(1 - \mu) + (1 - p_{t-1})\nu. \tag{6.1}$$

Solving the equation to give p_t in terms of p_0, the initial frequency of *A*, yields

$$p_t = \frac{\nu}{\mu + \nu} + \left(p_0 - \frac{\nu}{\mu + \nu}\right)(1 - \mu - \nu)^t. \tag{6.2}$$

To give intuitive meaning to this expression, let's consider two cases. When *t* is small, less than 100 generations, and p_0 is small, the term $(1 - \mu - \nu)^t$ will be approximately 1. The frequency of the mutant allele *a* will increase nearly linearly with time, with slope approximately equal to ν; however, ν is small, so the absolute rate of change is small. On the other hand, when *t* is large, say, thousands of generations, the $(1 - \mu - \nu)^t$ term will be approximately 0. In this case p_t will attain a value that remains the same generation after generation. This equilibrium value, $\hat{p}$, is given as

$$\hat{p} = \frac{\nu}{\mu + \nu}. \tag{6.3}$$

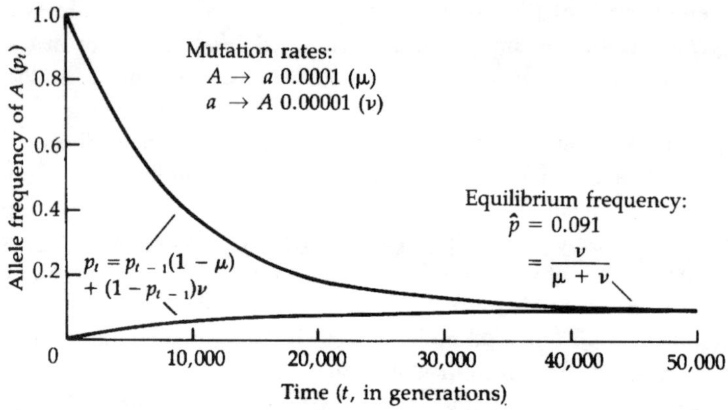

Figure 6.2 Change in allele frequencies under the pressure of reversible mutation. Attainment of equilibrium frequencies takes thousands of generations. Figure adapted from Hartl (1988).

Therefore, in the absence of other evolutionary forces (migration, selection, or drift), the frequency of alleles is related to the forward and backward mutation rates. The convergence of p_t toward its equilibrium value is shown in Figure 6.2. Mutation pressure is usually very weak in changing allele frequency. It takes thousands of generations for mutant allele frequencies to reach equilibrium under mutation pressure alone.

6.4 Interplay of Mutation and Other Evolutionary Forces

A key question for the field of population genetics is how genetic variability is maintained in populations. The mutation pressure model presented above considered mutation as an evolutionary force in isolation from other evolutionary forces. In the context of our discussion of mutation, we can define the question of genetic variability in terms of how many alleles can be maintained in a population due to mutation pressure. If loci exhibit more alleles than can be explained by mutation pressure alone, then other forces must be operating to accumulate alleles. If loci exhibit fewer alleles than expected, then other forces must be working to eliminate alleles.

6.4.1 *Interplay of Mutation and Drift*

Under the **infinite allele model** of mutation (Kimura and Crow 1964), we assume that every time a mutation occurs it creates a new allele, one that does not already exist in the population. We also assume that the population is finite because in an infinite population alleles could accumulate forever. Because the population is finite, variation will be lost through the process of random genetic drift (Chapter 9). The question becomes, for what number of alleles is the creation of new alleles, by mutation, balanced by the loss of alleles by random drift? Because of the assumption that every mutation is new, each allele arises in the population only once, and any individuals homozygous for the allele are **autozygous;** that is, their alleles at a particular locus are identical by descent from both sides of the pedigree. **Autozygosity**, then, is a measure of the frequency (or expectation) that two alleles are identical by descent. Autozygosity is the consequence of related individuals mating and is equal to the fixation index or coefficient of inbreeding,

F (see Chapter 10). The equilibrium value of the fixation index, $\hat{F}$, when mutation and drift are balanced can be approximated as

$$\hat{F} = \frac{1}{4N\mu + 1}, \tag{6.4}$$

where N is the effective population number (equals N_e, Chapter 9) and μ is the mutation rate. The number of selectively neutral alleles increases under mutation pressure until F equals $\hat{F}$. A complication arises because any number of allele frequencies can result in the same homozygosity value (Hartl 1988:99). This complication can be sidestepped if we assume that all alleles are equally frequent. The number of equally frequent alleles is called the **effective number of alleles,** often symbolized as n_e. In a biological sense, n_e is the number of equally frequent alleles in an ideal population (Chapter 9) that would be required to produce the same homozygosity as in an actual population. The n_e helps to quantify the "richness" or evenness of alleles and diminishes the weight of rare alleles. At equilibrium, then,

$$n_e = 4N\mu + 1. \tag{6.5}$$

In the context of discussing interplay of mutation and random drift, we might also note that in the absence of migration, mutation is the one mechanism by which loss of alleles through random drift may be reversed (Wallace 1981:186). A population that contains only A can regain a by mutation, or one containing only a can regain A. The rates of fixation and loss of genes from a large population are very small, occurring at a rate of ½N in segregating populations, so only a small rate of mutation is required to counterbalance the loss of variation though random drift. In small populations, loss of variation due to drift will exceed the rate of gain due to mutation.

6.4.2 *Interplay of Mutation and Selection*

Consider a locus with alleles A and a (the latter a mutant allele) with frequencies p and q, respectively. The fitness, W, of genotypes AA, Aa, and aa are W_{AA}, W_{Aa}, and W_{aa}, respectively. The average fitness of a recessive mutation is

$$\overline{W} = 1 - sq^2, \tag{6.6}$$

where $\overline{W}$ is the average fitness for the population, s is the coefficient of selection against the recessive genotype, and q is the frequency of recessive allele a in the population. For an incompletely recessive mutation, average fitness is

$$\overline{W} = 1 - 2hsq, \tag{6.7}$$

where h equals the degree of dominance at the locus. In this case, the average fitness of the population is

$$\overline{W} = p^2W_{AA} + 2pqW_{Aa} + q^2W_{aa}, \tag{6.8}$$

clearly a function of selection coefficients and allele frequencies. In our context, however, it is interesting to note that in the cases of both recessive and incompletely recessive mutations, stable allele frequencies can result from the interplay of mutation and selection (Wallace 1981:290). In the case of recessive mutations,

$$sq^2 = \mu \text{ or } q = (\mu/s)^{1/2}. \tag{6.9a}$$

In the case of incompletely recessive mutations,

$$hsq = \mu \text{ or } q = \mu/hs. \tag{6.9b}$$

Haldane (1937) showed that because of these relationships, the average fitness of a population can be expressed in terms of mutation rates. For recessive mutations,

$$\overline{W} = 1 - sq^2 = 1 - \mu. \tag{6.10}$$

For incompletely recessive mutations,

$$\overline{W} = 1 - 2hsq = 1 - 2\mu. \tag{6.11}$$

Hence, the average fitness of a population at equilibrium is reduced by an amount either equal to or twice as great as the mutation rate, depending on the recessivity of the alleles. The above relationships are for a single locus. For all loci combined, the average fitness for complete recessives is

$$\overline{W} = 1 - U, \tag{6.12}$$

and for incomplete recessives is

$$\overline{W} = 1 - 2U, \tag{6.13}$$

where U is the combined mutation rate for all loci. The Haldane principle is remarkable in showing that the impact of mutation on the population depends not on the severity of individual mutations but on the mutation rate. This concept can be grasped intuitively by noting that a lethal mutation causes one death, eliminating the gene from the population. On the other hand, a mutation posing 1 chance in 100 of causing death will persist in the population for an average of 100 generations before causing a death. At the population level, one individual exposed to a 100% chance of death poses the same cost to fitness as 100 individuals exposed to a 1% risk of death (Crow 1992). Hence, the fitness cost of mutation is equal to the mutation rate.

Muller observed that mutations are a common phenomenon. As discussed earlier, most mutations convey a disadvantage rather than an advantage, but in many cases this disadvantage is very small. Therefore, mutations can accumulate, and the total disadvantage of carrying many mutations imposes a "load" on the population. Muller (1950) introduced the terms genetic load and load of mutations to describe the effect of mutant alleles on populations. Utilizing a series of estimates concerning mutation rates and numbers of genes, he estimated that 10% or more of all germ cells carry a newly arisen mutation and that 20% of all human deaths have as their basis a genetic cause (although the genetic basis may not be recognized as the cause of death). Crow (1958) built on Muller's concept, defining **genetic load** as the proportional amount by which the average fitness of the population is depressed for genetic reasons below that of the genotype with maximum fitness. He expressed it mathematically as

$$\text{Genetic load} = \frac{W_{max} - \overline{W}}{W_{max}}, \tag{6.14}$$

where W_{max} is the fitness of the optimum genotype. If W_{max} is assigned the conventional, arbitrary value of 1.00, then

$$\text{Genetic load} = 1 - \overline{W}. \tag{6.15}$$

Doubling the mutation rate doubles the eventual mutational load, with consequential impacts on the fitness of the population. Muller (1950) brought forward the concepts of genetic load and load of mutations in part to alert people to the dangers of ionizing radiation because of the heightened frequency of mutations it induces. Anthropogenic sources of mutagens in the environment are expected to have similar effects on gene pools of aquatic organisms.

Although originally used to describe the effect of mutation, the term genetic load came to have a broader meaning (Crow 1992), referring to all fitness-reducing alleles in a population, whether maintained by mutation **(mutation load),** by selection on alleles that are favored in heterozygotes but not in homozygotes **(segregation load),** by mismatch between allele frequencies and environment **(dysmetric load),** or by several other mechanisms. When reading the literature, one must be clear regarding the intent of the term "genetic load."

To put the interplay of mutation and selection into a broad perspective, although mutation is a source of reduced fitness currently, it also is the source of necessary new variability for future evolution. Mutation load is the price a species pays for the privilege of evolution; without mutation there would be no evolution (Crow 1992).

6.4.3 *Interplay of Mutation, Selection, and Drift*

The interplay of mutation, selection, and random drift has proven a contentious issue among evolutionary geneticists. The key issue is whether selection or random drift (Chapters 8 and 9) is the primary force driving the fate of new mutations. The classical **selectionist** viewpoint is that the key driving force of evolution is Darwinian natural selection operating upon advantageous mutants. The **neutralist** viewpoint (Kimura 1968, 1983) is that the great majority of evolutionary changes, at least at the DNA level, result from random fixation of neutral or nearly neutral mutants through random drift, caused by random sampling of breeders or gametes. Neutralism does not deny the role of natural selection in determining the course of adaptive evolution but assumes that only a small amount of molecular-level changes are adaptive (Kimura 1992). Emphasis, then, is on mutation pressure and random drift as explanatory factors in molecular evolution. The selectionist viewpoint would be that most polymorphisms are adaptive and maintained by balancing selection. The proposal of the neutral theory has led to controversy, termed the selectionist–neutralist debate, and has stimulated much research in both molecular and population genetics.

■ 6.5 Mutation and Adaptedness

Mutation is a difficult process to study in the field. Hence, the rate and significance of mutations traditionally have been studied using visible traits in model organisms, notably the fruit fly *Drosophila* spp., leading to development of an extensive literature. Clearly, many mutations are deleterious to their carriers. How, then, could mutation provide the raw material for adaptive evolution? Dobzhansky (1970:79–82) pointed out that a mutation is neither useful nor harmful in the abstract, only in the context of a particular environment. He distinguished between two relevant types of environment. The first is the more obvious type of environment, the external environment. Within the context of a given external environment, a given mutation can prove adaptive, neutral, or maladaptive. Consider the scalelessness trait of chickens; would this not seem a maladaptive trait? Abbot et al. (1962), however, showed that scaleless chickens can survive with less of the amino acids methionine and cysteine in their diet, underlining the notion that the adaptiveness of a mutation must not be assessed simply as we perceive it but within an ecological context. The second type of environment is internal, that is, the genetic environment within which the mutation is expressed. The adaptiveness of the mutation depends on what *other* genes the individual carries. For example, certain mutant *Drosophila* alleles that are adaptive in some backgrounds prove maladaptive or even lethal in others.

Dobzhansky (1970:80) pointed out that the genes and proteins of a living species already have been subjected to selection and that, at least, the frequently occurring mutations have had time to become incorporated into the genotype. The adaptedness of a species is fairly close to the maximum achievable in its present environment. Hence, it is not surprising that most mutations are not clear-cut improvements given the ecological or genetic environments of the species. He adds that to improve the chances of observing a favorable mutation, the organism should be placed in a novel environment. Domesticated animals and plants are good subjects for detection of useful mutants. Although the agricultural environments provided are sheltered in many ways, they are different from the environments faced by wild ancestors and hence demand genetic adaptation (see also Box 6.1).

■ 6.6 Perspective: A Fisheries Manager s View of Mutation

In his widely used textbook, *Topics in Population Genetics*, Wallace (1968:131) wrote, "Molecular evolution at the moment does not greatly concern the population geneticist." Admitting to a change in perspective in a subsequent textbook, Wallace (1981:186) wrote, "During the intervening years, I have realized that an understanding of the types of mutations and of mutational events which are possible (and do occur) at the molecular level is important." Fisheries managers might relate readily to both his early view and to his later change of heart. There is little appreciable effect of mutation over the ecological time scale at which we observe the dynamics of population genetics processes in fisheries populations. This partially explains why fisheries geneticists infrequently study mutation as a process. However, given that it is the source of all genetic variation, exerting a great effect over evolutionary time, it is important that fisheries scientists understand the implications of the process of mutation.

For fisheries managers concerned with genetic diversity within and among populations, there might be an inclination to dismiss efforts to conserve or protect allelic diversity because the ongoing process of mutation will surely replenish any lost diversity. We need to be cautious of such logic because mutation is a slow, accumulating process (compared with time frames associated with resource management) and because the nondeterminative nature of mutations ultimately may lead to an undesired outcome. Further, the loss of allelic diversity over the short term increases the risk that a

■ Box 6.1 Genomic Control of Mutability

The classical view is that genetic change results from small, random mutations in individual genes. However, recent data collected from a wide range of species suggest that mutations are neither as small nor as rare as biologists have long assumed (reviewed by Pennisi 1998). Furthermore, these mutations may not be randomly located. Some sequences of DNA are more likely to be duplicated or moved to another place than are other sequences. The enzymes that copy and maintain DNA introduce more changes in some parts of the genome than in others, creating hotspots of mutation. These findings have resulted in a paradigm shift, whereas formerly it was assumed that genomes evolve to minimize mutation rates and prevent random genetic changes, under the new view, the most successful genomes may be those that have evolved to be able to change quickly if necessary.

Mutational Hotspots

The finding that the rate of mutation differs throughout the genome is illustrated by variation in toxins among *Conus* marine snails (B. Olivera, University of Utah, personal communication cited by Pennisi 1998). The 500 species of predatory *Conus* species use venom-laden harpoons to immobilize their invertebrate or fish prey. Each *Conus* species has its own toxin of 50–200 peptides. The genetic variation responsible for toxin diversity seems to be concentrated at a particular mutational hotspot, the third exon in the DNA sequence encoding the toxin peptide.

There could be several explanations for such mutational hotspots, including variation in the efficacy with which mutational errors that arise in the genome are repaired. In vitro studies have shown that the error rate for repair of mistakes made during DNA replication can vary from 99 to 3% (T. Kunkel, National Institute of Environmental Health Sciences, personal communication cited by Pennisi 1998). Further, certain sequences are more prone to error at the replication stages than are others. In particular, DNA microsatellite arrays, composed of tandem repeats of one- to six-base pair motifs, prove particularly prone to **replication slippage.** That is, the newly replicated strand may slip relative to the strand being copied, so that it ends up shorter or longer than the original, depending on whether it slipped forward or backward.

Mutation and Adaptation

Whereas many mutations have been associated with major phenotypic effects in their carriers (Chapter 1), most have more subtle effects. For example, mutability of microsatellite arrays in some cases has been linked to onset of disease (e.g., Huntington's disease in human). In contrast, recent work has suggested that mutability of microsatellite arrays can affect the expression of genes in a quantitative way, allowing the fine-tuning of gene expression (Kashi et al. 1997; King et al. 1997; Kashi and

continued on next page

■ Box 6.1 *continued*

Soller 1998). If the number of tandem repeats affects gene expression, some variants may be favored over others under particular environmental conditions. The number of tandem repeats at or near gene regulatory elements may affect the level of activity of a downstream gene and has been shown to affect expression of a transcription factor that, in turn, controls the expression of other genes.

Another mutational process that has been putatively linked to an adaptive response of genomes is provided by transposable elements. Following Barbara McClintock's demonstration that transposons control the expression of kernel coloration in Indian corn, geneticists tended to regard transposons as "rogue" DNA sequences that could land anywhere in the genome, disrupting gene expression. However, by affecting coloration or other fitness-related traits, transposable elements can provide evolutionarily relevant variation. By affecting the remote predecessors of antibody genes, transposons also may have played a key role in generating the immune system's wide variety of antibodies (D. Shatz, Yale University, and M. Gellert, National Institute of Diabetes and Digestive and Kidney Diseases, personal communications in Pennisi 1998). Transposons also may capture genes and move them to other parts of the genome (a process termed **DNA shuffling**), possibly to be expressed in a different regulatory context or a new role.

The evolutionary contribution of mutations may increase when a species faces new selective pressures. For example, mutation rates in *Escherichia coli* increased genome wide when the bacterium faced stress caused by environmental change (J. Cairns, Virginia Polytechnic Institute and State University, cited by Pennisi 1998). It seems that the bacterium activates mechanisms that give rise to higher rates of mutation, activating different repair enzymes and promoting shuffling of DNA (S. Rosenberg, Baylor College of Medicine, personal communication cited by Pennisi 1998).

Hence, the paradigmatic view of mutation is changing. In her 1983 address accepting the Nobel Prize, Barbara McClintock said that the genome is "a highly sensitive organ of the cell that in times of stress could initiate its own restructuring and renovation."

population will become extinct (Chapter 18), precluding the future chances of regaining diversity.

■ 6.7 References

Abbot, U. K., V. S. Asmundsen, and K. R. Shortridge. 1962. Nutritional studies with scaleless chickens: I. The sulfur-containing amino acids. Genetical Research 3:181–195.

Crow, J. F. 1958. Some possibilities for measuring selection intensities in man. Human Biology 30:1–13.

Crow, J. F. 1992. Genetic load. Pages 132–136 *in* E. F. Keller and E. A. Lloyd, editors. Keywords in evolutionary biology. Harvard University Press, Cambridge, Massachusetts.

Dobzhansky, T. 1970. Genetics of the evolutionary process. Columbia University Press, New York.

Duvernell, D. 1998. Population genetics of Death Valley pupfishes (Cyprinodontidae: *Cyprinodon* spp.) and the identification of a new retrotransposable element family. Doctoral dissertation. Virginia Polytechnic Institute and State University, Blacksburg.

Federoff, N.V. 1983. Controlling elements in maize. Pages 1–65 *in* J. A. Shapiro, editor. Mobile genetic elements. Academic Press, New York.

Haldane, J. B. S. 1937. The effect of variation fitness. American Naturalist 71:337–349.

Hartl, D. 1988. A primer of population genetics, 2nd edition. Sinauer Associates, Sunderland, Massachusetts.

Ives, P.T. 1950. The importance of mutation rate genes in evolution. Evolution 4:236–252.

Kashi, Y., D. King, and M. Soller. 1997. Simple sequence repeats as a source of quantitative genetic variation. Trends in Genetics 13:74–78.

Kashi, Y., and M. Soller. 1998. Functional roles of microsatellites and minisatellites. Pages 10–23 *in* D. D. Goldstein and C. Schlotterer, editors. Microsatellite evolution and application. Oxford University Press, Oxford, UK.

Kimura, M. 1968. Evolutionary rate at the molecular level. Nature (London) 217:624–626.

Kimura, M. 1983. The neutral theory of molecular evolution. Cambridge University Press, Oxford, UK.

Kimura, M. 1992. Neutralism. Pages 225–230 *in* E. F. Keller and E. A. Lloyd, editors. Keywords in evolutionary biology. Harvard University Press, Cambridge, Massachusetts.

Kimura, M., and J. F. Crow. 1964. The number of alleles that can be maintained in a finite population. Genetics 49:725–738.

King, D. G., M. Soller, and Y. Kashi. 1997. Evolutionary tuning knobs. Endeavor 21:36–40.

Lambert, M. E., J. F. McDonald, and I. B. Weinstein, editors. 1988. Eukaryotic transposable elements as mutagenic agents. Cold Spring Harbor Laboratory, Cold Spring Harbor, New York.

McClintock, B. 1948. Mutable loci in maize. Carnegie Institute of Washington Year Book 47:155–169, Carnegie Institute of Washington, Washington, D.C.

Muller, F. 1950. Our load of mutations. American Journal of Human Genetics 2:111–176.

Nei, M. 1975. Molecular population genetics and evolution. American Elsevier, New York.

Pennisi, E. 1998. How the genome readies itself for evolution. Science 281:1131–1134.

Shapiro, J. A., editor. 1983. Mobile genetic elements. Academic Press, New York.

Stebbins, G. L. 1976. Chromosome, DNA and plant evolution. Evolutionary Biology 9:1–34.

Thompson, P. E. 1962. Asynapsis and mutability in *Drosophila melanogaster*. Genetics 47:337–349.

Wallace, B. 1968. Topics in population genetics. Norton Press, New York.

Wallace, B. 1981. Basic population genetics. Columbia University Press, New York.

7

Migration

Anthony J. Gharrett and Lev A. Zhivotovsky

7.1 Introduction

Geneticists often use the term migration or immigration to mean gene flow, also referred to as gene migration. This use of the term migration differs from the fisheries and behavioral sciences uses, which describe a variety of geographic movements of individuals or groups, such as the anadromous migrations of Pacific salmon *Oncorhynchus* spp. or the catadromous migrations of the American eel *Anguilla rostrata*. **Migration** in geneticists' parlance implies movement from one spawning population to another *and* successful contribution of genetic material to the recipient population by the individual who moved. Inclusion of genetic material from immigrants can alter the local gene pool. In this chapter, we will consider first gene flow in general by describing some of its genetic indicators and by examining some models that predict the effect of gene flow and other models that estimate natural gene flow rates. Then we will use the models to consider the apparent extent of gene flow in natural fish population systems and the influence of gene flow on population structure as well as the possible consequences to wild populations of gene flow from cultured or transplanted stocks.

Gene flow is one of the forces that can strongly influence the genetic composition and structure of populations, which span from the differences in allelic frequencies among populations to multilocus linkages that may form coadapted complexes important in local adaptation. Whether or not gene flow is beneficial (increases average fitness) depends on the roles that contributed genes play in the adaptation of recipient populations to their environments. The effect of genes received through migration may range from providing novel, adaptively important genetic information to being maladaptive. In most marine and anadromous fish populations, gene flow between populations occurs naturally and persistently. The genetic compositions of individual populations linked by gene flow may coevolve as one complex unit; one term for this unit is a **metapopulation** (Gilpin and Hanski 1991). Metapopulations can be enormous; for example, the genetic similarity among eastern Bering Sea populations of Pacific herring *Clupea pallasi* suggests strongly that high levels of gene flow link those populations (Grant and Utter 1984). The genetic composition of each population in a metapopulation reflects a balance between its own local dynamics (i.e., selection) and introduced genes. If the propensity for returning to natal grounds to spawn **(homing)** or failing to home **(straying)** is heritable, straying may result in gene flow, and the balance between homing and straying may be regulated by natural selection (Quinn 1984). In response to the spawning success of returning fish, the balance between homing and straying of members of a population or metapopulation may evolve, and these characteristics may contribute to the overall genetic stability of the metapopulation.

Gene flow also can result from human activities, such as transplantation and artificial culture, that involve stocking, ocean ranching, or fish escaping from pens (see Chapters 14, 15). The process itself is neither inherently good nor bad. However, the results of gene flow between populations that have not been linked recently by genetic exchange are unpredictable and, therefore, may lower the fitness of the recipient population. The risk involved may be manifested in the recipient by poorer adjustment to the local environment and ultimately as decreased productivity (see Chapter 11 on local adaptation and outbreeding depression).

7.2 Indicators of Gene Flow

Theory predicts and data confirm several effects of gene flow on gene and genotypic frequencies of populations. These effects could influence fitness in populations of fishes.

7.2.1 *Change in Allelic Frequency*

The simplest consequence of gene flow (between genetically dissimilar populations) is a change in allelic frequencies in a recipient population. Immigrants from a genetically different source will tend to alter the allelic frequencies of the recipient population toward those of the source population (Figure 7.1). With continual gene flow, the recipient population eventually will resemble the donor population genetically, unless some other force (e.g., selection, random genetic drift, or mutation) affecting allelic frequencies is acting with sufficient strength to counter the effects of gene flow. Genetic exchange among populations tends to homogenize allelic and genotypic frequencies among the populations (Figure 7.1).

Even one or two transplantation events can modify the gene pool in the recipient population. For example, walleye *Stizostedion vitreum* possessing an *MDH* (malate dehygrogenase, enzyme number 1.1.1.37; IUBMBNC 1992) allozyme variant not observed in the recipient population were introduced into West Blue Lake, Canada. Alleles of these genetically distinct fish dominated the 1971 year-class in catches in 1973, 1974, and 1975 (Schweigart et al. 1977). Murphy et al. (1983) observed a similar result from supplemental stocking of walleye into Claytor Lake, Virginia.

7.2.2 *Wahlund Effect*

Another characteristic of populations that receive immigrants is that they tend to have an excess of homozygous genotypes (a deficit of heterozygous types) relative to Hardy–Weinberg expectations (Chapter 1) prior to spawning. The observed deviation, referred to as the **Wahlund effect,** can be observed in any mixture of individuals from two (or more) populations that differ in genotype frequencies and occurs whether or not these fish eventually interbreed. A numerical example of the effect is presented in Table 7.1. In this example, the proportion of *Aa* heterozygotes (0.24) observed in the mixture on the spawning ground (or in a mixed-stock fishery) is less than would be

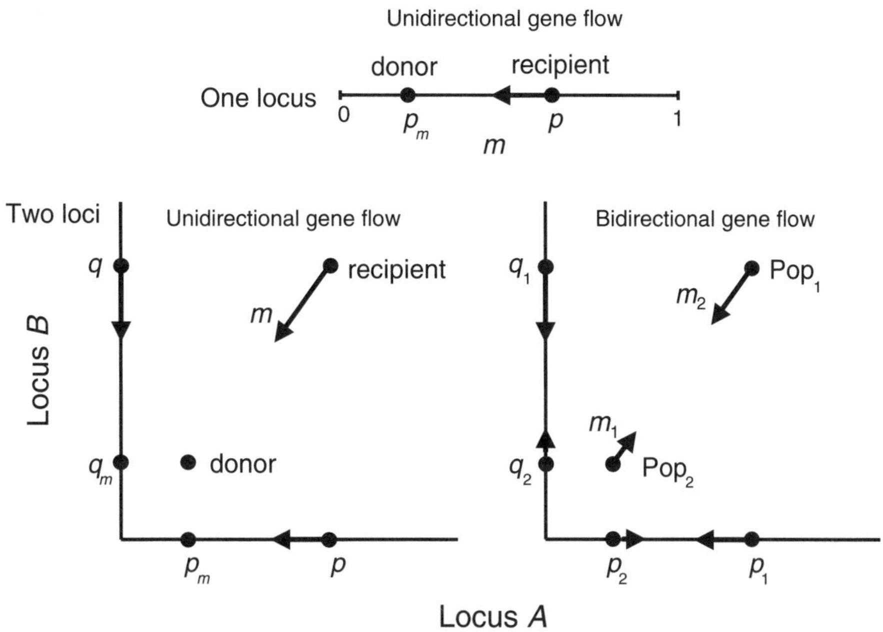

Figure 7.1 The results of gene flow on the allelic frequencies of a recipient population in the absence of other frequency-altering influences. The arrows show the direction of allelic frequency changes in the populations. Shown are unidirectional, recurrent gene flow at a single locus (top) and at two loci (bottom left) and bidirectional, recurrent exchange between two populations for two loci (bottom right). In the one-locus model, p_m is the frequency of an allele in the donor population. Immigrants will alter the frequency of that allele in the recipient population (p); the direction of the change will be toward the frequency of the donor, and the magnitude of the change will be proportional to the rate of immigration (m). Frequencies of alleles at two (or more) loci in the recipient population will also change toward that of the donor. The combined effect is plotted, but the result at each locus is shown on the axes; the magnitude of change is proportional to m. If two populations (e.g., Pop$_1$ and Pop$_2$) exchange immigrants, allelic frequencies of both will change to intermediate frequencies. The change in Pop$_1$ is proportional to the immigration from Pop$_2$ and vice versa.

expected in a randomly mating population (0.32). Similarly, the proportions of both homozygous genotypes observed, 0.68 for *AA* and 0.08 for *aa*, exceed Hardy–Weinberg expectations of 0.64 and 0.04, respectively. Although the example is illustrative of a particular mixture, it can be demonstrated algebraically that the Wahlund effect occurs in all mixtures. Of course, should a single generation of random mating occur, Hardy–Weinberg frequencies will be restored in the absence of natural selection for autosomal loci in species that have discrete generations.

The Wahlund effect, which reflects a mixture of population sources within the unit sampled, is analogous to the increased homozygosity resulting from inbreeding and can be analyzed in the same way with inbreeding coefficients (see Chapter 10). In particular, Wright's inbreeding coefficient F (Wright 1978; Hartl and Clark 1989), which is often designated F_{IS} when presented in population-by-population analyses, quantifies the excess (or deficit) of homozygotes (relative to Hardy–Weinberg expectations) within a population unit of interest and can be used to test for conformance to Hardy–Weinberg expectations (Table 7.1). Using Wright's concept, the Wahlund effect or inbreeding observed at different hierarchical levels of populations can be extended to descriptions

Table 7.1 Demonstration of the Wahlund effect (an increase in the proportion of homozygous individuals and a decrease in heterozygous individuals) for a prespawning mixture consisting of 20% immigrant individuals ($m = 0.2$) and 80 % native individuals ($1 \quad m = 0.8$). Shown also are the relationship of the Wahlund effect to Wright s inbreeding coefficient, F_{is} (for which H_{exp} and H_{obs} are the expected [Hardy Weinberg] and observed heterozygosities, respectively), Pearson s chi-square statistic, X^2, for a goodness-of-fit test of observed genotypic frequencies to Hardy Weinberg expectations, and the relationship of Wright s F_{is} to the expected and observed genotypic frequencies in the mixture (cf to inbreeding coefficient in Chapter 10).

Allele, genotype, or statistic	Native	Immigrant	Mix	
		Allele frequencies		
A	$p_1 = 0.9$	$p_2 = 0.4$	$p_{mix} = (1 - m)*p_1 + m*p_2 = 0.8*0.9 + 0.2*0.4 = 0.8$	
a	$q_1 = 0.1$	$q_2 = 0.6$	$1 - p_{mix} = q_{mix} = 0.2$	
		Genotype frequencies		
			Observed	Expected (Hardy–Weinberg)
AA	$p_1^2 = 0.81$	$p_2^2 = 0.16$	$0.8*0.81 + 0.2*0.16 = 0.68$	$p_{mix}^2 = 0.64$
Aa	$2p_1q_1 = 0.18$	$2p_2q_2 = 0.48$	$0.8*0.18 + 0.2*0.48 = 0.24$	$2p_{mix}\,q_{mix} = 0.32$
aa	$q_1^2 = 0.01$	$q_2^2 = 0.36$	$0.8*0.01 + 0.2*0.36 = 0.08$	$q_{mix}^2 = 0.04$
		Calculations showing the Wahlund effect		
Wright's F_{is}	$F_{is} = 1 - H_{obs}/H_{exp} = 1 - 0.24/0.32 = 0.25$			
Hardy–Weinberg test (for a sample of $N = 100$ fish)	$X^2 = NF_{is}^2 = N(0.25)^2 = 100*0.625 = 6.25 > 3.841 = \chi^2_{(1\,df,\,\alpha\,=\,0.05)}$			
Relationships to genotypic frequency changes	$AA_{obs} = AA_{exp} + p_{mix}\,q_{mix}\,F_{is} = 0.64 + 0.8*0.2*0.25 = 0.68$ $Aa_{obs} = Aa_{exp} + 2p_{mix}\,q_{mix}\,F_{is} = 0.32 - 2*0.8*0.2*0.25 = 0.24$ $aa_{obs} = aa_{exp} + p_{mix}\,q_{mix}F_{is} = 0.04 + 0.8*0.2*0.25 = 0.08$			

of the population structure by quantifying the variation in allelic frequencies within and among populations as in the gene diversity analysis presented later (section 7.3.3.2).

The chances of detecting a Wahlund effect in a spawning population is remote for natural populations because small effects are difficult to detect statistically and large effects would not ordinarily be expected in natural, presumably equilibrated, systems. To find examples, we must examine situations that involve mixtures that have not yet interbred or that are obviously not in equilibrium, such as mixed-stock fisheries for Pacific salmon. An extreme example of the Wahlund effect was observed in a small Swedish lake for two reproductively isolated but sympatric populations of brown trout *Salmo trutta* that were taken in the same collections. Only the two homozygous allozyme phenotypes were observed at the *LDH-1** (L-lactate dehydrogenase, 1.1.2.7) locus (**100/*100* and **240/*240*) in the fish; additional data showed that the two groups differed in frequency of alleles at five other allozyme loci and had very different growth rates (Ryman et al. 1979). It is likely that these populations diverged in isolation during the Pleistocene and recently came back into contact but have maintained reproductive isolation.

Another example of a Wahlund effect in a mixture is found in cod of the Baltic Sea and Danish belt sea (Sick 1965). Collections in three regions showed large allelic frequency differences. However, the two flanking regions had genotypic frequencies consistent with Hardy–Weinberg equilibrium, whereas the intervening region had intermediate allelic frequencies and a huge heterozygote deficit. An interpretation of this observation is that these cod have separate spawning areas but share some feeding areas. Probably stock mixing should be taken into account in managing fisheries in this area. The Wahlund effect should be examined where mixed-stock fisheries are suspected. However, the ability to detect a significant excess of homozygous types depends on the extent of allelic frequency differences between contributors to the mixture and the power of the test. Consequently, failure to detect a Wahlund effect does not demonstrate the absence of a mixture.

7.2.3 Gametic Disequilibrium

Another indicator of gene flow is gametic disequilibrium, a form of linkage disequilibrium (Chapter 1). Gametic disequilibrium is an extension of the Wahlund effect to multiple loci. The idea behind the term is that individuals contribute gametes, not genotypes, to the next generation and that gametes carry a single allele from each locus. In an equilibrium population, gametic frequencies can be calculated from the product of the frequencies of alleles for that population. Gametic equilibrium exists in a population if the alleles at different loci are distributed independently in the gamete pool produced by the population; if alleles at different loci are not distributed independently, there is **gametic disequilibrium.**

Gametic disequilibrium can result from selection, random drift, or population mixing. In mixed populations, disequilibrium results primarily from the Wahlund effect extended to multiple loci. Even divergent populations that are each in equilibrium produce mixtures that are in gametic disequilibrium because gametes of the mixture carry alleles with frequencies determined by the population from which those alleles were derived rather than by frequencies of the alleles in the mixture as a whole. Genetically different populations have different gametic frequencies. An example of the changes in genotypic frequencies involving two loci is presented in Figure 7.2. Unlike the single locus Wahlund effect, however, multilocus genotypic frequencies are not restored to frequencies predicted by allelic frequencies after a single generation of random mating. The approach to equilibrium (expected equilibrium gametic frequencies equal the product of allelic frequencies) is asymptotic; the rate of approach depends on the number of loci and the map distance (centiMorgans) separating them. Unlinked loci (>50 centiMorgans) approach equilibrium most rapidly.

The brown trout populations examined by Ryman et al. (1979) exhibited gametic disequilibrium for several pairs of loci. Another example of gametic disequilibrium in a mixture occurs in the Glenarm River in northern Ireland (Crozier 1993), where escaped farmed Atlantic salmon *Salmo salar* ascend the river and spawn with wild fish.

7.3 Models

In population genetics, models provide a framework for understanding a process. By reducing the process to mathematical terms, a model focuses attention on the crucial and quantifiable aspects of the process. Some simple models can provide an amazingly accurate picture of complicated processes. However, because most models are developed under simplifying assumptions about the process, care must be exercised in interpreting the results of a model. With that caveat, we will examine some models relating gene flow to allelic frequencies in populations and then incorporate the effect of other factors that

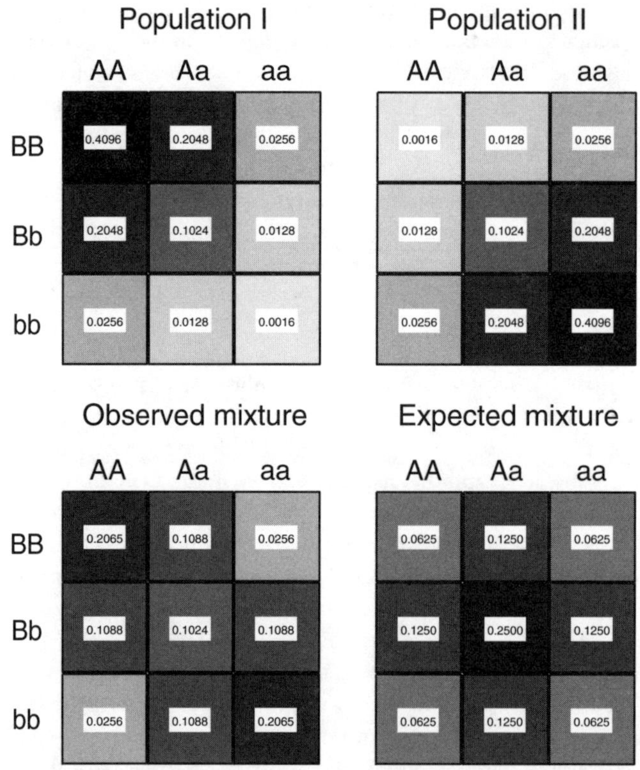

Figure 7.2 Gametic disequilibrium resulting from mixing two genetically distinct populations. Two loci are followed. In population I (top left), the allelic frequencies are $p_A = 0.8$ and $p_B = 0.8$; in population II (top right), $p_A = 0.2$ and $p_B = 0.2$. Each of the populations is at gametic equilibrium, so that the frequency of each gamete is the product of the corresponding allelic frequencies. Genotypic frequencies are presented both numerically and graphically (by the extent of stippling). In the resultant mixture, $p_A = 0.5$ and $p_B = 0.5$. Multilocus genotypic frequencies of the equal-part mixture (bottom left) are not equilibrium frequencies (bottom right), nor will they attain equilibrium after a single generation of random mating.

influence genetic structure; that is, we will begin with a very simple model and then incorporate terms that reduce the number of assumptions.

Both long and short time frames of gene flow are of interest to fisheries scientists. The contemporary native population structure of a species evolved over very long time periods (perhaps thousands of generations or more) in response to evolutionary forces (forces that genetically alter populations). The genetic differences among current populations reflect those historic influences and also provide biological information that can be important to management and conservation of the species. Such information provides details of the structure of and relationship among populations, information that can guide management decisions and may be critical to successful rehabilitation efforts. In stark contrast, recently (in an evolutionary context) many fish populations have been exposed to strong anthropogenic influences such as harvest, stocking, habitat loss, and pollution. The time frame for many of these activities extends back in terms of years or decades. Unfortunately, the results of these influences may not be apparent immediately and may take longer than a fisheries scientist's career or life span to become obvious.

Genetic effects of migration can be observed both in evolutionary and contemporary time frames. The same theory and resultant models can be used to predict results, although different applications of the theory may be appropriate, as discussed below.

7.3.1 Simple Gene Flow Model

The simplest population genetics models consider a single Mendelian locus (see Chapter 1 on Mendelian inheritance) in a random-mating population (see Chapter 1 on Hardy–Weinberg equilibrium) and relax one of the assumptions underlying Hardy–Weinberg equilibrium at a time. Therefore, we will relax only the assumption of no immigration into the population but will retain the rest of the assumptions. Among other things, this implies that neither the immigrant nor its genes will be discriminated against via selection or assortative mating in its contribution to the population. Following immigration, the genetic composition or gene pool will reflect the proportionate contributions of the native fish and the immigrants. Because a population attains Hardy–Weinberg equilibrium in a single generation of random mating, we can characterize the gene pool with frequencies of the alleles at the locus we are following. The frequency of an allele after one generation of immigration is the average of the frequencies in the donor and recipient populations weighted by their proportionate contributions:

$$q_1 = (1 - m)q_0 + mq_m, \tag{7.1}$$

where q_0 is the frequency of allele a in the population before immigration, q_1 is its frequency in the generation following immigration, q_m is its frequency in the immigrants, m is the proportion of immigrants in the spawning population, and $1 - m$ is the proportion of natives in the spawning population (Figure 7.1). This equation can be rearranged algebraically to solve for the proportion of migrants (per generation) that have entered a population if we know the allelic frequencies before and after migration. If a single introduction occurs in which the donor fish have an *LDH-B2*100* allele frequency of 0.8, the recipient population an initial frequency of 0.4, and the next generation the frequency is 0.5, one can solve for m:

$$m = \frac{q_1 - q_0}{q_m - q_0} = \frac{0.5 - 0.4}{0.8 - 0.4} = 0.25,$$

which indicates that the introduction contributed 25% of the gene pool. One would expect the postmigration frequency to persist in the absence of further migration and other evolutionary forces. This is a direct method for estimating migration rate; it was used to estimate the relative contribution of genetically marked pink salmon *Oncorhynchus gorbuscha* to Auke Creek as part of the evaluation of a genetic marking experiment (Lane et al. 1990).

If the immigration rate (m) and the allelic frequency (q_m) in the donor population are constant from generation to generation, one expression for the frequency of allele a after several (t) generations is

$$q_t \approx (q_0 - q_m)e^{-mt} + q_m, \tag{7.2}$$

where e^{-mt} approximates $(1 - m)^t$, the solution for discrete generations. For more complex situations (say an age-structured population), e^{-mt} remains a good approximation. In the previous hypothetical example, if identical introductions were made each generation for five generations, the frequency of the *LDH-B2*100* allele would increase to

$$q_t \approx (q_0 - q_m)e^{-mt} + q_m = (0.4 - 0.8)e^{-0.25*5} + 0.8 = 0.685.$$

With repeated introductions, the allelic frequencies in the recipient population will asymptotically approach the frequencies of the donor population (Figure 7.3). This equation can be rearranged to solve for the number of generations required to achieve a specified change in allelic frequency:

$$t \approx \frac{1}{m}\log_e\left(\frac{q_0 - q_m}{q_m - q_0}\right). \tag{7.3}$$

If recurrent gene flow occurred into a population, such as would occur as a result of repeated introductions, this equation allows us to predict the change in allelic frequencies over time.

7.3.2 Models of Gene Flow and Random Genetic Drift

The model for simple gene flow assumes that the populations under consideration are so large (infinite) that no random deviations in allelic frequencies occur from generation to

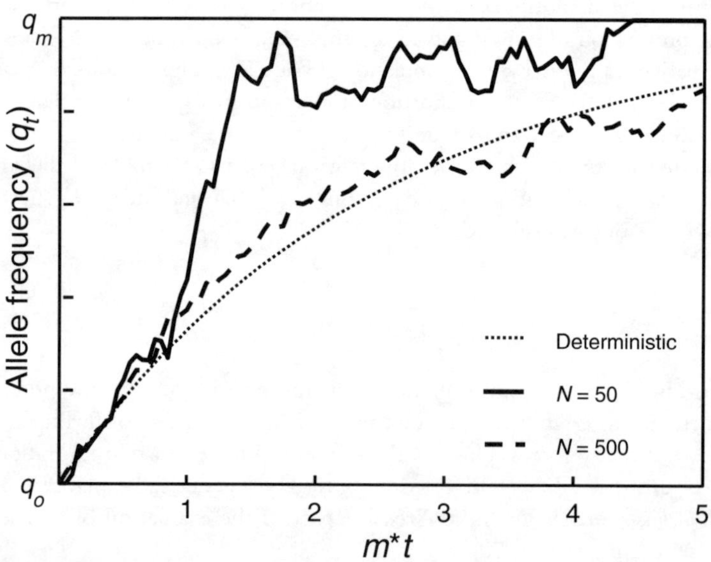

Figure 7.3 Effects of population size on allelic frequency changes due to gene flow. The allelic frequency change in a large (infinite) recipient population resulting from recurrent immigration is depicted by the dotted line (deterministic). If the recipient population has a finite size, random drift will produce fluctuations in allelic frequency. The trajectories for population sizes (N) of 50 and 500 are single examples of the random paths that might occur. Gene flow generally determines the direction of change, and smaller populations tend to exhibit greater fluctuations from the predicted trajectory. The figure is generalized to consider two arbitrary initial allelic frequencies (q_m in donor population and q_0 in the recipient) and combinations of immigration rate and time in generations ($m*t$).

generation. Real populations, however, are finite, and allelic frequencies deviate randomly from expectations as a result of sampling error in each generation. This effect is entirely a sampling (statistical) phenomenon unrelated to selection or other evolutionary forces.

In finite populations, allelic frequencies may change from generation to generation (Waples 1989). The expected magnitude (of the variance of the changes) is inversely proportional to the effective population size (N_e) (Chapter 9). Such random changes in allelic frequency are termed **random genetic drift** (Chapter 9). Because genetic drift is random, it typically results in divergence among populations whereas gene flow tends to make them similar. Examples of the effect that finite population size can have on gene flow is shown in Figure 7.3 for recipient populations of 50 and 500 individuals. Note that the curves shown are only single examples of the random effects of genetic drift; the actual results are unpredictable, although in general the recipient population will become increasingly similar to the donor population even in the face of random drift.

7.3.2.1 **Island Model.** Models that consider both random drift and gene flow depend on the geometric pattern of movements between the populations; that is, the paths the migrants follow and their numbers as well as population sizes. One conceptual model is the **island model** (Wright 1943), which considers populations as discrete units analogous to a number of islands among which there are exchanges of individuals (Figure 7.4). A migrant from an island is equally likely to be an immigrant to any other island; distance among islands is not a direct factor. For fish, an example is migration between spawning shoals within a lake. We mentioned previously that gene flow and random drift affect divergence among populations in opposite directions. The equation describing one generation of this process on one island is similar to equation (7.1), except that the allelic frequency of the immigrants is the average of all the other islands ($\bar{q}$) and there is a stochastic element (ε) that results from the finite population size of the island:

$$q_t = (1 - m)q_{t-1} + m\bar{q} + \varepsilon_{t-1}. \tag{7.4}$$

Therefore, the extent of divergence will eventually reach an equilibrium. If the rate of gene flow is large, the populations will remain similar in genetic composition. If there is little gene flow, allelic frequencies will diverge among populations. In the extreme, allelic frequencies may become fixed (allelic frequency $q = 0$ or 1), and the populations will have different allele complements.

The rate of random genetic drift (and divergence of populations) is inversely related to the N_e of the populations. Very large populations diverge very slowly under random drift in the absence of gene flow, whereas small populations diverge rapidly on average. As a consequence, relatively low rates of gene flow are required to arrest divergence among larger populations compared with those required among smaller populations and the divergence that results in the island model depends (approximately) on the absolute number of immigrants each generation. For a model in which all the islands have the same N_e and receive the same proportion of immigrants (m),

$$F_{ST} \approx \frac{1}{4N_e m + 1}, \tag{7.5}$$

where at equilibrium F_{ST} is the fixation index, which in this equation represents the proportion of the variation at a locus attributable to divergence among populations (islands), and the product $N_e m$ is the *number* of individuals that immigrate to each island

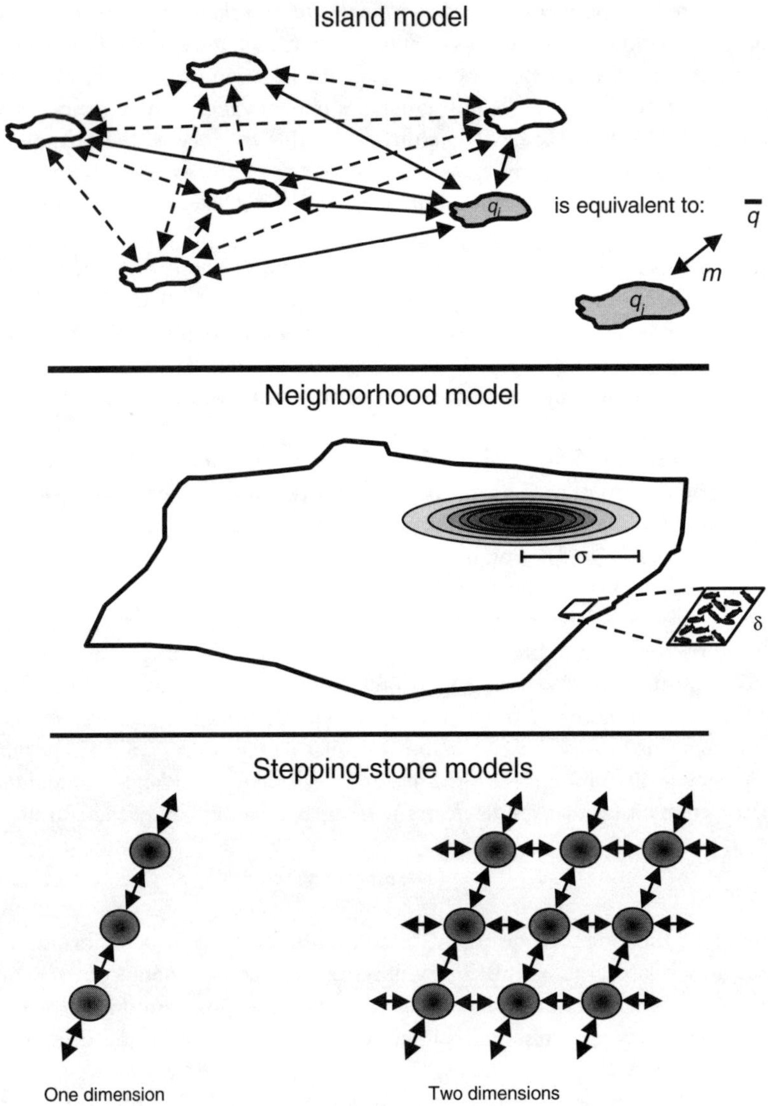

Figure 7.4 Models combining gene flow and random genetic drift. For the island model (top), immigrants to an island are equally likely to come from any other island. The total proportion of immigrants (and emigrants) in each population each generation is m. At time t the allelic frequency of a population of interest (and its emigrants) is q_t and the average allelic frequency of immigrants (and the populations of other islands) is $\bar{q}$. The two-dimensional neighborhood model (middle) uses the distribution of movement between birth and reproduction (characterized by the standard deviation, σ) and the density of breeding individuals, δ, to estimate the effective population size, N_e. Stepping-stone models (bottom) allow gene flow between adjacent populations.

each generation. It is usually very difficult to estimate m directly. Consequently gene flow is often estimated indirectly using $N_e m$, which is the number of immigrant individuals rather than the proportion of immigrant alleles (m). Clearly, by explaining the model in terms of the rate of divergence implies that time is a factor in these models. In fact, these models assume that sufficient time has elapsed that an equilibrium has obtained; therefore, interpretation of the results of these models is appropriate only for situations in which equilibrium or near equilibrium exists.

7.3.2.2 **Neighborhood Models.** In contrast to the island model, the **neighborhood models** assume that individuals are distributed over a continuum (Figure 7.4). Structure within the continuum occurs if the average distance that individuals move between birth and reproduction is small relative to the species range (Malécot 1950). The ranges of movement constitute a **neighborhood.** The result is that gene flow throughout the range is retarded and structure can develop. The N_e of a neighborhood for a two-dimensional distribution of a species can be estimated by

$$N_e = 4\pi\delta\sigma^2, \tag{7.6}$$

where δ is the number of breeding individuals per unit area and σ^2 is the variance of the distance between birth and breeding sites. In this expression, the standard deviation, σ, is analogous to the radius of a circle; therefore, N_e is the area of the circle times the density. At equilibrium, the relationship between migration and drift in the two-dimensional neighborhood model approximates the relationship derived from the island model.

7.3.3.3 **Stepping-Stone Models.** The **stepping-stone models** are a third type of migration and random drift model (Kimura and Weiss 1964). This type of model spatially structures populations so that primary sources of immigrants into a population are the adjacent populations (adjacent is defined relative to the particular structure of the model) and immigration from more distant populations is much less prevalent. These models can be one-dimensional (linear), such as along a coastline or river, or two- or three-dimensional. For a linear stepping-stone model, the equation describing one generation of the process on one island is similar to equation (7.4), except that the allelic frequency of the immigrants is the average of the two flanking islands:

$$q_{i,t} = (1-m)q_{i,t-1} + \frac{m}{2}q_{i-1,t-1} + \frac{m}{2}q_{i+1,t-1} + \varepsilon_{i,t-1}\,, \tag{7.7}$$

where i's index the population and t's index the generation of the process. The two- or three-dimensional models produce relationships between gene flow and random drift that are generally comparable to the island model. The linear stepping-stone model, however, can serve as a much stronger spatial and temporal filter for gene flow because each population can receive immigrants from only two adjacent populations and divergence among populations can be much greater than predicted from the island model. In fact, the extent of divergence depends on the number of stepping-stones (populations) in the system of populations.

7.3.3 *Results and Application of Gene Flow and Drift Models*

Both island and stepping-stone models, as well as the more complicated models, provide rules for generating mathematical and computer formulations that theoreticians can use to examine the combined effects of gene flow and random drift. In these models, populations can be conceptualized as being organized in a hierarchical structure: for example, streams feeding a tributary, tributaries within a drainage, drainages within a geographical region, and so forth.

7.3.3.1 **Wright s *F*-Statistics.** Wright (1978) characterized population hierarchies by partitioning the variation of an allele at a locus on the basis of the reduction in heterozygosity at each level of hierarchy that results from departures from random mating, for example from inbreeding or geographic structure (Wahlund effect). Locally isolated or partially isolated populations can create a hierarchical structure of "apparently" inbred units or, looked at as a whole, a mixture. In a system of multiple populations (one hierarchical level), the **inbreeding coefficient,** F_{IS} (the average of F_{is} over populations), measures the average departure from Hardy–Weinberg expectations in terms of the average deficit of heterozygotes within populations, and the **fixation index,** F_{ST}, measures the reduction in total expected heterozygosity of the entire system of populations due to random drift among isolated populations. The total inbreeding coefficient for the system of populations, F_{IT}, is related to the other *F*-statistics by

$$\left[(1-F_{IT})=\left(\frac{H_I}{H_T}\right)\right]=\left[(1-F_{IS})(1-F_{ST})=\left(\frac{H_I}{H_S}\right)\left(\frac{H_S}{H_T}\right)\right], \tag{7.8}$$

where H_I is the average heterozygosity observed within populations, H_S is the Hardy–Weinberg expectations of heterozygosity averaged among the populations, and H_T is the heterozygosity expected in the entire population system if there were no subdivisions and mating occurred at random. The analysis of population hierarchies can be extended to additional levels.

7.3.3.2 **Gene Diversity Analysis.** A multiallelic, multilocus treatment (Nei 1973; Chakraborty and Leimar 1987) of Wright's analysis of population subdivision using *F*-statistics is gene diversity analysis. It can be shown (Chakraborty and Leimar 1987) that the ultimate extent of gene differentiation among populations, G_{ST}, is related to both the immigration rate into a population (m) and N_e:

$$G_{ST}=\frac{1}{1+4N_em}, \tag{7.9}$$

where the term $N_e\,m$ is the number of immigrants per generation. The G_{ST} is determined by partitioning the genetic variation (heterozygosity) observed in the system of populations into two components, the relative proportion existing *within* populations and the proportion attributable to differences *among* the populations. The G_{ST} quantifies the divergence among populations and is an extension of Wright's F_{ST} ($G_{ST} = \bar{F}_{ST}$) for multiple loci and $G_{ST} = F_{ST}$ for loci with more than two alleles.

Estimates of gene diversity are based on the average heterozygosities at *many* loci (Waples 1998). If all the loci in a population are fixed (i.e., they have no variation), their heterozygosities are zero. If populations are fixed (only one allele occurs at a locus in a population) for different alleles, all the variability is interpopulation variability, that is, G_{ST} equals 1 (Table 7.2, example I), which may occur at extremely low migration rates (or under strong selection if selection acts). If all populations have the same allelic frequencies at their loci, there is no interpopulation variability, only intrapopulation variability, so G_{ST} equals 0 (Table 7.2, example II). A low level of intrapopulation variability may appear at high migration rates. The structure of most population hierarchies will be intermediate to these extremes (e.g., Table 7.2, example III). In simple gene diversity analysis, the average heterozygosity (based on Hardy–Weinberg expectations) for all loci and populations (H_S) is computed, and then the allele frequencies from all the populations are averaged over all loci to estimate total gene diversity (H_T). The relative proportion of variation within populations is H_S/H_T; the remainder, $1 - H_S/H_T$, is the relative proportion of variation among the populations, or G_{ST}.

The simple examples of gene diversity analysis in Table 7.2 are useful for describing population structure, especially if the population is the unit of concern, but the method is not useful for statistical inference. These simple analyses do not take into account biases in estimates that result from finite sample sizes and from subsampling all possible populations (Waples 1998). Although Chakraborty and Leimar (1987) modify the analysis to correct for biases, a more rigorous statistical analysis of gene diversity based on analysis of the variance of allelic frequencies has been developed by Weir and Cockerham (1984). A gene diversity analysis for a simple one-level population hierarchy is worked out in Table 7.3, and the parallel analysis of variance (ANOVA) for *F*-statistics (θ_S is used to denote total among-population variation and θ_P denotes variation among groups of populations—the second level of hierarchy) is in Table 7.4.

We used the results of the gene diversity analysis (Table 7.3) and ANOVA (Table 7.4) to estimate gene flow in terms of the number of immigrants per generation (N_em) at each level of the hierarchy (Table 7.5). The actual structure of the hierarchy is important in this extension and affects N_em estimates. We treated the data in two different ways, assuming two different models for sampling the hierarchy. If the hierarchy includes a large number (infinite in theory) of subunits at each level, the populations sampled are presumably random samples of that large hierarchy (model A; Figure 7.5). Alternatively, if the structure is known and includes a relatively small number of subunits, an adjustment is made in the estimate (model B; Figure 7.5). These estimates of gene flow show that the bias in the simple gene diversity analysis tends to overestimate gene flow by a small amount. In contrast, if one failed to recognize the finite nature of the hierarchy in model B, and treated the results as if the populations and regions were sampled from a large number, using model A might greatly overestimate the amount of gene flow. For example, compare the among-group gene flow estimated as 2.38 individuals per generation in model A but only 0.60 in model B.

Random drift can be overwhelmed by gene flow, just as weak gene flow may have little influence on small populations, which experience strong drift. Crude rules of thumb predict that when N_em exceeds four individuals per generation, the set of populations behaves as a randomly mating unit; when N_em is much less than one, populations will diverge (Hartl and Clark 1989). However, even with gene flows of 10 individuals and in the absence of differential selection, populations can diverge to the point that they will be statistically heterogeneous in allozyme studies that use practical ($N = 100$) sample sizes (Gharrett 1994). Also for many real populations, equilibrium may not have obtained (Zhivotovsky et al. 1994). Processes such as postglacial colonization or reestablishment of population structure following catastrophes, such as the 1964 earthquake that radically altered salmon spawning habitats in Prince William Sound, Alaska, as well as many anthropogenic influences, have occurred too recently to expect that an equilibrium between random drift and gene flow has been achieved. The time required to reach equilibrium can be roughly estimated (Li 1976):

Table 7.2 Examples of gene diversity analysis. In example I, there is no intrapopulation variation, consequently all the diversity exists among populations. In example II, there is variation within each population but no differences among them, so all diversity exists within populations. In example III, there is variation within each population and differences among the populations; 2.7% of gene diversity exists among populations and 97.3% of gene diversity exists within populations. Frequencies of two alleles at a locus *(k)* in population *j* are designated p_{jk} and q_{jk} $(p_{jk} + q_{jk} = 1)$.

	Locus *A*			Locus *B*			
	p_{jA}	q_{jA}	Expected heterozygosity for locus *A* in population *j* $H_{jA} = 2p_{jA}q_{jA} = 1 - p_{jA}^2 - q_{jA}^2$	p_B	q_B	Expected heterozygosity for locus *B* in population *j* $H_{jB} = 2p_{jB}q_{jB} = 1 - p_{jB}^2 - q_{jB}^2$	Average expected heterozygosity by population $H_{j.} = (H_{jA} + H_{jB})/2$
Example I: All loci fixed in a given population, but populations are fixed for different alleles							
Data							
Population							
Population 1	1	0	0	1	0	0	$H_{1.} = 0$
Population 2	0	1	0	0	1	0	$H_{2.} = 0$
Population 3	1	0	0	1	0	0	$H_{3.} = 0$
Data Analysis							
Statistic							
Average expected heterozygosity by locus							
$H_{.k} = (H_{1k} + H_{2k} + H_{3k})/3$			$H_{.A} = 0$			$H_{.B} = 0$	
Totals							
$p_{.k} = (p_{1k} + p_{2k} + p_{3k})/3$	$p_{.A} = 0.67$	$q_{.A} = 0.33$		$p_{.B} = 0.33$	$q_{.B} = 0.67$		
$H_{Tk} = 2p_{.k}q_{.k} = 1 - p_{.k}^2 - q_{.k}^2$			$H_{TA} = 1 - 0.67^2 - 0.33^2 = 0.44$			$H_{TB} = 1 - 0.33^2 - 0.67^2 = 0.44$	
Average expected heterozygosity		$H_S = (H_{.A} + H_{.B})/2 = (H_{1.} + H_{2.} + H_{3.})/3 = (0 + 0 + 0)/3 = 0$					
Average total expected heterozygosity		$H_T = (H_{TA} + H_{TB})/2 = (0.44 + 0.44)/2 = 0.44$					
Average intrapopulation variation		$H_S/H_T = 0/0.44 = 0$					
Average interpopulation variation		$G_{ST} = 1 - H_S/H_{T.} = 1 - 0 = 1$					
Interpretation		∴ **All gene diversity exists *among* populations**					
Example II: All populations have the same allelic frequencies							
Data							
Population							
Population 1	0.4	0.6	0.48	0.6	0.4	0.48	0.48

Population 2	0.4	0.6	0.48	0.6	0.4	0.48	0.48
Population 3	0.4	0.6	0.48	0.6	0.4	0.48	0.48

Data analysis			
Statistic			
$H_{.k} = (H_{1k} + H_{2k} + H_{3k})/3$		$H_{.A} = 0.48$	$H_{.B} = 0.48$
Totals			
$p_{.k} = (p_{1k} + p_{2k} + p_{3k})/3$	$p_{.A} = 0.4$ $q_{.A} = 0.6$	$p_{.B} = 0.6$ $q_{.B} = 0.4$	
$H_{Tk} = 2p_kq_k = 1 - p_k^2 - q_k^2$		$H_{TA} = 0.48$	$H_{TB} = 0.48$
Average expected heterozygosity	$H_S = (0.48 + 0.48 + 0.48)/3 = 0.48$		
Average total expected heterozygosity	$H_T = (0.48 + 0.48)/2 = 0.48$		
Average intrapopulation variation	$H_S/H_T = 0.48/0.48 = 1$		
Average interpopulation variation	$G_{ST} = 1 - 1 = 0$		
Interpretation	**∴ All gene diversity exists *within* populations**		

Example III: An intermediate case, alleles are not fixed and their frequencies vary among populations

Data							
Population							
Population 1	0.6	0.4	0.48	0.4	0.6	0.48	0.48
Population 2	0.5	0.5	0.5	0.5	0.5	0.5	0.5
Population 3	0.4	0.6	0.48	0.6	0.4	0.48	0.48

Data analysis			
Statistic			
$H_{.k} = (H_{1k} + H_{2k} + H_{3k})/3$		$H_{.A} = 0.487$	$H_{.B} = 0.487$
Totals			
$p_{.k} = (p_{1k} + p_{2k} + p_{3k})/3$	$p_{.A} = 0.5$ $q_{.A} = 0.5$	$p_{.B} = 0.5$ $q_{.B} = 0.5$	
$H_{Tk} = 2p_{.k}q_{.k} = 1 - p_{.k}^2 - q_{.k}^2$		$H_{TA} = 0.5$	$H_{TB} = 0.5$
Average expected heterozygosity	$H_S = (0.48 + 0.5 + 0.48)/3 = 0.487$		
Average total expected heterozygosity	$H_T = (0.5 + 0.5)/2 = 0.5$		
Average intrapopulation variation	$H_S/H_T = 0.487/0.5 = 0.973$		
Average interpopulation variation	$G_{ST} = 1 - 0.973 = 0.027$		
Interpretation	**∴2.7 % of gene diversity exist *among* populations**		
	∴97.3% of gene diversity exists *within* populations		

Table 7.3 An example of gene diversity analysis for a system with two hierarchical levels (populations and groups). This example partitions variability within and among populations of each of two groups of populations and between the groups of populations (Chakraborty and Leimar 1987). It is assumed that the population is the unit of interest; therefore, allele frequency estimates are not weighted by sample size. Examples of each calculation are shown. Note that calculations at the second level of hierarchy are weighted by the number of populations in each group. This analysis can be extended to higher-order hierarchies; such analyses are weighted by the number subdivisions (e.g., population or groups) at each level of hierarchy. There are two alleles *(l)* at each locus *(k)* in each population *(j)* of each group of populations *(i)* ($p_{ijk1} + p_{ijk2} = 1$). Calculations of heterozygosities at different levels of hierarchy are designated by *H* s and proportionate estimates of variation are denoted by *G* s. Averages across indices are indicated by dots; for example, $p_{2.B1}$ is the average allele frequency across populations in group 2 for the first allele at locus *B*, $H_{S(1..)}$ is the average expected heterozygosity of all loci in all populations in group 1, and $H_{G(2.)}$ is the average heterozygosity of all loci in group 2.

		Locus *A*			Locus *B*			Summary	
	N	p_{ijA1}	p_{ijA2}	$H_{S(ijA)}$	p_{ijB1}	p_{ijB2}	$H_{S(ijB)}$	values	Example of calculation
Group 1									
Data									
Population 1a	100	0.5	0.5	0.5	0.2	0.8	0.32		$H_{S(1aB)} = 1 - p_{1aB1}^2 - p_{1aB2}^2$ $= 1 - 0.2^2 - 0.8^2 = 0.32$
Population 1b	100	0.5	0.5	0.5	0.3	0.7	0.42		
Population 1c	100	0.5	0.5	0.5	0.4	0.6	0.48		
Data Analysis									
$p_{1.kl}$		0.5	0.5		0.3	0.7			$p_{1.B2} = (p_{1aB2} + p_{1bB2} + p_{1cB2})/3$ $= (0.8 + 0.7 + 0.6)/3 = 0.7$
$H_{S(1.k)}$				0.5			0.407		$H_{S(1.B)} = [H_{S(1aB)} + H_{S(1bB)} + H_{S(1cB)}]/3$ $= (0.32 + 0.42 + 0.48)/3 = 0.407$
$H_{G(1k)}$				0.5			0.42		$H_{G(1B)} = 1 - p_{1.B1}^2 - p_{1.B2}^2$ $= 1 - 0.3^2 - 0.7^2 = 0.42$
$H_{S(1..)}$								0.453	$H_{S(1..)} = [H_{S(1.A)} + H_{S(1.B)}]/2$ $= (0.5 + 0.407)/2 = 0.453$
$H_{G(1.)}$								0.460	$H_{G(1.)} = [H_{G(1A)} + H_{G(1B)}]/2$ $= (0.5 + 0.42)/2 = 0.460$
$H_{S(1..)} / H_{G(1.)}$								0.985	$H_{S(1..)} / H_{G(1.)} = 0.453/0.460$
$G_{SG(1)}$								0.014	$G_{SG(1)} = 1 - 0.985$

Group 2									
Data									
Population 2a	100	1.0	0.0	0.00	1.0	0.0	0.00		
Population 2b	100	0.9	0.1	0.18	0.6	0.4	0.48		
Population 2c	100	0.8	0.2	0.32	0.4	0.6	0.48		
Population 2d	100	0.7	0.3	0.42	0.0	1.0	0.00		
Data Analysis									
$p_{2.kl}$		0.85	0.15		0.5	0.5			
$H_{S(2.k)}$				0.23			0.24		
$H_{G(2k)}$				0.255			0.5		
$H_{S(2..)}$								0.235	
$H_{G(2.)}$								0.378	
$H_{S(2..)} / H_{G(2.)}$								0.623	
$G_{SG(2)}$								0.377	
Total gene diversity analysis									
$p_{..kl}$		0.7	0.3		0.414	0.586			$p_{..B2} = (p_{1aB2} + p_{1bB2} + \ldots + p_{2dB2})/7$ $= (0.8 + 0.7 + \ldots + 1.0)/7 = 0.586$
$H_{S(..k)}$				0.346			0.311		$H_{S(..B)} = [3{\star}H_{S(1.B)} + 4{\star}H_{S(2.B)}]/7$ $= (3{\star}0.407 + 4{\star}0.24)/7$
$H_{T(k)}$				0.42			0.485		$H_{T(B)} = 1 - p_{..B1}^{2} - p_{..B2}^{2}$ $= 1 - 0.414^2 - 0.586^2 = 0.485$
$H_S = H_{S(...)}$	Average expected heterozygosity within populations							0.329	$H_S = [H_{S(..A)} + H_{S(..B)}]/2$ $= (0.346 + 0.311)/2$
$H_G = H_{G(..)}$	Average expected heterozygosity of groups							0.413	$H_G = [3{\star}H_{G(1.)} + 4{\star}H_{G(2.)}]/7$ $= (3{\star}0.460 + 4{\star}0.378)/7$
H_T	Total average expected heterozygosity							0.453	$H_T = [H_{T(A)} + H_{T(B)}]/2$ $= (0.42 + 0.485)/2$
H_S/H_T	Proportion of variation due to within-population variation							0.726	$H_S/H_T = 0.329/0.453$
G_{SG}	Proportion of variability due to variation among populations within groups							0.185	$G_{SG} = (H_G - H_S)/H_T$ $= (0.413 - 0.329)/0.453$
G_{GT}	Proportion of variability due to variation among groups							0.088	$G_{GT} = (H_T - H_G)/H_T$ $= (0.453 - 0.413)/0.453$
G_{ST}	Proportion of variability due to variation among populations							0.274	$G_{ST} = 1 - H_S/H_T$ $= 1 - 0.726$

Table 7.4 Analysis of variance (ANOVA) for *F*-statistic analysis of allele frequencies in Table 7.3 to estimate gene diversities (Weir and Cockerham 1984; Weir 1996). In parentheses are the analogous values estimated in Table 7.3 based on average heterozygosities. Note that the unbiased estimates obtained from the ANOVA generally exceed estimates based on heterozygosities at the different levels of population structural hierarchy. Why bother with the estimates based on heterozygosities? They are most intuitive and much simpler to calculate. Subscripts *i* and *u* index group and locus, respectively; other symbols are as in Table 7.3. Weir s (1996) notation is used here; consequently *p* and *q* correspond to p_{ijk1} and p_{ijk2} in Table 7.3. Assuming Hardy Weinberg equilibrium, Weir s (1996) *F* equals θ in the one-level hierarchy and *F* equals θ_S in the two-level hierarchy. The θ s are estimated from composite statistics S s and R s, which are estimated from mean squares and corrected sample sizes (n_c).

Source of variance	Mean square	Expected mean squares	Group 1				Group 2			
			Locus *A*	Locus *B*	df	Total	Locus *A*	Locus *B*	df	Total
			One-level hierarchy model[a]							
Among populations within groups	MSP	$pq[(1-\theta)+2n_c\theta]$	0.0000	2.0000	2		3.3333	34.6667	3	
Individuals in populations	MSG	$pq(1-\theta)$	0.2525	0.2054	297		0.1162	0.1212	396	
$\theta_{iu} = S_{1iu}/S_{2iu}$			−0.0050	0.0421			0.1222	0.5888		
$\theta_{Si} = (S_{1A}+S_{1B})/(S_{2A}+S_{2B})$						0.0166 [$G_{SG(1)}$ = 0.014]				0.4443 [$G_{SG(2)}$ = 0.377]
			Two-level hierarchy model[b]							
Between groups	MSP	$pq[(1-\theta_S) + 2n_{c1}(\theta_S-\theta_P) + 2n_{c2}\theta_P]$	42.0000	13.7143	1					
Among populations within groups	MSS	$pq[(1-\theta_S) + 2n_{c3}(\theta_S-\theta_P)]$	2.0000	21.6000	5					
Individuals in populations	MSG	$pq(1-\theta_S)$	0.1746	0.1573	1,393					
$\theta_{Su} = R_{3u}/R_{2u}$			0.2797	0.3795						
$\theta_{Pu} = R_{1u}/R_{2u}$			0.2419	−0.0456						
$\theta_S = (R_{3A}+R_{3B})/(R_{2A}+R_{2B})$						0.3307 (G_{ST} = 0.274)				
$\theta_P = (R_{1A}+R_{1B})/(R_{2A}+R_{2B})$						0.0949 (G_{GT} = 0.088)				

[a] Each group analyzed separately. S_1 = MSP − MSG; S_2 = MSP + $(2n_c - 1)$MSG; and n_c = 100.000.

[b] Groups analyzed together. $R_1 = [(MSP - MSG)/2n_{c2}] - [n_{c1}(MSS - MSG)/2n_{c2}n_{c3}]$; $R_2 = [(MSP - MSG)/2n_{c2}] + [(n_{c2} - n_{c1})(MSS - MSG)/2n_{c2}n_{c3}] + MSG$; $R_3 = [(MSP - MSG)/2n_{c2}] + [(n_{c2} - n_{c1})(MSS - MSG)/2n_{c2}n_{c3}]$; n_{c1} = 100.000; n_{c2} = 342.8571; and n_{c3} = 100.0000.

Table 7.5 Gene flow (numbers of individuals, $N_e m$) estimated from gene diversity analysis (Table 7.3) and *F*-statistics analyses (Table 7.4). Two models (Figure 7.5 A, B) are analyzed. The first column of each analysis (model A) considers a two-level hierarchical model from which populations and groups are sampled from a much larger set of populations and groups. For the second column of each analysis (model B), data collected from a defined hierarchy of populations (n_1 in group 1, n_2 in group 2, and a total of k) and groups (g of them) are analyzed. In the first analysis, i indexes the group analyzed. The migration rate between groups is m_G.

	Estimates from gene diversity analysis			Estimated from analysis of variance		
Level	Gene diversity	Model A	Model B	*F*-statistic	Model A	Model B
	G_{SG_i}	$N_{e_i} m_i \approx \frac{1}{4}\left(\frac{1}{G_{SG_i}} - 1\right)$	$N_{e_i} m_i \approx \frac{1}{4}\left(\frac{1}{G_{SG_i}} - 1\right)\left(\frac{n_i - 1}{n_i}\right)^2$	θ_{S_i}	$N_{e_i} m_i \approx \frac{1}{4}\left(\frac{1}{\theta_{S_i}} - 1\right)$	$N_{e_i} m_i \approx \frac{1}{4}\left(\frac{1}{\theta_{S_i}} - 1\right)\left(\frac{n_i - 1}{n_i}\right)^2$
Group 1	0.014	17.0	7.56	0.0166	14.8	6.58
Group 2	0.377	0.41	0.23	0.4443	0.31	0.18
	G_{GT}	$N_{G_e} m_G \approx \frac{1}{4}\left(\frac{1}{G_{ST}} - 1\right)$	$N_{G_e} m_G \approx \frac{1}{4}\left(\frac{1}{G_{GT}} - 1\right)\left(\frac{g - 1}{g}\right)^2$	θ_P	$N_{G_e} m_G \approx \frac{1}{4}\left(\frac{1}{\theta_P} - 1\right)$	$N_{G_e} m_G \approx \frac{1}{4}\left(\frac{1}{\theta_P} - 1\right)\left(\frac{g - 1}{g}\right)^2$
Among groups	0.088	2.59	0.65	0.0949	2.38	0.60
	G_{ST}	$N_e m \approx \frac{1}{4}\left(\frac{1}{G_{ST}} - 1\right)$	$N_e m \approx \frac{1}{4}\left(\frac{1}{G_{ST}} - 1\right)\left(\frac{k - 1}{k}\right)^2$	θ_S	$N_e m \approx \frac{1}{4}\left(\frac{1}{\theta_S} - 1\right)$	$N_e m \approx \frac{1}{4}\left(\frac{1}{\theta_S} - 1\right)\left(\frac{k - 1}{k}\right)^2$
Among populations	0.274	0.66	0.49	0.3307	0.51	0.37

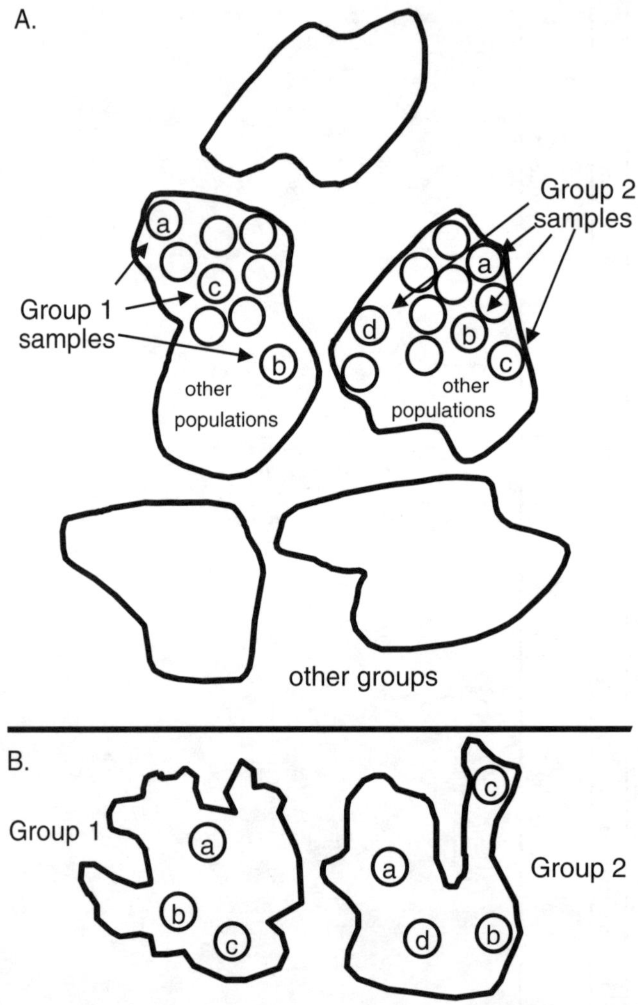

Figure 7.5 Hierarchical organization of the populations analyzed in Table 7.5. Shown are **(A)** populations and groups of populations sampled at random from a large number of possible groups and populations, and **(B)** sampling from a finite, known set of groups and populations.

$$\frac{1}{t} \ll \frac{1}{2N_e} + 2m . \tag{7.10}$$

A very small N_e or a high rate of gene flow (m) can dominate the expression.

Gene diversity has been used to analyze dispersal of marine fishes in southern Californian and Baja Californian waters. An allozyme examination of 10 species representing a wide range of larval life spans and prevalent at several different sampling sites showed that the amount of gene flow (N_em) correlated directly with the time spent as larvae (Waples 1987). It is presumed that longer-lived larvae tend to disperse over a wider range than do shorter-lived larvae. For example, gene flow in the live-bearing black perch *Embiotoca jacksoni* was low (about 0.3 individuals per generation), whereas the kelp bass *Paralabrax clathratus*, whose larvae have a life span of several months, exhibited high gene flow (over 150). Species with intermediate larval life spans had intermediate levels of gene flow. The relationships between genetic distances (Chapter 13) and geographic distribution for several species indicated that dispersion patterns of diverse species may be similar, which suggests that physical factors such as currents and temperature may influence different species similarly (Waples and Rosenblatt 1987). This analysis assumed that N_e is similar among species, or at least uncorrelated with m, because correlation between N_e and time spent as larvae would produce comparable results. A summary of gene diversity analyses for Pacific salmon is presented in Box 7.1

7.3.4 Gene Flow and Selection

Another assumption underlying Hardy–Weinberg equilibrium is that no selection is acting. Populations experiencing similar selection regimes will tend to resist the effects of random genetic drift, even if there is little gene flow. Of course, the relative magnitudes of selection and drift (proportional to $1/2N_e$) drive the process. Differential selection, on the other hand, tends to cause the populations to diverge and works in opposition to gene flow. If the rate of gene flow (m) exceeds the selection coefficient (s) for a particular allele, the selection process will have little net effect (Figure 7.6). If s greatly exceeds m, selection will dominate the process, but even when s is 10 times m, gene flow will noticeably alter the allelic frequencies. In situations in which local selection regimes differ, the combination of gene flow and selection ultimately will achieve an equilibrium; that is, there will be no net allelic frequency change in a generation. Figure 7.7 (left) depicts the combined influences of s and m. The magnitude of the vectors representing the net effect of the two forces decreases in the second generation, whereas the angles between the forces increase. At equilibrium, the vectors for s and m will be in opposite directions and of equal magnitude; the resultants will be zero. It may be difficult to resolve the combined effects of gene flow and the different selection regimes experienced by different populations (local adaptation). Deterministic models examining these combined effects have been developed for only some special cases (Zhivotovsky and Feldman 1993). Models that combine the effects of gene flow, heterogeneous selection, and random drift (Figure 7.7, right) are still unavailable.

Allelic frequency gradients or clines can develop from two opposing mechanisms. First, genetic clines can develop as a result of differential selection in a gradient of environmental conditions (e.g., temperature, pH, or elevation) (see also Chapter 8). Second, a cline may develop between two genetically different populations connected by gene flow through a chain of populations. In reality, both mechanisms may act together. Clines along a linear distribution of populations, such as a stream or coastline, are most often apparent (Figure 7.8; Box 7.2). For example, the American eel along the Atlantic seaboard of North America exhibit a cline in frequencies of the allozyme allele *Gpi-2** (glucose-6-phosphate isomerase, 5.3.1.9) (Williams et al. 1973; Koehn and Williams 1978) even though they presumably originate from a panmictic population (Avise et al. 1986). Other examples include mummichog *Fundulus heteroclitus,* with a cline in the frequency of *LDH-B** allozymes (Place and Powers 1979; Figure 8.4). We saw earlier (section 7.3.3.3) that, for a given immigration rate (m), linear stepping-stone models exhibit greater divergence than do models with more dimensions, so it is likely that clines in one-dimensional population assemblages will be more readily detected, particularly if the populations are distributed along an environmental gradient.

■ Box 7.1
Gene Diversity and Gene Flow in Pacific Salmon

Pacific salmon offer an opportunity to apply gene diversity analysis to estimates of gene flow because they home with high, but not perfect, precision to natal streams and have opportunity for gene flow in every generation. Because we do not have knowledge of the effective population sizes (N_e) of Pacific salmon, we are unable to estimate the migration rate (m), but we can estimate the number of immigrants per generation ($N_e m$) from a measure of genetic divergence among populations, G_{ST}. Of course if equilibrium has not obtained, the estimates of gene flow will be overestimates.

A summary of gene diversity analyses of anadromous Pacific salmon based on electrophoretic data indicates that the extent of gene flow may be species dependent (see table below). These results are not inconsistent with independent estimates of straying in Pacific salmon (Quinn 1984). Differences among the species are striking. Pink salmon and chum salmon *(Oncorhynchus keta),* which emigrate soon after emerging from the gravel, show much higher rates of gene flow than do coho *(O. kisutch),* sockeye *(O. nerka),* or chinook *(O. tshawytscha)* salmon, which generally grow in freshwater for several months or more prior to their seaward migration. Explanations of these differences may be related to the abundance and close proximity of pink and chum salmon populations to other conspecific populations or to the longer residence time and potentially stronger imprinting in most chinook, coho, and sockeye salmon populations. Quinn (1984) has suggested that straying and the resultant gene flow may be related to the stability of the system in which the fish spawn. If a spawning or rearing system is unstable because of environmental variation such as fluctuating flows, ice scour, or changes in stream access or location, a population can best ensure its persistence by straying. The genetic structure of pink salmon in some regions such as Asia (Zhivotovsky et al. 1989) and the Aleutian Islands (Gharrett et al. 1988) suggests high levels of gene flow. Glubokovsky and Zhivotovsky (1986) used the term "fluctuating stocks" to emphasize the importance of gene flow to the population structure of a region. If homing has genetic determinants, which is likely, then the precision of homing may respond to natural selection.

continued on next page

7.3.5 *Other Factors and Combinations of Factors*

Factors other than selection, gene flow, and random drift influence allelic frequencies in populations. However, the effects of other factors are generally weaker, difficult to quantify, or special cases of population breeding structure that generally are subsumed under effective population size (N_e) and related to random drift. In gene flow models that include mutation (μ), the combined influence is usually their sum ($m + \mu$), and gene flow usually predominates, particularly in the short time frame relevant to fisheries management. It is interesting that the ultimate source of genetic variation, mutation, is generally inconsequential in gene flow models. Because mutation rates for most molecular markers are quite small (approximately 10^{-4} to 10^{-6} per generation), the migration term dominates their combined effect unless it is very small (e.g., $\ll$1% of the breeders).

■ Box 7.1 *continued*

Gene diversity analyses of Pacific salmon based on studies that used the most loci and largest geographic distribution.

Species	Region	G_{ST}	Number of loci	Number of locations	N_em	Source
Oncorhynchus nerka	British Columbia	0.162	33	80	1.3	Wood et al. 1994
	SE Alaska	0.130	31	25	1.7	C. M. Guthrie[a]
	W Alaska	0.045	55	28	5.3	Varnavskaya et al. 1994
	Kamchatka	0.034	55	6	7.1	Varnavskaya et al. 1994
Oncorhynchus keta	Washington, S British Columbia	0.028	39	105	8.7	Phelps et al. 1994
	SE Alaska, N British Columbia	0.027	40	61	9.0	Kondzela et al. 1994
	W Alaska	0.036	24	30	6.7	Wilmot et al. 1994
	Asia	0.038	29	29	6.3	Winans et al. 1994
Oncorhynchus gorbuscha						
Odd year	S British Columbia, Puget Sound	0.018	22	22	13.6	Data from Beacham et al. 1985
	Alaska	0.037	23	22	6.5	Data from McGregor 1983; Gharrett, unpubished
	Asia	0.011	26	27	22.5	Hawkins et al. 2002
Even year	S British Columbia	0.015	12	3	16.4	Data from Beacham et al. 1985
	Alaska	0.025	23	22	9.8	Data from McGregor 1983; Gharrett et al. 1988; Gharrett, unpublished
	Asia	0.020	19	17	12.2	Noll et al., 2001
Oncorhynchus kisutch	California	0.158	45	27	1.3	Bartley et al. 1992
	S British Columbia	0.086	15	43	2.7	Wehrhahn and Powell 1987
Oncorhynchus tshawytscha	California	0.177	53	35	1.2	Bartley and Gall 1990
	Oregon, Washington, S British Columbia	0.123	25	65	1.8	Utter et al. 1989
	Alaska	0.059	28	37	4.0	Gharrett et al. 1987

[a] National Marine Fisheries Service Auke Bay Laboratory, personal communication.

Regardless of the evolutionary forces acting on populations, gene flow tends to reduce genetic differences between donor and recipient populations. If the extent of gene flow between populations is inversely related to the geographic distance between them, one would expect to detect clines across the geographic distribution of a species. The relationship between distance and gene flow can be measured directly by plotting the estimated number of migrants (or its logarithm) between two populations each generation ($N_e m$) against the geographic distance (or its logarithm) separating them (Slatkin 1993). A significant regression for such a plot indicates that gene flow is related to the geographic distance separating two populations.

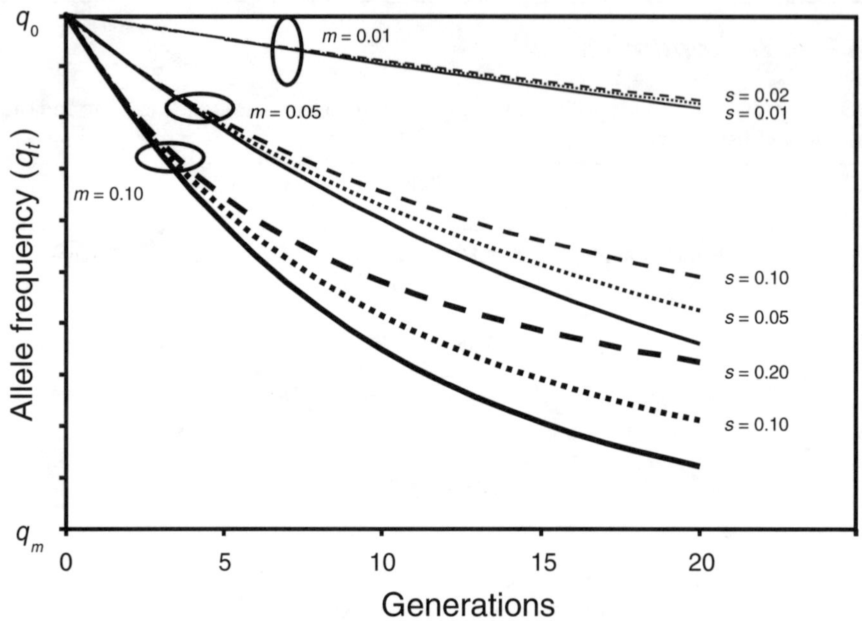

Figure 7.6 Combined effects of selection and gene flow for three levels of gene flow (m = 0.01, 0.05, and 0.10) each at three levels of selection (s = 0 [solid line], 1*m [dotted line], and 2*m [dashed line]). The figure is generalized to consider two arbitrary allelic frequencies (q_m, in the donor population, and q_0, in the recipient population at generation 0). Selection coefficients much less than m have much less effect on the gene flow trajectory. The model for selection assumes selection against the allele a (no dominance), but other models yield roughly comparable results.

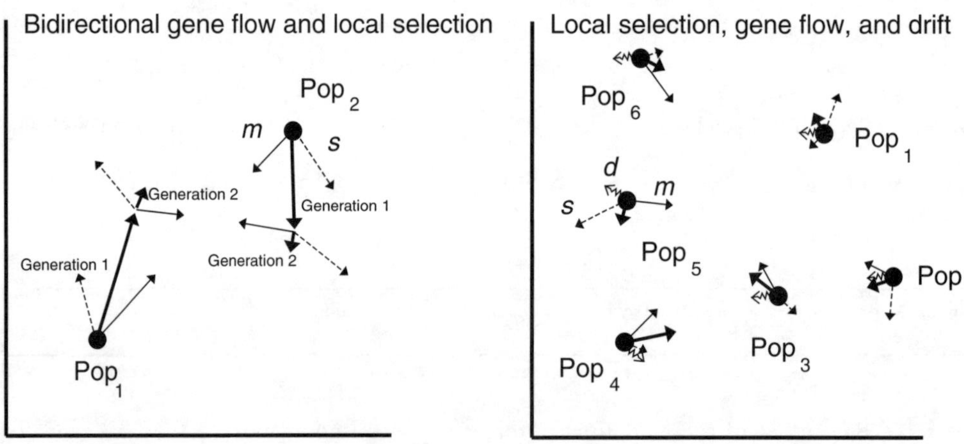

Figure 7.7 Conceptual depictions of allelic frequency changes in populations. (Left) The combined allelic frequency changes (thick arrow) from gene flow, m (narrow arrow), and natural selection, s (dashed arrow), between two populations that have experienced different selection regimes. (Right) The combined allelic frequency changes from local selection, gene flow, and random drift, d (zigzag arrow), on a collection of interacting populations. Note that the direction of the allelic frequency changes due to gene flow is always toward the overall system mean. Populations are plotted in "allele frequency space" transformed to two dimensions.

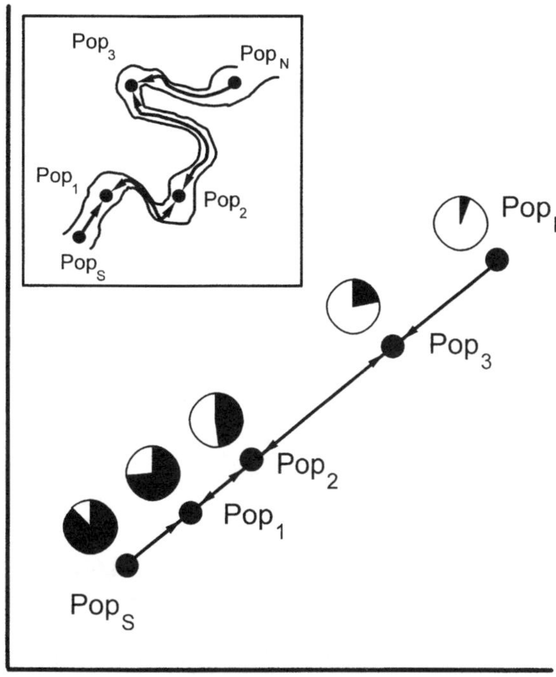

Figure 7.8 Example of a one-dimensional system of populations along which a cline in allelic frequencies might be detected. The spatial relationships between populations within a river system are transformed to a linear relationship and the arrows indicate that gene flow occurs mostly between adjacent populations. The small pie diagrams show the frequencies of two alleles at a locus for each population.

7.3.6 *Applications to DNA Sequence-Based Data*

The development of DNA sequence-based population genetics methods has modified some of the applications of models and has introduced new approaches. Polymerase chain reaction (PCR) based methods make it possible to amplify and scrutinize specific nucleotide sequences. Two of the most frequently used sources of DNA variation are microsatellite loci and mitochondrial DNA (mtDNA) sequences.

7.3.6.1 **Microsatellite DNA Loci.** Microsatellites are repeats of very short sequences distributed throughout the genome (see Chapter 5). The genetic variation that occurs at microsatellite loci results from differences in the number of copies of the repeat and is detected from the size or sequence of amplification products. Many microsatellite loci have numerous alleles (in excess of 40). The highly variable microsatellite loci provide an array of potent tools for population analysis. However, a consequence of the high variability is that many of the alleles may be observed only once in a population or a group of populations. To analyze such data it is necessary to use nonparametric methods based on permutations or Monte-Carlo methods (e.g., Genepop, Raymond, and Rousset 1995; Arlequin, Schneider et al. 2000). Another challenge of microsatellites is that mutations that prevent annealing of the PCR primers (and failure of the amplification), so called nulls, are relatively frequent. Null alleles produce an apparent departure from Hardy–Weinberg expectations (an excess of

■ Box 7.2
Isolation by Distance in Yukon River Chinook Salmon Populations

R. L. Wilmot, ***National Marine Fisheries Service, Alaska Fisheries Science Center, Juneau***

The Yukon River originates in the coastal mountains of British Columbia and flows for 3,250 km through the Yukon Territory and Alaska to the Bering Sea. Chinook salmon contribute to both commercial and subsistence fisheries in the Yukon River. More than 100 chinook salmon populations are distributed throughout the drainage. Chinook salmon enter the Yukon River continuously from late May or early June through mid-July and migrate to spawning grounds as far as 3,000 km from the river mouth. The Alaskan catch accounts for approximately 90% of the chinook salmon harvest and includes fish of both Alaskan and Canadian origin. The United States and Canada began negotiations in 1985 to arrive at an equitable allocation of Yukon River salmon. The primary information needed to develop a treaty was information on the national origins of fish harvested in Alaskan waters. One of the methods used to estimate the origins of catch was based on genetic differences among chinook salmon populations (see Chapter 13).

The baseline of allozyme frequencies developed to address allocation questions provides information on population structure of Yukon River chinook salmon. The distribution of populations along the Yukon River approximates a linear stepping-stone distribution of populations. The relationship between distance and gene flow (Slatkin 1993) was examined by plotting log(N_em) against the log(river distance) separating each population pair. This plot has a strong correlation (r^2 = 0.40), which clearly indicates isolation by distance. One interpretation of this result is that chinook salmon home with high fidelity and that strays to nearby populations are more likely than are strays to more distant populations.

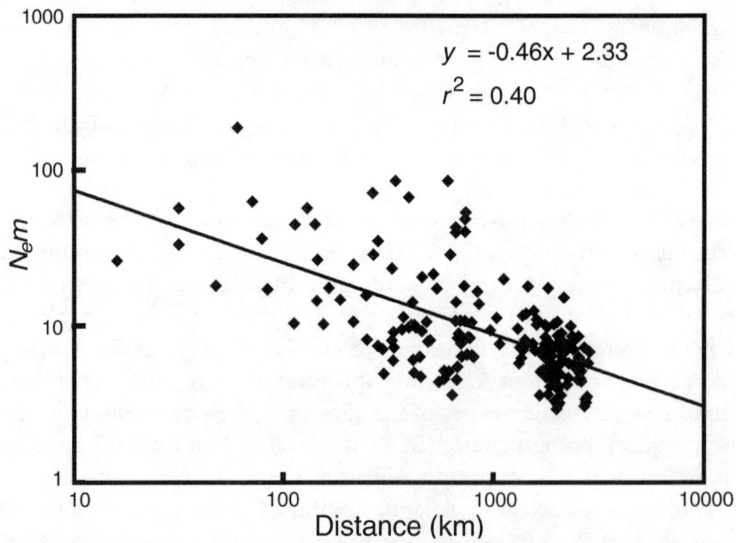

Regression of $\log_{10}(N_em)$ against the $\log_{10}$(river distance) separating pairs of spawning sites of chinook salmon populations in the Yukon River system.

homozygotes), although many of the apparent homozygotes are actually heterozygotes between a successfully amplified allele and an unamplified allele. Null alleles of the White Sands pupfish *Cyprinodon tularosa* were sequenced using alternative primers that annealed to other sites (Jones et al. 1998). Five different size-classes were discovered among the nulls as well as a four-base-pair deletion. Correct interpretation of microsatellite data that include null alleles is difficult, and applying such data to models is inappropriate. Unfortunately, rather than distinguishing between true homozygous pairs and null heterozygotes, researchers often discard data from such loci or, even worse, actually use it.

Microsatellites were used to examine the population structure of coastal cutthroat trout *Oncorhynchus clarki clarki* in the Puget Sound and along the Pacific coast of Washington (Wenburg et al. 1998). The effective number of fish exchanged between populations within a region (N_em) was estimated at about one. Although coastal cutthroat trout often have small spawning populations, they have high levels of variation relative to many other Pacific salmon species. It was hypothesized that a combination of iteroparity and gene flow (albeit low) are involved in maintaining the variation that survived Wisconsin glaciation.

The large amount of variation observed at microsatellite loci results from relatively high mutation rates at loci that probably have little influence on fitness. A consequence of this is that models combining migration (m) and mutation (μ) are not as heavily dominated by migration, and it is prudent to consider the mutation term in population genetics models. Because a microsatellite allele is characterized in terms of its length, the number of repeats allows us to consider microsatellite variation as quantitative. On this ground, additional population statistics have been introduced: R_{ST} as an analog of F_{ST} and G_{ST} statistics (Slatkin 1995), genetic distance (Goldstein et al. 1995), and higher-order statistics (Zhivotovsky and Feldman 1995). Migration also can be incorporated into these models (Feldman et al. 1999; Zhivotovsky 2001).

7.3.6.2 **Mitochondrial DNA.** The mitochondria of eukaryotes carry a set of sequences that is a convenient source of nucleotide sequence data for population genetic study of animals (Brown 1985; Chapter 4). Mitochondrial DNA is convenient for study because (1) each cell possesses multiple copies of mtDNA (roughly 1,000), (2) it can be easily isolated from nuclear DNA, (3) it is relatively small (about 16,500 nucleotide pairs or so), (4) it has been completely sequenced in several fish species, and (5) it is inherited independently from the nuclear DNA, primarily (Gyllensten et al. 1991) as a maternal haploid clone. There is also evidence that mtDNA sequences may diverge (evolve) more rapidly than do most nuclear sequences (Brown et al. 1982). As a consequence of being maternally inherited and haploid, the effective number of mitochondrial genomes in a population is about one-fourth of the effective size of diploid complements for autosomal, diploid loci.

One of the problems in estimating gene flow (N_em) among populations is that it is rarely possible to distinguish divergence among populations caused by a contemporary gene flow–drift (–selection) equilibrium from divergence introduced by historical events, for example, by colonization. Even if a small F_{ST} (and a resultant large N_em) is estimated, there may be no contemporary gene flow. Mitochondrial DNA and other sequence-based data, including both nucleotide sequence and restriction sites, carry information that can be used to infer evolutionary relationships among alleles or haplotypes; that is, gene genealogies, which have an implicit temporal framework. Genealogies of mtDNA haplotypes are especially advantageous because there is no recombination between mtDNA haplotype lineages, so the haplotype displays its mutational record relative to other lineages. Defining a genealogy is an ordered series of mutational events that can be interpreted as a temporal ordering of haplotypes, whether or not the temporal scale follows a regular clock. Analysis of the shape of the genealogy in a geographical context is referred to as **phylogeography** (e.g., Avise 1994, 2001). Populations that have similar haplotype compositions either exchange individuals or are isolated but have not diverged to fixation of different haplotypes. Populations that do not share haplotypes do not exchange individuals, and the extent of divergence of their haplotypes is an index of the length of time they have been isolated. For example, three sunfish species (spotted sunfish *Lepomis punctatus*, redear sunfish *L. microlophus*, and warmouth *L. gulosus*) and the bowfin *Amia calva* in the southeastern United States exhibit strong geographic differences in their mtDNA haplotype compositions, presumably reflecting distinct refugia isolated during the Pleistocene (Avise 1994).

Phylogeographic analysis has been extended by Templeton and colleagues (1995) who have developed a rigorous procedure for testing geographical associations of haplotypes and an inference key for interpreting the causes of association. Different demographic and migration processes leave different signals in the genealogies of the different haplotypes in a gene tree. Periods of population expansion are accompanied by the appearance of new haplotypes; population crashes reduce the number of haplotypes. Gene flow tends to homogenize populations, whereas isolation leads to divergence. At the periphery of a genealogy are the newest haplotypes (unless the ancestral root traces through one of them). The nested clade analysis is conducted by collapsing the genealogy in steps and analyzing the geographic patterns of haplotypes at each step. Demographic signals can be resolved from the geographic distribution of the haplotype and collapsed groups of haplotypes (nested clades) in the genealogy. Inferences can be made from these analyses about the historic demography and influences of gene flow on the contemporary population structure, inferences that are not possible using data from unordered alleles such as allozymes. Templeton's key (1998) is based on statistically significant patterns that are used to distinguish among causes of phylogeographic associations, among which are (1) the null hypothesis of no geographic structure, (2) historic reductions of population size that fragmented the species, (3) historic contiguous range expansions, (4) contemporary restricted gene flow resulting in isolation by distance, (5) historic long-distance colonization, and (6) contemporary restricted gene flow with some long-distance dispersal. Examples of four haplotype composition patterns consistent with phylogeographic patterns 1, 2, 3, and 4 are simply illustrated in Figure 7.9. In interpreting the patterns, keep in mind that haplotype A is ancestral to B and C. The null hypothesis tested is that there is no geographic structure. All other examples exhibit statistically significant structure. Historic fragmentation, which results from historic isolation (for example, during ice sheet advances) followed by mutation and random drift, produces qualitative differences in haplotype compositions. In contrast, an historic contiguous range expansion followed by drift can produce similarities in nonadjacent populations. Limited gene flow resulting in isolation by distance is the only model that is appropriate for F_{ST} ($N_e m$) analysis, although phylogeographic patterns 3–6 can have similar values of F_{ST}. Templeton's key (1998) also identifies patterns 5 and 6 as well as other associations for which too little data are available for interpretation.

Results of different branches of a gene tree may be in agreement, but as the genealogy is collapsed, analyses may reveal patterns that arose from earlier events (e.g., Templeton and Georgiadis 1996). Numerous events that shaped contemporary population structure may be resolved.

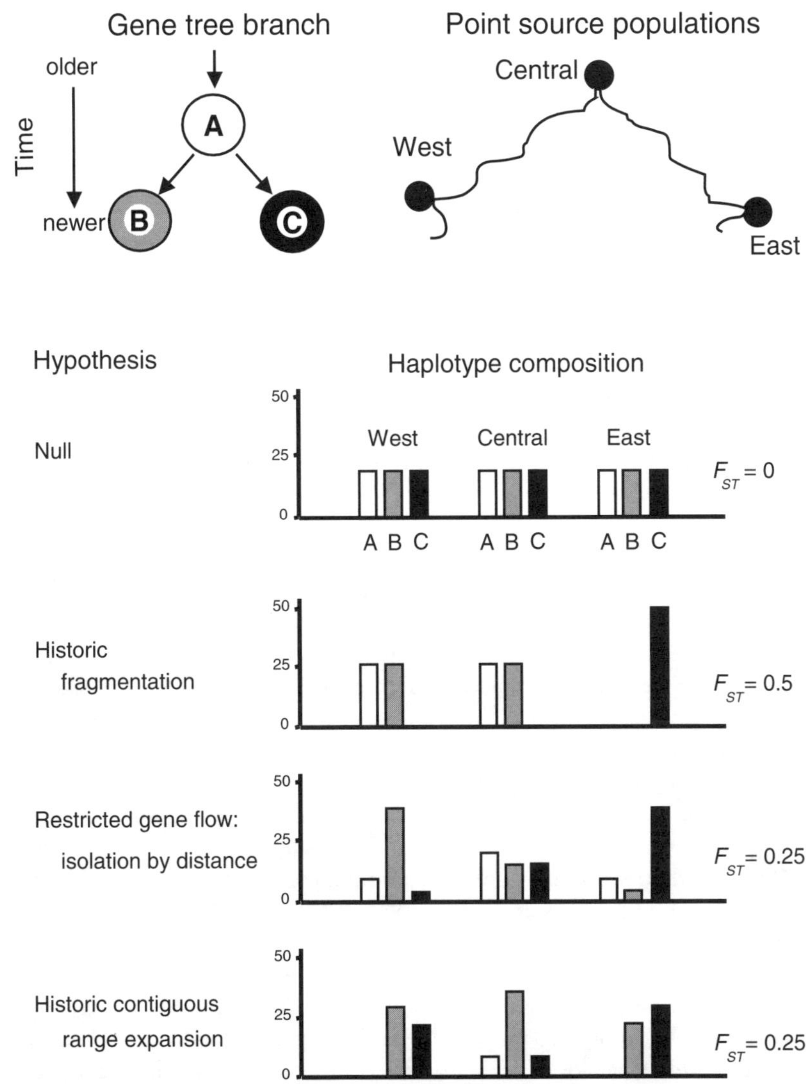

Figure 7.9 Phylogeographic examples of the events and processes that lead to divergence among three populations and the haplotype compositions consistent with those processes. There are no intervening populations. Note that two of the processes have the same F_{ST} values, although in one there is no contemporary gene flow.

■ 7.4 Perspective: Beyond the Models

The models presented here that examine the influences of gene flow on population structure generally predict the effect at a single locus. The results of random drift affect each independent locus differently. Moreover, each locus may behave differently depending on the influence of selection. One result is that the genetic structure and presumed relationships among populations described using neutral or nearly neutral alleles may differ substantially from structures based on other (unlinked) loci that are subject to different local selection regimes (Gillespie 1991; Gharrett 1994). Practically, this means that selectively neutral or nearly neutral loci (possibly some allozyme loci) may provide estimates of the pattern and level of gene flow between populations but not the full extent of genetic divergence among them.

Most of the adaptive characters of interest to fisheries biologists are more complex traits than are the Mendelian traits of population genetics models and result from the combined expression of many loci (quantitative traits). Quantitative traits are probably regulated by natural (and possibly human) selection. Traits such as fecundity and growth rate, which contribute to parameters critical to the formulation of population dynamics expressions, are rarely determined by a single locus. Although the models we have examined are applicable to allelic frequencies at each of the contributing loci, the number of loci involved and genetic interactions between loci generally increase the complexity of response to gene flow by quantitative traits. As a result, the effects of altered gene flow rates or of gene flow from introductions on local populations are unpredictable but can range from benign to severely disruptive. The latter effect is termed **outbreeding depression** (see Chapter 11), which describes the loss in fitness (and consequently productivity) resulting from hybridization between genetically dissimilar populations of a species (Shields 1982; Templeton 1986; Emlen 1991). The decrease in fitness of immigrants and their hybrids with natives depends on dissimilarities in both allelic frequencies and gametic disequilibria between the locally adapted recipient population and the donor population (Zhivotovsky and Christiansen 1995). Ample evidence for local adaptation in fishes (Taylor 1991; Carvalho 1993) and the consequences of altered or increased gene flow from aquacultural sources (Hindar et al. 1991) suggest that managers and culturists should avoid actions that alter natural levels of gene flow, particularly ones that will increase gene flow between populations with obvious adaptive differences.

An example of local adaptive differences is the precision of homing to spawning grounds in migratory fish. Observations of hybrids between populations of pink salmon suggest a heritable component in homing (Bams 1976; but see Thrower and Smoker 1995) in addition to imprinting; the homing mechanism (part of which is the imprinting process) must have some genetic determination in order for homing and straying to be acted on by natural selection. The evidence for adaptive differences and for genetic influences on migratory movements provides the components necessary to explain homing and straying in an evolutionary context (Quinn 1984). For relatively stable local environments, local adaptation will tend to favor native individuals over immigrants and their hybrids. If the local environments are less stable, there is less benefit to homing, and selection against immigrants is reduced. In nature, metapopulations are complex, and dynamic population structures involve balances between homing and straying, local adaptation, and genetic drift in the context of an environment that is fluctuating and changing (Gilpin and Hanski 1991). The simple genetic models we have examined can provide considerable insight into the biology of fish, and in combination with population dynamics models may enable us to develop better ways of predicting abundance. We can gain an understanding of the process by examining the theory; however, there still are very few data available that we can use to model natural populations. The behavior of the models depends on genetic parameters (e.g., m, N_e, and s) and on demographic parameters (e.g., δ and σ). In real life we can guess these values or "realistic" ranges, but virtually no direct estimates of these genetic parameters exist for wild populations. Numerous studies have made indirect estimates of N_em. Those estimates depend on a number of assumptions including a migration–mutation–selection–drift equilibrium. Although gene flow can be crudely estimated from fin clipping or physical tagging experiments, the values estimated indirectly or using physical tags have not been validated with direct estimates of migration rates (m) obtained by tracking alleles between populations.

In applying results of these models, one must be aware that although the models often provide amazing insights into population genetics processes, the results can also be over- or misinterpreted (Templeton et al. 1995; Waples 1998). Such errors are particularly pertinent in studies of species that have relatively high levels of gene flow, for instance, many marine species. Wrong conclusions about the genetic structure of a species result from inadequate or inappropriate samples and from misuse of the model (from inappropriate assumptions, for example). Other problems include bias in estimates that result from finite samples of individuals or populations. Finally, if a hypothesis is tested using a model, it is easy to make type II errors because alternative hypotheses are often not obvious or available. Also, one must remember that failure to reject a hypothesis does not necessarily mean the hypothesis is true. Conservation and management of fisheries resources will benefit from genetic analyses, but the interpretations must be made carefully.

7.5 References

Avise, J. C. 1994. Molecular markers, natural history, and evolution. Chapman and Hall, New York.

Avise, J. C. 2001. Phylogeography: the history and formation of species. Harvard University Press, Cambridge, Massachusetts.

Avise, J. C., G. S. Helfman, N. C. Saunders, and L. S. Hales. 1986. Mitochondrial DNA differentiation in North Atlantic eels: population genetic consequences of an unusual life history pattern. Proceedings of the National Academy of Science, USA 83:4350–4354.

Bams, R. A. 1976. Survival and propensity for homing as affected by presence or absence of locally adapted paternal genes in two transplanted populations of pink salmon *(Oncorhynchus gorbuscha)*. Journal of the Fisheries Research Board of Canada 33:2716–2725.

Bartley, D. M., B. Bentley, P. G. Olin, and G. A. E. Gall. 1992. Population structure of coho salmon *(Oncorhynchus kisutch)* in California. California Fish and Game 78:88–104.

Bartley, D. M., and G. A. E. Gall. 1990. Genetic structure and gene flow in chinook salmon populations of California. Transactions of the American Fisheries Society 119:55–71.

Beacham, T. D., R. E. Withler, and A. P. Gould. 1985. Biochemical genetic stock identification of pink salmon (*Oncorhynchus gorbuscha)* in southern British Columbia and Puget Sound. Canadian Journal of Fisheries and Aquatic Sciences 42:1474–1483.

Brown, W. M. 1985. The mitochondrial genome in animals. Pages 95–130 *in* R. J. MacIntyre, editor. Molecular evolutionary genetics. Plenum, New York.

Brown, W. M., E. M. Prager, A. Wang, and A. C. Wilson. 1982. Mitochondrial DNA sequences of primates: tempo and mode of evolution. Journal of Molecular Evolution 18:225–239.

Carvalho, G. R. 1993. Evolutionary aspects of fish distribution: genetic variability and adaptation. Journal of Fish Biology 43 (Supplement A):53–73.

Chakraborty, R., and O. Leimar. 1987. Genetic variation within a subdivided population. Pages 89-120 *in* N. Ryman and F. M. Utter, editors. Population genetics and fisheries management. University of Washington, Washington Sea Grant Program, Seattle.

Crozier, W. W. 1993. Evidence of genetic interaction between escaped farmed salmon and wild Atlantic salmon (*Salmo salar* L.) in a northern Irish river. Aquaculture 113:19–29.

Emlen, J. 1991. Heterosis and outbreeding depression: a multilocus model and an application to salmon production. Fisheries Research 12:187–212.

Feldman, M. W., J. Kumm, and J. K. Pritchard. 1999. Mutation and migration in models of microsatellite evolution. Pages 98–115 *in* D. G. Goldstein and C. Schlotterer, editors. Microsatellites: evolution and applications. Oxford University Press, New York.

Gharrett, A. J. 1994. Genetic dynamics of a small population system: a model applicable to interactions between hatchery and wild fish. Aquaculture and Fisheries Management 24 (Supplement 2):79–92.

Gharrett, A. J., S. M. Shirley, and G. R. Tromble. 1987. Genetic relationships among populations of Alaskan chinook salmon *(Oncorhynchus tshawytscha)*. Canadian Journal of Fisheries and Aquatic Sciences 44:765–774.

Gharrett, A. J., C. Smoot, A. J. McGregor, and P. B. Holmes. 1988. Genetic relationships of even-year northwestern Alaskan pink salmon. Transactions of the American Fisheries Society 117:536–545.

Gillespie, J. H. 1991. The causes of molecular evolution. Oxford University Press, New York.

Gilpin, M., and I. Hanski. 1991. Metapopulation dynamics: empirical and theoretical investigations. Academic Press, San Diego, California.

Glubokovsky, M. K., and L. A. Zhivotovsky. 1986. Population structures of pink salmon: systems of fluctuating stocks. Biologiya Morya 2:39–44 (English translation).

Goldstein, D. B., A. R. Linares, L. L. Cavalli-Sforza, and M. W. Feldman. 1995. An evaluation of genetic distances for use with microsatellite loci. Genetics 139:463–471.

Grant, W. S., and F. M. Utter. 1984. Biochemical population genetics of Pacific herring *(Clupea pallasi)*. Canadian Journal of Fisheries and Aquatic Sciences 41:856–864.

Gyllensten, U., D. Wharton, A. Josefsson, and A. C. Wilson. 1991. Paternal inheritance of mitochondrial DNA in mice. Nature (London) 352:255–257.

Hartl, D. L., and A. G. Clark. 1989. Principles of population genetics, 2nd edition. Sinauer Associates, Inc. Sunderland, Massachusetts.

Hawkins, S. L., N. V. Varnavskaya, E. A. Matzak, V. V. Efremov, G. M. Guthrie III, R. L. Wilmot, H. Mayama, F. Yamazaki, and A. J. Gharrett. 2002. Population structure of odd-broodline Asian pink salmon and its contrast to the even broodline structure. Journal of Fish Biology 60:370–388.

Hindar, K., N. Ryman, and F. Utter. 1991. Genetic effects of cultured fish on natural fish populations. Canadian Journal of Fisheries and Aquatic Sciences 48:945–957.

IUBMBNC (International Union of Biochemistry and Molecular Biology, Nomenclature Committee). 1992. Enzyme nomenclature, 1992. Academic Press, San Diego, California.

Jones, A. G., C. A. Stockwell, D. Walker, and J. C. Avise. 1998. The molecular basis of a microsatellite null allele from the White Sands pupfish. The Journal of Heredity 89:339–341.

Kimura, M., and G. H. Weiss. 1964. The stepping stone model of population structure and the decrease of genetic correlation with distance. Genetics 49:561–576.

Koehn, R. K., and G. C. Williams. 1978. Genetic differentiation without isolation in the American eel, *Anguilla rostrata*. II. Temporal stability of geographic patterns. Evolution 32:624–637.

Kondzela, C. M., and five coauthors. 1994. Genetic relationships among chum salmon populations in southeast Alaska and northern British Columbia. Canadian Journal of Fisheries and Aquatic Sciences 51 (Supplement 1):50–64.

Lane, S., A. J. McGregor, S. G. Taylor, and A. J. Gharrett. 1990. Genetic marking of an Alaskan pink salmon population, with an evaluation of the mark and the marking process. Pages 395–406 *in* N. C. Parker and five coeditors. Fish-marking techniques. American Fisheries Society, Symposium 7, Bethesda, Maryland.

Li, W. H. 1976. Effect of migration on genetic distance. American Naturalist 110:841–847.

Malécot, G. 1950. Quelques schémas probabilistes sur la variabilité des populations naturelles. Annales de l'Université de Lyon Sciences, Section A 13:37–60.

McGregor, A. J. 1983. A biochemical genetic analysis of pink salmon *(Oncorhynchus gorbuscha)* from selected streams in northern Southeast Alaska. Master's thesis. University of Alaska-Juneau, Juneau.

Murphy, B. R., L. A. Nielsen, and B. J. Turner. 1983. Use of genetic tags to evaluate stocking success for reservoir walleyes. Transactions of the American Fisheries Society 112:457–463.

Nei, M. 1973. Analysis of gene diversity in subdivided populations. Proceedings of the National Academy of Sciences, USA 70:3321–3323.

Noll, C., and 12 coauthors. 2001. Analysis of contemporary genetic structure of even-broodyear populations of Asian and western Alaskan pink salmon, *Oncorhynchus gorbuscha*. Fishery Bulletin 99:123-130.

Phelps, S. R., L. L. LeClair, S. Young, and H. L. Blakenship. 1994. Genetic diversity patterns of chum salmon in the Pacific Northwest. Canadian Journal of Fisheries and Aquatic Sciences 51 (Supplement 1):65–83.

Place, A. R., and D. A. Powers. 1979. Genetic variation and relative catalytic efficiencies: lactate dehydrogenase B allosymes of *Fundulus heteroclitus*. Proceedings of the National Academy of Sciences, USA 76:2354–2358.

Quinn, T. P. 1984. Homing and straying in Pacific salmon. Pages 357-362 *in* J. D. McCleave, G. P. Arnold, J. J. Dodson, and W. H. Neill, editors. Mechanisms of migration in fishes. Plenum Press, New York.

Raymond, M., and F. Rousset. 1995. GENEPOP (version 1,2): population genetics software for exact tests and ecumenicism. Journal of Heredity 86:248–249.

Ryman, N., F. W. Allendorf, and G. Ståhl. 1979. Reproductive isolation with little genetic divergence in sympatric populations of brown trout *(Salmo trutta)*. Genetics 92:247–262.

Schweigart, J. F., F. J. Ward, and J. W. Clayton. 1977. Effects of fry and fingerling introductions on walleye *(Stizostedion vitreum vitreum)* production in West Blue Lake, Manitoba. Journal of the Fisheries Research Board of Canada 34:2142–2150.

Schneider, S., D. Roessli, and L. Excoffier. 2000. Arlequin ver. 2.000: a software for population genetics data analysis. Genetics and Biometry Laboratory, University of Geneva, Switzerland.

Shields, W. M. 1982. Philopatry, inbreeding, and the evolution of sex. State University of New York Press, Albany.

Sick, K. 1965. Haemoglobin polymorphism of cod in the Baltic and Danish Belt Sea. Hereditas 54:19–48.

Slatkin, M. 1993. Isolation by distance in equilibrium and non-equilibrium populations. Evolution 47:264–279.

Slatkin, M. A. 1995. A measure of population subdivision based on microsatellite allele frequencies. Genetics 139:457–462.

Taylor, E. B. 1991. A review of local adaptation in Salmonidae, with particular reference to Pacific and Atlantic salmon. Aquaculture 98:185–207.

Templeton, A. R. 1986. Coadaptation and outbreeding depression. Pages 105-116 *in* M. E. Soulé, editor. Conservation biology: the science of scarcity and diversity. Sinauer Associates, Sunderland, Massachusetts.

Templeton, A. R. 1998. Nested clade analyses of phylogeographic data: testing hypotheses about gene flow and population history. Molecular Ecology 7:381–397.

Templeton, A. R., and N. J. Georgiadis. 1996. A landscape approach to conservation genetics: conserving evolutionary processes in the African Bovidae. Pages 398–430 *in* J. C. Avise and J. L. Hamrick, editors. Conservation genetics: case histories from nature. Chapman and Hall, New York.

Templeton, A. R., E. Routman, and C. A. Phillips. 1995. Separating population structure from population history: a cladistic analysis of the geographical distribution of mitochondrial DNA haplotypes in the tiger salamander, *Ambystoma tigrinum*. Genetics 146:767–782.

Thrower, F. P., and W. W. Smoker. 1995. Homing propensity in transplanted native chum salmon. Pages 575–576 *in* H. L. Schramm, Jr., and R. G. Piper, editors. Uses and effects of cultured fishes in aquatic ecosystems. American Fisheries Society, Symposium 15, Bethesda, Maryland.

Utter, F., G. Milner, G. Ståhl, and D. Teel. 1989. Genetic population structure of chinook salmon, *Oncorhynchus tshawytscha*, in the Pacific Northwest. U.S. National Marine Fisheries Service Fishery Bulletin 87:239–264.

Varnavskaya, N. V., C. C. Wood, and R. J. Everett. 1994. Genetic variation in sockeye salmon *(Oncorhynchus nerka)* populations in Asia and North America. Canadian Journal of Fisheries and Aquatic Sciences 51 (Supplement 1):132–146.

Waples, R. S. 1987. A multispecies approach to the analysis of gene flow in marine shore species. Evolution 41:385–400.

Waples, R. S. 1989. Temporal stability of allele frequencies: testing the right hypothesis. Evolution 43:1236–1251.

Waples, R. S. 1998. Separating the wheat from the chaff: patterns of genetic differentiation in high gene flow species. Journal of Heredity 89:438–450.

Waples, R. S., and R. H. Rosenblatt. 1987. Patterns of larval drift in southern California marine shore fishes inferred from allozyme data. U.S. National Marine Fisheries Service Fishery Bulletin 85:1–11.

Wehrhahn, C. F., and R. Powell. 1987. Electrophoretic variation, regional differences, and gene flow in coho salmon *(Oncorhynchus kisutch)* of southern British Columbia. Canadian Journal of Fisheries and Aquatic Sciences 44:822–831.

Weir, B. S., and C. C. Cockerham. 1984. Estimating *F*-statistics for the analysis of population structure. Evolution 38:1358–1370.

Weir, B. S. 1996. Genetic data analysis II. Sinauer Associates, Sunderland, Massachusetts.

Wenburg, J. K., P. Bentzen, and C. J. Foote. 1998. Microsatellite analysis of population genetic structure in an endangered salmonid: the coastal cutthroat trout *(Oncorhynchus clarki clarki).* Molecular Evolution 7:733–749.

Williams, G. C., R. K. Koehn, and J. B. Mitton. 1973. Genetic differentiation without isolation in the American eel, *Anguilla rostrata.* Evolution 27:192–204.

Wilmot, R. L., and five coauthors. 1994. Genetic stock structure of western Alaska chum salmon and a comparison with Russian Far East stocks. Canadian Journal of Fisheries and Aquatic Sciences 51 (Supplement 1):84–94.

Winans, G. A., P. B. Aebersold, S. Urawa, and N. V. Varnavskaya. 1994. Determining the continent of origin of chum salmon *(Oncorhynchus keta)* using genetic stock identification techniques: status of allozyme baseline in Asia. Canadian Journal of Fisheries and Aquatic Sciences 51 (Supplement 1):95–113.

Wood, C. C., B. E. Riddell, D. T. Rutherford, and R. E. Withler. 1994. Biochemical genetic survey of sockeye salmon *(Oncorhynchus nerka)* in Canada. Canadian Journal of Fisheries and Aquatic Sciences 51 (Supplement 1):114–131.

Wright, S. 1943. Isolation by distance. Genetics 28:114–138.

Wright, S. 1978. Evolution and the genetics of populations: a treatise in four volumes. Volume 4, Variability within and among natural populations. The University of Chicago Press, Chicago.

Zhivotovsky, L. A. 2001. Estimating divergence time with use of microsatellite genetic distances: impacts of population growth and gene flow. Molecular Biology and Evolution 18:700–709.

Zhivotovsky, L. A., and F. B. Christiansen. 1995. The selection barrier between populations subject to stabilizing selection. Evolution 49:490–501.

Zhivotovsky, L. A., and M. Feldman. 1993. Heterogeneous selection in subdivided populations. Journal of Mathematical Biology 31:747–759.

Zhivotovsky, L. A., and M. W. Feldman. 1995. Microsatellite variability and genetic distances. Proceedings of the National Academy of Sciences USA 92:11549–11552.

Zhivotovsky, L. A., A. J. Gharrett, A. J. McGregor, M. K. Glubokovsky, and M. W. Feldman. 1994. Gene differentiation in Pacific salmon (*Oncorhynchus* sp.): facts and models with reference to pink salmon *(O. gorbuscha).* Canadian Journal of Fisheries and Aquatic Sciences 51 (Supplement 1):223–232.

Zhivotovsky, L. A., and 12 coauthors. 1989. Genetic differentiation of pink salmon. Genetika 25:1261–1274 (English translation).

8

Natural Selection

Eric Hallerman

8.1 Introduction

Natural selection, the differential perpetuation of genes and gene complexes (Dobzhansky 1970:97), is the evolutionary force mediating adaptation of populations to their environment. Charles Darwin and A. Russell Wallace deduced natural selection from Malthus's principle that populations increase in numbers at a geometric rate, sooner or later facing limitations of key resources. Only a part, perhaps only a small part, of progeny survive. Those carrying better-adapted genotypes survive more often than do those carrying other genotypes. The frequencies of the better-adapted genotypes will increase, and those of other genotypes will decrease, from generation to generation. Yet, natural selection can occur even if all progeny survive in some generations. In principle, differential reproductive success is as powerful a selective agent as is differential survival or mortality.

To understand the action of natural selection, a working understanding of the concept of fitness is needed. **Fitness,** often denoted as W, is the ability to survive to reproductive age and leave viable offspring. The viability and reproductive components of fitness may be defined in either absolute or relative terms.

— **Absolute viability** is the probability of an individual surviving from fertilization to maturity, which ranges from zero to one. The related concept of **relative viability** is the absolute viability of a particular genotype or phenotype standardized relative to that of the fittest genotype or phenotype, which is assigned the value of 1.00. For example, the relative viability of a particular genotype may be 0.90 relative to that of the fittest genotype.

— Reproductive fitness also may be quantified either in absolute or relative terms. However, because **reproductive success,** as defined by the number of viable offspring parented, can vary from zero to an undefined upper bound, the theory and associated modeling become more complex.

Population and evolutionary geneticists frequently are concerned with estimating the relative fitness of individuals expressing different genotypes (Section 8.3).

8.2 Types of Selection

Natural selection is a common name for several distinct processes. The types of selection and their effects upon genetic variation are major topics for both theoretical and experimental studies in population genetics.

8.2.1 Modes of Selection

Three modes of selection (Figure 8.1) have been described in the population genetics literature.

Directional selection favors phenotypes *at one end* of the population's distribution (Figure 8.1A). It changes the **adaptive norm** of the population; that is, it changes the mean phenotypic value for the trait at issue. The clearest example of directional selection in aquatic organisms comes from artificial selection (Chapter 12), for example, as practiced to increase growth rate in cultured fishes. Only the fastest-growing individuals are retained as broodstock; assuming that growth rate is heritable, the mean growth rate of progeny of the selected broodstock will exceed that of the progeny of randomly mated individuals from the same stock. Hence, the adaptive norm has been changed. As an example of directional selection in natural populations of aquatic organisms, a cogent argument can be made that size-selective fisheries have changed the size at maturity in highly exploited stocks (Box 8.1). Directional selection enhances the frequencies of some alleles or allele combinations and reduces the frequencies of others (Dobzhansky 1979:96), generally reducing the genetic variation in a population.

Stabilizing selection, sometimes termed **normalizing selection,** favors *intermediate* phenotypes (Figure 8.1B). There are many examples of stabilizing selection in nature. Clutch size in birds is subject to stabilizing selection because of the trade-off between the number of young hatched and the likelihood of an individual surviving to fledge. Birth weight in humans clearly is subject to stabilizing selection; lower birth weight directly affects the likelihood that the child will survive, whereas higher birth weight affects the likelihood that the mother will survive to raise the child. A number of ecological factors render the timing of a run in a population of salmon subject to stabilizing selection. To deposit gametes too early is to risk superimposition of other redds upon one's own, and the young may emerge before suitable food resources are available the following spring. To deposit gametes too late will put the young at a competitive disadvantage for exploiting food resources upon emergence. Further, for purposes of choosing the fittest mate, the number of potential mates is greatest in the middle of the run. Stabilizing selection is believed by many population geneticists to be the form of selection most commonly operating in nature. Several forms of selection are subsumed under

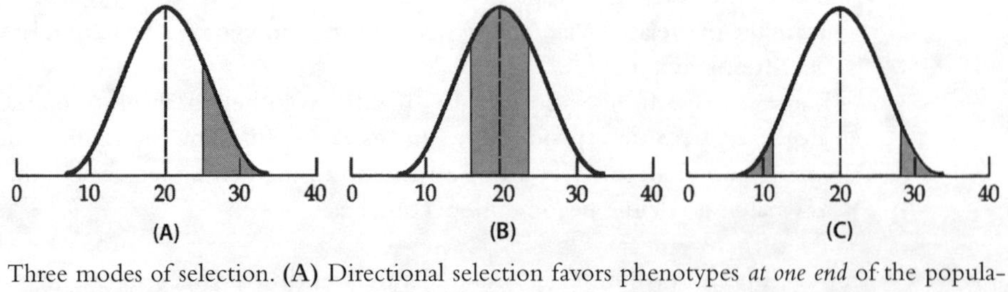

Figure 8.1 Three modes of selection. **(A)** Directional selection favors phenotypes *at one end* of the population's distribution. **(B)** Stabilizing selection favors *intermediate* phenotypes. **(C)** Disruptive selection favors extreme phenotypes *at both ends* of the distribution.

■ Box 8.1
Harvest as a Selective Force

Loren M. Miller

The hypothesis that harvest acts as a selective force has intuitive appeal. It is known that many forms of exploitation preferentially capture fish with certain characteristics. For example, most commercial fishing gear selectively catches fish according to size. Other traits on which fishing may impose selection include age and size at maturity, spawning time, and behavior. In general, any trait that affects the catchability of fish and has a heritable component may be selected upon by human harvest. Traits that are genetically correlated with those selected by a fishery also may be changed. For example, if size and fecundity are positively correlated, selection against large size will also decrease fecundity.

Despite the widespread belief that fisheries may be imposing selection, there are few documented cases. Both a selection differential (that is, a difference between the average quantitative character in the population as a whole and the average among those selected, Chapter 12) and an estimate of heritability (the proportion of phenotypic variance that has a genetic basis, Chapter 12) that correspond to observed degrees of change must be obtained to establish selection definitively as the cause of changes in a trait. Nelson and Soulé (1987) suggested that the lack of evidence for harvest-mediated selection is due to the difficulty of documenting selection. Therefore, the few defensible examples may be just the tip of the iceberg.

Studies by William Ricker and colleagues in British Columbia often are cited as evidence of harvest-mediated selection on fish size. They examined historical commercial fishing records for several species of Pacific salmon *Oncorhynchus* spp. (summarized in Ricker 1981). The most straightforward examination was for pink salmon *Oncorhynchus gorbuscha* (Ricker et al. 1978), which does not have overlapping generations. They concluded that a decrease in average individual size of as much as 34% from 1951 to 1975 was due to size-selective fishing gear leaving smaller fish to spawn each season. Estimates of heritability, h^2, were obtained by applying the equation $h^2 = R/S$. The response to selection, R, was observed as declines in mean individual size of fish each generation. Selection differentials, S, were estimated by comparing the catch to selective gill nets and nonselective seines. They obtained reasonable estimates of h^2 ranging from 0.22 to 0.40.

The conclusions of Ricker and associates have been questioned. Healy (1986) concluded that the Ricker studies did not definitively document the effect of selection because many of the observed declines were not sustained after 1975, even though size-selective fishing continued. Other factors, such as changes in oceanographic conditions or increased competition within or between species, could have accounted for the observed trends. Nelson and Soulé (1987) also noted that interpretation of the data set could not rule out interpopulation selection as opposed in intrapopulation selection. If different populations have distinct growth rates, then harvest of larger fish may cause the replacement of a mixed-population group of fish by a single, slower-growing population that has not undergone genetic change.

continued on next page

■ Box 8.1 *continued*

The Ricker studies were weakened, in part, by the use of historical records that were not collected to test their hypotheses. McAllister et al. (1992), therefore, proposed a large-scale experimental design to test for a genetic effect of size-selective fishing on pink salmon. They determined that h^2 could be estimated with reasonable statistical power by using an experimental design with temporal and spatial replication (see figure below). Pink salmon all mature at 2 years of age. Because of this life history (1) 2 years are required to estimate the effect of one generation of selection, and (2) one stream may provide two distinct populations, an odd- and an even-year stock. Three to six stocks would be selectively fished and an additional matching number of stocks would be nonselectively fished as controls for natural variation. Simulations showed that continuing this experiment for 5 to 10 years would provide sufficient statistical power to estimate h^2 with small standard error. Estimating h^2 is important because it, with a measurable selection differential, can be used to predict the genetic contribution to changes in mean individual fish size.

The discrete generations maintained by pink salmon simplified the genetic analysis, but discrete generations are not typical of most species. Miller and Kapuscinski (1994) estimated selection differentials caused by gill-net fishing for a species with multiple age-classes. Because many fish species continue to grow throughout life, sizes of fish of different ages cannot be compared directly. Comparison of size among individuals at a common age is necessary. Relying on the measurement of scale annuli to age fish, the authors compared estimated lengths to a common annulus and found that a selection differential was imposed on the Red Lake, Minnesota, population of walleye *Stizostedion vitreum*. In spawning year 1955, a selection differential of −0.111 mm was imposed on the length to the third annulus, while in 1958, a selection differential of 0.042 mm was imposed. The authors suggested that changes in intensity and direction of selection were due to the nature of gill-net selectivity. It may be possible to combine this experimental design with that of McAllister et al.'s (1992) large-scale experimental design to estimate h^2 in species with multiple age-classes.

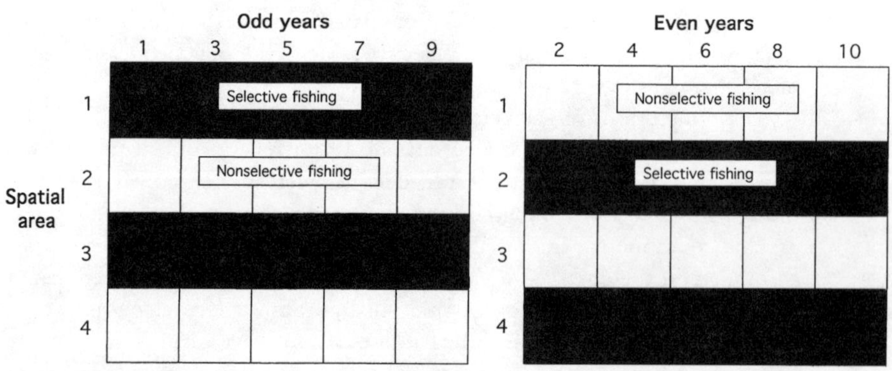

Diagram of experimental design to test for the effect of selection imposed by size-selective fishing of both odd- and even-year stocks of pink salmon. This design encompassed 10 years and four spatial areas. Figure adapted from McAllister et al. (1992).

the name stabilizing selection. A common feature, however, is that they maintain genetic heterogeneity or polymorphisms within a population at frequencies determined by selection (Dobzhansky 1979:96).

Disruptive selection favors extreme phenotypes *at both ends* of phenotypic distribution (Figure 1C). Aquatic organisms provide several good examples of disruptive selection. Young eastern oysters *Crassostrea virginica* are susceptible to juvenile oyster disease; however, it is the largest and smallest members of a population are most likely to survive. Many fishes for which males establish breeding territories exhibit alternative male life history strategies subject to disruptive selection (Box 8.2). For example, male salmon may return to spawn either as jacks or as full-size salmon. There is no fit intermediate alternative.

■ Box 8.2
Alternative Male Reproductive Strategies

For an individual to realize its fitness and contribute its genes to the next generation, it must both survive and reproduce. Survival rates to maturity are low for fishes, and males compete for mates. Also, individuals within a population of fish can exhibit considerable variation regarding age and size at maturity. Male mating success will, in many cases, be a function of body size, with large males favored in aggressive interactions to gain matings. Males adopt behavioral tactics to suit their situations. Large males will choose to fight, whereas small males will choose to sneak matings. If the behavioral decision is reversible, an individual may change tactics to suit a given situation. If the decision is irreversible, as in the case of precocious versus delayed maturity, the alternatives represent alternative male reproductive strategies (Gross 1991). Given the high stakes, survival or extinction of one's genes, it is not surprising that life history traits are subject to natural selection and that alternative male reproductive strategies have persisted in species facing uncertain selective regimes. The alternative reproductive strategies reflect different trade-offs between the likelihood of survival and of successful reproduction for males of different sizes. Consideration of alternative male reproductive strategies among species will illustrate the influence of disruptive, sexual, and frequency-dependent selection.

The life cycle of Pacific salmon is typified by that of coho salmon *Oncorhynchus kisutch* (Gross 1991). The young hatch and typically spend 1 year growing in freshwater before migrating to the sea. Male coho salmon exhibit two alternative reproductive strategies. **Hooknose males** usually mature at a large size after approximately 18 months in the ocean and defend reproductive territories. However, some males, known as **jacks,** mature after just 6 months at sea at a much smaller size than that attained by hooknose males. The physiological mechanism by which a juvenile commits irreversibly to a reproductive strategy is not entirely understood. However, the commitment seems to be made at the fry stage, with growth rate as a key determinant. Upon maturation, adults return to their natal streams to reproduce and die. In mate competition, a hooknose male will employ the behavioral tactic of fighting, whereas a jack will employ the tactic of sneaking. A jack will lurk beside a male guarding a female as she builds her redd. Upon release of

continued on next page

■ Box 8.2 *continued*

eggs, a jack, avoiding the attack of the large male, will intrude into the redd to release milt and gain fertilizations.

The expected fitness (in terms of lifetime reproductive success) from choosing either strategy will depend on two ecological factors. The jack life history offers a survivorship advantage over the hooknose life history because exposure to predation and other sources of mortality in the ocean is reduced by 1 year. The greater the survivorship advantage, the greater the relative advantage of the jack life history to a juvenile. The second factor is breeding success, as determined by competition within the behavioral morphs. The more juveniles that become jacks, the greater the competition among them, and thus, their mean breeding success declines (Gross 1984). On the other hand, as more juveniles become hooknose males, there will be greater competition among fighters. Hence, each reproductive strategy is more successful when rare, and the relative fitness of the jack and hooknose life histories is frequency dependent. An equilibrium frequency among the two life histories in a given population will reflect the trade-off between survivorship and reproductive success among the morphs in the population (see panel A in figure).

Although the reproductive strategy for an individual male is irreversible, there is considerable flexibility at the population level in response to shifts in ecological conditions. Changes in relevant ecological conditions will elicit a response at the population level in the frequency of males returning as jacks. For example, commercial fisheries with size-selective gear result in selective harvest of hooknose males. With the intensification of the fishery, the likelihood for survival as a parental male decreases, and selection increasingly favors early maturation and reproduction as a jack (panel B). The change in the selective regime, in this case, imposition of directional selection, has led to upward shifts in the frequency of early maturing males in many exploited coho salmon populations. Changes in selective equilibria among hooknose and jack life histories also are anticipated from human-mediated changes in the breeding and freshwater rearing environments of coho salmon (Gross 1991).

Other fishes also exhibit alternative male reproductive strategies (Gould and Gould 1989) that are responsive to the forces of disruptive, sexual, and frequency-dependent selection. Atlantic salmon *Salmo salar* exhibit an alternative reproductive strategy analogous to the jack strategy of Pacific salmonids; **grilse** are males that return to the spawning grounds after only 1 year at sea. However, Atlantic salmon exhibit an additional alternative male reproductive strategy. **Precocious parr** are males that never migrate to sea, instead wintering in their freshwater stream and becoming reproductively mature. These parr have testes that constitute 20% of body weight, as opposed to 4% in **parental males**, that is, full-sized males that establish and defend breeding territories. Precocious parr enter unnoticed into the redd of a spawning pair, shed milt, and frequently gain fertilizations. In some European rivers, male Atlantic salmon are nearly 100% precocious parr (Gould and Gould 1989).

Another system of alternative male reproductive strategies has been observed in the bluegill *Lepomis macrochirus* (Gould and Gould 1989). In northern populations, parental males begin building nests at age 7, competing actively with other males for the best sites and for access to females. Females are attracted to the nest. After a

continued on next page

■ Box 8.2 *continued*

courtship display, a female will release her eggs while the male fertilizes them. He will guard the eggs and yolk sac fry until they disperse after swim-up. There are two alternatives to the parental male strategy. Commitment to an alternative strategy is irreversible, although males may move from the sneaker to the satellite strategy as they grow. About 20% of the males become committed to sneaking fertilizations (via either alternative strategy) at the nests of parental males. Males adopting the alternative reproductive strategy mature at age 2 instead of age 7 and might reproduce for as many as five seasons, a span of reproductive potential much like that of the parental males. Small **sneaker males** hide near an active nest and wait for a pair to start spawning. When the eggs are released, they dart into the nest and release gametes, obtaining fertilizations. Once a sneaker male becomes too large to hide effectively, he may become a **satellite male.** Expressing the coloration pattern and outward behavior of a female, a satellite will wait near the water surface near an active nest. When a spawning female begins to release eggs, the satellite will enter the nest. Instead of releasing eggs, of course, the satellite male will release milt. The genetic basis by which male bluegill enter one reproductive strategy or the other is not known. However, parental males tend to father more parental males than do nonparental males, and nonparental males father more nonparental males than do parental males.

A satellite strategy also has been observed in oyster toadfish *Opsanus tau* (Bass 1996). Males establish territories and tend nests. Satellites sneak fertilizations as in bluegill, with satellite males imitating females by expressing coloration and making vocalizations like those of females.

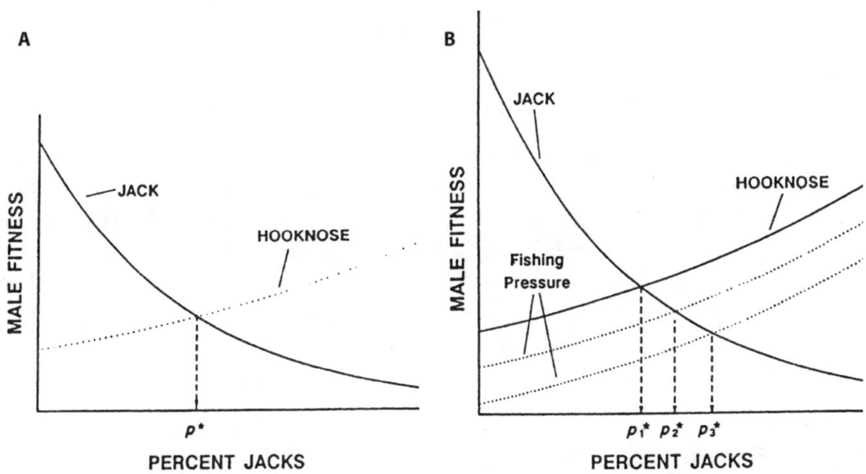

(A) The relative fitness of hooknose and jack male coho salmon is determined by the probability of their survival to return to the spawning grounds times their reproductive success, which is frequency dependent. The relative fitness of the two tactics is equal at $p^\star$, and the frequency of jacks in a population of stable size should reach equilibrium there. **(B)** Selective harvest of hooknose males decreases their relative fitness and favors an evolutionary increase in the frequency of jacks. With no harvest, $p_1^\star$ represents the equilibrium; $p_2^\star$ and $p_3^\star$ represent equilibria at increasing levels of size-selective harvest. Figure adapted from Gross (1991).

8.2.2 Frequency-Dependent Selection

In **frequency-dependent selection,** the frequency of a phenotype determines the fitness of the carrier. Among ecological mechanisms giving rise to frequency-dependent selection in nature is the selective effect of predation on a prey species. The most frequently cited example involves frequencies of banding coloration morphs in the land snail *Cepaea nemoralis* (Clarke 1962). Bird predators develop a search image for common morphs and take high proportions of individuals from the common morphs and only low proportions from the rare ones. Rarer types enjoy a selective advantage until they become sufficiently common that predators switch search images and begin to prey selectively on them. The coexistence of distinct morphs and the cycling of phenotypic frequencies in populations of land snails are maintained in this way.

As an ecological counterpoint, consider the frequency-dependent selective effect caused by prey species on their predator. Scale-eating cichlid fish of the genus *Perissodus* exhibit a mouth torqued to one side (Figure 8.2A). These fish attack their prey from behind, snatching scales from the victim's side; the mouth is skewed to one side to favor taking of scales from one side of prey. The direction of the mouth opening in *P. microlepis* is determined by a one locus, two allele system (Hori 1993). The relative frequencies of the left- and right-opening mouth morphs fluctuate around unity (Figure 8.2B), effected by frequency-dependent selection exerted by prey alertness. As prey become wary to attack from a particular side, the associated mouth morph is selected against, and the alternative morph becomes favored.

Another mechanism giving rise to frequency-dependent selection is the **rare male advantage,** in which females select as their mates males of an uncommon phenotype. For example, for *Drosophila melanogaster* (Petit 1958) and *D. pseudoobscura* (Ehrman et al. 1965), when two types of males are present in a confined space, the rare type has an advantage in breeding. Similarly, individuals expressing less common alternative male reproductive strategies are likely to enjoy a fitness advantage (Box 8.2).

8.2.3 Hard and Soft Selection

Wallace (1981) recognized distinctions between what he termed hard and soft selection. In **hard selection,** phenotype is the absolute criterion determining the fitness of an individual, independent of other environmental or genetic factors. An example of hard selection is the expression of **lethal genes,** that is, genes whose expression in a homozygote results in the death of the carrier. In **soft selection,** the fitness of a phenotype is related to kind, frequency, and distribution of phenotypes in the population. Examples of soft selection are those imposed by ecologically mediated mechanisms, such as frequency-dependent selection. Most selection in nature is of the "soft" type.

8.2.4 Sexual Selection

Among animals with mating systems involving choice of mates, sexual selection frequently has given rise to **sexual dimorphism,** whereby the sexes differ in size or morphology, or **sexual dichromatism,** whereby the sexes differ in coloration. Under **sexual selection,** a mechanism first proposed by Charles Darwin, there is competition among males for mates, and characters enhancing the reproductive success of the carriers are perpetuated irrespective of their survival value (King and Stansfield 1990). Sexual selection often is subdivided into intrasexual and epigamic selection.

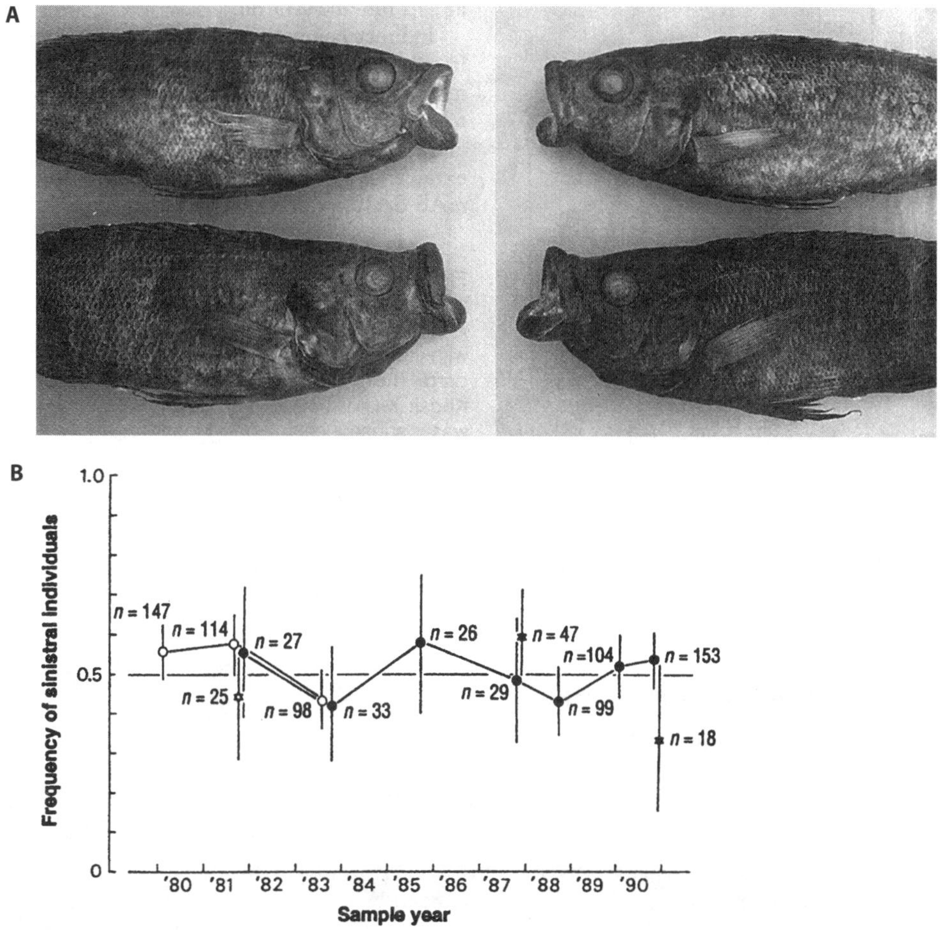

Figure 8.2 (A) Handedness of mouth opening in the Lake Tanganyikan cichlid fish *Perissodus microlepis* exhibiting mouths torqued to the right (above) and left (below). **(B)** Frequencies of the left- and right-torqued mouth morphs in two populations of *P. microlepis* (Luhanga coast, open circles; Bemba coast, closed circles). The relative frequencies of the two morphs fluctuate around unity, mediated by frequency-dependent selection exerted by prey alertness. Reprinted with permission from Hori, M. Frequency-dependent natural selection in the handedness of scale-eating cichlid fish. Science 260:216–219. Copyright 1993. American Association for the Advancement of Science.

Intrasexual selection is the mechanism described by Darwin, and it involves adaptation that confers the ability of a male to dominate other males in battle. In egg-laying fishes in which the males are territorial during the breeding season, such as salmonids, the males often are larger than the females (Moyle and Cech 1996:121). An example of a morphological adaptation in response to intrasexual selection is the highly modified kype, or hooked mouth, of male salmon, an adaptation for fighting. A more common example, found in at least 25 families of fishes (Wiley and Collette 1970), is the expression of breeding tubercles and contact organs during the breeding season. These keratinized structures are found on the male body parts that are likely to come in contact with females or other males and assist the male in courtship or in defense of territory and nest.

Although intrasexual selection classically is regarded as occurring among males (Box 8.2), it also has been shown to occur among females. Observing breeding competition in the wild, van den Berge and Gross (1989) demonstrated intrasexual selection among female coho salmon, quantifying selection on body size and kype size. Body size contributed to female fitness through (1) increased initial biomass of egg production, (2) ability to acquire a high-value territory for egg incubation, and (3) success in nest defense. These factors together resulted in as much as a 23-fold fitness advantage for the largest females in the population. Fleming and Gross (1989) examined morphology, investment in egg production, and breeding competition in 13 wild and 5 hatchery populations. Female secondary sexual characters, including body coloration and length of kype, increased significantly with the degree of breeding competition within wild populations. Egg size and investment in egg production decreased as competition increased, probably as a result of life history trade-offs. In hatcheries, where breeding competition was relaxed, characters associated with breeding competition and spawning performance were reduced and egg size was increased.

Under **epigamic selection,** the female is the selective agent, choosing among a field of genetically variable males of her species, for example, on the basis of the elaborate courtship displays in certain birds. Among fishes, examples of epigamic selection are provided by female guppies *Poecilia reticulata* and African cichlids (Box 8.3) choosing their mates on the basis of body and tail coloration.

8.2.5 Kin Selection

Altruism is the expression of behaviors that increase the fitness of other individuals at the expense of one's own fitness. There are many well-accepted examples of altruism in many taxa in the animal kingdom. Social insects provide the most extreme example; in many species, one individual female in a colony will contribute all the reproduction through her reproductive lifetime, and the workers will support this reproductive effort altruistically. In a number of territorial birds, nonbreeding, "helper" individuals will attend the nest of a breeding pair, assisting in the rearing of the nestlings. Many animals living in groups will provide the group with a warning signal when a predator or other danger is sighted, for example, the beaver slapping the surface of the water with its tail on the approach of a human. Given that every individual in a population would want to realize his or her own fitness, that resources are limited, and that individuals in a population frequently are in a competitive relationship, how would evolutionary theory explain altruism?

Resolution of the apparent conundrum starts from the observation that relatives have genes in common, that is, alleles that are identical-by-descent from a common ancestor. Altruistic behaviors are favored by selection because the altruist is increasing his or her **inclusive fitness.** That is, the altruist increases the likelihood of survival or reproduction of alleles carried by relatives as opposed to those same alleles carried by his or her self. The "altruistic allele" can increase in frequency if the benefit in terms of increased fitness of others bearing one's genes offsets the cost to one's own fitness. For example, if helper birds are last year's offspring of the same parental birds at the nest, they share half their genes with the current nestlings, the same relatedness they would have with their own young. The helpers, therefore, are increasing their inclusive fitness by assisting in rearing. I know of no examples of altruism in fishes. A complete and colorful exploration of altruism is presented by Dawkins (1976).

■ Box 8.3
Disruptive Selection, Sexual Selection, and Radiation of the Lake Victoria Cichlids

The explosive radiation of cichlid fishes in the African Great Lakes has long intrigued evolutionary biologists. How can the rapid diversification and speciation of halplochromine cichlids be explained in terms of ecological and evolutionary processes? Galis and Metz (1998) reviewed findings suggesting that sexual and disruptive selection are key forces underlying the diversity of the cichlid assemblage of Lake Victoria.

Interest in evolution of the Lake Victoria cichlid community intensified when geological data recently revealed that the lake was completely dry during the most recent Ice Age (Johnson et al. 1996). This suggests that the approximately 500 haplochromine cichlid species there must have evolved within the last approximately 12,400 years, perhaps from a single ancestral species. An old hypothesis—that allopatric speciation had occurred in temporarily isolated satellite lakes around Lake Victoria—has fallen out of favor. Seehausen et al. (1997) hypothesized that sexual selection could be the driving force behind speciation of Lake Victoria haplochromines. Females of a given species or morph have a strong preference for males of a particular color. Mate choice of females for differently colored males maintains reproductive isolation between sympatric species and color morphs. Males of sympatric, closely related species always differ in color, one species having blue males, the other red or yellow males. The dichotomy of blue versus red or yellow males also holds for sympatric, conspecific color morphs. Females usually are inconspicuously colored. Changes in coloration of the males happen relatively frequently. Evidence for the importance of sexual selection for the maintenance of reproductive isolation is provided by behavioral experiments in monochromatic light, which masks color differences. Other evidence comes from the observation that increased turbidity in Lake Victoria is causing a breakdown in reproductive barriers and heightened frequency of hybridization in parts of the lake.

Although the sexual selection hypothesis explains how new evolutionary lineages arise, it does not explain how they are maintained. Niche differentiation will be needed for a large number of new lineages to coexist and persist. Liem (1973) argued that morphological changes in the pharyngeal apparatus have great evolutionary importance in the adaptive radiation of cichlids. The pharyngeal apparatus of cichlids has a large number of independent elements that can easily be modified. Rather small morphological and behavioral changes allow cichlids to specialize on different food items. Disruptive selection on feeding specializations promotes the divergence of incipient species, and the resulting niche shifts promote their coexistence. This line of argument is supported by the observation that sibling species are characterized by differences in feeding morphology and behavior.

Hence, the adaptive radiation of Lake Victoria cichlids has taken place because of two selective processes acting together. Species are splitting repeatedly due to sexual selection, and adaptive radiation is occurring due to disruptive selection on feeding apparatus.

■ 8.3 Demonstration and Measurement of Selection

Many data sets have been collected on the occurrence and frequencies of genetic polymorphisms, but only a subset directly or indirectly offer insight on the adaptive significance of the polymorphism at issue, that is, on its relationship to natural selection. The adaptive significance of genetic variation is a major topic of discussion for population and evolutionary geneticists (Koehn and Hilbish 1987).

Dobzhansky et al. (1966 and references therein, summarized in Dobzhansky 1970:134–137) were among the first to relate genetic polymorphism to selective processes, in their case, driven by ecological and climatic variation. Examining chromosomal variation in *Drosophila pseudoobscura,* they tracked the frequencies of the Arrowhead, Chiracahua, and standard karyotypes. They observed regular and cyclical variability in frequencies of the karyotypes, paralleling the changes of seasons. They inferred that changes in karyotypic frequencies reflected the changing fitness of genotypes with environmental factors.

There are numerous publications in the scientific literature demonstrating changes of allele frequencies that parallel variation in environmental factors. Among aquatic organisms, examples include variation at esterase loci of barnacles *Balanus amphitrite* (Nevo et al. 1977) and of suckers (family Catostomidae, Koehn and Rasmussen 1967) that paralleled variation in environmental temperature regimes. The inference reached in these and many studies of similar design is that the pattern of allelic variation at the loci screened was the consequence of natural selection. However, it is difficult to prove that the environmental parameter measured drove natural selection at the marker locus. Among the questions that could be posed are, did selection indeed operate at this locus or at a tightly linked locus, and, could other mechanisms have given rise to the observed pattern of genetic variation? These considerations, in turn, raise the question, how is selection demonstrated?

A classical approach to demonstrating selection involves testing of a one-locus model; the aim of the experiment is to attach a fitness value to each possible genotype at the locus at issue. After measuring allele frequencies at the locus in the parental generation, the Hardy–Weinberg model (Chapter 1) is used to determine the expected genotypic frequencies among the progeny. After observing genotype frequencies among the progeny after selection presumably occurred, departures from Hardy–Weinberg expectation are attributed to the effect of selection. Alternatively, variation at a locus among populations can be attributed to the effects of differential selective pressures operating over the range of the species. Though perhaps simplistic, these approaches have been applied frequently to document selection and estimate its intensity. Especially when combined with data gathered using model systems under controlled though simplified conditions, the findings can lead to insight on processes that apply in ecological contexts.

Demonstration and measurement of selection in natural populations is a complex undertaking. The single-locus approach does not reflect the complexity of adaptation in the real world. There are many fitness-related loci relevant to most aspects of ecological adaptation, and variation at such loci is subject to both selective and nonselective forces (e.g., mutation, migration, and drift). Hence, there is a need for corroborating evidence, hopefully showing the mechanism by which selection acted on the genotypes at the locus or loci at issue. However, in the words of Koehn and Hilbish (1987), "It has been frustratingly difficult to identify the specific aspects of the environment that may be acting as a source of natural selection." Hence, definitive demonstration and quantification of selective forces in nature pose a difficult line of research.

Because of the importance of selection as an adaptive process, much effort has been expended in characterizing its action. The next two sections explore the effects of natural selection within the context of the adaptive significance of polymorphisms considered over either time or space.

8.3.1 Changes in Allele Frequency over Time

8.3.1.1 **Models of Selection at a Locus.** Although selection operates on phenotypes, we must consider the genotypes at a locus in terms of how their expression gives rise to the relevant phenotype. In the absence of selective forces, the Hardy–Weinberg model predicts that carriers of genotypes *AA*, *Aa*, and *aa* contribute, on average, equal numbers of gametes to the gene pool of the next generation (Dobzhansky 1970: 101–102). Should selection be operative, however, there will be differential survival or reproductive success among the *AA* and *aa* genotypes. Suppose the carriers of *AA* genotypes contribute 100 gametes to the gene pool of the next generation while carriers of *aa* genotypes contribute only 90. The relative fitness of the *AA* genotype, W_{AA}, by convention is taken to equal 1. The relative fitness of the *aa* genotype, W_{aa}, equals $1 - s$, where s is the **selection coefficient,** that is, the degree of selection against the homozygous *aa* genotype. In the above example, s equals 0.10. Measurement of the intensity of selection entails estimation of the selection coefficient.

Because selection operates on phenotypes, we must consider how phenotype relates to the mode of gene expression, that is, to the dominance relationships among alleles at a locus. Dominance effects determine the phenotype upon which selection will operate, in other words, which genotypes will be selected for, which against, and at what intensity. Dominance of one allele over another is not always complete. Hence, we must consider the degree of dominance of an allele, denoted as h. The degree of dominance of an allele determines the fitness of the heterozygote. Several cases can be distinguished: no dominance (W_{Aa} is the midpoint between W_{AA} and W_{aa}, often termed an **additive gene effect**), **partial dominance** (W_{Aa} is above the midpoint between W_{AA} and W_{aa}), **complete dominance** (W_{Aa} is equal to W_{AA}), and **overdominance** (W_{Aa} is greater than W_{AA}).

To model the effect of selection at a locus, let's assume that there are two alleles at the locus, *A* and *a*, that *A* is dominant to *a* with a coefficient of dominance equal to h, and that the *AA* genotype is the most fit. The frequency of *A* is denoted p and the frequency of *a* is denoted q. Under the Hardy–Weinberg model, the fitness and frequencies of each genotype before and after selection are as follow (Dobzhansky 1970:103).

Genotype	*AA*	*Aa*	*aa*
Fitness	1	$1 - hs$	$1 - s$
Initial frequency	p^2	$2pq$	q^2
Frequency after selection	p^2	$2pq(1 - hs)$	$q^2(1 - hs)$

The rate of change of allele frequencies also can be modeled as a function of dominance effects and the current allele frequencies at the locus. In terms of dominance effects, three cases can be distinguished: (1) *A* is the favored, completely dominant allele ($h = 0$), (2) *A* is the favored allele with an additive gene effect relative to an h equal to 0.5, and (3) *A* is the favored, recessive allele ($h = 1$). The frequencies of the respective genotypes and the phenotypes upon which selection will act are functions of the current allele frequency. The interacting effects of dominance effects and current allele fre-

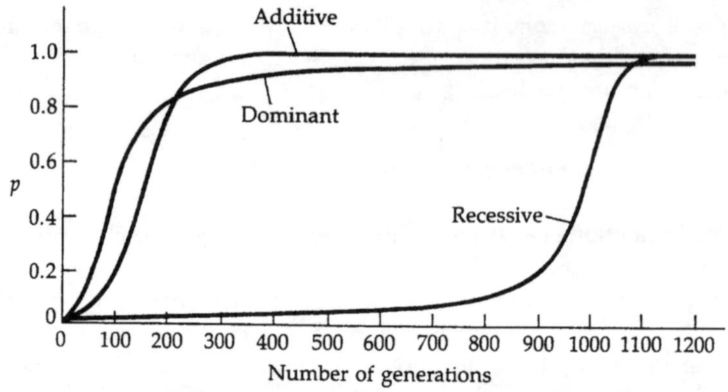

Figure 8.3 The rate of change of allele frequencies *(p)* as a function dominance effects and current allele frequencies for cases where *A* is the favored, dominant allele ($h = 0$), *A* is the favored allele with additive effect ($h = 0.5$), and *A* is the favored recessive allele ($h = 1$). The difference in relative fitness between the contrasting homozygous genotypes is assumed to be 5%. Figure adapted from Hartl (1988).

quency are shown in Figure 8.3 (Hartl 1988). Note that a favored dominant allele will increase rapidly in frequency when it is rare because selection will favor both homozygous and heterozygous genotypes; however, dominant alleles will increase in frequency only slowly when they are common, for the same reason. For a favored additive allele, the initial increase in frequency is slower than it is for a favored dominant allele; however, the additive allele will proceed more quickly to fixation because selection continues to distinguish between homozygous and heterozygous genotypes. Increase in the frequency of favored recessive alleles is slow when they are rare, as the recessive phenotype will be expressed only in the rare homozygote. As the recessive allele becomes more common, recessive phenotypes will become more common and progress toward fixation will be rapid.

At a practical level, the model above poses important implications for fisheries managers. Introduction of nonnative stocks can introduce new alleles or change allele frequencies in a locally adapted population. Should maladaptive alleles be introduced into a population, it will take many generations for selection to purge them from the population (Chapter 15). To the degree that allele frequencies are the product of local adaptation, changes may shift the population from a selective equilibrium.

8.3.1.2 **Models of Selective Equilibria at a Locus.** Certain models of the effects of selection predict that allele frequencies tend toward equilibrium values, which prove stable for generations. Indeed, persistence of allele frequencies at variable loci is observed more or less frequently in natural populations. Such findings support inference of the existence of **selective equilibria.** Three kinds of selective equilibria have been distinguished in terms of what happens when frequencies deviate from the equilibrium value. For a **stable equilibrium,** once displaced, an allele frequency converges once again to its equilibrium frequency. For an **unstable equilibrium,** once displaced, an allele frequency progresses away from its equilibrium value. With directional selection, that is, selection favoring one end of the phenotypic distribution, frequencies of alleles tend to become fixed at 0 or 1.

How, then, are selective equilibria produced and maintained? The occurrence of stable equilibria often is explained by selection for heterozygotes at given loci. The classic and often-repeated example is that of the alleles at the hemoglobin locus, where genotypes determine susceptibility to disease. The wild-type homozygote, ++, is malaria susceptible but not anemic. The alternative homozygote, *ss*, is resistant to malaria but suffers from sickle cell anemia. The heterozygote, +*s*, is favored because he or she is resistant to malaria and does not suffer from sickle cell anemia. Selection against heterozygotes gives rise to unstable equilibria. This model of selective equilibrium has not always proven satisfactory, for a number of reasons. Few good examples of heterozygote superiority have been demonstrated at particular loci. Further, fitness is determined by the action of *many* loci. Fitness of a given genotype varies over space and through time. Fitness not only of individuals but also of populations is at issue. Some population geneticists question how frequently or infrequently selective equilibria are achieved in nature. Hence, other models have been developed to explain selective equilibria, as detailed in next two sections.

8.3.2 Changes in Allele Frequency over Space

A **cline,** in the sense relevant to us, is a gradual spatial variation in allele frequencies exhibited by a species along a line of environmental or geographic transition. Clinal variation in allele frequencies occurs rather frequently in natural populations, including populations of aquatic organisms such as marine mussels (Gartner-Kepkay et al. 1980), fathead minnow *Pimephales promelas* (Merritt 1972), bluegill (Avise and Smith 1974), and largemouth bass *Micropterus salmoides* (Philipp et al. 1983). What causes and maintains these clines? Clinal variation in allele frequencies is the consequence of the dynamic, interacting effects of migration (Chapter 7) and selection.

A single-locus model for the establishment of clinal variation was developed by Nagylaki (1975). Selection intensity is allowed to vary by location. There are regions where a particular allele is advantageous or deleterious; these are separated by a belt where either allele is selectively neutral. The length of the cline is a consequence of the length of the driving environmental gradient and of the relative fitness of heterozygotes at the locus at issue. If heterozygotes are more fit than are homozygotes through a large portion of the relevant environmental gradient, there is a long cline. On the other hand, if heterozygotes are less fit than homozygotes, a narrow cline results. There is no cline if either migration is too strong or selection is too weak.

Place and Powers (1979) observed clinal variation in allozyme frequency at the *LDH-B** (L-lactate dehydrogenase, enzyme number 1.1.1.27; IUBMBNC 1992) locus in mummichog *Fundulus heteroclitus* along the Atlantic coast of North America (Figure 8.4). The *LDH-B*b* allele occurred at high frequencies in the north and the *LDH-B*a* allele in the south, with a cline of gradual transition in allele frequency from Georgia through Nova Scotia. The observation of clinal variation in allele frequencies suggests, but does not in itself prove, the interacting effects of migration and selection. Alternatively, clinal variation could have arisen from the dispersal of populations from two centers of distribution, for example, following dispersal from Pleistocene glacial refugia. Hence, corroborating lines of evidence are useful for supporting the hypothesis of selection as the driving force for the observed cline. In the case of the mummichog, corroborating evidence was provided by data on the thermal optima for catalytic activity of the respective allozymes. At a low pyruvate concentration and a pH below 8.0, the *LDH-B* b/b* allozyme exhibited a greater reaction rate at lower temperatures, such as 10°C. At temperatures above 25°C and pH values between 6.5 and 7.0, the *LDH-B*a/a* allozyme exhibited a higher catalytic rate. These observations suggest why selection might favor heightened frequencies of the respective allozymes at the northern and southern ends of

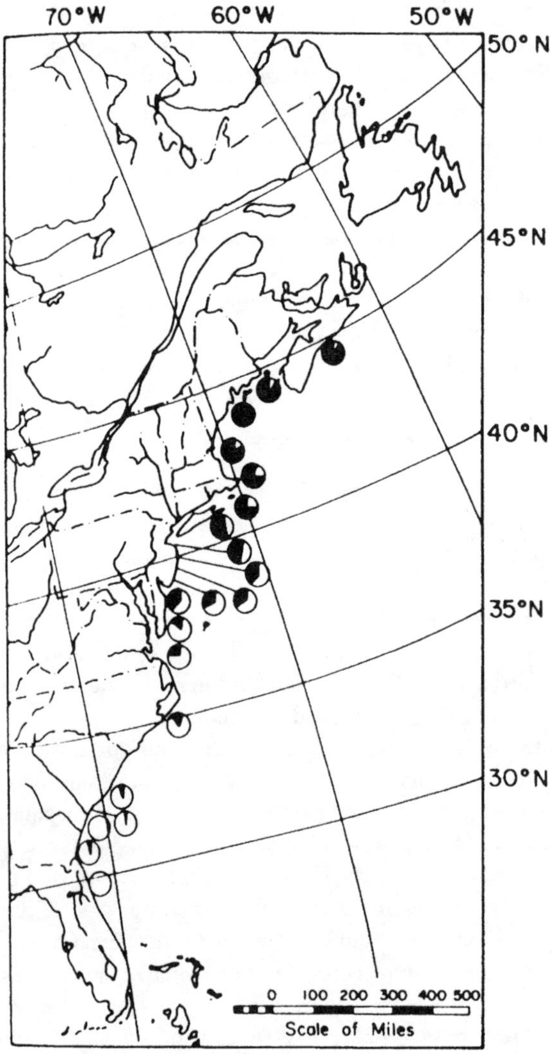

Figure 8.4 Clinal variation in allozyme frequency at the *LDH-B* locus in mummichog. Dark portions of the pie diagrams represent frequencies of the *LDH-B*b* allele and light parts those of the *LDH-B*a* allele. Figure adapted from Place and Powers (1979).

the species' distribution. Expression of both gene products would confer the ability to efficiently catalyze the biochemical reaction at issue through the broad range of environmental temperatures characterizing the center of the species' range. It can be inferred, then, that selection plays an important role in maintenance of this cline.

■ 8.4 Models of Selection at Many Loci and at the Population Level

To this point, we have considered selection in terms of the fitness of individuals. To the degree that we have considered selection at the population level, populations could be regarded simply as collections of individuals who did or did not survive or reproduce.

Fitness and the effects of selection, however, also can be considered at the population level (Wright 1982).

Let's begin by considering allelic variation at a particular fitness-related locus. The population, living as it does in an ecological setting, faces a changing environment. Hence, in many cases, the fitness of the population is low if it is fixed for one allele at the fitness-related locus, maximized at some intermediate allele frequency, and low if fixed for the alternative allele. This is easy to depict graphically in two dimensions: with allele frequency at the locus on the horizontal axis and fitness on the vertical axis, the relationship might appear like an upside-down U.

However, fitness is actually the consequence of allelic segregation at *many* loci. Extending our two-dimensional depiction of fitness at one locus to many loci results in an *n*-dimensional surface with pinnacles, pits, saddles, and other topographic features. To depict our analysis in an understandable fashion, let's limit our consideration to allelic variation at two fitness-related loci. Let's depict allele frequencies at fitness-related loci *A* and *B* as the two axes of the graph lying in the plane of the page (Figure 8.5A). We'll depict fitness as a third dimension going up from or down into the page, much as elevation is depicted on a conventional topographic map. This model for considering the fitness of populations is termed the **adaptive landscape.** Different combinations of allele frequencies at the respective loci relate to different levels of fitness for a population. Combinations of allele frequencies relating to high fitness are labeled with a plus sign and those relating to low fitness with a minus sign. The peaks on the adaptive landscape represent locally stable selective equilibria.

Evolutionary processes operating on the gene pool of a population change the population's position on the adaptive landscape. Natural selection acts upon the allele frequencies of a population, moving those frequencies toward a more optimal combination for that ecological context, increasing the population's fitness, and thereby moving the population toward the nearest peak on the adaptive landscape. However, the action of natural selection is such that a population may become "stuck" on a peak on the landscape, which may or may not be the highest peak on the landscape. To reach a higher peak, other evolutionary processes will have to act. For example, random genetic drift can affect the allele frequencies of a population in a nonadaptive sense, moving the population from a peak to a valley. From there, under the action of selection, the population can ascend another peak, perhaps a higher peak. The action of a variety of evolutionary processes on a population on the adaptive landscape is shown in Figure 8.5B. Let's consider the effect of evolutionary processes operating on a population that is situated on the encircled peak of the landscape in the upper left of Figure 8.5A. The dashed circle containing the plus sign represents the range of allele frequencies that the population may possess that are selectively optimal. The effects of different processes or combinations of processes on the composition of the gene pool of the population can be depicted by the position of the population on the landscape, the area of the landscape enclosing optimal allele frequencies, or the topography of the landscape itself.

— Under a regime of increased mutation pressure (large $4N\mu$) or reduced selection pressure (small $4Ns$), the selective equilibrium encompasses a broader range of possible allele frequencies at fitness-related loci. This is depicted in panel a of Figure 8.5B as a shaded area that is larger than that originally occupied by the population in Figure 8.5A.

— Under the opposite regime, with increased selective pressure (large $4Ns$) or reduced mutation pressure (small $4N\mu$), the selective equilibrium will be narrower–only a

Figure 8.5 Fitness of populations as depicted by Wright (1932, again presented by Provine 1989:284) in his shifting balance theory. **(A)** The adaptive landscape. The axes of the figure are allele frequencies at two fitness-related loci; the topographic dimension of the figure is fitness (+ represents high fitness, – represents low fitness). **(B)** The effects of given evolutionary processes operating on the gene pool of a population on that population's position on the adaptive landscape. **(a)** Increased mutation rate or reduced selection intensity; **(b)** increased selection intensity or reduced mutation rate; **(c)** qualitative change in the environment (effects of mutation and selection large); **(d)** close inbreeding (effects of mutation and selection small); **(e)** slight inbreeding (effects of mutation and selection medium); and **(f)** division into local demes (effects of migration medium). Figure adapted from Provine, W. B. 1999. Sewall Wright and evolutionary biology. University of Chicago Press, by permission of the University of Chicago.

narrow range of allele frequencies will prove stable. This is depicted in panel b of Figure 8.5B as a shaded area that is smaller than that originally occupied by the population in Figure 8.5A.

— Populations live and adapt to a changing environment. As the environment changes, so too do selective processes and consequent selective equilibria. Environmental change with associated changes in selective regime can be depicted as change in topography in the selective landscape. Provided that the population has the adaptive resources to respond to the environmental change (e.g., both $4N\mu$ and $4Ns$ are large), its array of allele frequencies at fitness-related loci will change such that the population will remain on a peak in the adaptive landscape (panel c, Figure 8.5B).

— Should the population undergo an episode of close inbreeding (when both $4N\mu$ and $4Ns$ are small), the genetic variability of the population will be decreased, as shown by a decrease in the shaded area of panel d in Figure 8.5B. The population's array of allele (and genotypic) frequencies will depart from optimality, and the population will sink from the peak to a lower area of the adaptive landscape.

— Should the population undergo an episode of light inbreeding ($4N\mu$ and $4Ns$ both medium), the diminution of genetic variability will be less severe, and the adaptive impact less profound (panel e, Figure 8.5B).

— Should the population become divided into local demes, connected by a moderate level of migration ($4Nm$ medium), then each deme will exhibit less genetic variation than did the original population. Allele frequencies of many demes will not be selectively optimal (panel f, Figure 8.5B). This situation approximates the concept of a **metapopulation,** a system of local demes that are linked by migration. Any of them may become extinct, perhaps because of inability to respond to selective pressure, with their habitat to be recolonized by migration from a nearby deme. Consideration of this case sets the stage for discussion of a selective process suggested by Sewall Wright.

The adaptive landscape was developed by Wright (1932) to support explanation of his **shifting balance theory.** In his theory, **demes,** or isolated reproductive units, "explore" the adaptive topography, seeking to maximize their fitness. The process has three phases. In the first, exploratory phase, random genetic drift moves small populations over the topographic surface. In the second, under a regime of **mass selection,** favorable genotypes increase in frequency within a deme. In the third phase, under a regime of **interdeme selection,** interplay of migration and selection can give rise to fusion of demes, loss of low-fitness populations, and spread of high-fitness populations. Focusing here on the selective aspects, demes containing certain genotypes or combinations of genotypes are more likely to become extinct than are other demes. Other, more fit demes are more likely to colonize the habitat vacated by the now-extinct demes.

The shifting balance theory has proven controversial, notably because the process of interdeme selection is a type of group selection. Interdeme selection remains largely untested. However, Wade and Goodnight (1991) imposed interdemic selection on large arrays of populations of the flour beetle *Tribolium castaneum* in the manner described by Wright, with differential dispersal of individuals from demes of high fitness into demes of low fitness. A significant increase in average fitness was observed in the experimental populations as compared with control populations with equivalent but random migration rates. The results indicated that the interdemic selection phase of shifting balance theory can increase average fitness.

8.5 Perspective

Natural selection, the force driving adaptive evolution, is one of the most fascinating processes at issue in population genetics. Its impact on fish populations has been highlighted in this chapter. However, a subject as well explored as natural selection can only be introduced in a chapter of this scope. To broaden and deepen their understanding of natural selection, students will have to read beyond the boundaries of fisheries science. For example, many of the classical demonstrations of natural selection on natural populations examined populations of fruit flies *Drosophila* spp., notably the work of Dobzhansky and colleagues. Their classical studies are presented as a collection (Lewontin et al. 1981) and summarized within a broad evolutionary context in Dobzhansky's *Genetics of the Evolutionary Process* (1970). The elegant models and predictions of natural selection theory perhaps are presented best by Williams (1966, 1975). A more in-depth overview of selection theory, including discussion of both general and ecological models, is presented by Hedrick (1983).

8.6 References

Avise, J. C., and M. H. Smith. 1974. Biochemical genetics of sunfish. I. Geographic variation and subspecific intergradation in the bluegill, *Lepomis macrochirus*. Evolution 28:42–56.

Bass, A. H. 1996. Shaping brain sexuality. American Scientist 84:352–363.

Clarke, B. 1962. Balanced polymorphism and the diversity of sympatric species. Systematics Association Publications 4:47–70.

Dawkins, R. 1976. The selfish gene. Oxford University Press, Oxford, UK.

Dobzhansky, T. 1970. Genetics of the evolutionary process. Columbia University Press, New York.

Dobzhansky, T., W. W. Anderson, and O. Pavlovsky. 1966. Genetics of natural populations. XXXVIII. Continuity and change in populations of *Drosophila pseudoobscura* in the western United States. Evolution 20:418–427.

Ehrman, L., B. Spassky, O. Pavlovsky, and T. Dobzhansky. 1965. Sexual selection, geotaxis, and chromosomal polymorphism in experimental populations of *Drosophila pseudoobscura*. Evolution 19:337–346.

Fleming, I. A., and M. R. Gross. 1989. Evolution of adult female life history and morphology in a Pacific salmon (coho: *Oncorhynchus kisutch*). Evolution 43:141–157.

Galis, F., and J. A. J. Metz. 1998. Why are there so many cichlid species? Trends in Ecology and Evolution 13:1–2.

Gartner-Kepkay, K. E., L. M. Dickie, K. R. Freeman, and E. Zouros. 1980. Genetic differences and environments of mussel populations in the Maritime Provinces. Canadian Journal of Fisheries and Aquatic Sciences 37:775–782.

Gould, J. L., and C. G. Gould. 1989. Sexual selection. Freeman, New York.

Gross, M. R. 1984. Sunfish, salmon, and the evolution of alternative reproductive strategies and tactics in fishes. Pages 55–75 *in* G. Potts and R. Wotton, editors. Fish reproduction: strategies and tactics. Academic Press, London.

Gross, M. R. 1991. Salmon breeding behavior and life history evolution in changing environments. Ecology 72:1180–1186.

Hartl, D. 1988. A primer of population genetics, 2nd edition. Sinauer Associates, Sunderland, Massachusetts.

Healy, M. C. 1986. Optimum size and age at maturity in Pacific salmon and effects of size-selective fisheries. Pages 39–52 *in* D. J. Meerburg, editor. Salmonid age at maturity. Canadian Special Publication in Fisheries and Aquatic Sciences 89.

Hedrick, P. W. 1983. Genetics of populations. Science Books International, Boston, and Van Nostrand Reinhold, New York.

Hori, M. 1993. Frequency-dependent natural selection in the handedness of scale-eating cichlid fish. Science 260:216–219.

IUBMBNC (International Union of Biochemistry and Molecular Biology, Nomenclature Committee). 1992. Enzyme nomenclature, 1992. Academic Press, San Diego, California.

Johnson, T. C., C. A. Scholz, M. R. Talbot, K. Kelts, R. D. Ricketts, G. Ngobi, K. Beuning, I. Ssemmanda, and J. W. McGill. 1996. Late Pleistocene dessication of Lake Victoria and rapid evolution of cichlid fishes. Science 273:1091–1093.

King, R. C., and W. D. Stansfield. 1990. A dictionary of genetics, 4th edition. Oxford University Press, New York.

Koehn, R. K., and T. J. Hilbish. 1987. The adaptive importance of genetic variation. American Scientist 75:134–141.

Koehn, R. K., and D. I. Rasmussen. 1967. Polymorphic and monomorphic serum esterase heterogeneity in catostomid fish populations. Biochemical Genetics 1:131–144.

Lewontin, R. C., J. A. Moore, W. B. Provine, and B. Wallace, editors. 1981. Dobzhansky's genetics of natural populations I–XLIII. Columbia University Press, New York.

Liem, K. F. 1973. Evolutionary strategies and morphological innovations: cichlid pharyngeal jaws. Systematic Zoology 22:425–441.

McAllister, M. K., R. M. Peterman, and D. M. Gillis. 1992. Statistical evaluation of a large-scale fishing experiment designed to test for a genetic effect of size-selective fishing on British Columbia pink salmon *(Oncorhynchus gorbuscha)*. Canadian Journal of Fisheries and Aquatic Sciences 49:1294–1304.

Merritt, R. B. 1972. Geographic distribution and enzymatic properties of lactate dehydrogenase allozymes in the fathead minnow, *Pimephales promelas*. American Naturalist 106:173–184.

Miller, L. M., and A. R. Kapuscinski. 1994. Estimation of selection differential from fish scales: a step towards evaluating genetic alteration of fish size in exploited populations. Canadian Journal of Fisheries and Aquatic Sciences 51:774–783.

Moyle, P. B., and J. J. Cech, Jr. 1996. Fishes: an introduction to ichthyology, 2nd edition. Prentice Hall, Upper Saddle River, New Jersey.

Nagalaki, T. 1975. Conditions for the existence of clines. Genetics 80:595–615.

Nelson, K., and M. Soulé. 1987. Genetical conservation of exploited fishes. Pages 345–368 *in* N. Ryman and F. Utter, editors. Population genetics and fishery management. University of Washington Press, Seattle.

Nevo, E., T. Shimony, and M. Libni. 1977. Thermal selection of allozyme polymorphisms in barnacles. Nature (London) 267:699–701.

Petit, C. 1958. Le determinisme genetique et psycho-physiologique de la competition sexuelle chez *Drosophila melanogaster*. Bulletin Biologiques France Belgique 92:248–329.

Philipp, D. P., W. F. Childers, and G. S. Whitt. 1983. Biochemical genetics of largemouth bass. Transactions of the American Fisheries Society 121:1–20.

Place, A. R., and D. A. Powers. 1979. Genetic variation and relative catalytic efficiencies: lactate dehydrogenase B allozyme of *Fundulus heteroclitus*. Proceedings of the National Academy of Sciences USA 76:2354–2358.

Provine, W. B. 1989. Sewall Wright and evolutionary biology. University of Chicago Press, Chicago.

Ricker, W. E. 1981. Changes in the average size and average age of Pacific salmon. Canadian Journal of Fisheries and Aquatic Sciences 38:1636–1656.

Ricker, W. E., H. T. Bilton, and K. V. Aro. 1978. Causes of the decrease in size of pink salmon *(Oncorhynchus gorbuscha)*. Fisheries and Marine Service Technical Report 820.

Seehausen, O., J. J. M. van Alphen, and F. Witte. 1997. Cichlid fish diversity threatened by eutrophication that curbs sexual selection. Science 277:1808–1811.

van den Berghe, E. P., and M. R. Gross. 1989. Natural selection resulting from female breeding competition in a Pacific salmon (coho: *Oncorhynchus kisutch*). Evolution 43:125–140.

Wade, M. J., and C. J. Goodnight. 1991. Wright's shifting balance theory: an experimental study. Science 253:1015–1018.

Wallace, B. 1981. Hard and soft selection. Chapter 16 *in* Basic population genetics. Columbia University Press, New York.

Wiley, B., and B. Collette. 1970. Breeding tubercles and contact organs in fishes: their occurrence, structure, and significance. Bulletin of the American Museum of Natural History 143:145–216.

Williams, G. C. 1966. Adaptation and natural selection. Princeton University Press, Princeton, New Jersey.

Williams, G. C. 1975. Sex and evolution. Princeton University Press, Princeton, New Jersey.

Wright, S. 1932. The role of mutation, inbreeding, crossbreeding, and selection on evolution. Proceedings of the Sixth International Congress of Genetics 1:356-366. Reprinted, pages 161–171 *in* W. B. Provine, editor. 1986. Sewall Wright: evolution, selected papers. University of Chicago Press, Chicago.

Wright, S. 1977. Evolution and genetics of populations, volume 3: experimental results and evolutionary deductions. University of Chicago Press, Chicago.

Wright, S. 1982. The shifting balance theory and macroevolution. Annual Review of Genetics 16:1–19.

9

Random Genetic Drift

Eric Hallerman

9.1 Introduction

When a population propagates itself, genetic information in the form of allele frequencies is passed from one generation to the next. Even in the absence of mutation, migration, or selection, allele frequencies in a finite population may be changed by random sampling and consequent sampling error. **Random genetic drift** is the term applied to several nonselective processes driven by this sampling error that result in allele frequency changes. In this chapter, I will characterize the processes collectively referred to as random drift and relate random drift to issues of importance to fisheries scientists.

The concept of random drift itself has evolved over the years (Provine 1986; Beatty 1992), and its importance has been debated, sometimes heatedly. Varieties of random drift first were articulated in the late nineteenth and early twentieth centuries in opposition to the neo-Darwinian notion that natural selection is the all-sufficient agent driving evolution (Beatty 1992). Actually, in the first edition of *On the Origin of Species*, Darwin (1859:81) acknowledged fluctuations in the frequency of variations that have no adaptive significance or that appear equally fit. Darwin did not suggest what process(es) might be driving such fluctuations.

Gulick (1905:209) was the first to describe the random **sampling of breeders,** the "indiscriminate destruction or failure to propagate" of part of a population, as a process giving rise to changing allele frequencies among generations. He invoked the example of a volcanic eruption that kills all nearby members of a species indiscriminately, resulting by chance in a change in the frequency of the various types in the species. This is an extreme example of a mechanism driving differential reproductive success among individuals. A more typical example is provided by the results of Gross and Kapuscinski (1997), who studied nesting success of smallmouth bass *Micropterus dolomieu* in Lake Opeongo, Ontario. They found that 86 of 202 nests produced no surviving fry. A sample of juvenile fish taken in the fall included 19 individuals produced by one mating, whereas no survivors were found for 60 of the nests that had fry in the spring.

A second mechanism underlying random drift is random **sampling of gametes,** first described by Hagedoorn and Hagedoorn (1921). Given that heterozygous organisms may leave homozygous offspring, and by chance may leave only offspring homozygous for one of the alleles in question, then an allele may become lost. At the population level, the finite number of *Aa* parents in one generation may contribute, say, more *A*'s than *a*'s to their offspring. The frequencies of alleles *A* and *a* may change from generation to generation until one is eliminated from the population. The change often is referred to as a random walk; the direction and magnitude result from chance.

Gulick (1905) also noted that a group newly split off from a larger group, especially if small, would be unlikely to have all the heritable variations, and certainly not in the same proportions, as the original group. This phenomenon later was termed the **founder effect.** The founder effect differs from the two previously described mechanisms of random drift. It concerns changes in allele frequencies *among* populations, whereas sampling of breeders or gametes causes changes in allele frequencies *within* populations. However, the founder effect, like all causes of sampling error, is a random process mediated, nonadaptive mode of evolution.

Though the processes driving genetic drift are rather straightforward, drift frequently is misunderstood by students of genetics. Two frequent misunderstandings are as follow. First, random drift, as we shall see, can be a powerful force acting on the gene pool of a small population; the term drift should not be taken necessarily to mean gradual but rather *undirected*. Second, inbreeding and random drift share the feature of being most important in small populations but are distinct processes. Inbreeding directly affects genotype frequencies but, absent selection against a particular genotype, does not change allele frequencies. Random drift directly affects allele frequencies. The only instances in which drift and inbreeding effects are identical is for haploid genomes or for clonal organisms that do not experience meiotic recombination (J. Epifanio, Illinois Natural History Survey, personal communication).

■ 9.2 The Idealized Population and Effective Population Number

The process of genetic drift is influenced greatly by the number of breeders and their makeup in the population. As a standard for comparison, population geneticists often consider a theoretically idealized population against which a real population can be tested. Generally, there are five conditions that describe the ideal population:

— an unlimited number of breeders,
— equal numbers of males and females breeding,
— equal probability of reproductive success for each breeder (i.e., random variation in family size),
— nonoverlapping generations, and
— absence of migration, mutation, and selection.

Because populations of fish, shellfish, or any other organism rarely if ever meet these conditions, allele and genotype frequencies often depart from ideal expectations. We can examine the dynamics of actual populations and model random genetic drift.

To assess the effect of small population size on the gene pool of a population, Wright (1931, 1938) developed the concept of the effective population number, N_e. The **effective population number** can be defined either in terms of the amount of increase in homozygosity (inbreeding effective number, Chapter 10) or the amount of allele frequency drift (variance effective number, this chapter) relative to an idealizd population. Under many circumstances, these are the same, but this is a special case. The inbreeding effective number depends primarily on the number in the parent generation, whereas the variance effective number is more directly related to the number in the progeny generation (Kimura and Crow 1963). Therefore, when population size is changing, the two effective sizes tend to differ. Over the longer term, population size must be fairly stable, so the two N_e values will be approximately the same. A third way of defining N_e is in terms of the rate of loss of heterozygosity (eigenvalue effective number, Ewens 1982), but this number is less commonly used.

When a population is "ideal," its N_e equals the number of reproductively mature individuals. However, natural populations exhibit departures from idealized conditions. The N_e of an actual population is defined as the size of an idealized population with the same amount of inbreeding or random drift exhibited by the actual population. Although real populations' census sizes (i.e., the count of all individuals) are generally greater than N_e's, this is not necessarily so, as we will see later (section 9.3.2).

Generally, there is a relationship between the size of a population (both in its present condition and in its history) and the rate and magnitude of drift effects. A population that has multiple alleles across many loci can be thought of as possessing genetic variance. Drift erodes this variance (in the absence of mutation) such that the expectation, E, of the change in genetic variance, σ_g^2, is inversely proportional to the population size, N, in an otherwise ideal population (Epifanio, personal communication):

$$E(\delta\sigma_g^2) = \tfrac{1}{2}N. \tag{9.1}$$

The change in genetic variance is low for large populations (zero for an infinite population) and higher for small populations. The change in variance is relative to the variance of the previous generation; therefore, drift can have a cumulative effect through time. It also is possible for the effects of severe drift to persist for many generations.

9.3 Demographic Methods for Calculation of Effective Population Number

Determination of N_e when there is departure from the idealized model has been discussed extensively (e.g., Wright 1938; Haldane 1939; Crow 1954; Crow and Morton 1955; Kimura and Crow 1963). Many of these models address specific departures from idealized conditions, presenting derivations of mathematical expressions to estimate the effect of inbreeding or random drift. Three departures from idealized conditions especially relevant to fisheries genetics (Gall 1987) are described below.

9.3.1 *Unequal Numbers of Male and Female Breeders*

There are a number of relevant contexts in which the numbers of males and females reproducing are unequal; for example,

— species for which a dominant male mates with a harem, for example, ungulates such as elk;
— species, including many fishes (e.g., centrarchids), in which males establish and defend breeding territories and breed with multiple females;
— size-selective fisheries (e.g., the walleye *Stizostedion vitreum* fishery on Red Lake, Minnesota) taking more individuals of the larger sex than of the smaller sex; and

— hatcheries where reproductively ready females are more limited in number than are males, or vice versa.

In such cases, N_e is related directly to the sex ratio. We then can estimate N_e as

$$N_e = \frac{4N_m N_f}{N_m + N_f}, \tag{9.2}$$

where N_m is the number of breeding males and N_f is the number of breeding females. Should the interest be in estimating N_e for purposes of estimating the rate of inbreeding, the one-generation change in inbreeding, δF, can be calculated as

$$\delta F = \tfrac{1}{2}N_e = \frac{1}{8N_m} + \frac{1}{8N_f}. \tag{9.3}$$

Note that if N_m equals N_f, then N_e is the number of breeding adults. Also, note the similarity of this equation with that used to estimate the expectation of change in genetic variance.

For purposes of illustration, let's consider the familiar situation for elk, for which a few dominant males establish harems and dominate all matings for the population. Hence, if two males gain matings with 50 females, N_m is 2, N_f is 50, and N_e is 7.7. Even though 52 animals are breeding, the skewed sex ratio limits the amount of genetic information that is transmitted to the progeny to roughly equal that of four males and four females breeding as pairs. In this context, it is important to note that rapid turnover of dominant male elk among years will bring new genetic variation into subsequent cohorts. Still, the key point of this example remains: N_e is very sensitive to the number of breeders of the less numerous sex.

Fisheries managers generally do not have the opportunity to affect sex ratios among spawners in the wild. However, the impact of skewed sex ratios on N_e should not be lost upon aquaculturists, who can control matings in the hatchery. Aquaculturists should, whenever practical, equalize N_m and N_f in their propagation plans and actions (Chapter 14) and consider how different mating designs (e.g, single-pair breeding or factorial breeding) potentially can affect the rate of drift in a population.

9.3.2 *Nonrandom Family Size*

There are many (nonselective) reasons why reproductive success might vary among individuals, why some leave many progeny and some few. Among the many nongenetic factors driving differences in reproductive success among individuals are the following examples.

— In a common aquaculture situation, some female broodstock are truly ready at spawning time and yield high-quality eggs whereas some females who are induced to spawn hormonally yield poorer-quality gametes; such differences in gamete quality results in differences in survival rates of their young.
— Some females are larger than others and, therefore, are more fecund.

— In the wild, some spawners secure more favorable nest locations than do others. For example, success of redds of salmonids may be affected by superimposition of later nests on earlier ones.
— In the wildlife context, some breeding pairs secure better foraging territories than do others, affecting recruitment of their young.

Under idealized conditions (Gall 1987), each parent has an equal opportunity to spawn, the contribution of progeny by individual parents is random and families vary in size, survival of progeny to breeding age is random, and generations do not overlap (i.e., all parents are replaced each generation). Random family size follows the binomial distribution, which is well approximated by the Poisson distribution, which in this case will have a mean and variance of 2. Such a population of parents will contribute numbers of progeny according to the following proportions.

Expected number of progeny	Proportion of parents
0	0.135
1	0.271
2	0.271
3	0.180
4	0.090
5	0.036
6	0.012
7	0.003
8 or more	<0.001

However, in most populations, variance in family size will be greater than that in the Poisson distribution, and the rate of random change in allele frequencies will be greater. If N is constant, and mating is random, N_e as a function of variance in family size can be estimated as

$$N_e = \frac{4N}{2+V_k}, \tag{9.4}$$

where V_k is variance in family size. Note that N_e equals N if V_k equals 2. In many contexts, family size is likely to be different for the sexes, in which case we might use the following expression from Hill (1979).

$$N_e = \frac{8N}{4+V_{km}+V_{kf}}, \tag{9.5}$$

where V_{km} is the variance in family size among males and V_{kf} is variance in family size among females. The N_e is a measure of the evenness of family sizes among parents, that is, of the variance in progeny number in an ideal population that displays the same amount of drift as the population under examination. Variation in family size greater than that expected under idealized conditions frequently causes N_e to be less than N.

The N_e of a population of given N can be maximized by equalizing family size. Aquaculturists frequently have the ability to equalize family size in their controlled breeding context (Chapter 14). The fecundity of fishes is such that successful spawning of few females can meet production goals. Further, more eggs can be held than can fingerlings approaching stocking size. Because some fish at younger life stages are likely to be disposed of, culturists would do well to discard extra members of the largest families, while keeping most or all members of smaller families, thereby equalizing family size. The large impact of *not* equalizing family size on the N_e of a hatchery stock is demonstrated well in the results of Simon et al. (1986), who estimated N_e's much smaller than census sizes in a hatchery stock of coho salmon *Oncorhynchus kisutch*. On the other hand, when differences in family size are controlled properly, N_e can, in fact, surpass the census size of the population (Allendorf 1993). If reproduction is restricted to equal-sized full-sib families with each male and female parenting a single family, V_k will equal 0 and the N_e will be nearly double the census population size. Ultimately, the variance in family size is limited by the lifetime fecundity of females in a population. For mammals, lifetime fecundities range in the tens, whereas in sturgeon (Acipenseridae), fecundity can well exceed a million. The chances for a sizable proportion of the population coming from a single sturgeon is much greater than that chance is for mammals, increasing the likelihood of random drift through variance in family size.

Family sizes can and do vary widely in the wild. Perhaps the best example is offered by marine organisms, which exhibit high reproductive potential and are subject to high mortality, especially in the early life stages. Successful reproduction depends on timing of reproduction to match oceanographic conditions conducive to fertilization, larval survival, and successful recruitment. For such species, the variance of reproductive success may be orders of magnitude greater than binomial or Poisson variance. This huge variance leads to discrepancies of several orders of magnitude between effective and actual population numbers. For example, Hedgecock (1994) showed that genetic drift yielded a 10,000-fold discrepancy between N_e and N for Pacific oyster *Crassostrea gigas* in Dabob Bay, Washington.

9.3.3 *Different Numbers of Parents in Successive Generations*

Different numbers of parents in successive generations can have a profound effect on the amount of genetic information passed from generation to generation. If loss of variability occurs consequent to a restricted number of individuals colonizing new habitat, it is referred to as a founder effect. If loss of variation occurs due to a restricted number of individual reproducing within an existing population, it is termed a **bottleneck effect.** The effect of a reduction in the number of breeders on the N_e is estimated using the expression

$$\frac{1}{N_e} = \frac{1}{t}\left(\frac{1}{N_{et=1}} + \frac{1}{N_{et=2}} + \ldots + \frac{1}{N_{et=t}}\right), \tag{9.6}$$

where t is the number of generations at issue and $N_{et=n}$ is the effective size of the population in the generation specified. Because N_e appears in the denominator of the fractions, it follows that small values of N_e make a large impact on the mean N_e over a period of generations. Suppose, for example, that over a period of four generations, N_e's for a given population were 50, 10, 50, and 50. Were mean N_e calculated as the arithmetic mean, it would be 40, with but a minor impact of the bottleneck in the second generation. However, because mean N_e is calculated as the harmonic mean, that is as

$$\frac{1}{N_e} = \frac{1}{4}\left(\frac{1}{50} + \frac{1}{10} + \frac{1}{50} + \frac{1}{50}\right) = 0.04,$$

the mean N_e equals 25, and the impact of the bottleneck in the second generation is rather dramatic. It follows that the genetic character of the population could have been changed by the effect of drift at the time of the bottleneck.

The practical implication of these considerations is not to allow a managed population to become bottlenecked. In a fisheries context, managers must ensure escapement of a sufficiently large number of spawners that a genetic bottleneck does not occur. In an aquacultural context, sufficient numbers of breeding pairs must be used so that random drift does not drive loss of genetic diversity (Chapter 14). The importance of using sufficiently large numbers of broodstock for propagating cultured lines can be inferred from the results of Allendorf and Phelps (1980), Cross and King (1983), Vuorinen (1984), Hallerman et al. (1986), and others, who have shown loss of considerable genetic variation after few generations of captive propagation.

9.3.4 Other Departures from Idealized Conditions

I have presented three cases in which departures from idealized conditions affect allele frequencies in populations via random drift. Other departures from idealized conditions also might prove important in certain contexts. For example, N_e in species with life histories exhibiting age-specific survivorship and reproduction can be estimated using the model of Giesel (1971). In species with overlapping generations, N_e can be estimated using models developed by Hill (1972, 1979). Other models (Ryman et al. 1981; Lacava and Hughes 1984; Reed et al. 1986; Lande and Barrowclough 1987; Lemkuhl 1987; Wood 1987; Crow and Denniston 1988) are available to estimate N_e under other demographic situations. Still others might be developed to meet the needs of modeling the genetic effects of the demographics of any species of interest.

9.4 Effects of Random Drift on Genetic Variability

Random drift affects genetic variability both within and between populations. Generally, the effect is to reduce the allelic diversity *within* populations. Alternately, the main effect of drift *among* populations is to cause the arrays of allele frequencies among those populations to diverge. We will examine each case separately.

9.4.1 Effect of Random Drift within Populations

Whenever a population reproduces, there is sampling of the genetic variation within the population. Whether by sampling of breeders, sampling of gametes, or founder effect, random changes in allele frequency may be caused by sampling error. Were this sampling to go forward in replicated fashion a sufficiently large number of times, the mean frequency of an allele in the progeny generation would equal that in the parental population, but there would be a variance about this mean. Hence, random drift can be expressed in terms of the variance of the change in allele frequency (Tave 1993) as follows:

$$\sigma^2_{\delta q} = \frac{pq}{2N_e}, \tag{9.7}$$

where $\sigma^2_{\delta q}$ is the variance of the change in allele frequency and p and q are frequencies of alleles at a given locus. It should be clear, then, that the effect of drift at a locus is dependent on the frequencies of the respective alleles and on the N_e. We also see that σ^2 is inversely related to N_e; that is, drift has a greater effect in smaller populations than in larger ones.

The result of genetic drift within a population, then, is random change of allele frequencies the direction of which is not predetermined and the magnitude of which reflects population size and the current frequency of alleles. The relationships of allele frequency, population size, and likelihood of an allele becoming lost are shown in Table 9.1. Probabilities of losing an allele via random drift are higher for rare alleles or small N_e and lower for common alleles or large N_e. When random drift is strong, rare alleles may be lost, whereas common alleles may become **fixed,** that is, their frequency may become 1.0. Among the many populations for which N_e has been estimated (Frankham 1995), few have been populations of aquatic species. For those aquatic populations, N_e frequently was but a fraction of *N*, as discussed below. It is reasonable to infer that many aquatic populations are sufficiently small or experience bottlenecks such that uncommon alleles are likely to be lost through random drift.

Against this background, three points must be emphasized. Although the term drift suggests a gentle process, under certain circumstances random drift can have a profound effect on the array of allele frequencies in a population's gene pool. Loss of rare alleles is of interest because some of these alleles could have adaptive value, if not under present ecological conditions then perhaps later. Last, the direct effect of drift on allele frequencies differentiates it from inbreeding, a process with which it often is confused; inbreeding operates directly on genotype frequencies by reducing the proportion of heterozygotes but does not directly affect allele frequencies. Therefore, small population sizes do not affect the rates of inbreeding except in cases in which small population size increases the chance of alleles that are identical by descent coming together in an individual through mating among relatives (see Chapter 10).

Investigating the effects of random drift on genetic variability within populations, Nei et al. (1975) carried out a computer simulation to model its effects under various conditions. They worked with heterozygosity values, which vary from 0 to 1, and which are more readily given to modeling than is allelic diversity. Although they used data on *Drosophila pseudoobscura*, their findings can be generalized to other organisms. Starting with an average heterozygosity of 0.138 in generation 1, Nei et al. (1975) followed changes in heterozygosity when the population size was reduced in simulated bottlenecks. The bottlenecks were simulated by having the computer randomly draw given numbers of individuals from the data set; then those individuals' genotypes were used to generate genotypes for the next generation. Nei et al. modeled the effect of bottlenecks of 2 or 10 individuals and the effect of how quickly the population size grew after the bottleneck. Their results, shown in Figure 9.1, illustrate three key points regarding the effects of a genetic bottleneck on within-population genetic variability. First, the decrease in average heterozygosity was directly dependent on the N_e at the time of the bottleneck—note that the lines for an *N* of 2 are lower than those for an *N* of 10. Second, the effect of the bottleneck is dependent on how quickly population number recovered after the bottleneck—note that after a bottleneck of a given severity, the loss of heterozygosity is greater for populations with a lower intrinsic rate of increase than for those with a higher one. Third, in the absence of migration, with only mutation as a source of new variability, the effect of random drift is very long lasting—the original level of hererozygosity is attained only after millions of generations of mutation.

As is the example above, many studies of the loss of genetic variation caused by population bottlenecks use heterozygosity as the measure of genetic variation. Allendorf

Table 9.1 Single-generation probabilities of losing an allele via genetic drift for eight allele frequencies at various effective population sizes (N_e). Table adapted from Tave 1993.

	Allele frequency							
N_e	0.5	0.4	0.3	0.2	0.1	0.05	0.01	0.001
2	0.06250	0.12960	0.24010	0.40960	0.65610	0.81451	0.96060	0.99601
3	0.01562	0.04666	0.11765	0.26214	0.53144	0.73509	0.94148	0.99402
4	0.00391	0.01680	0.05765	0.16778	0.43047	0.66342	0.92274	0.99203
5	0.00098	0.00605	0.02825	0.10737	0.34868	0.59874	0.90438	0.99004
6	0.00024	0.00218	0.01384	0.06872	0.28243	0.54036	0.88638	0.98807
7	0.00006	0.00078	0.00678	0.04398	0.22877	0.48768	0.86875	0.98609
8	0.00002	0.00028	0.00332	0.02815	0.18530	0.44013	0.85146	0.98412
9		0.00010	0.00163	0.01801	0.15009	0.39721	0.83451	0.98215
10		0.00004	0.00080	0.01153	0.12158	0.35849	0.81791	0.98019
15			0.00002	0.00124	0.04239	0.21464	0.73970	0.97043
20				0.00013	0.01478	0.12851	0.66897	0.96077
25				0.00001	0.00515	0.07694	0.60501	0.95121
30					0.00180	0.04607	0.54716	0.94174
35					0.00063	0.02758	0.49484	0.93236
40					0.00022	0.01652	0.44752	0.92308
45					0.00008	0.00989	0.40473	0.91389
50					0.00003	0.00592	0.36603	0.90479
55						0.00354	0.33103	0.89578
60						0.00212	0.29938	0.88687
65						0.00115	0.26537	0.87628
70						0.00076	0.24487	0.86930
75						0.00046	0.22145	0.86064
80						0.00027	0.20028	0.85208
85						0.00016	0.18113	0.84359
90						0.00010	0.16381	0.83520
95						0.00006	0.14814	0.82688
100						0.00004	0.13398	0.81865
125							0.08106	0.77870
150							0.04904	0.74071
175							0.02967	0.70456
200							0.01795	0.67019
225							0.01086	0.63748
250							0.00657	0.60638
275							0.00398	0.57679
300							0.00240	0.54863
325							0.00146	0.52188
350							0.00088	0.49641
375							0.00053	0.47219
400							0.00032	0.44915
425							0.00019	0.42723
450							0.00012	0.40639
475							0.00007	0.38656
500							0.00004	0.38656

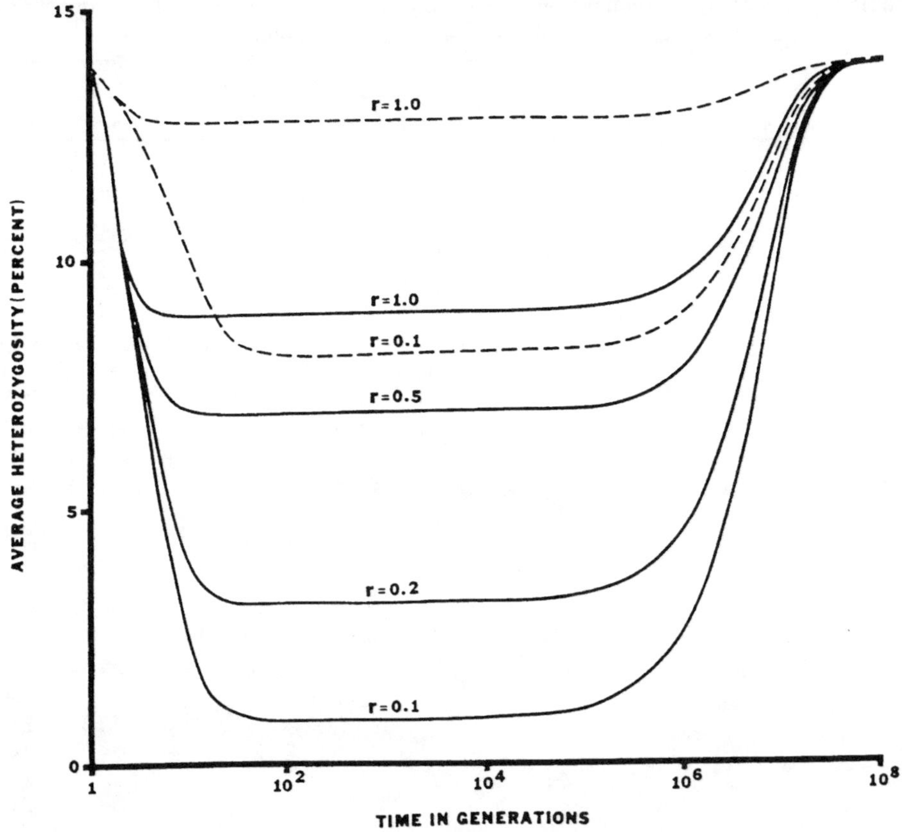

Figure 9.1 Changes in average heterozygosity when a population goes through a genetic bottleneck of given size. Solid lines refer to cases in which the bottleneck size (N_0) is 2 individuals and broken lines to an N_0 of 10. Population growth rate is assumed to be logistic, and *r* is the intrinsic rate of population growth. Original population heterozygosity is 0.138. Time in generations is presented on a logarithmic scale. Figure adapted from Nei et al. (1975).

(1986) compared the effects of bottlenecks on the loss of alleles, as well as on reduction of heterozygosity. He showed that population bottlenecks of short duration will have little effect on heterozygosity but will reduce severely the number of alleles present (Figure 9.2). Heterozygosity provides a good measure of the capability of a population to respond to selection immediately following a bottleneck. However, the number of alleles remaining is important for the population's long-term response to selection.

9.4.2 *Effect of Random Drift among Populations*

Having considered the effect of random drift within a population, let's move on to consider its effect among populations. We will assume that the effects of mutation, migration, and selection are negligible. Simulations have shown that even if a group of populations start with the same array of allele frequencies and N_e, random drift will cause their allele frequencies to diverge over time (Figure 9.3). Close inspection of the figure will show that allele frequencies at a given locus will tend to 0 or 1 in a particular population.

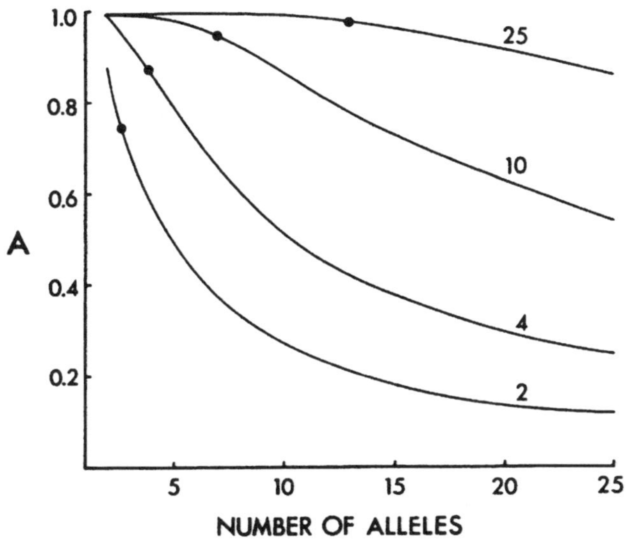

Figure 9.2 Allelic diversity *(A)* remaining versus original number of alleles after a bottleneck of a single generation of sizes (N_0) 2, 4, 10, or 25. All alleles are assumed to be equally frequent. The solid circle represents the proportion of heterozygosity expected to be retained (1 – ½*N*). Figure adapted from Allendorf, F. W. 1986. Genetic drift and the loss of alleles versus heterozygosity. Zoo Biology 5:181–190. Reprinted by permission of Wiley-Liss, Inc., a subsidiary of John Wiley and Sons, Inc.

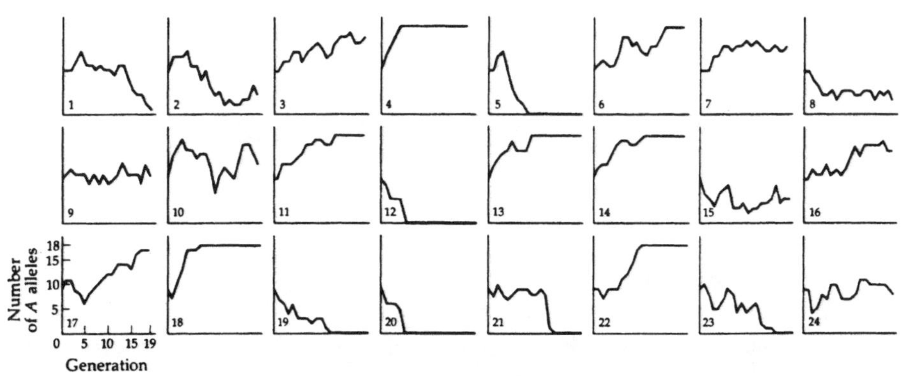

Figure 9.3 Effect of random drift on genetic divergence of populations. Shown are simulated allele frequencies at a given locus initially with two alleles at equal frequencies for 24 different populations of equal size (*N* = 9). The process of divergence would occur at all segregating loci. Figure adapted from Hartl (1988).

The effects of random drift on systems of populations in the wild will prove more complicated. Because drift is a random process, it will tend to have different effects on the arrays of allele frequencies for different populations. The differing effect follows as a consequence of the populations having differing allele frequencies and N_e at given times and of the random nature of the process of drift. Still, in the absence of mutation, migration, or selection, random drift will cause allele frequencies to diverge. The effects of drift on allele frequencies in the presence of the other evolutionary forces are highly dynamic and are discussed in Chapter 8. A discussion of using differing allele frequencies to assess genetic relationships among populations is presented in Chapter 13.

9.5 Genetic Methods for Estimating Effective Population Size

To this point in the chapter, I have discussed classical, demographics-based models for estimating N_e and inferring any effects of random drift. However, the data requirements for demographics-based methods of estimating N_e frequently restrict their usefulness for fisheries applications. Moreover, demographic methods require numerous assumptions regarding the relative success of breeders. Only in exceptional instances—notably in captive or controlled breeding contexts—do we have access to data on who mated with whom and how successfully. Such data for fishes in the wild are virtually absent. Based on the notion that genetic drift affects the faithfulness with which genetic information is passed from one generation to the next, alternative approaches have been developed for using genetic data to obtain indirect estimates of N_e.

Some of these models estimate N_e by means of inferences based on the correlation of genetic variation with population size. Large populations generally will exhibit greater genetic variation; further, alleles at unlinked loci will be randomly associated. Smaller populations will exhibit less variation, and because of the limited number of allelic combinations occurring in a small population, some unlinked alleles will be inherited together by chance alone. Applying this line of reasoning, Hill (1981) estimated N_e by assessing linkage disequilibrium *(D)* and correlation of alleles at different loci *(r)* in a sample drawn from a population. Both D and r will be 0 in an ideal, infinite, randomly mating population. However, both D and r will depart from 0 in real, finite populations due to drift, migration, selection, and linkage. The assumptions needed to estimate N_e by assessing the linkage disequilibrium and correlation of alleles are (1) alleles are selectively neutral, (2) there is no migration, (3) there is no subpopulation structure, that is, one panmictic population is sampled, and (4) there is random sampling of the entire population. Box 9.1 elaborates on the approach and shows the results of Bartley et al. (1992), who applied the modified Hill (1981) model to estimate N_e for four fish populations. Bartley et al.'s results suggest that the approach yields a reliable estimate of N_e for a well-sampled population.

Other models (e.g., Nei and Tajima 1981; Pollack 1983) use temporal variation in allele frequencies to estimate an average N_e per generation in the examined time interval. That is, these models use direct measurement of the effects of random drift to infer what population size could have caused the observed change in allele frequencies. The approach is particularly robust over intervals of 2 to 10 generations and when N_e is small (Waples 1989). Modifications of such models for estimating N_e of salmonid populations were made by Waples (1989, 1990), Waples and Teel (1990), and Tajima (1992). Jorde and Ryman (1995) showed how to estimate N_e for populations of brown trout *Salmo trutta* with overlapping generations. Temporal variation models have been shown useful in estimating N_e in closed, hatchery populations of salmon (Waples and Teel 1990) and shellfish (Hedgecock and Sly 1990; Hedgecock et al. 1992) and more recently for natural populations. For example, Miller and Kapuscinski (1997) used the temporal variation approach to reconstruct historical genetic changes in a natural population of northern pike *Esox lucius* in Lake Escanaba, Wisconsin. They examined genetic variation at three time intervals over a period of 32 years by sampling fish scales from a historical collection. Variability was assessed by extracting DNA from epithelial cells that adhered to the scales and by using polymerase chain reaction to amplify seven microsatellite DNA loci (Chapter 4). Based on temporal changes in genetic variation, they estimated N_e for two time periods. Estimates of N_e for the time intervals 1961–1977 and 1977–1993 were 35 and 72, respectively. Corresponding ratios of N_e/N ranged from

■ Box 9.1
Use of Genetic Data to Estimate Effective Population Size

Models (e.g., Hill 1981) have been developed to relate physical linkage of alleles versus chance co-inheritance to the effective size of a population (N_e). These models have been modified for application to fish populations. Bartley et al. (1992) demonstrated use of genetic data to estimate N_e for four different fish populations. Their work showed utility of the approach both within the particular contexts examined and suggested its use more generally.

Using isozyme markers, Bartley et al. (1992) estimated allele frequencies for polymorphic loci for each population. They then estimated *r*, the correlation among alleles, as

$$r = \frac{D}{\left[p(1-p)+q(1-q)\right]^{1/2}},$$

where *p* is the frequency of allele *A* at locus 1, *q* is the frequency of allele *B* at locus 2, and *D* is Burrow's composite measure of equilibrium, a measure of linkage disequilibrium ($D^\star$ in Campton 1987). They then generated a matrix of correlation values. Then $N_e(D)$, the N_e as estimated from linkage disequilibrium data, was estimated as

$$N_e(D) = \frac{1}{3\left(r^2 - \frac{1}{s}\right)},$$

where *s* is the sample size. This yielded an estimate of $N_e(D)$ for each polymorphic locus. To obtain a single value of $N_e(D)$, they took the arithmetic mean value of the r^2s and the harmonic mean of the *s*'s and substituted them into the equation. Equations were presented for calculating confidence intervals (CI) and the variance of $N_e(D)$. Using this procedure, Bartley et al. (1992) examined four sets of allozyme genotype data representing four fish populations.

Case 1 was an analysis of a hatchery enhancement program for white seabass *Atractoscion nobilis*. The individuals sampled were juveniles produced by 25 adults spawning in a tank year after year. Although this number of spawners potentially contributed to reproduction, hatchery workers reported that it appeared that one to two females were fertilized by one to three males each year. Six samples of 21 to 72 individuals were taken and analyzed; the samples each exhibited three to six polymorphic loci. Unexpectedly, the estimated $N_e(D)$ values were negative for five of six groups, and the 95% CI ranged from negative to 18. These negative estimates for N_e suggested that either an infinite population had been sampled—clearly not the case—or a sampling error had yielded spurious correlations of allele frequencies. Bartley et al. (1992) then combined data between sampling dates, yielding greater sample size, more polymorphic loci, and more precise results. For 1985, $N_e(D)$ was 7 (95% CI = 2.5–15), and for 1986, $N_e(D)$ was 10 (95% CI = 6–16). In each case,

continued on next page

■ Box 9.1 *continued*

the ratio of $N_e(D)/N$ was depressed below 1.0 because of an unequal sex ratio among spawners and an unequal contribution of adults to spawn. Given that one of the goals of the project was to preserve the genetic diversity of the species, it became clear that the operational design of the program was ineffective.

Case 2 was an estimate of N_e for offspring of chinook salmon *Oncorhynchus tshawytscha* spawned at the Coleman National Fish Hatchery, California. These chinook salmon are a semelparous population with the potential to lose genetic variation due to bottlenecks and inbreeding. Through the duration of the 1988 fall run, over 10,000 adults returned to the hatchery. Twenty-one polymorphic loci were observed among 100 individuals screened. The estimated $N_e(D)$ was only 132 (95% CI = 67 – 355). The authors noted that their surprisingly low estimate of the N_e agreed with an estimate obtained using the method of Waples and Teel (1990). Reasons that the N_e was so much smaller than the number of fish returning to the hatchery could include a small number of adults actually spawning, large variance in family size, and sampling error in estimation of $N_e(D)$.

Case 3 estimated N_e for a wild population of juvenile chinook salmon collected from the 1987 winter run in the Sacramento River, California. This population is protected as a threatened species and was a candidate stock for an enhancement program. Sampled individuals were collected as out-migrating smolts in a run with an estimated size of 2,000. Twenty polymorphic loci were observed among 94 individuals screened. The estimated $N_e(D)$ was 89 (95% CI = 45–265). The estimated ratio of N_e:N was just 0.04. Run sizes for other years in that period of time were 2,422 (1986), 548 (1989), and 440 (1990). Assuming that the ratio N_e:N is constant, then N_e for 1990 was approximately 20. The conservation implication is that the genetic resources of this population are indeed threatened.

Case 4 examined N_e in the rainbow trout *Oncorhynchus mykiss* stock at Shasta Hatchery, California, the object of an intensively managed hatchery breeding program. In this highly structured program, family data, sex ratios, and genetic contributions are monitored. At the time, the program was in its eighth generation. The program was modest in size, involving the offspring of 17 to 30 single-pair matings; N equaled 40 adults. Seven polymorphic loci were observed among 160 individuals screened. The estimated $N_e(D)$ was 36 (95% CI = 13–113). Hence, $N_e(D)$ was found to approximate N. This finding argues for the good breeding practice used at the hatchery and also for the validity of the method used for estimating N_e.

Considering the results of the study overall, it was found that to obtain useful estimates of $N_e(D)$, the number of samples should be on the order of 100 or more and the number of polymorphic loci should be at least six. The method gives a good estimate of N_e for the *sampled* population. Hence, geneticists may need to sample repeatedly to estimate N_e for the whole population, for example, to sample repeatedly over time through the spawning period, over year-classes for migratory salmon, and among all rearing units in a hatchery. The same life stage should be sampled each time. It also was noted that genetic methods for estimating N_e yielded more precise estimates, in terms of narrower CIs, for small populations.

0.03 to 0.14. The conservation implication of this finding is that over the course of these two time intervals, the population could have lost as much as 8% of its heterozygosity.

9.6 Perspective: Interplay of Selection and Random Drift

Although all population geneticists agree that random drift is an evolutionary force, there has been considerable argument about how often population sizes and selection intensities are small enough that genetic drift proves important (Beatty 1992, and citations therein). During the 1930s and 1940s, random drift frequently was invoked to account for many nonadaptive patterns of variation. However, certain of these apparent cases of random drift later were reinterpreted in terms of evolution by natural selection. Collectively, these reinterpretations were taken by many, who came to be called **selectionists,** as support for the overwhelming importance of natural selection. Selectionists tended to wonder whether any invocation of drift would stand up to rigorous inspection or whether these cases were simply instances for which we did not know the subtle ways in which selection worked.

More recently, the importance of nonselective, random processes has become more widely recognized. The rise of molecular genetics played a large role in the founding of a worldview emphasizing the prevalence of selectively neutral variation. Beginning in the 1950s and 1960s, it was recognized that different codons of three nucleotides can encode the same amino acid. Therefore, the degeneracy of the code suggested that certain base pair substitutions (see Chapter 6) could be functionally equivalent and, therefore, selectively neutral substitutions. Moreover, it was found that proteins could differ in some of their amino acids and still prove functionally equivalent. With the development of isozyme methods (Chapter 2), it became clear that many populations carried far more variation than had been expected. If all this variation represented differentially fit genotypes, how could the large amount of suboptimally fit variation be maintained? The large amount of variation could be explained by regarding it as (largely) selectively neutral. This worldview came to be called the **neutralist** school (Kimura 1983).

The past decade has seen a moderation of the debate among selectionist and neutralist viewpoints. Beatty (1992) cites Kimura (1983) for two key points that tend to resolve much of the difference between the selectionist and neutralist worldviews. First, the essential part of the neutral theory is not so much that molecular mutants are selectively neutral in the strict sense but rather that their fate largely is determined by random drift. Second, the random drifting of frequencies of neutral genes accounts for variation at the molecular level, whereas natural selection is more important for explaining variation at the phenotypic level. However, echoes of the selectionist–neutralist debate still resound today.

The effect of genetic drift over an evolutionary time scale is considered in Box 9.2.

9.7 References

Allendorf, F. W. 1986. Genetic drift and the loss of alleles versus heterozygosity. Zoo Biology 5:181–190.

Allendorf, F. W. 1993. Delay of adaptation to captive breeding by equalizing family size. Conservation Biology 7:416–419.

Allendorf, F. W., and S. R. Phelps. 1980. Loss of genetic variation in a hatchery stock of cutthroat trout. Transactions of the American Fisheries Society 109:537–543.

Barrowclough, G. F., and G. F. Shields. 1984. Karyotypic evolution and long-term effective population size of birds. Auk 101:99–102.

■ Box 9.2
Effective Population Sizes over Evolutionary Time

Through most of this chapter we have been considering estimation of effective population sizes, N_e, within an *ecological* time frame. What about estimating N_e in an *evolutionary* time frame? Estimates of N_e over evolutionary time would yield insight regarding the relative importance of random drift as an evolutionary force.

Lande (1979) developed a method for estimating the long-term N_e that is based on the assumption that individuals heterozygous for chromosomal mutations are at a selective disadvantage. Hence, a new mutation can become fixed only through the process of random drift. The probability of the mutation becoming fixed in a population is dependent on N_e. Thus, it's possible to work backwards and to estimate N_e from the observed rate of fixation of chromosomal changes:

$$N_e = -\left(\frac{1}{s}\right)\left[\ln\left(\frac{R}{\mu}\right) - 1\right],$$

where s is the selection coefficient against heterozygotes, μ is the rate of mutation, and R is the rate of fixation.

Barrowclough and Shields (1984) applied the method to published data. The selection coefficient, s, is relatively well known, approximately 0.10. The mutation rate, μ, also is relatively well known, approximately 10^{-4} for chromosomal mutations. The fixation rate, R, varies widely and is known for certain taxa. Long-term N_e's for such taxa are presented below. Barrowclough and Shields (1984) argued that their results were rather robust against inaccuracy in estimation of parameters. The notable observation to be made on their results is that estimated long-term N_es are small, at least among vertebrates.

Taxon	Long-term N_e
Drosophila	200–800
Mammals	30–200
Birds	
Rattites (kiwis, flightless birds)	>100
Galliformes (grouse, quail, turkey)	~102
Columbina (pigeon)	~64
Other nonpasserines	~106
Passerines (song birds)	~87

Over an ecological time scale, N_e is unlikely to be constant, fluctuating with habitat and environmental conditions. At any moment in time, the N_e might be much larger than the long-term estimates presented above, likely on the order of several hundreds or thousands for many species. However, many populations go through occasional bottlenecks, which, as we saw earlier in this chapter, can have a large and lasting effect on the genetic variability of a population. These bottleneck events are important with regard to interpopulation differentiation and arguably to speciation. Hence, random drift effects are important in vertebrate taxa over the evolutionary time scale. Random drift and chromosomal modes of speciation seem important in birds and some mammals but not in dipterans. Such analyses have not been carried out for fishes.

Bartley, D., M. Bagley, G. Gall, and B. Bentley. 1992. Use of linkage disequilibrium data to estimate effective size of hatchery and natural fish populations. Conservation Biology 6:365–375.

Beatty, J. 1992. Random drift. Pages 273–281 *in* E. F. Keller and E. A. Lloyd, editors. Keywords in evolutionary biology. Harvard University Press, Cambridge, Massachusetts.

Campton, D. 1987. Natural hybridization and introgression in fishes: methods of detection and genetic interpretations. Pages 161–192 *in* N. Ryman and F. Utter, editors. Population genetics and fishery management. University of Washington Press, Seattle.

Cross, T. F., and J. King. 1983. Genetic effects of hatchery rearing in Atlantic salmon. Aquaculture 33:33–40.

Crow, J. F. 1954. Breeding structure of populations. II. Effective population number. Pages 543–556 *in* Statistics and mathematics in biology. Iowa State College Press, Ames.

Crow, J. F., and C. Denniston. 1988. Inbreeding and variance effective population numbers. Evolution 42:482–495.

Crow, J. F., and N. E. Morton. 1955. Measurement of gene frequency drift in small populations. Evolution 9:202–214.

Darwin, C. 1859. On the origin of species, a facsimile of the 1st edition, with an introduction by Ernst Mayr. Published 1964, Harvard University Press, Cambridge, Massachusetts.

Ewens, W. J. 1982. On the concept of effective population size. Theoretical Population Biology 3:87–112.

Frankham, R. 1995. Effective population size/adult population size ratios in wildlife: a review. Genetical Research (Cambridge) 66:95–107.

Frankham, R. 1996. Relationship of genetic variation to population size in wildlife. Conservation Biology 10:1500–1508.

Gall, G. A. E. 1987. Inbreeding. Pages 47–87 *in* N. Ryman and F. W. Utter, editors. Population genetics and fishery management. University of Washington Press, Seattle.

Giesel, J. T. 1971. The relations between population structure and inbreeding. Evolution 25:491–496.

Gross, M. L., and A. R. Kapuscinski. 1997. Reproductive success of smallmouth bass estimated and evaluated from family-specific DNA fingerprints. Ecology 78:1424–1430.

Gulick, J. T. 1905. Evolution, racial and habitudinal. Carnegie Institution of Washington, Washington, D.C.

Hagedoorn, A. L., and A. C. Hagedoorn. 1921. On the relative value of the processes causing evolution. Nijhoff, The Hague.

Haldane, J. B. S. 1939. The equilibrium between mutation and random extinction. Annals of Eugenics 9:400–405.

Hallerman, E. M., R. A. Dunham, and R. O. Smitherman. 1986. Selection or drift— isozyme allele frequency changes among channel catfish selected for rapid growth. Transactions of the American Fisheries Society 115:60–68.

Hartl, D. 1988. A primer of population genetics, 2nd edition. Sinauer Associates, Sunderland, Massachusetts.

Hedgecock, D. 1994. Does variance in reproductive success limit effective population sizes of marine organisms? Pages 122–134 *in* A. Beaumont, editor. Genetics and evolution of aquatic organisms. Chapman and Hall, London.

Hedgecock, D., V. Chow, and R. S. Waples. 1992. Effective population numbers of shellfish broodstocks estimated from temporal variance in allele frequencies. Aquaculture 108:215–232.

Hill, W. G. 1972. Effective population size with overlapping generations. Theoretical Population Biology 3:278–289.

Hill, W. G. 1979. A note on effective population size with overlapping generations. Genetics 92:317–322.

Hill, W. G. 1981. Estimation of effective population size from data on linkage disequilibrium. Genetical Research 38:209–216.

Jorde, P. E., and N. Ryman. 1995. Temporal allele frequency change and estimation of effective size in populations with overlapping generations. Genetics 139:1077–1090.

Kimura, M. 1983. The neutral theory of molecular evolution. Cambridge University Press, Cambridge, UK.

Kimura, M., and J. F. Crow. 1963. The measurement of effective population number. Evolution 17:279–288.

LaCava, J., and J. Hughes. 1984. Determining minimum viable population levels. Wildlife Society Bulletin 12:370–376.

Lande, R. 1979. Effective deme sizes during long-term evolution: estimates from rates of chromosomal rearrangement. Evolution 33:234–251.

Lande, R., and G. G. Barrowclough. 1987. Effective population size, genetic variation, and their use in population management. Page 87–124 *in* M. E. Soulé, editor. Viable populations for conservation. Cambridge University Press, Cambridge, UK.

Lemkuhl, J. F. 1984. Determining size and dispersion of minimum viable populations for land management planning and species conservation. Environmental Management 8:167–176.

Miller, L. R., and A. R. Kapuscinski. 1997. Historical analysis of genetic variation reveals low effective population size in a northern pike, *Esox lucius,* population. Genetics 147:1249–1258.

Nei, M., T. Maruyama, and R. Chakraborty. 1975. The bottleneck effect and genetic variability in populations. Evolution 29:1–10.

Nei, M., and F. Tajima. 1981. Genetic drift and estimation of effective population size. Genetics 98:625–640.

Pollack, E. 1983. A new method for estimating the effective population size from allele frequency changes. Genetics 104:531–548.

Provine, W. B. 1986. Sewall Wright and evolutionary biology. University of Chicago Press, Chicago.

Reed, J. M., P. D. Doerr, and J. R. Walters. 1986. Determining minimum population sizes for birds and mammals. Wildlife Society Bulletin 14:255–261.

Ryman, N., R. Baccus, C Reuterwall, and M. H. Smith. 1981. Effective population size, generation interval, and potential loss of variation in game species under different hunting regimes. Oikos 36:257–266.

Simon, R. C., J. D. McIntyre, and A. R. Hemmingsen. 1986. Family size and effective population size in a hatchery stock of coho salmon *(Oncorhynchus kisutch).* Canadian Journal of Fisheries and Aquatic Sciences 43:2434–2442.

Tajima, F. 1992. Statistical method for estimating the effective population size in Pacific salmon. Journal of Heredity 83:309–311.

Tave, D. 1993. Genetics for fish hatchery managers, 2nd edition. Van Nostrand Reinhold, New York.

Vuorinen, J. 1984. Reduction of genetic variability in a hatchery stock of brown trout, *Salmo trutta.* Journal of Fish Biology 24:339–348.

Waples, R. S. 1989. A generalized method for estimating population size from temporal changes in allele frequency. Genetics 121:379–391.

Waples, R. S. 1990a. Conservation genetics of Pacific salmon. II. Effective population size and the rate of loss of genetic diversity. Journal of Heredity 81:267–276.

Waples, R. S. 1990b. Conservation genetics of Pacific salmon. III. Estimating effective population size. Journal of Heredity 81:277–289.

Waples, R. S., and D. J. Teel. 1990. Conservation genetics of Pacific salmon. I. Temporal changes in allele frequency. Conservation Biology 4:144–156.

Wood, J. W. 1987. The genetic demography of the Gainj of Papua New Guinea. 2. Determinants of effective population size. American Naturalist 129:165–187.

Wright, S. 1931. Evolution in Mendelian populations. Genetics 16:97–159.

Wright, S. 1938. Size of populations and breeding structure in relation to evolution. Science 87:430–431.

10

Inbreeding

Eric Hallerman

10.1 Introduction

The genotypic structure of a population is determined not only by the frequencies of alleles and the forces affecting allele frequency—mutation, migration, selection, and random drift—but also by the population's mating system (Gall 1987). There may be **random mating,** whereby each individual in the population has an equal chance of mating with any other individual. However, the mating systems of most populations deviate from random mating. If mates are chosen on the basis of greater than average phenotypic resemblance, for example regarding morphological traits, the result is **positive assortative mating** (Chapter 1). Should mates be more closely related to one another than are individuals chosen at random, inbreeding results. **Inbreeding** is defined as the mating of related individuals. The consequence of inbreeding is an increase in homozygosity and a corresponding decrease in heterozygosity relative to expectations under random mating. Inbreeding occurs in natural and cultured populations due to matings among relatives and positive assortative mating when phenotypic resemblance is based on heritable traits. In the context of fisheries management or aquaculture, inbreeding can result from mate selection in the natural environment or from crossings by humans of related individuals in hatcheries. This chapter will cover a range of topics regarding inbreeding, including estimation of inbreeding, inbreeding depression, hypotheses explaining inbreeding depression, and implications of inbreeding in natural populations and cultured fish stocks.

The concept of inbreeding often is misunderstood by fisheries managers and the general public. Frequently, the term inbreeding is used erroneously to explain any genetics-related reason for poor performance of a population in a given situation. That is, inbreeding may be invoked to explain the failure of a population to respond to a selective challenge when the actual reason may be lack of genetic variation due to random genetic drift (Chapter 9), that is, a limited number of individuals breeding rather than matings among relatives. Although both inbreeding and drift may act upon the gene pool of a small population (Chapter 18), their effect on genetic variability is different. Inbreeding acts directly upon genotype frequencies, increasing the frequency of homozygotes while decreasing that of heterozygotes, with no direct effect on allele frequencies. In contrast, random genetic drift acts directly upon allele frequencies.

10.2 Estimation of Inbreeding

Two related individuals share one or more common ancestor(s). These related individuals may carry alleles in common because they inherited the same alleles from the common ancestor(s). Should these individuals mate, their offspring will carry at some loci identical copies of alleles received from both parents that originated from the common ancestor. The **level of inbreeding,** then, is defined as the probability that the two alleles at a given locus are identical copies of a single ancestral allele (Gall 1987). The estimated level of inbreeding is quantified by the **inbreeding coefficient,** denoted as *F*. Inbreeding is quantified in a relative fashion by estimating the increase in homozygosity, or decrease in heterozygosity, relative to some ancestral or **reference or base population.** The reference population is assumed to be noninbred, that is, its *F* equals 0. The reference population may be the original source population used to establish the population of interest or an intervening past generation from which the increase in inbreeding level is estimated.

Estimating the level of inbreeding requires that distinction be made among two classes of homozygotes. The process of gamete formation is such that, in the absence of mutation, offspring will carry exact copies of alleles transmitted from the parent. Such alleles are **identical by descent** (IBD) from a common ancestor. In contrast, alleles that are identical but are not recent copies of an ancestral allele are said to be **alike in state** (AIS). All homozygotes in the reference population are considered AIS homozygotes. An inbred population, however, will contain both AIS and IBD homozygotes. The inbreeding coefficient, *F*, is the probability that two alleles at a locus are IBD, or a measure of the increase in homozygosity due to inbreeding. The inbreeding coefficient, *F*, also can be defined as the percent increase in homozygosity in a lineage over the population mean homozygosity, that is, the percent increase in homozygosity above that which would have occurred had breeding been random.

10.2.1 *Calculating Inbreeding Coefficients from Pedigree Information*

An individual is inbred if one or more individuals appears on both the maternal and paternal sides of its pedigree. An analytical technique called path analysis (Gall 1987; Tave 1993) is used to identify the lines of descent from a common ancestor to parents of an individual of interest and to estimate the contribution of each line of descent to the resulting level of inbreeding of the individual.

10.2.1.1 Evaluating Inbreeding in Simple Pedigrees. To illustrate estimation of inbreeding for a simple case, let's evaluate the inbreeding coefficient, F_I, of individual I produced by a mating of cousins (Figure 10.1A). The first step in the analysis is to identify individuals in the pedigree that are common ancestors to both the sire and dam of an individual in question. The pedigree for individual I shows that both grandparents, A and B, are common ancestors to I's parents, G and H. The second step in the analysis is to use the pedigree to draw a path diagram (Figure 10.1B) showing the paths of inheritance from the common ancestor(s) to the inbred individual. Arrows are used to identify the direction of inheritance for each path. This diagram allows us to trace the paths by which an IBD allele in either grandparent A or B could be inherited by individual I and to estimate the probability of that occurrence. Because there are two common ancestors in this case, A and B, there will be two paths, one including common ancestor A and one including common ancestor B (Figure 10.1C). Using the path diagram, count the number of individuals in the path from the individual being examined, I, moving against the arrows

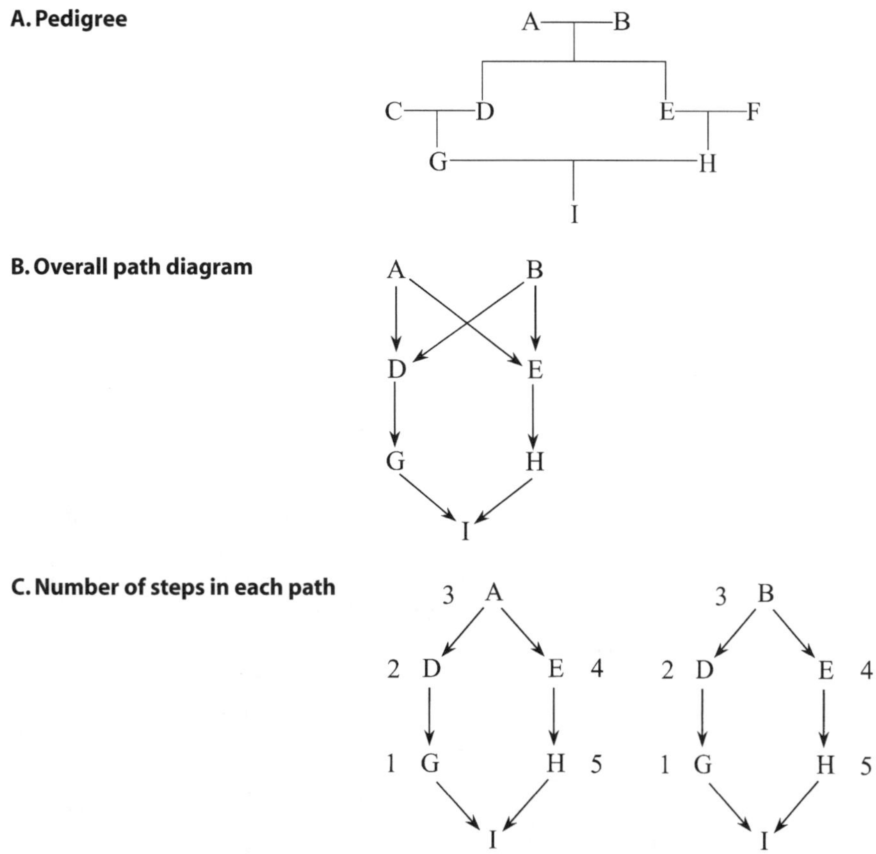

D. Calculation of estimated inbreeding coefficient

For the path through common ancestor A:
$F = (½)^5 (1 + 0) = 0.03125$

For the path through common ancestor B:
$F = (½)^5 (1 + 0) = 0.03125$

F_I equals the summation of paths for all common ancestors:

$F_I = 0.03125 + 0.03125$
$= 0.0625$

Figure 10.1 Estimating the inbreeding coefficient *(F)* for offspring of a mating of cousins. The numbers in (C) count the number of steps in the path from one parent of I, through the common ancestor, to the other parent.

from one parent to the common ancestor, then moving with the arrows back through the other parent. The paths are G ← D ← **A** → E → H and G ← D ← **B** → E → H. The number of individuals in each path is five (Figure 10.1C).

Next, to calculate the inbreeding coefficient (F_x) of an individual, make appropriate substitutions into the expression

$$F_x = \Sigma\, [(½)^n(1 + F_A)], \tag{10.1}$$

in which F_x is the inbreeding coefficient for individual x. The number of individuals in a given path is given by n, and the value ½ is raised to the nth power because in each generation, the likelihood is ½ that a particular ancestral allele will be transmitted from a parent to individual offspring. In equation (10.1), F_A is the inbreeding coefficient for the common ancestor. If this number is not specified, it is assumed that F_A is 0. This assumption generally is reasonable for wild populations, although it may not be true for small populations or for cultured stocks of small to moderate size. Summation (Σ) indicates that the contributions to inbreeding for each path are independent and summed to estimate the overall inbreeding coefficient for the individual in question.

In the example for individual I, there are two five-membered paths, one through each grandparent. Each path contributes $(\frac{1}{2})^5$ to I's inbreeding coefficient. Hence, $F_I = (\frac{1}{2})^5 + (\frac{1}{2})^5 = \frac{1}{32} + \frac{1}{32} = \frac{1}{16} = 0.0625$ (Figure 10.1D).

10.2.1.2 Evaluating Inbreeding in More Complex Pedigrees. Matings among relatives can result in more complex relationships than that in the example above. However, the procedure for estimating the degree of inbreeding in a more complex pedigree follows that outlined above (Gall 1987; Tave 1993: 221).

1. Identify ancestors common to both sides of an individual's pedigree.
2. Construct a path diagram linking the individual in question with each common ancestor. If more than one common ancestor exists, a path diagram is constructed for each. If more than one line of descent links an ancestor with a descendant, there will be more than one path between them. However, an intermediate ancestor can occur only once in each pathway.
3. Estimate the degree of inbreeding associated with each path estimated as $(\frac{1}{2})^n$. Should the common ancestor itself be inbred, the effect of that inbreeding on the inbreeding of the individual in question can be accounted for by substituting the ancestor's inbreeding coefficient, F_A, into the term $(1 + F_A)$ in expression (10.1). However, inbreeding coefficients of all individuals in the path is ignored except for that of the common ancestor(s).
4. Calculate the contribution of each path to the degree of inbreeding independently and add it to the total.

The procedure for estimating the inbreeding coefficient for an individual in a complex pedigree can best be understood by means of an example. Let's assume that the simple pedigree considered above is but a part of the more complex pedigree shown in Figure 10.2A. In this pedigree, common ancestor A now is assumed to be inbred, with F_A is 0.0625. Let's estimate the inbreeding coefficient for individuals O (F_O) and Q (F_Q).

We start by estimating the degree of inbreeding for individual O. Examining the pedigree (Figure 10.2A), we identify two common ancestors, A and B. The next step is to draw the path diagram from O through parent I to A and B and returning to parent N (Figure 10.2B). Four paths contribute to the inbreeding coefficient of individual O (Figure 10.2C). The contributions of the four paths are calculated (Figure 10.4D). Note that paths 1 and 3 with common ancestor A at the apex include the nonzero term for F_A, increasing the calculated inbreeding value associated with that path. Summing across paths, the estimated inbreeding coefficient for individual O is 0.032.

Let's move on to estimate the inbreeding coefficient for individual Q. Examining its pedigree, we note although individual O is inbred, it is **outcrossed,** that is, mated to an unrelated individual, P. Hence, the progeny of O and P, including Q, are not inbred, and F_Q is 0.

The latter example illustrates a key point. Inbreeding can be reduced to zero by mating unrelated individuals. Consequently, if all individuals in a captive breeding program are tracked with pedigrees and matings are carefully planned, then inbreeding in captive

A. Pedigree

B. Overall path diagram

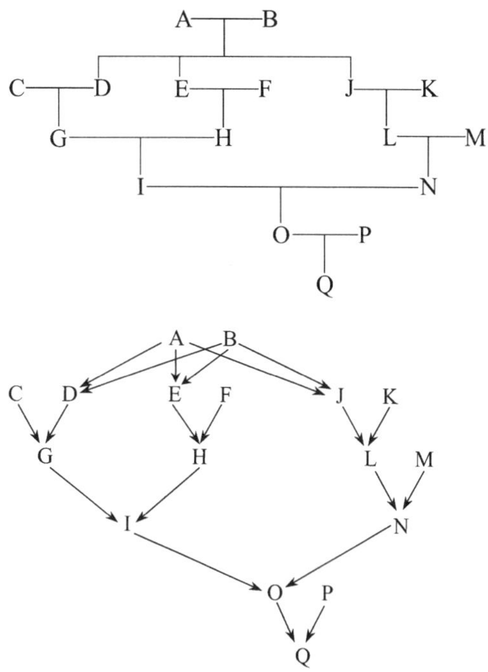

C. Number of steps in each path

$I \leftarrow G \leftarrow D \leftarrow \mathbf{A} \rightarrow J \rightarrow L \rightarrow N$ (1)

$I \leftarrow G \leftarrow D \leftarrow \mathbf{B} \rightarrow J \rightarrow L \rightarrow N$ (2)

$I \leftarrow H \leftarrow E \leftarrow \mathbf{A} \rightarrow J \rightarrow L \rightarrow N$ (3)

$I \leftarrow H \leftarrow E \leftarrow \mathbf{B} \rightarrow J \rightarrow L \rightarrow N$ (4)

D. Calculation of estimated inbreeding coefficient

For path 1: $F = (\frac{1}{2})^7 (1 + 0.0625) = 0.0083$
For path 2: $F = (\frac{1}{2})^7 (1 + 0) = 0.0078$
For path 3: $F = (\frac{1}{2})^7 (1 + 0.0625) = 0.0083$
For path 4: $F = (\frac{1}{2})^7 (1 + 0) = 0.0078$

$$F_O = \Sigma \text{ (all paths)} = 0.0322$$

Figure 10.2 Estimating the inbreeding coefficient *(F)* for offspring of a more complex pedigree system including multiple paths and an inbred ancestor.

populations can be eliminated by mating unrelated individuals. Note, however, that in small populations this is possible for only a few generations. However, planned matings can greatly reduce the rate of inbreeding accumulation.

Truly complex pedigrees will result from breeding within closed populations. With passage of a number of generations, breeding within closed populations can result in considerable inbreeding accumulation. A good example is provided by herds of cattle in the United States. The mean inbreeding coefficients for cows of five breeds born in the late 1990s ranged from 0.045 to 0.06 (Weigel 2001). The widespread use of a limited number of proven Holstein sires has resulted in particularly complex pedigrees, in which individuals may have inbreeding coefficients of 0.30 or more. Captive populations in zoos also can have highly complex pedigrees, as exemplified by Laikre and Ryman's (1991) study of blindness in captive wolves *Canis lupus*. Closed aquacultural stocks may have complex pedigrees, although regrettably, pedigrees rarely are kept. A number of computational methods have been developed for calculating inbreeding coefficients for large populations for which pedigrees are known (Emik and Terrill 1949; Cruden 1949 as cited by Gall 1987).

10.2.2 Estimating Inbreeding Coefficients without Pedigree Information

Calculating a discrete estimate of the inbreeding coefficient for an individual is based on knowledge of its pedigree. However, pedigrees for individuals in wild populations and for most cultured stocks are unknown. Even for a hatchery population for which the breeding system is well documented, estimation of inbreeding is complicated by lack of information on genotypic frequencies in the founder population, effects of selection practiced to obtain spawners, and differential survival of progeny (Gall 1987).

In the absence of pedigree information, we cannot calculate F for individuals, but we can estimate the mean rate of inbreeding for the population if we make a key simplifying assumption. Assuming that each individual contributed equally to the progeny generation, a mean F for the population can be estimated when the numbers of breeders of each sex used to propagate a population are known (Tave 1993: Table 4.14). Because this simplifying assumption is unlikely to be met fully in most situations, the resulting estimated rate of inbreeding per generation (Table 10.1) should be regarded as a lower estimate than that actually occurring in the population.

The rate of inbreeding for a population also can be estimated as a function of its inbreeding effective population size, N_e, as

$$F = ½N_e. \tag{10.2}$$

Procedures for estimation of N_e are considered in detail in Chapter 9.

10.3 Inbreeding Depression

Every individual carries recessive alleles that are not expressed because they are "masked" in the heterozygous state. Many of these alleles are deleterious, and were they expressed, they would negatively affect the fitness of the carrier, perhaps dramatically as

Table 10.1 Percent inbreeding per generation in a population with random mating, assuming each parent contributes an equal number of offspring. Table adapted from Tave (1990, 1993).

Number of females	Number of males: 1	2	5	10	25	100	250	Infinite
1	25.00	18.75	15.00	13.75	13.00	12.62	12.55	12.50
2	18.75	12.50	8.75	7.50	6.75	6.38	6.30	6.25
5	15.00	8.75	5.00	3.75	3.00	2.62	2.55	2.50
10	13.75	7.50	3.75	2.50	1.75	1.38	1.30	1.25
25	13.00	6.75	3.00	1.75	1.00	0.62	0.55	0.50
100	12.62	6.38	2.62	1.38	0.62	0.25	0.18	0.12
250	12.55	6.30	2.55	1.30	0.55	0.18	0.10	0.05
Infinite	12.50	6.25	2.50	1.25	0.50	0.12	0.05	0.00

an abnormal or lethal phenotype. Related individuals are more likely to share the same recessive alleles than are unrelated individuals. Inbreeding increases the rate of homozygosity by pairing alleles that are identical by descent, thereby increasing the likelihood that rare deleterious alleles will be expressed in the homozygous state. The probability of deleterious recessive alleles pairing increases as the relatedness of the parents increases. Expression of deleterious recessive alleles can affect the fitness of the carriers as decreased larval viability, survival through key life cycle events, growth rate, or reproductive ability. The frequency of abnormalities is increased. Decreased fitness due to inbreeding is referred to as **inbreeding depression.** Increasing levels of inbreeding tend to be associated with more pronounced inbreeding depression.

10.3.1 *Demonstrating Inbreeding Depression*

The landmark studies demonstrating inbreeding depression in fishes involved rainbow trout *Oncorhynchus mykiss.* Starting with a noninbred stock, Kincaid (1976a, 1976b, 1983) developed lines of rainbow trout with inbreeding ranging from 12.5% to 59.4%. He also maintained a noninbred control line from the same base stock. Expression of a variety of fitness-related traits was quantified, involving viability (hatch and survival rates), growth (weight and length at seven ages), and reproduction (weight of egg mass and percent deformed). The performance of the inbred lines for each trait was compared with that of the noninbred control line. Hence, inbreeding depression was reported as a negative value, indicating the percentage of performance decrease in an inbred line versus the control line. A positive value indicated the degree to which the mean value for a trait was larger in an inbred population. The findings (Table 10.2) illustrated several key points about inbreeding depression. First, at a relatively low level of inbreeding, 12.5%, viability traits were negatively affected but growth rate was modestly higher in the inbred line than in the control line. Low levels of inbreeding enhanced growth rates when favorable allele combinations were brought together in genotypes, but not so at higher levels of inbreeding. At higher levels of inbreeding, 18.5% and above, all traits except hatch rate were negatively affected. In addition, inbreeding depression was more pronounced at higher levels of inbreeding. The effect of inbreeding on growth rate is shown dramatically as size at age in Figure 10.3. Reproductive fitness, measured as fecundity, was profoundly affected by inbreeding, in the most highly inbred line decreasing to less than half that in the control line.

Kincaid's studies of rainbow trout are exceptional because inbred lines were produced for the purpose of investigating inbreeding depression. It should be noted that the level of inbreeding at which inbreeding depression becomes evident is not clearly defined and will vary among species. Inbreeding depression depends on the particular alleles and allele frequencies in a pedigree. Hence, the sensitivity to inbreeding can vary among lines and populations of a species (Box 10.1). Some degree of sensitivity to inbreeding depression seems common to virtually all organisms studied. Inbreeding depression has been shown for a number of traits in cultured finfish stocks, including survival and growth (Gjerde et al. 1983; Gjerde 1988) and abnormal body curvature (Aulstad and Kittelsen 1971) in rainbow trout, growth rate in brook trout *Salvelinus fontinalis* (Cooper 1961), and growth rate (Bondari 1984) and a number of economic traits (Bondari and Dunham 1987) in channel catfish *Ictalurus punctatus.* The effects of inbreeding in shellfish have proven inconsistent. Longwell and Stiles (1973) observed severely reduced larval viability and moderate growth rate reduction in larvae of brother–sister matings in eastern oyster *Crassostrea virginica.* However, Mallet and Haley (1983) noted higher survival

Table 10.2 Inbreeding depression in rainbow trout. Negative values indicate the percentage by which the trait s mean value is smaller in an inbred population as compared with the control, noninbred population ($F = 0.0\%$). Positive values indicate the degree to which the trait s mean values is larger in the inbred population. Table adapted from Kincaid 1983.

	Inbreeding coefficient, *F* (%)					
Trait	12.5	18.5	25.0	37.5	50.0	59.4
Hatch rate	–4.5	1.7	10.1	–5.5	9.0	
Survival rate	–2.5	–14.4	–7.7	–4.0	–11.0	
77-d weight	0.0	–15.4	–16.7	–23.0	–29.0	
91-d weight	4.4	–16.3	–17.5	–33.0	–30.0	
105-d weight	6.8	–14.2	–15.6	–32.0	–30.0	
126-d weight	3.8	–7.9	–15.7	–23.0	–20.0	
150-d weight	8.5	–1.8	–11.6	–21.0	–21.7	
1-year weight						
Males			–23.3	–16.4	–12.8	–34.8
Females			–20.9	–27.2	–21.9	–41.8
2-year weight						
Males			–26.2	–32.2	–33.7	–41.8
Females			–18.0	–38.1	–28.8	–51.2
2-year length						
Males			–9.2	–13.0	–15.2	–19.0
Females			–9.1	–16.8	–13.9	–21.8
Weight of egg mass			–18.1	–33.9	–40.3	–57.0
Percent deformed			–2.9	–10.0		

Figure 10.3 Comparison of size at age for inbred and noninbred rainbow trout strains derived from the same founder stock and reared under the same conditions. Photo provided courtesy of H. Kincaid.

■ Box 10.1
The Effects of Lineage and Environment on Inbreeding Depression

Certain aspects of inbreeding depression relevant to fisheries management and aquaculture prove difficult to study using managed aquatic organisms. However, experimental designs exploiting the favorable attributes of laboratory model systems—that is, ease of maintenance and short generation time—can offer insights useful to fisheries managers and aquaculturists. Studies using the red flour beetle *Triboleum castaneum* demonstrated that inbreeding depression is influenced by lineage and environmental factors.

Pray et al. (1994) measured fitness in two environments of a series of 35 inbred lineages of the red flour beetle derived from a single ancestral population. Humidity has major effects on the population growth and competitive ability of flour beetles *Triboleum* spp. Hence, one treatment was maintained at 70% relative humidity, and the other was allowed to vary (37.8% relative humidity, SD 10.8%). Both treatments were held at 29°C. Inbred lines were produced by five generations of full-sib mating, resulting in an inbreeding coefficient *(F)* of 0.672 relative to the outbred stock (control) population ($F = 0$). Inbreeding depression in each environment was measured by comparing the mean fitness of inbred and control lines. Test populations consisted of five males and five females. Nine individuals were homozygous *b*/*b* for black body coloration; one test individual, whose fitness was measured directly, was homozygous +/+ for red coloration. Progeny of the test individual were heterozygous +/*b* and phenotypically distinguishable. The test individual's fitness was measured as the proportion of brown offspring in the test population. For any particular lineage from which this individual was drawn, the difference between its mean fitness and that of the control line was taken as a measure of inbreeding depression.

Inbreeding depression was evident as significant differences in fitness among inbred and control groups (see table on next page). The effects of inbreeding depression were more pronounced for males than for females in both treatments. Both sexes exhibited more inbreeding depression in the variable than in the constant humidity environment, shown as a greater difference for the "control – inbred" column in the table.

Analysis of variance of the fitness data showed a statistically significant effect of lineage on the degree of inbreeding depression. Although all lineages were recently derived from the same founder stock, the random loss of genetic variation in the five generations of sib-matings resulted in different alleles becoming combined in the homozygous state in the respective inbred lines. By chance, some lines were affected at a greater proportion of fitness-related loci than were other lines. Hence, some lines were better adapted and some less well adapted than others to deal with the stresses of experimental conditions.

Lineages were ranked by fitness within each treatment for each sex. Changes in the relative rankings of each lineage were observed between the two humidity treatments. This genotype-by-environment interaction (Chapter 12) was statistically significant, showing that the fitness of some lineages was more sensitive to environment than that of others.

continued on next page

■ Box 10.1 *continued*

Table Mean proportion of offspring produced by 35 lineages of inbred and control red flour beetles reared in different humidity regimes. Table adapted from Pray et al. 1994.

Sex and humidity regime	Mean fitness (±SE) Inbred	Mean fitness (±SE) Control	Inbred vs. control[a]	Control – inbred
Female				
Constant	0.16 ± 0.06	0.20 ± 0.04	$P < 0.001$	0.04
Variable	0.17 ± 0.06	0.23 ± 0.05	$P < 0.0001$	0.06
Male				
Constant	0.11 ± 0.07	0.16 ± 0.07	$P < 0.0001$	0.05
Variable	0.16 ± 0.08	0.25 ± 0.07	$P < 0.0001$	0.09

[a]For females, data for both environments were tested with a one-tailed *t*-test. For males, data for the variable environment were tested with a one-tailed *t*-test; for the constant environment data were tested with a Welch analysis of variance for unequal variances.

The key findings of this study, that the effects of inbreeding can differ among lineages and environments and that some lineages are more sensitive to environmental effects than are others, pose implications beyond the model system to the practice of fisheries management. For example, the environmental dependency of inbreeding depression suggests that estimates of inbreeding depression for a population in a captive breeding setting may not predict the level of inbreeding depression in the wild. This poses a cautionary note for predicting the outcome of reintroduction of hatchery-produced endangered species into the wild, emphasized by the observation that inbreeding depression in this study was more severe in the variable environment than it was in the constant environment. Significant variation in the level of inbreeding depression among the closely related lineages in Pray et al.'s study suggests that estimates of inbreeding depression for one population cannot be extrapolated to another with a great degree of confidence, even within the same species. Quantification of a population's response to inbreeding requires scarce experimental resources. The best strategy for any propagation program, whether for fisheries management or for aquaculture, is to plan matings carefully so as to avoid inbreeding (Chapter 14).

with increased homozygosity in eastern oyster at the larval stage, no difference in survival of inbred and outbred eastern oysters from juvenile to adult phase, and lower growth rate through the adult stage in inbred oysters. Lannan (1980) found no evidence of inbreeding effects on survival of larvae of Pacific oyster *C. gigas* to metamorphosis.

Genetic uniformity is valued in stocks used as laboratory model systems because genetic variation then can be ruled out as a factor affecting the response to an experimental treatment. However, inbreeding depression, as shown in zebra danio *Danio rerio* (Piron 1978; Mrakovcic and Haley 1979), affects the suitability of a stock for experimental work.

The potential impacts of inbreeding depression on fisheries management and aquaculture are considerable. Within the context of fisheries management, inbreeding was shown to affect the recapture rate of stocked Atlantic salmon *Salmo salar* (Ryman 1970), presumably by acting upon survival rate (Box 10.2). In an evaluation of inbreeding in salmonid broodstocks at federal and state hatcheries, inbreeding rate ranged from 0 to 10.4% among reported broodstocks (Kincaid 1995); 79.2% of broodstocks had inbreeding rates less than 0.5% per generation, and 6.9% had inbreeding rates greater than 1.0% per generation. Within the context of production aquaculture, computer simulation of breeding programs typical of those used by many salmonid farmers (McKay et al. 1992)

■ Box 10.2
Effect of Inbreeding on Hatchery-Based Enhancement Programs

Stocking hatchery-produced fish is an important fisheries management practice. If a limited number of spawners is propagated in a closed stock, does this pose the prospect of inbreeding depression? If a fitness-related trait is affected, could inbreeding affect the effectiveness of the hatchery-based enhancement program?

Ryman (1970) examined recapture frequencies of Atlantic salmon marked at age 2 and released from a hatchery in Sweden from 1958 through 1965. Recapture frequencies of 44 families representing various degrees of inbreeding were compared with those of 351 presumably noninbred families. Family size varied from 100 to 1,700 individuals, with a mean of approximately 500. Mean recapture frequencies varied considerably between rivers and years, with an overall return rate of about 10%. To make comparisons among families, rivers, and years, the return rate, O, for each family was expressed as a deviation, measured in standard errors, from the mean for families stocked into the river in question for the year of release. The Z scores (i.e., the proportion of the normal distribution for observations more extreme than the observation under consideration) were calculated for each family as

$$Z = (O_{ijk} - O_{ij}) \,/\sigma(O_{ij}),$$

where the subscript i denotes river, j year, and k family.

Return rates for inbred families were lower than those for noninbred families (see table below). Differences in return rates for inbred and noninbred families were statistically significant (t = 2.45, 393 df). Recapture rates were more variable for inbred than for noninbred families.

Lower recapture rates for inbred families strongly suggest a lower survival rate for inbred fish. This finding poses important implications for planning matings in hatchery programs, underlining the importance of maintaining pedigrees in closed stocks and avoiding matings among relatives. Cognizance of the potential for inbreeding depression is an important part of hatchery operations (Chapter 14).

Table Recapture rates of adult Atlantic salmon from noninbred and inbred families. Table adapted from Ryman 1970.

	Noninbred (F = 0)	Inbred (F > 0)
Number of families	351	44
Z, mean transformed return rate	0.052	–0.418
Variance of Z	0.9125	1.5577

showed that inbreeding would increase by 3 to 5% per generation. This rate of inbreeding could cause inbreeding depression within three generations, affecting the efficiency and profitability of the enterprise. Systematic mating designs, notably rotational line crossing (Kincaid 1977), have been proposed as an approach for reducing the rate of inbreeding accumulation in managed broodstocks. Alternatively, many broodstock management plans call for marking and tracking individual broodstock replacement candidates to preclude matings among relatives. The high potential of detrimental impacts from inbreeding depression warrants careful planning for fish and shellfish propagation programs, especially for stocks intended for planting in the wild (Chapter 14).

Although many studies have addressed inbreeding in captive stocks, relatively little work has addressed inbreeding or inbreeding depression in natural populations of aquatic organisms (Box 10.3). Lack of genetic variation reduced the rate of population growth in western mosquitofish *Gambusia affinis* (Leberg 1990). Should such a finding prove general, it poses significant implications for conservation of small populations (Chapter 18). The possible importance of inbreeding and inbreeding depression warrants more work on natural populations of aquatic organisms.

■ Box 10.3
Inbreeding in Natural Populations

Whereas studies of captive or experimental animal populations suggest the ubiquity of inbreeding effects, relatively few studies have addressed inbreeding in natural populations. This is especially true for populations of aquatic organisms. Ralls et al. (1986) examined the inbreeding literature for natural populations of birds and mammals. Their findings and discussion have relevance for those interested in other taxa as well.

The question to be considered at the outset is, what is the frequency of matings between close relatives in natural populations? Direct evidence can be drawn from field studies based on identifiable individuals. Ralls et al. (1986) noted that parent–offspring or full-sib matings ranged from 0 to 6% among species observed. Considering that many of the populations observed were small, this would seem to be a low frequency of matings among relatives. However, it is difficult to apply statistical tests to determine whether such matings are less frequent than what would be expected by chance. Because it is difficult to determine the *expected* frequency of matings among relatives, it is problematic to test *departures* of data from a demonstrably correct null hypothesis. Assumptions regarding the effect of natal or reproductive dispersal can affect the test results. Ralls et al. (1986) felt that available evidence tended to support the inference that matings between close relatives occur less frequently than by chance.

A second question is, if matings between relatives are less common than what would be expected by chance, how is this mediated? One possible inbreeding avoidance mechanism involves reduction of opportunity for mating among relatives. Temporal or spatial isolation would reduce the likelihood that close relatives would encounter each other during the breeding season. The most common mechanism of inbreeding avoidance is sex bias in natal or breeding dispersal. This is especially common in species with promiscuous or polygynous mating systems. Juvenile females are

continued on next page

10.3.2 *Increasing the Frequency of Expression of Recessive Alleles through Inbreeding*

The phenomenon of inbreeding depression raises the question of its underlying mechanism. It frequently is assumed that inbreeding depression is the consequence of a heightened frequency of expression of deleterious recessive alleles. How, then, does the mating of related individuals give rise to this heightened frequency? The frequency of individuals in a population carrying the homozygous recessive *aa* genotype can be calculated (Hartl 1988:50) as

$$\text{frequency } (aa) = q^2(1 - F) + qF, \tag{10.3}$$

where F is the inbreeding coefficient and q is the frequency of allele a in the population. To see how a mating of relatives increases the likelihood that a maladaptive trait will be expressed, let's consider the example of phenylketonuria in humans (Research and Education Association 1985:321), in which expression of the recessive genotype results in severe mental retardation in untreated infants. The frequency of the recessive allele, q, is 0.01 in the general human population. The risk of phenylketonuria for children of unrelated parents, then, is $q^2 = (0.01)^2 = 0.0001$. What, then, would the risk be if the parents are first cousins? As shown section 10.2.1.1, the inbreeding coefficient for offspring of a

■ Box 10.3 *continued*

the predominant dispersers in birds and juvenile males in mammals. There is considerable controversy over proximate and ultimate causes of sex-biased dispersal. A key argument is whether such dispersal is primarily for inbreeding avoidance. Reports on sexual biases in dispersal of fishes are not available. Another mechanism minimizing opportunity for relatives to mate is monogamous mating and high annual mortality, as best demonstrated by passerine birds.

Another inbreeding avoidance mechanism, should relatives meet during the mating season and have the opportunity to mate, would have the mating not be realized. This mechanism depends on the ability of animals to recognize individuals likely to be close relatives and, on such recognition, to avoid such mating. Kin recognition is known to occur in many bird and mammal species. Cues include recognition of natal location (with the assumption that an individual there may be a relative), prior association with an individual, and smell (with recognition mediated by immunogenetic loci). The occurrence and extent of kin recognition in fishes is largely uninvestigated. In cases in which a potential mate is rejected, most frequently it is the female that rejects the male. This is not surprising, as the female has a greater physiological investment in the young.

Inbreeding in natural populations of fishes has not been well investigated. Fisheries scientists have tended to study the effects of inbreeding (e.g., heightened frequency of fluctuating asymmetries) as opposed to demonstrating its occurrence (e.g., by monitoring matings among physically or genetically marked individuals). We have much to learn about the occurrence, frequency, and fitness implications of inbreeding in natural populations of fishes.

first-cousin mating is 0.0625. The risk of the disease among children of first cousins can be found by making appropriate substitutions into equation (10.3). Hence,

$$\text{frequency } (aa) = (0.01)^2(1 - 0.0625) + (0.01)(0.0625)$$
$$= 0.000937 + 0.00625 = 0.0007.$$

Comparing this result with the frequency for expression of the trait in the general population, 0.0001, we see that risk of expression of the disease is increased sevenfold in a mating of cousins. Although the risk for this or any other genetic disorder caused by a recessive genotype at a single locus remains small, the risk or producing recessive genotypes at many loci across the genome is increased significantly in matings among relatives. Hence, matings among related individuals can appreciably increase the risk of some fitness-related trait being negatively affected. Recognition of this risk led to the taboo against cousin marriages in many human cultures. Loss of fitness due to inbreeding depression presumably selected for development of inbreeding avoidance mechanisms in many organisms (Box 10.3).

■ 10.4 Alternative Hypotheses Regarding Inbreeding Depression

Two rival hypotheses regarding the genetic basis for inbreeding depression are debated in the genetics literature (Charlesworth and Charlesworth 1987). The **dominance hypothesis** states that loss of fitness due to inbreeding results from the heightened frequency of expression of deleterious recessive alleles across the genome. Discussion of inbreeding to this point in this chapter has been based upon the dominance hypothesis. However, an alternative hypothesis is advanced by some geneticists. The **overdominance hypothesis** states that fitness is determined by genome-wide heterozygosity levels. Under this hypothesis, high heterozygosity is inherently advantageous; hence, it sometimes is called the heterozygous advantage hypothesis. Inbreeding depression, then, is the consequence of reduced heterozygosity throughout the genome due to matings between related individuals. Two mechanisms might explain the overdominance effect. When high fitness of the heterozygote is the consequence of variation at a single locus, often a genetic marker locus, it is termed **direct overdominance;** this generally is regarded as the exceptional case. When high fitness of the heterozygote is the consequence of variation at loci associated with the marker locus, it is termed **associative overdominance.** Associative overdominance is regarded as the more common phenomenon.

Several lines of evidence have been cited to support the relationship between heterozygosity and fitness-related traits (reviewed by Mitton and Grant 1984; Allendorf and Leary 1986). Examples of differential fitness of homozygotes and heterozygotes at protein or DNA loci in natural populations of aquatic organisms are presented in Table 10.3. Key fitness-related traits include the following.

— *Increased survival rate.* The most defensible association of heterozygosity with survival rate is by demonstration of temporal changes in the genetic composition of a cohort. More species exhibited an increase in heterozygosity after a selection event (e.g., a hard winter) than showed a decrease. Other studies have used comparisons among cohorts, age-classes, or size-classes to support the inference that heterozygosity confers a survival advantage, although such conclusions are open to alternative interpretations (Allendorf and Leary 1986).

Table 10.3 Examples of differential fitness of homozygotes and heterozygotes at genetic marker loci in natural populations of aquatic organisms. Fitness of heterozygote indicated as (+) heterozygous advantage and ( ) heterozygous disadvantage or absence of an effect. Table adapted from Allendorf and Leary (1986), with addition of recent citations.

Species	Trait	Fitness of heterozygote	Source
	Finfishes		
Bleak *Alburnus alburnus*	Body size	+	Handford 1963
Livebearing eelpout *Zoarces viviparus*	Survival	–	Christiansen et al. 1977
Mummichog *Fundulus heteroclitus*	Survival	+	Mitton and Koehn 1975
Plaice *Pleuronectes platessa*	Survival	+	Beardmore and Ward 1977
Sand shiner *Notropis stramineus*	Body size	+	Koehn et al. 1971
Skipjack tuna *Katsuwonus pelamis*	Body size	–	Fujino and Kang 1968
	Shellfishes		
Blue mussel *Mytilus edulis*	Growth rate	–	Koehn and Gaffney 1984
		–	Mallet et al. 1986
	Survival	+	Koehn et al. 1976
California mussel *Mytilus carpenteri*	Survival	+	Tracey et al. 1975
Dward surf clam *Mulinia lateralis*	Growth rate	+	Gaffney et al. 1990
		+	Koehn et al. 1988
Eastern oyster *Crassostrea virginica*	Growth rate	+	Koehn and Shumway 1982
		–	Foltz and Chatry 1986
	Metabolic rate	+	Rodhouse & Gaffney 1984
	Survival	+	Singh 1982
Marine bivalve *Baltic macoma*	Survival	+	Singh and Green 1984
Spisula ovalis	Survival	+	David and Jarne 1997
Northern quahog *Mercenaria mercenaria*	Growth rate	–	Dillon and Manzi 1988
	Survival	–	Adamkewicz et al. 1984
Pacific oyster *Crassostrea gigas*	Survival	+	Buroker 1979
		+	Fujio et al. 1979
Ribbed mussel *Guekensia demissa*	Survival	+	Chiasson et al. 1976
		+	Koehn et al. 1973
Scallop *Pecten maximus*	Survival	–	Wilkins 1978
Sea scallop *Placopectin magellanicus*	Survival	+	Foltz and Zouros 1984
	Growth rate	+	Pogsen and Zouros 1994

— *Increased growth rate*. In most cases, association of heterozygosity with growth has been approached by seeking correlation of the number of heterozygous loci per individual with growth rate. Many of these studies have involved marine bivalves. However, the association of growth rate with heterozygosity may be inconsistent, depending on age, developmental stage, and environmental conditions.

— *Developmental rate*. Associations between heterozygosity and developmental rate have been demonstrated in two fishes. Heterozygotes at a lactate dehydrogenase locus (enzyme number 1.1.1.27; IUBMBNC 1992) in mummichog exhibited a hatching time intermediate to the two homozygotes (DiMichele and Powers 1982), although it is important to note that the enzyme in question affects blood oxygen affinity. More highly heterozygous individual rainbow trout consistently hatched sooner than did more homozygous individuals (Danzmann et al. 1986).

— *Increased developmental stability*. Genetically determined developmental pathways are widely believed to be expressed more reliably in the phenotypes of highly heterozygous individuals. **Fluctuating asymmetry** is the occurrence of randomly distributed differences in size, shape, or number for bilateral traits among the sides of an individual (Box 10.4); it is an indicator of developmental instability. Higher levels of heterozygosity at isozyme loci have been correlated with lower levels of fluctuating asymmetry in several aquatic organisms, including the freshwater mussels *Elliptio complanata* and *Lampsilis radiata* (Kat 1982), the Mexican livebearer *Poeciliopsis monacha* (Vrijnhoek and Lerman 1982), rainbow trout (Leary et al. 1983, 1985), brook trout (Leary et al. 1984), and cutthroat trout *Oncorhynchus clarki* (Leary et al. 1984, 1985).

— *Improved disease resistance*. Heterozygotes in mammals and birds frequently exhibit higher disease resistance than do homozygotes (Allendorf and Leary 1986). This is especially well explained when heterozygosity at histocompatibility loci is demonstrated. Histocompatibility loci mediate the immunological recognition of self and not-self epitopes of molecules, affecting both recognition of pathogens and coordination of the immune response.

Despite the existence of a body of supporting data, the overdominance hypothesis has not, in a general sense, been proven. Both positive and negative effects of heterozygosity have been observed. Further, many of the studies supporting the hypothesis have been flawed because the effects of immigration and emigration were not excluded (Allendorf and Leary 1986), and the groups studied may not have been closed populations. Perhaps most important in a general sense, the mechanism (or mechanisms) by which heterozygosity would affect development, growth, survival, and other traits is not clear. One possible explanation is that heterozygosity confers an increased physiological ability to compensate for environmental variation. For example, the gene product of one allele might function well at high temperature and that of another allele at low temperature. Thus, a heterozygous individual might exhibit higher fitness for development, survival, or growth traits than would a homozygote under changing environmental conditions.

10.5 Perspective: Conservation Implications of Inbreeding Depression

From the conservation viewpoint, it is important to note that the dominance and overdominance hypotheses support different predictions about the relative sensitivity of populations to inbreeding (Lacy 1992). If inbreeding is the result of expression of deleterious recessive alleles, then selection should have removed most such alleles from populations with long histories of inbreeding, and such populations should prove resistant to further impacts of inbreeding. In such cases, mean heterozygosity at molecular marker loci may have little predictive value regarding a population's genetic well being. Species with a long evolutionary history of small, closed populations include pupfishes *Cyprinodon* spp., certain populations of which have lived isolated in small, desert springs for millennia and which show no evidence of inbreeding depression. Alternatively, if inbreeding depression is the consequence of low genome-wide heterozygosity, then previously inbred populations would exhibit reduced fitness upon further inbreeding.

■ Box 10.4
Fluctuating Asymmetry as a Measure of Inbreeding Depression

The hypothesis that heterozygosity stabilizes embryological development, first put forward by Lerner (1954), states that heterozygosity confers an increased ability to compensate for environmental variation during development. Hence, genetically determined developmental pathways are **canalized** that is, precisely expressed in spite of environmental perturbations, in the phenotypes of highly heterozygous individuals, who will have an increased tendency to express the characteristic morphology of the species. Much of the evidence supporting this hypothesis comes from studies of inbred and hybrid lines. Morphologically deformed individuals occur at heightened frequencies in inbred lines of animals. Although some deformities are due to the increased frequency of homozygotes for deleterious recessive alleles, other deformities do not have a Mendelian basis and apparently are caused by the general inability of highly homozygous individuals to progress normally through embryonic development (Wright 1934). Phenotypic similarity of the left and right sides of an individual has been used as a measure of developmental stability. Conversely, asymmetry between the left and right sides of an individual is taken as a measure of developmental instability.

Fluctuating asymmetry occurs when the difference between a character on the left and right sides of individuals is normally distributed about a mean of 0 (Van Valen 1962). There is a large body of data supporting the relationship between increased homozygosity levels and reduced developmental stability as measured by fluctuating asymmetry.

Heightened frequencies of fluctuating asymmetry have been observed in three salmonid fishes (rainbow trout, westslope cutthroat trout *Oncorhynchus clarki lewisi*, and brook trout) exhibiting high levels of homozygosity at allozyme loci. For example, Leary et al. (1984) found a negative correlation between heterozygosity and asymmetry in 12 of 13 salmonid populations, which was significant ($P < 0.05$) in 4 populations. Small differences among individuals in allozyme heterozygosity had a detectable effect on developmental stability. Increased frequencies of deformities have been associated with loss of heterozygosity caused by inbreeding in a number of fishes, including common carp *Cyprinus carpio* (Kirpichnikov 1981), rainbow trout (Aulstad and Kittelsen 1971; Kincaid 1976a, 1976b), and channel catfish *Ictalurus punctatus* (Bondari 1983). Fluctuating asymmetry has been advanced as a useful tool for measuring loss of heterozygosity in fish populations.

The relationship between heterozygosity loss and heightened fluctuating asymmetry poses important implications for fisheries management and aquaculture. For example, Leary et al. (1985) presented morphological evidence that loss of genetic variation in a hatchery stock of westslope cutthroat trout adversely affected embryological development of the fish. The Murray Springs Hatchery, Montana, stock was founded from a maximum of 60 fish collected in 1965 from a natural population in Hungry Horse Creek, Flathead County, Montana. Loss of genetic variation in the stock was demonstrated using allozyme markers (Allendorf and Phelps 1980). Leary et al. (1985) collected data on expression of five meristic traits in the 1983 year-class

continued on next page

■ Box 10.4 *continued*

of the stock. The bilateral traits at issue were presence versus absence of pectoral fins, number of mandibular pores, number of gill rakers in the upper arch, number of gill rakers in the lower arch, and number of pelvic fin rays. They examined 51 individuals from the Murray Springs Hatchery stock, and 27–51 individuals from five wild populations (see table below). None of the fish from wild populations were missing pectoral fins, but 13 of 51 hatchery fish were completely missing one or both pectoral fins. Many hatchery fish had a dorsal bump posterior to the head resulting from disarticulation of the first few thoracic vertebrae. Samples from four wild populations exhibited similar numbers of asymmetric traits per individual (see table below). The hatchery fish, however, exhibited a significantly ($P < 0.05$) higher mean number of asymmetries (2.24) than did the wild fish in Hungry Horse Creek (1.85), the stream from which they were derived. Occurrence of deformities, that is lack of pectoral fin or disarticulated vertebrae, was positively correlated with asymmetries, suggesting a general breakdown of developmental stability in affected individuals.

Presumably, inbreeding in the small, closed hatchery population reduced heterozygosity. The heightened level of homozygosity, by decreasing the degree of developmental canalization, increased the developmental instability that was manifested as heightened levels of deformity and fluctuating asymmmetry relative to wild stocks. Caution must be taken when interpreting the implications of the study, as hatchery and wild fish experienced embryological development under different conditions. However, the apparent effect of loss of genetic variation in the hatchery stock suggests the need for introducing new genetic variation to the stock and more careful attention to propagation practices (Chapter 14). Hatchery operations must be considered within the larger programmatic context of fish culture and fisheries management. Reduced fitness of the hatchery stock poses negative implications for the restoration and enhancement programs served by the hatchery program.

Monitoring frequencies of fluctuating asymmetries relative to a well-defined baseline stock provides a basis for estimating impacts of inbreeding as they occur in a selected aquaculture broodstock. Bilateral meristic traits might be chosen so that measurement does not involve sacrifice of the fish. This is particularly important for stocks of conservation concern or for valued aquaculture broodstocks. For example, Borkholder (1993) examined the numbers of scales along, above, and below the lateral line and the numbers of pectoral or pelvic fin rays in highly selected strains of goldfish *Carassius auratus.* Among the five traits monitored, a mean of 1.8 asymme-

Table Mean number of asymmetries of five bilateral meristic traits in five populations of westslope cutthroat trout (data from Leary et al. 1985).

Population (sample size)	Mean number of asymmetries (SE)
Murray Springs Hatchery (51)	2.24 (1.11)
Emery Creek (27)	1.93 (1.04)
Granite Creek (27)	1.86 (0.99)
Hungry Horse Creek (47)	1.85 (0.98)
O'Keefe Creek (51)	1.82 (1.10)

continued on next page

■ Box 10.4 *continued*

tries per individual was found in the comet strain (the classical orange goldfish), significantly fewer than the 2.5 per individual in fancy strains. This finding suggested that the high intensity of selection and inbreeding involved in propagation of the fancy strains had affected the developmental stability of the strains. In turn, the loss of fitness had affected the profitability of the aquaculture enterprise.

Whereas Leary and Allendorf (1989) advanced the potential of developmental stability as a measure of genetic or environmental stress upon a population, Clarke (1995) developed the case for developmental stability as an indirect measure of fitness. The ability of developmental stability to predict changes in fitness assumes that developmental stability will be affected before there are detectable changes in direct components of fitness, such as fecundity (Clarke et al. 1986; Clarke and McKenzie 1992). Due to its potential sensitivity and ease of measurement compared with more direct fitness estimates, measurement of developmental stability poses a number of practical applications for conservation. Such applications include identifying populations under genetic or environmental stress, predicting loss of fitness should stress persist, and monitoring the fitness of captive populations. Among studies of genetic stress in populations of fishes, a population of Gila topminnow *Poeciliopsis occidentalis* with no detectable allozyme variation experienced significant reduction of fitness, measured in terms of survival, growth, fecundity, and developmental stability, compared with genetically variable populations (Quattro and Vrijnhoek 1989). Angus and Schultz (1983) showed that inbred *Poeciliopsis lucida* were developmentally less stable than were outbred individuals. Interspecific hybridization also can disrupt developmental stability of fishes, as shown in centrarchids (Graham and Felley 1985) and salmonids (Leary et al. 1985).

Whatever the underlying mechanism, it is clear that different populations exhibit widely differing losses of fitness associated with inbreeding.

Definitive demonstration of a causal link between molecular genetic heterozygosity and population viability has proven elusive. Avise (1994:366) listed several reasons for using caution when interpreting the low molecular genetic heterozygosity reported for rare species.

- —Most reductions of heterozygosity have been the *result* of rather than the *cause* of population bottlenecks.
- —At least a few widespread and successful species also appear to have low heterozygosities, as estimated using the same methods as used for rare species.
- —In some threatened or endangered species (e.g., northern elephant seal, Bonnell and Selander 1974), low genetic variation does not appear to have inhibited population recovery from low levels.
- —The fitness cost of inbreeding differs widely among species (Ralls et al. 1988), with some taxa proving highly sensitive and others not.
- —Estimates of molecular genetic variability based on any single class of markers may not adequately characterize genome-wide heterozygosity.

Inability to identify and measure variability at fitness-related loci is a major shortcoming in population genetics. Heterozygosity measured at molecular genetic loci may not reflect variation at fitness-related loci. Successful application of genetic marker based methods for detection of fitness-related loci (Box 12.3) may help address this problem.

10.5 References

Adamkewicz, L., S. R. Raub, and J. R. Wall. 1984. Genetics of the clam *Mercenaria mercenaria*. II. Size and genotype. Malacologia 25:525–533.

Allendorf, F. W., and R. F. Leary. 1986. Heterozygosity and fitness in natural populations of animals. Pages 57–76 *in* M.E. Soulé, editor. Conservation biology: the science of scarcity and diversity. Sinauer Associates, Sunderland, Massachusetts.

Allendorf, F. W., and S. R. Phelps. 1980. Loss of genetic variation in a hatchery stock of cutthroat trout. Transactions of the American Fisheries Society 109:537–543.

Angus, R. A., and R. J. Schultz. 1983. Meristic variation in homozygous and heterozygous fish. Copeia 1983:287–299.

Aulstad, D., and A. Kittlesen. 1971. Abnormal body curvatures of rainbow trout *(Salmo gairdneri)* inbred fry. Journal of the Fisheries Research Board of Canada 28:1918–1920.

Avise, J. C. 1994. Molecular markers, natural history, and evolution. Chapman and Hall, New York.

Beardmore, J. A., and R. D. Ward. 1977. Polymorphism, selection, and multilocus heterozygosity in the plaice, *Platessa platessa* L. Pages 207–221 *in* S. Levin, editor. Measuring selection in natural populations. Lecture Notes in Biomathematics 19. Springer-Verlag, New York.

Bondari, K. 1983. Caudal fin abnormality and growth and survival of channel catfish. Growth 47:361–370.

Bondari, K. 1984. Growth comparison of inbred and randombred catfish at different temperatures. Proceedings of the Southeastern Association of Fish and Wildlife Agencies 35(1981):547–553.

Bondari, K., and R. A. Dunham. 1987. Effects of inbreeding on economic traits of channel catfish. Theoretical and Applied Genetics 74:1–9.

Bonnell, M. L., and R. K. Selander. 1974. Elephant seals: genetic variation and near extinction. Science 184:908–909.

Borkholder, B. D. 1993. Genetic marker frequency differences among strains of goldfish, *Carassius auratus*. Master's thesis, Virginia Polytechnic Institute and State University, Blacksburg, Virginia.

Buroker, N. E. 1979. Overdominance of a muscle protein (Mp-1) locus in the Japanese oyster, *Crassostrea gigas* (Ostreidae). Journal of the Fisheries Research Board of Canada 36:1313–1318.

Charlesworth, D., and B. Charlesworth. 1987. Inbreeding depression and its evolutionary consequences. Annual Review of Ecology and Systematics 18:237–268.

Christiansen, F. B., O. Frydenberg, and V. Simonsen. 1977. Genetics of *Zoarces* populations. X. Selection component analysis of the *Est*III polymorphism using samples of successive cohorts. Hereditas 87:129–150.

Clarke, G. M. 1995. Relationships between developmental stability and fitness. Application for conservation biology. Conservation Biology 9:18–24.

Clarke, G. M, G. W. Brand, and M. J. Whitten. 1986. Fluctuating asymmetry: a technique for measuring developmental stress caused by inbreeding. Australian Journal of Biological Sciences 39:145–153.

Clarke, G. M., and L. J. McKenzie. 1992. The use of fluctuating asymmetry as a quality control indicator for insect mass rearing processes. Journal of Economic Entomology 85:2045–2050.

Cooper, E. L. 1961. Growth of wild and hatchery strains of brook trout. Transactions of the American Fisheries Society 90:424–438.

Cruden, D. 1949. The computation of inbreeding coefficients. Journal of Heredity 40:248–251.

Danzmann, R. G, M. M. Ferguson, F. W. Allendorf, and K. L. Knudsen. 1986. Heterozygosity and developmental rate in a strain of rainbow trout *(Salmo gairdneri)*. Evolution 40:86–93.

David, P., and P. Jarne. 1997. Context-dependent survival differences among electrophoretic genotypes in natural populations of the marine bivalve *Spisula ovalis*. Genetics 146:335–344.

Dillon, R. T., and J. J. Manzi. 1988. Enzyme heterozygosity and growth rate in nursery populations of *Mercenaria mercenaria* (L.). Journal of Experimental Marine Biology and Ecology 116:79–86.

DiMichele, L., and D. A. Powers. 1982. LDH-B genotype-specific hatching times of *Fundulus heteroclitus* embryos. Nature (London) 296:563–565.

Emik, L. O., and C. E. Terrill. 1949. Systematic procedures for calculating inbreeding coefficients. Journal of Heredity 40:51–55.

Foltz, D. W., and M. Chatry. 1986. Genetic heterozygosity and growth rate in Louisiana oysters *(Crassostrea virginica)*. Aquaculture 57:261–269.

Foltz, D. W., and E. Zouros. 1984. Enzyme heterozygosity in the scallop *Plactopectin magellanicus* (Gmelin) in relation to age and size. Marine Biology Letters 5:255–263.

Fujino, K., and T. Kang. 1968. Transferrin groups in tunas. Genetics 59:79–91.

Fujio, Y., Y. Nakamura, and M. Sugita. 1979. Selective advantage of heterozygotes at catalase locus in the Pacific oyster, *Crassostrea gigas*. Japanese Journal of Genetics 54:359–366.

Gaffney, P. M., T. M. Scott, R. K. Koehn, and W. J. Diehl. 1990. Interrelationships of heterozygosity, growth rate, and heterozygote deficiencies in the coot clam, *Mulinia lateralis*. Genetics 124:687–699.

Gall, G. A. E. 1987. Inbreeding. Pages 47–87 *in* N. Ryman and F. Utter, editors. Population genetics and fishery management. University of Washington Press, Seattle.

Gjerde, B. 1988. Complete diallele cross between six inbred groups of rainbow trout, *Salmo gairdneri*. Aquaculture 75:71–87.

Gjerde, B., K. Gunnes, and T. Gjedrem. 1983. Effect of inbreeding on survival and growth in rainbow trout. Aquaculture 34:327–332.

Graham, J. H., and J. D. Felley. 1985. Genomic coadaptation and developmental stability with introgressed populations of *Enneacanthis gloriosus* and *E. obesus* (Pisces: Centrarchidae). Evolution 39:72–87.

Hartl, D. 1988. A primer of population genetics, 2nd edition. Sinauer Associates, Sunderland, Massachusetts.

IUBMBNC (International Union of Biochemistry and Molecular Biology, Nomenclature Committee). 1992. Enzyme nomenclature, 1992. Academic Press, San Diego, California.

Kat, P. W. 1982. The relationship between heterozygosity for enzyme loci and developmental homeostasis in peripheral populations of aquatic bivalves (Unionidae). American Naturalist 119:824–832.

Kincaid, H. L. 1976a. Effects of inbreeding on rainbow trout populations. Transactions of the American Fisheries Society 105:273–280.

Kincaid, H. L. 1976b. Inbreeding in rainbow trout (*Salmo gairdneri*). Journal of the Fisheries Research Board of Canada 33:2420–2426.

Kincaid, H. L. 1977. Rotational line crossing: an approach to the reduction of inbreeding accumulation in trout brood stocks. Progressive Fish-Culturist 39:179–181.

Kincaid, H. L. 1983. Inbreeding in fish populations used in aquaculture. Aquaculture 33:215–227.

Kincaid, H. L. 1995. An evaluation of inbreeding and effective population size in salmonid broodstocks in federal and state hatcheries. Pages 193–204 *in* H. L. Schramm, Jr., and R. G. Piper, editors. Uses and effects of cultured fishes in aquatic ecosystems. American Fisheries Society Symposium 15, Bethesda, Maryland.

Kirpichnikov, V. S. 1981. Genetic bases of fish selection. Springer-Verlag, New York.

Koehn, R. K., W. J. Diehl, and T. M. Scott. 1988. The differential contribution by individual enzymes of glycolysis and protein catabolism to the relationship between heterozygosity and growth rate in the coot clam, *Mulinis lateralis*. Genetics 118:121–130.

Koehn, R. K., and P. M. Gaffney. 1984. Genetic heterozygosity and growth rate in *Mytilis edulis*. Marine Biology 82:1–7.

Koehn, R. K., R. Milkman, and J. B. Mitton. 1976. Population genetics of marine pelecypods. IV. Selection, migration, and genetic differentiation of the blue mussel, *Mytilis edulis*. Evolution 30:2–32.

Koehn, R. K., J. E. Perez, and R. B. Merritt. 1971. Esterase enzyme function and genetical structure of populations of the freshwater fish *Notropis stramineus*. American Naturalist 105:51–69.

Koehn, R. K., and S. E. Shumway. 1982. A genetic/physiological explanation for differential growth rate among individuals of the American oyster, *Crassostrea virginica* (Gmelin). Marine Biology Letters 3:35–42.

Koehn, R. K., F. J. Turano, and J. B. Mitton. 1973. Population genetics of marine pelecypods. II. Genetic differences in microhabitats of *Modiolus demissus*. Evolution 27:100–105.

Lacy, R. C. 1992. The effects of inbreeding on isolated populations: are minimum viable population sizes predictable? Pages 277–296 *in* P. L Fiedler and S. K Jain, editors. Conservation biology. Chapman and Hall, New York.

Laikre, L., and N. Ryman. 1991. Inbreeding depression in a captive wolf *(Canis lupus)* population. Conservation Biology 5:33–40.

Lannan, J. E. 1980. Broodstock management of *Crassostrea gigas*. IV. Inbreeding and larval survival. Aquaculture 21:353–356.

Leary, R. F., and F. W. Allendorf. 1989. Fluctuating asymmetry as an indicator or stress: implications for conservation biology. Trends in Ecology and Evolution 4:214–217.

Leary, R. F., F. W. Allendorf, and K. L. Knudsen. 1983. Developmental stability and enzyme heterozygosity in rainbow trout. Nature (London) 301:71–72.

Leary, R. F., F. W. Allendorf, and K. L. Knudsen. 1984. Superior developmental instability of enzyme heterozygotes in salmonid fishes. American Naturalist 124:540–551.

Leary, R. F., F. W. Allendorf, and K. L. Knudsen. 1985. Developmental instability as an indicator of reduced genetic variation in hatchery trout. Transactions of the American Fisheries Society 114:230–235.

Leberg, P. L. 1990. Influence of genetic variability on population growth: implications for conservation. Journal of Fish Biology 37 (Supplement A):193–195.

Lerner, I. M. 1954. Genetic homeostasis. John Wiley and Sons, New York.

Longwell, A. C., and S. Stiles. 1973. Gamete cross incompatibility and inbreeding in the commercial oyster, *Crassostrea virginica* Gmelin. Cytologia 38:521–533.

Mallet, A. L., and L. E. Haley. 1983. Effects of inbreeding on larval and spat performance in the American oyster. Aquaculture 33:229–235.

Mallet, A. L., E. Zouros, K. E. Gartner-Kapkay, and K. R. Freeman. 1986. Genetics of growth in blue mussels: family and enzyme heterozygosity effects. Marine Biology 92:475–482.

McKay, L. R., I. McMillan, S. E. Sadler, and R. D. Moccia. 1992. Effects of mating system on inbreeding levels and selection response in salmonid aquaculture. Aquaculture 100:100–101.

Mitton, J. B., and M. C. Grant. 1984. Associations among protein heterozygosity, growth rate, and developmental homeostasis. Annual Review of Ecology and Systematics 15:479–499.

Mitton, J. B., and R. K. Koehn. 1975. Genetic organization and adaptive response of allozymes to ecological variables in *Fundulus heteroclitus*. Genetics 79:97–111.

Mrakovcic, M., and L. E. Haley. 1979. Inbreeding depression in the zebra fish, *Brachydanio rerio*. Journal of Fish Biology 15:323–327.

Piron, R. D. 1978. Spontaneous skeletal deformities in the zebra danio *(Brachydanio rerio)* bred for fish toxicity tests. Journal of Fish Biology 13:79–83.

Pogson, G. H., and E. Zouros. 1994. Allozyme and RFLP heterozygosities as correlates of growth rate in the scallop *Placopecten megellanicus*: a test of the associative overdominance hypothesis. Genetics 137:221–231.

Pray, L. A., J. M. Schwartz, C. J. Goodnight, and L. Stevens. 1994. Environmental dependency of inbreeding depression: implications for conservation biology. Conservation Biology 8:562–568.

Quatro, J. M., and R. C. Vrijnhoek. 1989. Fitness differences among remnant populations of the endangered Sonoran topminnow. Science 245:976–978.

Ralls, K., J. Ballou, and A. Templeton. 1988. Estimates of lethal equivalents and the cost of inbreeding in mammals. Conservation Biology 2:185–193.

Ralls, K., P. H. Harvey, and A. M. Lyles. 1986. Inbreeding in natural populations of birds and mammals. Pages 35–56 *in* M. E. Soulé, editor. Conservation biology: the science of scarcity and diversity. Sinauer Associates, Sunderland, Massachusetts.

Research and Education Association. 1985. The genetics problem solver. Research and Education Association, Piscataway, New Jersey.

Rodhouse, P. G., and P. M. Gaffney. 1984. Effect of heterozygosity on metabolism during starvation in the American oyster *Craassostrea virginica*. Marine Biology 80:179–187.

Ryman, N. 1970. A genetic analysis of recapture frequencies of released young salmon (*Salmo salar* L.). Hereditas 5:159–160.

Singh, S. M. 1982. Enzyme heterozygosity associated with growth at different developmental stages in oysters. Canadian Journal of Genetics and Cytology 24:451–458.

Singh, S. M., and R. G. Green. 1984. Excess of allozyme heterozygosity in marine molluscs and its possible significance. Malacolgia 25:569–581.

Tave, D. 1990. Effective breeding number and broodstock management: I. How to minimize inbreeding. *In* R. O. Smitherman and D. Tave, editors. Proceedings of the Auburn symposium on fisheries and aquaculture. Alabama Agricultural Experiment Station, Auburn University, Alabama.

Tave, D. 1993. Genetics for fish hatchery managers, 2nd edition. Van Nostrand Reinhold, New York.

Van Valen, L. 1962. A study of fluctuating asymmetry. Evolution 16:125–142.

Vrijnhoek, R. C., and S. Lerman. 1982. Heterozygote developmental stability under sexual and asexual breeding systems. Evolution 36:768–776.

Weigel, K. A. 2001. Controlling inbreeding in modern breeding programs. Journal of Dairy Science 64 (Supplement E):E177–E184.

Wilkins, N. P. 1978. Length correlated changes in heterozygosity at an enzyme locus in the scallop (*Pecten maximus* L.). Animal Blood Groups and Biochemical Genetics 9:69–77.

Wright, S. 1934. On the genetics of subnormal development of the head (otocephaly) in the guinea pig. Genetics 19:471–505.

11

Coadaptation and Outbreeding Depression

Eric Hallerman

11.1 Introduction

One of the most controversial fisheries management issues through the 1990s was the appropriate role of hatchery-supported fish stocking programs (Philipp et al. 1993; Schramm and Piper 1995). Some of the most critical arguments pertain to possible genetic effects of fish introductions (Chapters 14, 15). In large measure, the possibility of genetic effects turns on the likelihood of negative consequences from introgression of the gene pool of an nonnative stock into the gene pool of a native population. This possibility provides the context for this chapter. Coadaptation is a controversial area not only for fisheries geneticists but also for population geneticists concerned with a wide range of taxa. This chapter discusses the process and implications of coadaptation. The reader must recognize that subtle genetic processes may be involved and that this chapter cannot begin to settle all controversial points.

11.2 The Related Concepts of Coadaptation and Outbreeding Depression

Recognition that allelic variation at *single* loci affects the fitness of individuals expressing different genotypes is fundamental to the concept of selection and a foundation for population genetics. However, allelic variation at single loci does not completely explain the genetic basis for fitness. *Combinations* of alleles at multiple loci also may affect the fitness of the carrier via epistasis. If favorable allele combinations arise by chance, they are likely to become spread through the population and maintained by selection. The process is termed **coadaptation,** and the groups of alleles so produced and maintained are termed **coadapted gene complexes.**

Recognition of the concept of coadaption is attributed (Wallace 1991) to Dobzhansky (1937), who noted that if we assemble a broad collection of individuals from a particular species, the phenotypic variation observed is not a continuous distribution; rather, it is a multitude of separate, discrete distributions. The peaks of the distributions relate to the allele combinations that render their carriers fit in their particular environment. The classic demonstration of the effect of coadaptation is also attributed to Dobzhansky (1948). Geographic populations of the fruit fly *Drosophila pseudoobscura* were shown to have different sets of alleles. When populations expressing these different gene combinations were hybridized, the result was reduced fitness. The phenomenon of loss of fitness upon breakup of coadapted gene complexes is termed **outbreeding depression.** In general, demonstration that groups of genes affect fitness has proven difficult. Hence, the frequency and relative importance of coadaptation have proven contentious within the field of population genetics.

11.3 Processes That Give Rise to Coadaptation

11.3.1 *Local Adaptation*

Two processes may give rise to coadaptation (Templeton 1986). The first process, having to do with a population responding to extrinsic, or environmental, factors is termed **local adaptation.** Many species are distributed over a broad geographic range, presenting diverse ecological conditions that challenge an organism's ability to survive and reproduce. Local populations, subject to selection, adapt to their local environment. Their array of allele frequencies approaches and then tracks the optimum for that local environment, corresponding conceptually to a particular peak in Wright's (1931, 1932, 1982) adaptive landscape (Chapter 8, Figure 8.5A). The process of local adaptation is most likely if dispersal of individuals among local populations is limited. Hybridization among differentially adapted local populations can give rise to outbreeding depression.

A well-cited example of outbreeding depression consequent to breakup of local adaptation is provided by the goatlike mammal ibex *Capra ibex ibex* in central Europe, where overhunting led to extinction of the Czechoslovakian herd. There was a successful introduction of ibex of the same subspecies from Austria (Greig 1979). However, there also were later introductions of *C. i. aegagrus* from Turkey and *C. i. nubiana* from the Sinai peninsula. Hybrids of *C. i. ibex* and later-introduced stocks rutted in fall instead of winter; the resulting kids were born in February, only to freeze; and the population went extinct (Turcek 1951; Greig 1979). As do many putative examples of outbreeding depression, this example proved controversial, as quantitative data were not presented (Frankham 1995).

Local adaptation also occurs among fishes (Hindar et al. 1991; Taylor 1991, Carvalho 1993), perhaps the best examples of which occur in salmonids. Reproductive isolation and variation in selective regimes are characteristic of salmonid species, many of which are distributed as a collection of semi-isolated populations over a broad geographic area. There is strong evidence that spawning populations of anadromous salmonids exhibit highly specific local adaptations for a number of different traits (Allendorf and Waples 1996). To complete its complex life history successfully, an anadromous salmonid must develop as an embryo at an appropriate rate so as to emerge at a time appropriate for its natal ecosystem, forage effectively and avoid predators as a juvenile, find cool water for rearing or have the physiological capability of dealing with relatively high water temperatures, complete the physiologically complex process of smoltification at an appropriate time for emigration, migrate to oceanic regions where food supplies are adequate for survival and growth, re-enter freshwater at an appropriate time while undergoing physiological adaptation to freshwater, home accurately to the natal ecosystem, find a mate, and if female, deposit eggs at an appropriate time, place, and depth to promote the survival of the offspring. Disruption of any aspect of the life history leads to drastic loss of fitness.

A specific example of adaptation to local conditions is provided by the time of spawning among populatons of sockeye salmon *Oncorhynchus nerka* in the Fraser River system of British Columbia (Brannon 1987). Timing varies little from year to year at a given spawning location. However, there are large but consistent differences among spawning populations because each spawns at a time favorable for successful incubation, timing of emergence, and juvenile feeding under local conditions (Burgner 1991). The timing of spawning for local populations is strongly related to the temperature regime at the spawning site, with later spawning in warmer incubation environments. The differences in spawning times among populations results in similar emergence times of progeny the following spring because of the faster rate of development at higher temperatures.

It is important to note that local adaptation is not necessarily attributable to coadaptation. For example, in England the peppered moth *Biston betularia* exhibits **industrial melanism,** the tendency of individuals in polluted areas to be darkly colored. This phenomenon is due to selection upon a single major gene (Bishop and Cook 1980) and, therefore, is not the result of coadaptation across loci.

11.3.2 *Intrinsic Coadaptation*

The second process giving rise to coadapted gene complexes, relating to factors internal to the gene pool of the population, is termed **intrinsic coadaptation.** Intrinsic coadaptation occurs when a karyotype or genetic complex evolves in response to the state of other chromosomes or genes (Templeton 1986). Two underlying mechanisms have been recognized.

Karyotypic coadaptation results from certain combinations of chromosomal variants conferring fitness on their carriers whereas other combinations do not. In general, the process of gamete formation in meiosis requires a matched set of chromosomes (excepting sex chromosomes if present in the species). If the chromosome sets match, meiosis proceeds as normal. However, if the chromosome sets differ in number or structure, abnormal meiosis and fertility problems are possible. Hence, individuals may require two morphologically identical sets of chromosomes in order to prove fit. An example of breakdown of karyotypic coadaptation is provided by fertility problems observed in captive owl monkeys *Aotus trivirgatus* (De Boer 1982). After karyotypes of the individuals were examined, distinct groups carrying different chromosomal variants were recognized. Pairings among chromosomally similar individuals led to enhanced reproductive success (Cicmanec and Campbell 1977), supporting the hypothesis that karyotypic coadaptation had been at issue.

Genic coadaptation results from interactions among alleles at different loci that confer fitness on their carriers. The simplest case (Wallace 1991) would involve two loci, A and B, where A_1B_1 and A_2B_2 are well-adapted allele combinations whereas A_1B_2 and A_2B_1 are not. Let us assume that 1 and 2 refer to two geographically (or otherwise) isolated populations and the allelic combinations A_1B_2 and A_2B_1 arose through interpopulation hybridization followed by subsequent hybrid and backcross generations. If combinations A_1B_1 and A_2B_2 confer high fitness and A_1B_2 and A_2B_1 do not, an excess of A_1B_1 and A_2B_2 individuals will be expected within a population of mature individuals. Unless the relative fitnesses of A_1B_1 and A_2B_2 individuals are nicely balanced, the population rapidly will become uniformly one or the other.

Although genic coadaptation arguably may be rather common, it proves difficult to demonstrate. A clear example of genic coadaptation is provided by results of Templeton's (1979) study of *Drosophila mercatorum*. Although the species normally exhibits sexual reproduction, isolated, virgin females can reproduce via **parthenogenesis.** In the absence of fertilization, the haploid egg divides mitotically, and the cleavage nuclei fuse to form a diploid nucleus, which then initiates embryogenesis. Because diploidy is restored by fusion of identical haploid nuclei, the resulting parthenogenetic fly is completely homozygous. Because parthenogenetic lines are diploid and retain the ability to have normal meiosis, they can be crossed to males and reproduce sexually (Templeton 1983). Templeton produced parthenogenetic, homozygous lines from the same founder stock In order to remain viable, these lines presumably expressed fitness-conferring combinations of alleles. Templeton hybridized two distinct lines (using males from the founder stock) and used the resulting females for further parthenogenetic reproduction. The decline in fitness observed was attributed to breakdown of genic coadaptation. The experiment was repeated, yielding the same results. Because all lines were reared under identical conditions in a laboratory, differential local adaptation could not have been at issue. Hence, the interpretation reached was that in each case, the respective parthenogenetic lines each expressed different coadapted gene complexes that were incompatible with other lines.

Intrinsic coadaptation is most likely in cases in which the likelihood of creation of new allelic combinations is restricted, perhaps by meiotic crossover suppression (e.g., the chromosomal inversion in Dobzhansky's *Drosophila pseudoobscura*), population subdivision, small population size, or inbreeding. On the other hand, intrinsic coadaptation is unlikely in populations for which outcrossing frequencies are high or effective population size (N_e) is large.

The issue of whether or not genic coadaptation is common is debated by population geneticists, notably including fisheries geneticists. Arguing that the integrated function of the individual, not variation at individual genes, is the target of selection, Mayr (1963) posed that *most* genes in a species are coadapted. According to this line of thought, the formation of coadapted gene complexes is the heart of the process of speciation, an assertion that also has proven controversial.

11.4 Fitness Implications of Coadaptation and Outbreeding Depression

In a situation in which two distinct populations are mixed, two counteracting population genetic processes, inbreeding (Chapter 10) and coadaptation, might be at issue. If a population is small, there may be loss of alleles due to random genetic drift (Chapter 9). **Gametic phase disequilibrium,** that is, transmission of given combinations of alleles by chance as opposed to by linkage (Chapter 1) or by migration (Chapter 7), also may develop as a consequence of small population size (Ohta 1982). If a population is inbred, heterozygosity declines and, similarly, any heterotic advantage. Crossing of such small populations or inbred strains may result in re-establishment of lost alleles and combinations of alleles, bringing about hybrid vigor and thereby reversing inbreeding depression. On the other hand, however, if parental stocks are adapted to different locales or simply have different coadapted gene complexes, such a cross can lead to outbreeding depression.

The fitness implications of outbreeding depend in part on the number of generations since hybridization. The crossing of stocks can give rise to poor fitness in the F_1 generation; fitness in the F_1 intermediate to the parental strains; or heterosis—fitness higher

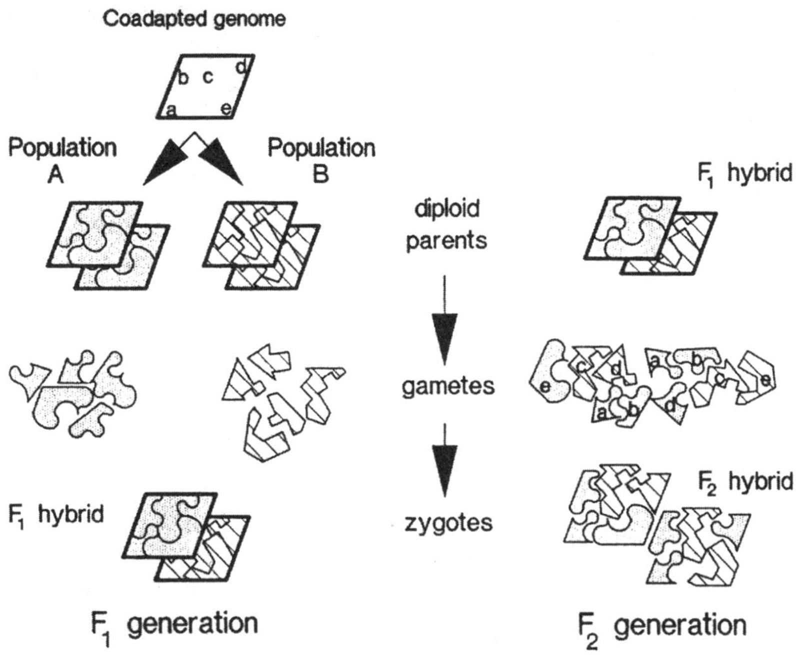

Figure 11.1 A conceptual representation of outbreeding depression resulting from disrupted coadapted gene complexes. A particular set of interacting (epistatic) alleles at multiple loci (a–e) are selected jointly to produce a well-adapted phenotype (coadapted gene complex) for a particular local environment or series of environments encountered by a population. Alternative coadapted gene complexes may be possible (populations A and B). If two populations possessing different coadapted gene complexes are hybridized, the F_1 hybrid individuals will receive two complete sets of coadapted alleles, which may or may not affect the fitness of the F_1 population. However, independent assortment of alleles in the F_2 and later hybrid generations disrupts coadapted complexes of alleles. Disruption of coadapted gene complexes by outbreeding may reduce population fitness and result in outbreeding depression. Reprinted from Fisheries Research, Volume 18, A. J. Gharrett and W. W. Smoker, A Perspective on the Adaptive Importance of Genetic Infrastructure in Salmon Populations to Ocean Ranching in Alaska, Pages 45–58, 1993, with permission from Elsevier Science.

than either parent—in the F_1 generation. In the last case, heterosis in the F_1 can mask any underlying outbreeding depression, which will not become apparent until the F_2 and subsequent generations (Figure 11.1). Loss of fitness in later-generation hybrids occurs because segregation will cause the breakup of gene combinations that the F_1 had inherited intact from the parents. Recombination among genes that occurred upon meiosis in the F_1 changes the genetic environment in which the genes are expressed. Outbreeding depression occurs because locally advantageous alleles are lost or favorable epistatic gene interactions that were fixed in the parental stocks recombine to produce unfavorable epistatic interactions.

The occurrence of outbreeding depression has been well demonstrated using fruit flies *Drosophila* spp. as a model system (e.g., Dobzhansky 1948, 1950; Brncic 1954; Wallace and Vetukhiv 1955). For example, in *D. pseudoobscura*, F_1 interpopulation hybrids were consistently larger than the mean of the two parental strains, whereas the F_2 hybrids were consistently smaller than were their F_1 parents (Anderson 1968). To make the point that outbreeding depression is not inevitable, in *D. subobscura*, neither F_1 heterosis nor F_2 breakdown was observed (MacFarquhar and Robertson 1963). However, outbreeding depression has been shown in a wide variety of organisms (e.g., Lacy et al. 1993; Waldman and McKinnan 1993; Wasser 1993).

A key determinant of whether inbreeding or outbreeding depression might result from a particular cross among members of different populations is the degree of genetic divergence among the parental stocks. If the parents are relatives, inbreeding depression is possible. If the parents are unrelated but share coadapted gene complexes, their progeny would be expected to exhibit high fitness. If the parents are so unrelated that they have different coadapted gene complexes, outbreeding depression may result. Hence, the fitness of F_1 hybrids of diverse parents can be expected to rise and then fall as the genetic difference between parental stocks increases (Figure 11.2). A degree of relatedness among outcrossing parents falling between the levels giving rise to either inbreeding or outbreeding depression provides "optimal outcrossing," maximizing the fitness of the progeny.

What lines of evidence, then, would support the inference that coadaptation exists in a given population? The most obvious—and defensible—would be a breakdown in fitness or some component of fitness following recombination (Wallace 1991). Such lines of evidence provided the basis for most inferences of coadaptation in the scientific literature (see section 11.5). A second line of evidence concerns observation of linkage disequilibrium among genes (Chapter 1), a finding that can be interpreted as evidence that genic coadaptation has occurred. Gametic phase disequilibrium can be demonstrated and quantified by estimating the coefficient of linkage disequilibrium (Weir 1979), the multilocus organization index (Brown et al. 1980), the variance of disequilibrium (Brown and Feldman 1981), or other metrics. However, unless linkage is tight and the advantage of the allele combination large, the favored combinations of alleles are easily disrupted. If the linkage is indeed tight, the combinations also can be attributed to chance events that occurred during past population bottlenecks (Ohta 1982) as well as to coadaptation. Hence, Wallace (1991) concludes that linkage disequilibrium in and of itself is a poor means of detecting coadaptive interactions.

A point pertinent to both theoretical and applied perspectives of coadaptation is the rate at which it might occur. Wallace (1991) contrasts the view of Bush and Howard (1986), who implied that coadaptation arises slowly over many generations, with that of Dobzhansky and Pavlovsky (1953), who documented the origin of new coadapted gene arrangements within but a few generations of fruit flies. The issue of how long it takes a coadapted complex to arise is in part a function of the mechanism of its formation. The chromosomal inversion at issue in the Dobzhansky and Pavlovsky study arose through a particular mutation event. Other mechanisms of coadaptation would take longer, depending on the number of loci and alleles involved, the intensity of selection, and the size and degree of isolation of the population at issue. Predictive models are lacking.

If a coadapted gene complex in a population has been disrupted, in the absence of further disruption (e.g., from further immigration or deliberate stocking), selection will lead to recovery of fitness in time. How long, then, will it take for new coadapted gene complexes to arise? When relatively few genes are involved in coadaptation, outbreeding depression "should" be temporary, some argue perhaps on the order of a few generations. When many genes are involved in coadapted gene complexes, it will take longer for similarly elaborated new complexes to arise after disruption. The new coadapted complexes may exhibit higher or lower fitness than did the complex that was disrupted. A key point in the context of conservation of genetic resources is that outbreeding depression, even if temporary, can increase the likelihood of a population's extinction, especially if outbreeding depression is severe or the population at issue is already small.

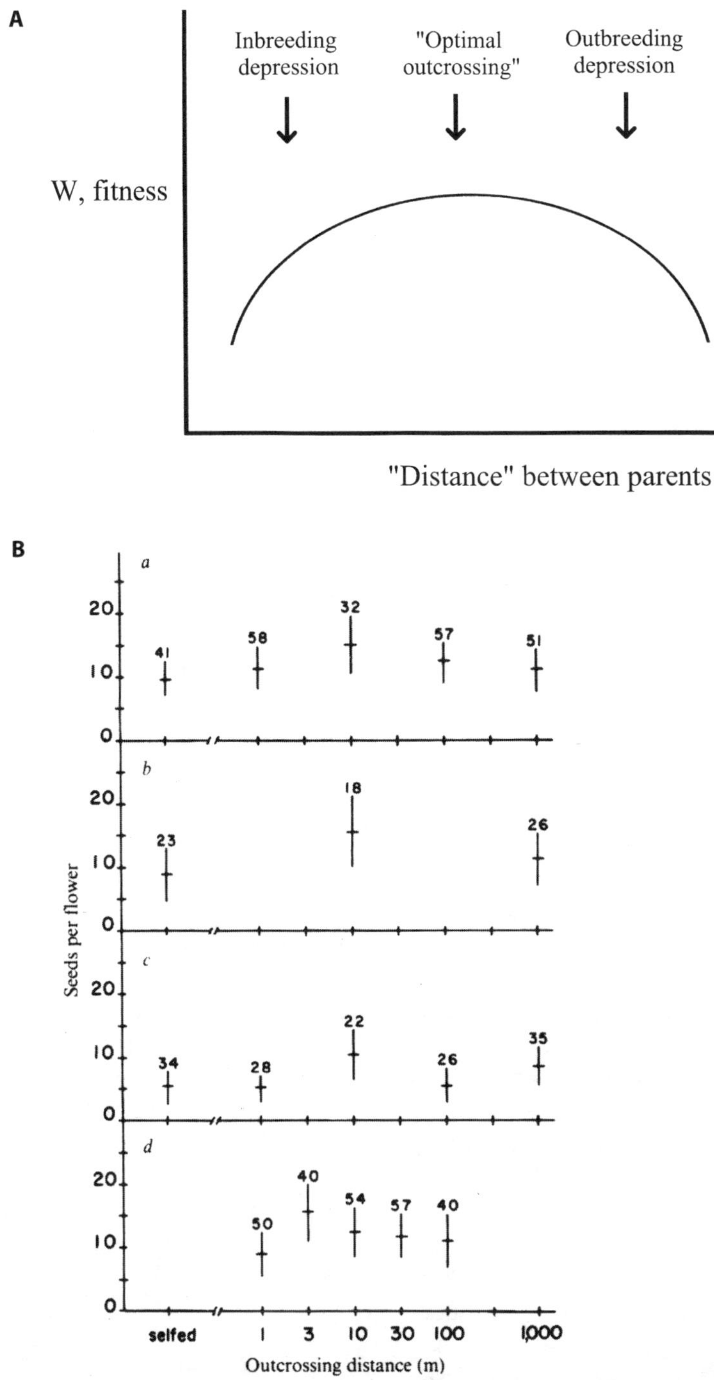

Figure 11.2 Optimal outcrossing. **(A)** Conceptual depiction of how inbreeding and outbreeding effects give rise to an optimal level of outcrossing. **(B)** Effect of outcrossing distance on seed set, a measure of reproductive fitness, in hand-pollinated *Delphinium nelsoni*, a mountain flower. Each panel of the figure represents a different field × year combination. Values are means ± 95% confidence intervals, with sample sizes shown. Figure from Price and Wasser (1979), reprinted with permission from *Nature* (277:294–297). Copyright 1979, Macmillan Magazines Ltd.

11.5 Coadaptation in Aquatic Organisms

Most studies of the process of coadaptation have been carried out using model organisms, and many of the most widely accepted examples of the phenomenon are based on laboratory studies of fruit flies. A considerable number of studies (e.g., Price and Wasser 1979) have provided widely accepted demonstrations of coadaptation in plants. In populations of these taxa, dispersal is limited and, as discussed above, the likelihood for coadaptation relatively great. For animals with relatively high dispersal ability, as in many fishes, the likelihood of coadaptation is relatively small in the absence of population isolating mechanisms and strong selective regimes. Hence, examples of coadaptation in more vagile animals, including fishes, are less numerous and frequently less clear-cut.

A number of laboratory and field studies have addressed coadaptation in fishes or other aquatic organisms. Table 11.1 lists examples of studies demonstrating or suggesting coadaption in aquatic animals. Many of the studies cited concern populations of

Table 11.1 Presumptive evidence for coadaptation in aquatic and marine organisms.

Presumptive evidence	Source
Invertebrata	
Development time of F_2 hybrids among local populations of the intertidal marine copepod *Tigriopus californicus* were as much as 50% longer than those of the parental stocks or the F_1 hybrid. Parental-type homozygotes exhibited higher viability than did nonparental homozygotes.	Burton (1987)
Dispersal is limited and mating success is proximity-dependent in the colonial ascidian *Botryllus schlosseri*.	Grosberg (1987)
Salmonidae	
Returns of pink salmon *Onchorynchus gorbuscha* to stream of release and to fishery were greater for hybrids of native males and nonnative females than for fish entirely of introduced genetic background.	Bams (1976)
Nine of 14 rainbow trout *O. mykiss* interstrain hybrids developed more slowly than did their maternal parent, consistent with the hypothesis of regulatory divergence between the hybridized strains.	Ferguson et al. (1985)
Reduction in resistance to the myxospiridian parasite *Ceratomyxa shasta* was observed in hybrids between resistant and nonresistant stocks of coho salmon *O. kisutch*.	Hemmingsten et al. (1986)
Transplantation of chum salmon *O. keta* from the Kurilka River in the Kuril Islands to the Kalininka River on Sakhalin Island in the Russian Far East led to disruption in run timing and population crash.	Altukhov and Salmenkova (1987)
Distance transferred from the natal stream was negatively related to recovery rate for hatchery-reared coho salmon.	Reisenbichler (1988)
Egg size and development rate are mismatched in hybrid sockeye × kokanee (lacustrine sockeye salmon) alevins, compromising viability. Other incompatibilities concern seawater adaptability, swimming performance, and age at maturation.	Foote et al. (1989); Wood and Foote (1990); Taylor and Foote (1991); Wood and Foote (unpublished)[a]

continued on next page

salmonids. This is not particularly surprising because life histories of salmonids are well studied, and population differentiation and local adaptation are salient features of salmonid ecology. Local adaptation, however, is but one mechanism underlying coadaptation. Box 11.1 focuses on demonstration of coadaptation in pink salmon based on an experimental design for which local adaptation is not at issue, and genic coadaptation may have been uncovered. Although coadaptation is less apparent in nonsalmonid taxa, it should not be concluded that coadaptation is not potentially at issue. Box 11.2 presents results of studies demonstrating coadaptation in largemouth bass. More study of coadaptation clearly is needed both in salmonid and nonsalmonid taxa.

The literature on coadaptation in fishes comprises a collection of frequently cited studies. Although certain studies are well designed to uncover and demonstrate coadaptation, others might more correctly be taken as suggestive of coadaptation. Taken together, this body of research shows that coadaptation does occur in populations of

Table 11.1 *continued.*

Presumptive evidence	Source
Reduced return rate was observed in inter-brood-year F_2 hybrid pink salmon.	Gharrett and Smoker (1991); Gharett et al. (1999); Box 1.1
Expression of alleles affecting upper thermal tolerance in rainbow trout varies among genetic backgrounds, suggesting that upper thermal tolerance is a coadapted trait.	Danzmann et al. 1999
Poeciliidae	
Hybrids between populations of southern platyfish *Xiphophorus maculatus* developed abnormal blotches of melanophores across the sides of the body.	M. Gordon and Smith (1938); M. Gordon (1947); H. Gordon and M. Gordon (1957)
Genetically based differences in life history traits (growth rate, fecundity, age and weight at maturity, interbrood interval, weight of offspring, and longevity) were observed among six stocks of western mosquitofish *Gambusia affinis* that shared common ancestors in 1905.	Stearns (1983)
Gasterosteidae	
Production of F_1 progeny among geographically isolated populations of threespine stickleback *Gasterosteus aculeatus* frequently results in alteration of the pattern of plate development.	Avise (1976)
Centrarchidae	
Hybrid populations of bluespotted sunfish *Enneacanthus gloriosus* and banded sunfish *E. obesus* exhibited more fluctuating asymmetry, a measure of developmental instability, than did populations of the parental species.	Graham and Felley (1985)
Reduced survival, growth and reproductive rates were observed in transplanted pure subspecies and hybrids of largemouth bass *Micropterus salmoides* in common garden experiments.	Philipp and Whitt (1991); Philipp and Claussen (1995); Box 11.2

[a] Cited in Wood and Foote (1990).

■ Box 11.1
Coadaptation and Outbreeding Depression in Pink Salmon: A Possible Example of Intrinsic Coadaptation

Pink salmon exhibit a strict 2-year life cycle. Hence, in any given spawning stream, there will be genetically distinct even- and odd-year stocks. These stocks are adapted to the same locale but may be differentially so adapted. By using cryopreserved milt, Gharrett and Smoker (1991) and Gharrett et al. (1999) hybridized even- and odd-year pink salmon from Auke Creek, Alaska, to test the hypothesis that the two sympatric stocks were differentially adapted to the same locale.

In the first experiment (Gharett and Smoker 1991), in the F_1 generation survival rates of hybrid and control fish were similar (see table on next page). The authors marked and released F_2 hybrids but were unable to release controls. The known return rate for wild-produced fish in Auke Creek, however, provided a useful reference value. The F_2 hybrids exhibited a very low return rate and an increased frequency of bilateral asymmetry. The results are consistent with an explanation of outbreeding depression.

The inter-brood-year cross experiment was repeated during the 1990s with the addition of a within-brood-year cross as a control for the F_2 generation. Return rates of F_1 hybrids and controls in 1994 (Gharrett et al.1999) did not differ significantly (see table). The F_2 hybrid and control crosses were made from 40 males and 40 females selected at random from each return group. Offspring were differentially marked and released. In 1996, return rates differed significantly between hybrids and controls. The low rate of return of control fish was similar to that of a larger group of tagged Auke Creek pink salmon and probably was not an artifact of the experiment. The observation of decreased survival in F_2 hybrids confirmed the findings of the earlier study and supported the inference that outbreeding depression for survival occurs in pink salmon.

continued on next page

fishes, especially in salmonids. However, the issues of how important, frequent, rapid, and universal a process is coadaptation in fishes remain controversial among fisheries geneticists. The importance of resolving these issues is discussed in the next section.

■ 11.6 Fisheries Management Implications of Coadaptation and Outbreeding Depression

Coadaptation and outbreeding depression have serious implications for the practice of fisheries management, particularly in the context of introducing hatchery-derived or nonnative fish into established native populations (Emlen 1991; Chapter 15). Sustaining the viability and productivity of wild populations of fishes requires minimization of any negative interactions between hatchery-derived and native populations of fishes. From the preceding section, it is clear that there is evidence that populations of salmonids, other fishes, and other aquatic organisms exhibit coadapted gene complexes for a range

■ Box 11.1 *continued*

The breeding design using between-year hybrids of pink salmon provided an extreme test for the hypothesis of outbreeding depression in the species. A less extreme test would involve hybridization of geographically distant populations. Ongoing research by the same group will evaluate performance of crosses between pink salmon from Kamchatka and Alaska and to follow up with crosses of geographically separated Alaskan populations to determine how closely located populations must be before outbreeding depression no longer is detectable (A. J. Gharrett, University of Alaska-Fairbanks, personal communication).

Table Numbers of released fry and returning adults for F_1 and F_2 hybrids and controls in tests of outbreeding depression in interyear crosses of pink salmon

Date	Action	Number released	Number returned	Percent returned
	Experiment 1 (Gharrett and Smoker 1991)			
1982	Milt cryopreserved			
1983	F_1 hybrid cross (1982 male × 1983 female)	5,483	144	2.63
	Control cross (1983 male × 1983 female)	5,492	139	2.53
1985	F_2 hybrid cross (returns from 1983 F_1 hybrids)	5,156	12	0.23
	Reference (wild, instream spawn)	59,666	2,430	4.07
	Experiment 2 (Gharrett et al. 1999)			
1990	Milt cryopreserved			
1991	Milt cryopreserved			
1992	F_1 hybrid cross (1991 male × 1992 female)	10,723	185	1.73
	Control cross (1990 male × 1992 female)	9,379	153	1.63
	Reference (wild, instream spawn)	237,073	22,167	9.35
1994	F_2 hybrid cross (returns from 1992 F_1 odd × even brood year hybrids)	14,309	34	0.24
	Control cross (returns from 1992 F_1 even × even brood year hybrids)	10,387	44	0.42
	Reference (independent experiment)	633,146	2,759	0.44

of fitness-related traits. These adaptations are the result of genetic differences between local populations at many loci, most often for quantitative traits (Chapter 12). Hence, we should expect it to be difficult to "replace" or "improve" a local population by introducing individuals from other populations. The highly variable success of fish stocking programs (e.g., Ricker 1972) and the impacts of stocking programs on local fish populations (Hindar et al. 1991; Chapters 14, 15) are such that fisheries managers would value the ability to predict the outcome of a proposed fisheries management action. In this context, conservation-minded fisheries managers want to know whether a stocking, stock transfer, supplementation, or other management action poses the risk of outbreeding depression to native fish populations.

■ Box 11.2
Coadaptation and Outbreeding Depression in Largemouth Bass: An Example of Local Adaptation

Although most inferences regarding coadaptation in fishes involve salmonids, it does not follow that the process is unique to salmonids and their complex life histories. Studies of other fishes (Table 11.1) have also led to inference of coadaptation. One of the best examples regards largemouth bass, the most popular sport fish in the eastern United States. Two subspecies of largemouth bass are recognized on the basis of meristics (Bailey and Hubbs 1949), differentiation that is supported by screening biochemical genetic markers (Philipp et al. 1983). The popularity of largemouth bass with the angling public has led to its widespread stocking. These stocking programs often have gone forward with little or no consideration given to their genetic or ecological consequences. Largemouth bass commonly are transported great distances to be stocked into lakes that have environmental conditions very different from the stock's native habitat. In many cases, largemouth bass are stocked into lakes that have native populations that have very different genetic backgrounds. Posing particular concern are fisheries management actions that involve the introduction of the Florida largemouth bass *M. s. floridanus* into native populations of northern largemouth bass *M. s. salmoides*. Presumed positive effects of these programs follow from harvest of trophy largemouth bass from the natural intergrade zone of the subspecies in northern Florida and from intergrade populations artificially produced in Texas and California. However, the genetic and ecological effects of stocking practices generally are unknown.

A number of studies (e.g., Cichra et al. 1982; Guest 1985; Philipp et al. 1985; Fields et al. 1987; Koppelman et al. 1988) have shown thermophysiological differences between the subspecies of largemouth bass. Philipp and Whitt (1991) proposed that introduction of a stock of largemouth bass from one environment into a population of largemouth bass in another environment (i.e., a stock transfer) and subsequent interbreeding of the two stocks would result in disruption of coadapted gene complexes that had evolved in both stocks. They further proposed that this disruption would result in reduction of the fitness of the new, resulting intergrade stock relative to the original, locally adapted native stock. They compared key components of fitness—survival, growth, and reproductive rates—among the two subspecies and the subspecies reciprocal F_1 hybrids in experimental ponds in the northern environment of central Illinois. The genetic composition of each of the stocks was confirmed using allozyme techniques. Survival rates over each of three winters for northern largemouth bass was significantly ($P = 0.0293$) greater than that of Florida largemouth bass, and that of the two reciprocal hybrids was intermediate. The absolute survival rate of all stocks decreased as the severity of the winter increased, but the effect was most dramatic for the Florida subspecies. The northern subspecies also exhibited greater second- and third-year growth than did the Florida subspecies, and that of hybrids was intermediate. Reproductive success of the pure northern largemouth bass was greater than that of the Florida largemouth bass and both reciprocal F_1 hybrids (Philipp 1991). Because of a lack of assortative mating, the two gene pools

continued on next page

■ Box 11.2 *continued*

rapidly mixed, resulting in a mixture of individuals expressing the two genomes to different extents. In a larger experiment of similar design, involving transplantations and performance evaluations among Florida, Illinois, Minnesota, and Texas populations of largemouth bass, local populations exhibited highest survival rates at each location of origin (D. P. Philipp, Illinois Natural History Survey, personal communication). Results of these studies taken together indicate the potential negative effects that may arise when Florida largemouth bass or hybrids between it and the northern subspecies are introduced into the range of the northern subspecies.

The interpretation that the fitness of northern largemouth bass was superior to that of Florida largemouth bass due to the effect of coadapted gene complexes proved controversial. Examining Philipp and Whitt's (1991) data, Maceina and Murphy (1992) concluded that northern largemouth bass survived and grew better than did the Florida subspecies only in small (0.08 ha) ponds in central Illinois. They pointed out that observed performance differences may have been attributable, in part, to competition related to unrealistically high densities and standing crops. The results of comparative performance studies from pond studies might not prove applicable to larger lakes and reservoirs. Noting the extensive impoundment of lentic habitats during the past 75 years, Maceina and Murphy suggested that fisheries managers might do well to consider stocking fish adapted to lentic conditions, that is, the Florida subspecies. Philipp (1992) responded that even the small ponds offered temperature refugia of 4°C and the cause of greater observed mortality among the Florida and hybrid largemouth bass stocks was the decreased ability of those stocks to store or use energy stores efficiently for the winter. Confronted with an energy deficit during lengthy northern winters, largemouth bass adapted to southern conditions could not alter their metabolic strategy from one of maximizing somatic growth to one of allocating energy into storage reserves for winter use. Regarding any impact of density, Philipp cited data showing that when largemouth bass of the pure subspecies and hybrid stocks were allowed to spawn naturally in each of four ponds, northern largemouth bass again survived and grew better than did other stocks. Evidence for continued production and survival of Florida largemouth bass beyond 2 years was lacking in three of the four ponds. Philipp took issue with the interpretation that flow regime is the single dominant selective force shaping the fitness of largemouth bass populations, pointed out that lentic habitats once were common along the flood plains of northern rivers, and questioned whether biologists can recognize the agents that shape selection and successfully predict the success of introduced stocks.

It must be recognized that the studies cited above used stocks representing distinct subspecies that were drawn from geographically distant locations. Especially in the context of this chapter, it should be recalled that coadaptation is a population level phenomenon. For truly effective fisheries management, it is the local population, or stock (Chapter 13), that must be considered the operational unit for management concern. The question then becomes, how geographically close can two stocks be and yet still show demonstrable local adaptations? Philipp and Claussen (1995) tested whether there were performance and fitness differences among stocks of northern

continued on next page

■ Box 11.2 *continued*

largemouth bass from two river drainages in Illinois, the Fox in northern Illinois and the Big Muddy in southern Illinois. Broodstock were collected from these systems and used to produce stocks of fingerlings, which were weighed, measured, and physically marked. Equal numbers were stocked into 10 ponds in the north, central, and southern regions of the state. In the northern Illinois study ponds, survival of the northern Illinois largemouth bass (NILMB) stock was significantly greater than that of the southern Illinois largemouth bass (SILMB) stock (see figure on next page, panel A). The reverse was true in southern Illinois ponds, and no significant difference in survival was observed among stocks in central Illinois study ponds. During each of the 3 years sampled, growth of NILMB was significantly greater than that of SILMB in northern Illinois ponds, the reverse was true in southern Illinois ponds, and there were no consistent differences in central Illinois ponds.

Samples of young of the year were collected from each pond, diagnostic allozymes were scored to characterize the genetic background of individual fish, and genetic contributions made to each year-class by each of the stocks were estimated. Results (see figure, panel B) indicated that the fitness of NILMB was significantly greater than that of SILMB in northern Illinois, the reverse was true in southern Illinois, and no consistent differences were observed in central Illinois. Survival differences played a greater role in establishing differential fitnesses of the two stocks than did differences in reproductive success. Collectively, the results demonstrated that local adaptations can and do exist among different stocks within a relatively small geographic area.

continued on next page

In order to predict the likelihood and magnitude of outbreeding depression, how might we quantify the relationships among fitness, generations since hybridization, and genetic distance between parental stocks? A number of surrogate measures of adaptive "distance" might be considered. The most obvious and easily measured is spatial distance, which exhibited clear correlation with the fitness of offspring of hybrids of differentially coadapted populations in flowering plants (e.g., Price and Wasser 1979), coho salmon (Reisenbichler 1988), and pink salmon (Altukhov and Salmenkova 1987). Although geographical distance between the source habitats of populations works reasonably well as a first approximation, other sorts of adaptive distances also might matter. Ecological differences among native habitats for populations suggest the action of different selective pressures, which would give rise to different sets of coadapted genes. Populations that have been subjected to different population histories, even if they originally were drawn from the same founder stocks, may exhibit differential coadaptation. In terms of the shifting balance theory of Sewall Wright (Chapter 9), populations that started at the same point shifted to different peaks on the adaptive topography. Emlen (1991; see Box 11.3) defined distance as the decline in fitness following crossing of distinct genetic lines, as experimentally observed in the habitat of the native stock.

How to anticipate whether outbreeding depression is likely following a management action is an area of current, sometimes heated, debate in the fisheries science, wildlife science, and population genetics communities. Development of predictive models is needed. Once we achieve the ability to predict the likelihood or magnitude of out-

■ Box 11.2 *continued*

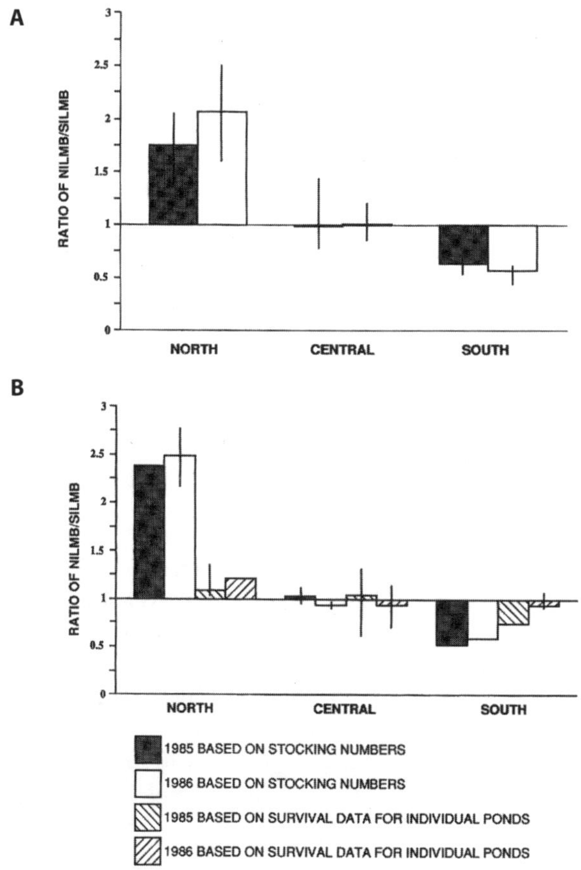

Comparative performance of northern Illinois largemouth bass (NILMB) and southern Illinois largemouth bass (SILMB) in common garden experiments in northern, central, and southern Illinois. **(A)** Ratios of relative abundance (NILMB/SILMB) in electrofishing collections for 1985 (shaded) and 1986 (open) in the three study regions. **(B)** Ratios of reproductive success for 1985 and 1986 based on (1) equal numbers of parents for each stock, as determined by equal numbers originally stocked, and (2) unequal numbers of parents as determined by survival (relative abundance data) of each stock in each pond. Figure from Philipp and Claussen (1995).

breeding depression following a proposed management action, we can better approach a range of important fisheries management questions (Emlen 1991):

— At what distance does differential, local adaptation begin to cause significant deterioration in performance for transplanted organisms? This question has important bearing on the design and execution of hatchery-based stocking programs posing the possibility of interbreeding of nonnative and native individuals.

■ Box 11.3
A Model for Estimating Outbreeding Depression and Guiding Decision Making by Fisheries Managers

It would be useful to predict the likelihood and magnitude of outbreeding depression posed by a proposed stocking or transplantation action. However, there is no widely accepted model for doing so. As described in the text, a key determinative factor is differential coadaptation among the two stocks that might become mixed. The question becomes one of quantifying adaptive divergence among the stocks.

In this context, Emlen (1991) developed a model for estimating the likelihood and magnitude of outbreeding depression, thereby guiding decision making by fisheries managers. The key parameter in the model is distance, D, the fitness difference between a transferred and a local population. For example, if stock A is transferred to site B and displays a fitness 10% less than that of native stock B, then D is 10%. However, D generally is not known, and its estimation will require time, money, and, perhaps, risk. The model considers a genetic subunit consisting of a structural gene and two epistatic loci and the change in fitness of different combinations of genotypes at the three loci along an ecological gradient. The fitness of the stock, W, is considered across all such genetic subunits. If the native population is replaced with another population, the matrix of fitness values and the fitness of the nonnative stock, W', is considered similarly. Simulations are run to consider the effects of mixing the gene pools. Zygote frequencies are calculated before and after selection, meiotic crossing-over is allowed to occur, and gametic frequencies for the next generation determined. Random mating will lead to new zygote frequencies, and the life cycle is repeated. The mixed population's fitness is found using the mean of fitness values over all zygotes weighted by their frequencies. Mathematical expressions of the model and procedures for estimating input parameters are presented by Emlen (1991).

For purposes of demonstrating use of the model, Emlen (1991) presented a hypothetical example based on coho salmon within the context of a typical fisheries management planning scenario. Faced with a decision regarding stocking or transplantation of coho salmon, a permissible drop in fitness would be identified. Proponents and opponents of the management action would provide estimates of model parameters, and the model would be run, yielding the families of curves shown in the figure below. If the model predicts a small drop of fitness, the population would remain above a critical fitness threshold. The management action would be regarded as

continued on next page

- Supposing that organisms bred in captivity (i.e., hatcheries, game farms, or zoos) experience selection pressures different from those operating in the wild (Chapter 14), can deterioration of fitness for life in the wild be mitigated by influx of genes from local, wild populations? If so, managers can hybridize captive and wild stocks to produce animals suitable for stocking. For example, Bams (1976) showed that poor homing ability of nonnative pink salmon was improved by crossing with the wild stock.
- Because straying of individuals from artificially reared stocks and their subsequent spawning with wild individuals can affects wild gene pools, what straying rate can be

■ Box 11.3 *continued*

"good," and managers would go ahead with the action. Should the predicted drop of fitness place the population within the grey area of the figure, the management action would be regarded as "bad," and the decision whether to go ahead is uncertain. Should the predicted drop in fitness place the population below a critical threshold, the management action is regarded as "ugly," and managers would not go forward with the management action.

The value of this model depends on two possible outcomes. First, how well does it predict the fitness of hybrids? Empirical tests of its predictions will be needed. Second, does the model inspire useful experiments or development of better models? The ability to predict the genetic implications of fisheries management actions would be highly valued.

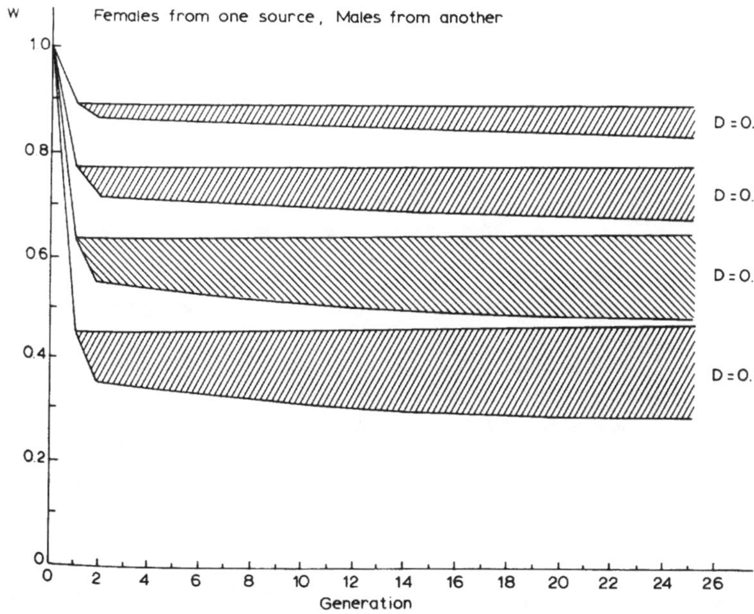

Changes in fitness *(W)* over time (in generations) when female parents come from one source and male parents from another as a function of *D*, parental distance. The hatched areas for each value of *D* delimit the good-bad-ugly envelope for decision making. Reprinted from Fisheries Research, Volume 12, J. M. Emlen, Heterosis and Outbreeding Depression: A Multilocus Model and an Application to Salmon Production, Pages 187–212, 1991, with permission from Elsevier Science.

tolerated? The answer to this question will affect the decision whether to go ahead with stocking of hatchery fish when valued, native populations of conspecifics reproduce nearby.

— When is it appropriate to carry out crossbreeding among populations of endangered species? In the context of supportive breeding, prospective broodstock are scarce and highly valued. Would crossbreeding reduce the effect of any inbreeding from previous supportive breeding work or from reproduction within a small population in the wild? Alternatively, would crossbreeding lead to outbreeding depression? In the context of a transplantation program, is it advisable to introduce individuals from a large population into a smaller population to boost the smaller population demographically and provide new genetic diversity?

11.7 Perspective

The concept of coadaptation has been discussed for more than 50 years and has itself evolved. The paucity of demonstrated instances of coadaptation in the wild has led some geneticists to dismiss coadaptation as a factor in evolutionary change. Kimura (1983), for example, regards "coadaptation" and "integration of gene pools" as vague terms lacking the precision needed for mathematical analysis. Wallace (1991), however, concludes that sound evidence supports coadaptation as a factor in the genetic change of populations.

Resolution of the frequency, speed, and importance of coadaptation, and successful prediction of the effects of outbreeding depression, would significantly improve the scientific basis for the practice of fisheries management. Much experimental and empirical work must be executed to improve our understanding of coadaptation in fishes and other aquatic organisms.

11.8 References

Allendorf, F. W., and R. S. Waples. 1996. Conservation and genetics of salmonid fishes. Pages 238–280 *in* J. C. Avise and J. L. Hamrick, editors. Conservation genetics: case histories from nature. Chapman and Hall, New York.

Altukhov, Y. P., and E. A. Salmenkova. 1987. Stock transfer relative to natural organization, management, and conservation of fish populations. Pages 333–343 *in* N. Ryman and F. W. Utter, editors. Population genetics and fishery management. University of Washington Press, Seattle.

Anderson, W. W. 1968. Further evidence for coadaptation in crosses between geographic populations of *Drosophila pseudoobscura*. Genetical Research 12:317–330.

Avise, J. C. 1976. Genetics of plate morphology in an unusual population of threespine sticklebacks *(Gasterosteus aculeatus)*. Genetical Research 27:33–46.

Bailey, R. M., and C. L. Hubbs. 1949. The black basses *(Micropterus)* of Florida, with description of a new species. University of Michigan Museum of Zoology Occasional Papers 516:1–40.

Bams, R. A. 1976. Survival and propensity for homing affected by presence or absence of locally adapted paternal genomes into two transplanted populations of pink salmon *(Oncorhynchus gorbuscha)*. Journal of the Fisheries Research Board of Canada 33:2716–2725.

Bishop, J. A., and L. M. Cook. 1980. Industrial melanism and the urban environment. Advances in Ecological Research 11:373–404.

Brannon, E. L. 1987. Mechanisms stabilizing salmonid fry emergence timing. Canadian Journal of Fisheries and Aquatic Sciences 96 (Supplement 1):120–124.

Brncic, D. 1954. Heterosis and the integration of the genotype in geographic populations of *Drosophila pseudoobscura*. Genetics 39:77–88.

Brown, A. H. D., and M. W. Feldman. 1981. Population structure of multilocus association. Proceedings of the National Academy of Sciences of the USA 78:5913–5916.

Brown, A. H. D., M. W. Feldman, and E. Nevo. 1980. Multilocus structure of natural populations of *Hordeum spontaneum*. Genetics 96:523–536.

Burgner, R. L. 1991. Life history of sockeye salmon *(Oncorhynchus nerka)*. Pages 1–117 *in* C. Groot and L. Margolis, editors. Pacific salmon life histories. University of British Columbia Press, Vancouver, BC.

Burton, R. S. 1987. Differentiation and integration of the genome in populations of the marine copepod, *Tigriopus californicus.* Evolution 41:504–513.

Bush, G. L., and D. J. Howard. 1986. Allopatric and non-allopatric speciation: assumptions and evidence. Pages 411–438 *in* S. Karlin and E. Nevo, editors. Evolutionary processes and theory. Academic Press, New York.

Carvalho, G. R. 1993. Evolutionary aspects of fish distribution: genetic variability and adaptation. Journal of Fish Biology 43 (Supplement A):53–73.

Cichra, C. E., W. E. Neill, and R. L. Noble. 1982. Differential resistance of northern and Florida largemouth bass to cold shock. Proceedings of the Annual Conference of the Southeastern Association of Fish and Wildlife Agencies 34:19–24.

Cicmanec, J. C., and A. K. Campbell. 1977. Breeding the owl monkey *(Aotus trivirgatus)* in a laboratory environment. Laboratory Animal Science 27:517.

Danzmann, R. G., T. R. Jackson, and M. M. Ferguson. 1999. Epistasis in allelic expression at upper temperature tolerance QTL in rainbow trout. Aquaculture 173:45–58.

De Boer, L. E. M. 1982. Karyological problems in breeding owl monkeys, *Aotus trivirgatus.* International Zoo Yearbook 22:119–124.

Dobzhansky, T. 1937. Genetics and the origin of species. Columbia University Press, New York.

Dobzhansky, T. 1948. Genetics of natural populations. XVIII. Experiments on chromosomes of *Drosophila pseudoobscura* from different geogoraphic regions. Genetics 33:588–602.

Dobzhansky, T. 1950. Genetics of natural populations. XIX. Origin of heterosis through natural selection in populations of *Drosophila pseudoobscura.* Genetics 35:288–302.

Dobzhansky, T., and O. Pavlovsky. 1953. Indeterminate outcome of certain experiments on *Drosophila* populations. Evolution 7:198–210.

Emlen, J. M. 1991. Heterosis and outbreeding depression: a multilocus model and an application to salmon production. Fisheries Research 12:187–212.

Ferguson, M. M., R. G. Danzmann, and F. W. Allendorf. 1985. Developmental divergence among hatchery strains of rainbow trout *(Salmo gairdneri).* II. Hybrids. Canadian Journal of Genetics and Cytology 27:298–307.

Fields, R., S. S. Lowe, C. Kaminski, G. S. Whitt, and D. P. Philipp. 1987. Critical and chronic thermal maxima of northern and Florida largemouth bass and their reciprocal F_1 and F_2 hybrids. Transactions of the American Fisheries Society 116:856–863.

Foote, C. J., C. C. Wood, C. Clarke, and J. Blackbourne. 1989. The saltwater adaptability of sympatric anadromous and nonanadromous sockeye salmon and their "hybrids." Aquaculture 82:377.

Frankham, R. 1995. Conservation genetics. Annual Review of Genetics 29:305–327.

Gharrett, A. J., and W. W. Smoker. 1991. Two generations of hybrids between even- and odd-year pink salmon *(Oncorhynchus gorbuscha):* a test for outbreeding depression? Canadian Journal of Fisheries and Aquatic Sciences 48:1744–1749.

Gharrett, A. J., and W. W. Smoker. 1993. A perspective on the adaptive importance of genetic infrastructures in salmon populations to ocean ranching in Alaska. Fisheries Research 18:45–58.

Gharrett, A. J., W. W. Smoker, R. R. Reisenbichler, and S. G. Taylor. 1999. Outbreeding depression in hybrids between odd- and even-broodyear pink salmon. Aquaculture 173:117–129.

Gordon, H., and M. Gordon. 1957. Maintenance of polymorphism by potentially injurious gene in eight natural populations of the platyfish, *Xiphophorus maculatus.* Journal of Genetics 55:1–44.

Gordon, M. 1947. Genetics of ocular tumor development in fishes (preliminary report). Journal of the National Cancer Institute 7:87–92.

Gordon, M., and G. M. Smith. 1938. The production of melanotic neoplastic disease in fishes by selective matings. IV. Genetics of geographical species hybrids. American Journal of Cancer 34:543–565.

Graham, J. H., and J. D. Felley. 1985. Genomic coadaptation and developmental stability within introgressed populations of *Enneacanthus gloriosus* and *E. obesus.* Evolution 39:104–114.

Greig, J. C. 1979. Principles of genetic conservation in relation to wildlife management in Southern Africa. South African Journal of Wildlife Research 9:57–78.

Grosberg, R. K. 1987. Limited dispersal and proximity-dependent mating success in the colonial ascidian *Botryllus schlosseri*. Evolution 41:372–384.

Guest, W. C. 1985. Temperature tolerance of Florida and northern largemouth bass: Effects of subspecies, fish size, and season. Texas Journal of Science 37:75–84.

Hemmingsen, A. R., R. A. Holt, R. D. Ewing, and J. D. McIntyre. 1986. Susceptibility of progeny from crosses among three stocks of coho salmon to infection by *Ceratomyxia shasta*. Transactions of the American Fisheries Society 115:494–495.

Hindar, K., N. Ryman, and F. M. Utter. 1991. Genetic effects of cultured fish on natural fish populations. Canadian Journal of Fisheries and Aquatic Sciences 48:945–957.

Kimura, M. 1983. The neutral theory of molecular evolution. Cambridge University Press, Cambridge, UK.

Koppelman, J. B., G. S. Whitt, and D. P. Philipp. 1988. Thermal preferenda of northern, Florida, and reciprocal F_1 hybrid largemouth bass. Transactions of the American Fisheries Society 117:238–244.

Lacy, R. C., A. Petric, and M. Warneke. 1993. Inbreeding and outbreeding in captive populations of wild animal species. Pages 352–374 *in* N. W. Thornhill, editor. The natural history of inbreeding and outbreeding: theoretical and empirical persepctives. University of Chicago Press, Chicago.

Maceina, M. J., and B. R. Murphy. 1992. Stocking Florida largemouth bass outside its range. Transactions of the American Fisheries Society 121:686–688.

MacFarquhar, A. M., and F. W. Robertson. 1963. The lack of evidence for coadaptation in crosses between geographical races of *Drosophila subobscura* Coll. Genetical Research 4:104–131.

Mayr, E. 1963. Animal species and evolution. Belknap Press, Cambridge, Massachusetts.

Ohta, T. 1982. Linkage disequilibrium due to random genetic drift in finite subdivided populations. Proceedings of the National Academy of Sciences of the USA 79:1940–1944.

Philipp, D. P. 1991. Genetic implications of introducing Florida largemouth bass, *Micropterus salmoides floridanus*. Canadian Journal of Fisheries and Aquatic Sciences 48 (Supplement 1):58–65.

Philipp, D. P. 1992. Stocking Florida largemouth bass outside its range. Transactions of the American Fisheries Society 121:688–691.

Philipp, D. P., W. F. Childers, and G. S. Whitt. 1983. A biochemical genetic evaluation of the northern and Florida subspecies of largemouth bass. Transactions of the American Fisheries Society 112:1–20.

Philipp, D. P., and J. E. Claussen. 1995. Fitness and performance differences between two stocks of largemouth bass from different river drainages within Illinois. Pages 236–243 *in* H. L. Schramm, Jr., and R. G. Piper. Uses and effects of cultured fishes in aquatic ecosystems. American Fisheries Society, Symposium 15, Bethesda, Maryland.

Philipp, D. P., J. E. Epifanio, and M. J. Jennings. 1993. Conservation genetics and current stocking practices: are they compatible? Fisheries 18(12):14–16.

Philipp, D. P., C. Kaminski, and G. S. Whitt. 1985. A comparison of the embryonic development of northern, Florida, and reciprocal F_1 hybrid largemouth bass in different thermal environments. Illinois Natural History Survey Bulletin 33:261–173.

Philipp, D. P., and G. S. Whitt. 1991. Survival and growth of northern, Florida, and reciprocal F_1 hybrid largemouth bass in central Illinois. Transactions of the American Fisheries Society 120:58–64.

Price, M. V., and N. M. Wasser. 1979. Pollen dispersal and optimum outcrossing in *Delphinium nelsoni*. Nature 277:294–297.

Reisenbichler, R. R. 1988. Relation between distance transferred from natal stream and recovery rate for hatchery coho salmon. North American Journal of Fisheries Management 8:172–174.

Ricker, W. E. 1972. Hereditary and environmental factors affecting certain salmonids. Pages 27–160 *in* R. C. Simon, editor. The stock concept in Pacific salmon. H. R. MacMillan Lectures in Fisheries, University of British Columbia, Vancouver.

Schramm, Jr., H. L., and R. G. Piper, editors. 1995. Uses and effects of cultured fishes in aquatic ecosystems. American Fisheries Society, Symposium 15, Bethesda, Maryland.

Stearns, S. C. 1983. The genetic basis of differences in life history traits among six populations of mosquitofish *(Gambusia affinis)* that shared ancestors in 1905. Evolution 37:618–627.

Taylor, E. B. 1991. A review of local adaptation in Salmonidae, with particular reference to Pacific and Atlantic salmon. Aquaculture 98:185–207.

Taylor, E. B., and C. J. Foote. 1991. Critical swimming velocities of juvenile sockeye salmon and kokanee, the anadromous and nonanadromous forms of *Oncorhynchus nerka* (Walbaum). Journal of Fish Biology 38:407–419.

Templeton, A. R. 1979. The unit of selection in *Drosophila mercatorum*. II. Genetic revolution and the origin of coadapted genomes in parthenogenetic strains. Genetics 92:1265–1282.

Templeton, A. R. 1983. Natural and experimental parthenogenesis. Pages 343–398 *in* M. Ashburner, H. L. Carson, and J. N. Thompson, editors. The genetics and biology of *Drosophila,* volume 3C. Academic Press, London.

Templeton, A. R. 1986. Coadaptation and outbreeding depression. Pages 105–116 *in* M. E. Soulé, editor. Conservation biology: the science of scarcity and diversity. University of Chicago Press, Chicago.

Turcek, F. J. 1951. Effect of introductions on two game populations in Czechoslavakia. Journal of Wildlife Management 15:113–114.

Waldman, B., and J. S. McKinnan. 1993. Inbreeding and outbreeding in fishes, amphibians, and reptiles. Pages 250–282 *in* N. W. Thornhill, editor. The natural history of inbreeding and outbreeding: theoretical and empirical persepctives. University of Chicago Press, Chicago.

Wallace, B. 1991. Coadaptaton revisited. Journal of Heredity 82:89–95.

Wallace, B., and M. Vetukhiv. 1955. Adaptive organization of the gene pools of *Drosophila* populations. Cold Spring Harbor Symposia in Quantitative Biology 20:303–310.

Wasser, N. M. 1993. Population structure, optimal outbreeding, and assortative mating in angiosperms. Pages 173–199 *in* N. W. Thornhill, editor. The natural history of inbreeding and outbreeding: theoretical and empirical persepctives. University of Chicago Press, Chicago.

Weir, B. S. 1979. Inferences about linkage disequilibrium. Biometrics 35:235–254.

Wood, C. C., and C. J. Foote. 1990. Genetic differences in the early development and growth of sympatric sockeye salmon and kokanee *(Oncorhynchus nerka)* and their hybrids. Canadian Journal of Fisheries and Aquatic Sciences 47:2250–2260.

Wright, S. 1931. Evolution in Mendelian populations. Genetics 16:97–159.

Wright, S. 1932. The roles of mutation, inbreeding, crossbreeding, and selection in evolution. Proceedings of the Sixth International Congress on Genetics 1:356–366.

Wright, S. 1982. The shifting balance theory and macroevolution. Annual Review of Genetics 16:1–19.

12

Quantitative Genetics

Eric Hallerman

12.1 Introduction

Anyone who has sampled populations of fishes will agree that there is great phenotypic variation in fishes. Conspicuous and economically important differences in growth rate, body size at maturity, and other traits frequently occur within and between populations of many fish species (Allendorf et al. 1987; Table 12.1). Although morphological characters cannot be compared directly among taxa, fish are phenotypically more variable than are other vertebrates (Mayr 1969:170). Coefficients of variation among individuals *within* populations (Table 12.1, upper panel) are typically less than 10% in other vertebrates but exceed that level for most morphological characters in fishes. Variation *among* conspecific populations of fishes (Table 12.1, lower panel) is generally larger than that in other vertebrates, often several times that among species in other vertebrate taxa. Such wide variation in quantitative characters raises a number of questions that fisheries science must strive to answer. How much of the observed variation is due to genetic factors? What is the adaptive and evolutionary significance of this variation? How should we consider quantitative genetic variation in fisheries management and conservation actions?

This chapter on quantitative variation might have been presented in the first section of this text along with those on molecular and chromosomal variation. However, I elected to defer discussion of quantitative genetics until the reader became familiar with population genetic processes. The intent of this chapter is to provide a foundation for understanding how quantitative variation is measured and its genetic basis inferred and then to discuss the implications of quantitative genetic variation for fisheries management and conservation.

12.1.1 *Recognizing the Nature of Quantitative Traits*

Quantitative traits exhibit a range of phenotypes that are measured as opposed to scored into a few discrete classes. Quantitative traits include such adaptively important traits as morphological, physiological, and behavioral characters. There are three general classes of quantitative traits, distinguished on the basis of the distribution of phenotypes.

Table 12.1 Phenotypic variation within and between populations for chosen characters in fishes. Adapted from Allendorf et al. (1987).

Variation within populations

Species	Character	Coefficient of variation (CV)[a]	Source
Rainbow trout	Body length (140 d–2.5 years)	10–19	Gjedrem (1983)
Oncorhynchus mykiss	Body weight (150 d–2.5 years)	17–56	
Atlantic salmon	Body length (190 d–3.5 years)	7–23	Gjedrem (1983)
Salmo salar	Body weight (190 d–3.5 years)	25–76	
Common carp	Body weight (adults)	22	Gjedrem (1983)
Cyprinus carpio			
Channel catfish	Body weight (juveniles)	46	Gjedrem (1983)
Ictalurus punctatus	Body weight (adults)	27	
Guppy	Body length (28–63 d)	10–12	Ryman (1972)
Poecilia reticulata	Body weight (42–63 d)	35–36	

Variation between populations

Species group	Character	Smallest measurement	Largest value as % of smallest	Source
Arctic char *Salvelinus alpinus*				
(15 Palearctic stocks)	Body length at 2 years	25 mm	800	Johnson (1980)
	Body length at 4 years	80 mm	398	
	Body length at 10 years	145 mm	417	
(10 nonmigratory Palearctic stocks)	Weight at maturity	23 g	4,213	Johnson (1980)
	Number of eggs	48	6,465	
Brown trout	Body length at 3 years	11.1 cm	270	Alm (1939)
Salmo trutta	Body length at 6 years	19.3 cm	337	
(11 European populations)				
European perch	Body length	9.8 cm	315	Alm (1946)
Perca fluviatilis				
(23 Swedish populations)				
Lake whitefish	Number of gill rakers	23	150	Lindsey (1981)
Coregonus clupeaformis				
(2 sympatric forms)				

[a] CV = (standard deviation/mean) × 100.

Continuous traits exhibit a continuum of phenotypes that are measured. There are infinitely many possible phenotypes, among which discrimination is limited only by our ability to measure them. Examples of continuous traits include growth rate, yield, and morphometric traits (e.g., head length:standard length, eye diameter:head length, and body depth:standard length).

Meristic traits exhibit phenotypes expressed in discrete elements that are counted to form integral classes. An individual's phenotype is characterized as the number of elements of the trait it exhibits. Examples of meristic traits in fishes are number of scales along the lateral line, number of fin rays, and number of pyloric caecae. When the number of elements exhibited becomes very large (e.g., fecundity), the distinction between continuous and meristic traits vanishes.

Threshold traits are discrete traits that are either present or absent in an individual. Generally, both genetic and environmental factors affect expression of the trait. Observation of the trait in a particular individual implies that the individual has a liability over the threshold for the trait's expression. This liability is not always directly observable; a geneticist would need to study an individual's pedigree to infer the likelihood that this individual may express the trait. The best-known and studied threshold traits are the human diseases diabetes, schizophrenia, and certain cancers. Exemplifying the genetic and nongenetic factors affecting expression of a threshold trait, the likelihood that a person with a genetic predisposition for diabetes will actually suffer the disease will also depend upon his or her diet, weight, and age. Threshold traits also are observed in fishes. Geneticists hypothesize that the tendencies of certain salmonids to undergo precocious maturation and to pursue anadromy are threshold traits.

12.1.2 Recognizing Qualities Common to All Classes of Quantitative Traits

It is not possible to infer from phenotype the action of segregation and single-locus-mediated gene expression because expression of quantitative traits can be controlled by many genes, a property termed **polygenic** trait determination. The **infinitesimal model** for expression of quantitative traits assumes a large number of loci each contributing a small and equal amount to expression of the trait. In reality, however, the contribution of individual loci to expression of the trait ranges from small to large. Should one gene play a large role in determining expression of a quantitative trait, it is termed a **major gene.** For example, major genes are known to affect expression of double muscling in Belgian Blue cattle and fertility in Beroola sheep and Meishan pigs. The number of genes affecting expression of certain quantitative traits has been estimated (Box 12.1). A particular gene may affect multiple traits, a condition referred to as **pleiotropy.** Because many loci generally are involved in the expression of a quantitative trait, all of which segregate independently (excepting linked loci), a large range of genetic variation is common.

Quantitative loci exhibit the same modes of inheritance as Mendelian, or qualitative, traits (Chapter 1) except that they are influenced by alleles at many loci. Each locus segregates in accordance to Mendelian expectation. Following meiosis (in a diploid organism), each gamete contains one of the two alleles carried by the parent. Alleles at a locus interact with each other according to Mendelian modes of phenotypic expression described in Chapter 1.

Expression of quantitative traits is affected by environmental (i.e., nongenetic) factors. Sensitivity to environmental effects will differ among loci. Polygenic trait determination and environmental influence give rise to continuous distributions for quantitative traits.

Because phenotypes are the consequence of multiple, interacting genetic and environmental effects, it is not possible to determine an individual's genotype from its external appearance. However, quantitative geneticists have developed a rich body of theory on how to apply statistical inference to quantify genetic and nongenetic factors underlying expression of quantitative traits. This body of theory is surveyed in the next section, preparing us for discussion of the implications of quantitative variation for adaptation and evolution.

■ Box 12.1
The Number of Genes Affecting Expression of a Quantitative Trait

It generally is assumed that expression of quantitative traits is controlled by many genes. For example, under the infinitesimal model, it is assumed that a large number of loci each contribute a small and equal amount to the variance for expression of the trait at issue. This paradigm is supported by ample experimental evidence for a variety of traits. A variety of statistical techniques have been proposed for estimating the effective, or minimum number of loci, contributing to the difference in a quantitative trait between two inbred lines grown in a common environment. Estimates of the **effective number of genes,** n_E, influencing the trait are obtained by comparing the phenotypic means and variances of the character in parental, F_1, F_2, and backcross populations. However, the requirement for inbred lines can lead to complications, notably in our context, precluding application of the approach to wild populations.

Lande (1981) outlined a procedure for estimating n_E between two populations that have diverged through artificial or natural selection. If certain criteria are met, notably that inheritance of the trait is largely additive, then by using an appropriate scale of measurement, n_E can be estimated by comparing the means and variances for the trait in the parental lines, hybrids, and backcrosses as

$$n_E = (\mu_{P2} - \mu_{P1})^2/(8\sigma_s^2),$$

where $(\mu_{P2} - \mu_{P1})$ is the difference in phenotypic expression of the trait among the parental lines, and σ_s^2 is the additional genetic variance in the F_2 population above that in the F_1 hybrids. Lande (1981) also presents methods for estimating the standard errors of the estimates and for determining whether underlying assumptions of the model are met.

Application of the method is illustrated by estimating the n_E affecting eye diameter in two species of cave- and surface-dwelling characins. Wilkens (1971) studied the

continued on next page

■ 12.2 Quantification and Partitioning of Quantitative Variation

12.2.1 *Quantifying Quantitative Variation*

Study of quantitative traits focuses on characterization of their genetic variation. In contrast to Mendelian traits, which are expressed as discrete classes, most quantitative traits are expressed in normal distributions. Hence, rather than consider frequencies of individuals in each phenotypic class, we consider quantitative traits in terms of a mean value (often expressed as μ, or $\bar{x}$) and its associated variance (σ^2 for a parameter or s^2 for a sample, $\Sigma[\{x_i - \bar{x}\}^2/\{n-1\}]$) or standard deviation (σ or s, that is $\sqrt{\sigma^2}$ or $\sqrt{s^2}$).

Although we generally assume that a normal distribution underlies our observations of a quantitative trait, often the underlying distribution is not normal. This nonnormality affects estimation of key quantitative trait parameters defined below, such as μ, a, d, and h. In such cases, the data sets must be transformed to normalize the observations or to prevent dependence of variation on the mean (Chapter 17 *in* Falconer and Mackay

Box 12.1 *continued*

inheritance of eye size in crosses between cave-dwelling *Anoptichthys hubbsi* and *Anoptichthys antrobius,* with highly reduced eye size, and the surface-dwelling Mexican tetra *Astyanax mexicanus,* with unreduced eyes. Mean eye diameters were measured among parental lines, F_1, F_2, and both types of backcross hybrids (see Table 1). Small variances were observed within the parental and F_1 populations and larger variances in F_2 and backcross populations. The estimates of n_E obtained based on variants of the model, n_1 to n_4, were reasonably consistent, ranging from 5.0 to 7.0, with reasonably small variances.

The number of genes affecting expression of a variety of traits has been estimated using this method (see Table 2). The minimum number of genes involved in producing large differences between populations on a quantitative trait frequently is estimated to be on the order of 5 to10, with much higher numbers for certain traits. Lande (1981) took this observation as evidence that large evolutionary changes usually occur by the accumulation of multiple genetic differences with relatively small effects.

Table 1 Estimation of the minimum number of genes, n_E, affecting eye size in hybrids of cave- and surface-dwelling characins. Abbreviations are parental (P), filial (F), and backcross (B) populations. Adapted from Lande (1981).

Population or model	N	μ	σ^2	Estimate of n_E
P_1	30	2.10	0.0264	
B_1	450	3.78	0.2190	
F_1	30	5.09	0.0571	
F_2	702	4.72	0.5628	
B_2	142	6.21	0.4691	
P_2	30	7.05	0.0220	
n_1				6.1 ± 0.4
n_2				5.9 ± 0.4
n_3				7.0 ± 1.3
n_4				5.0 ± 0.5

Table 2 Estimated minimum numbers of genes, n_E, affecting chosen quantitative traits. Compiled from Lande (1981) and Hartl (1988).

Trait	n_E
Eye diameter in characins	5–7
Coat color in lab rat	5–8
Skin color in human	4–6
Antigen recognition and graft rejection in mouse	40
Body weight in mouse	150
Fruit weight in tomato	9–11
Oil content in corn kernels	17–22
Female head shape in *Drosophila* spp.	6–9
Date of anthesis in goldenrod	6–7

1996). Key transformation functions include the arcsin ($\theta = \arcsin\sqrt{p}$, where p is a proportion) and the logarithm ($\theta = \log[x]$).

12.2.2 Partitioning Phenotypic Variance

Expression of a quantitative trait typically is the result of many, individually small genetic and environmental effects. Classical study of a quantitative trait partitions observed variation into components attributable to different underlying causes. The relative magnitude of each component determines the degree of resemblance between relatives and, consequently, the genetic properties of the population (Falconer and Mackay 1996:122; Chapters 5–7 *in* Lynch and Walsh 1998). Partitioning phenotypic variance into its components allows us to estimate the relative importance of each, particularly the relative importance of heredity and environment. The relative importance of a source of variation is the variance due to that source as a proportion of total phenotypic variance. The total phenotypic variance is the sum of the individual components, expressed as

$$V_P = V_G + V_E + V_{G\times E}, \tag{12.1}$$

where V_P = total phenotypic variance; V_G = genetic component of phenotypic variance; V_E = environmental component of phenotypic variance; and $V_{G\times E}$ = genotype × environmental interaction component of phenotypic variance.

The **genetic component of variance** includes variation with an underlying genetic basis. The relative importance of heredity in determining variation of a phenotype in a population is called the **heritability** of the character. The ratio V_G/V_P expresses the extent to which phenotypic variation between individuals is determined by genetic factors. This ratio is termed **broad-sense heritability** or the **degree of genetic determination,** often denoted h_b^2 (Falconer and Mackay 1996:123; see also Lynch and Walsh:170, ff). For historic reasons, the symbol h^2 stands for heritability itself and not for its square. Heritability ranges from 0, when all variation is caused by environmental factors, to 1, when all variation is genetically determined. In some cases, V_G can be estimated with simple experiments based on measuring phenotypic variation among individuals with identical genotypes and subtracting this estimate of V_E from V_P to estimate V_G. However, this experimental design proves difficult to apply in practice, especially for populations in the wild.

The **environmental component of variance** includes variation resulting from a variety of nongenetic factors (Falconer and Mackay 1996:134). The nature of environmental variation depends on the trait and the organism studied. To an investigator, environmental variance is a source of variation that reduces the precision of genetic variance component estimates. Investigators can reduce environmental variation by careful design and management of experiments. Nutritional and climatic factors are common sources of environmental variation—although they may be under experimental control for captive populations, they are not for wild populations. **Maternal effects,** the nongenetic effects of the mother on her young, are another source of environmental variation. In fish, maternal effects frequently are a consequence of how much yolk material was deposited in the egg, which affects survival through larval transition to exogenous feeding. Measurement error is another source of variation and usually is unimportant.

Although most of the variance terms above have some degree of intuitive meaning, the **genotype × environmental interaction** needs further explanation. In many cases, the genetic and environmental components of total phenotypic variance are independent of each other. There are cases, however, in which there is correlation between genotypic and environmental variation. One poignant example is human intelligence (Falconer and Mackay 1996:131): the phenotypic values of the parents affect the environment in which the children are reared to the extent that intelligence is increased or decreased from the level inherited. Therefore, a correlation is introduced between the genotype and the environment. In an aquacultural context, the genotype × environment interaction is at issue when a given genotype is superior to another in one environment but inferior in another. For example, Wohlfarth et al. (1975) showed differences in growth performance between Chinese and European races of the common carp, with the Chinese race growing faster in low-input culture systems and the European race in high-input systems.

12.2.3 *Partitioning Genetic Variance*

Partitioning a population's total phenotypic variance for a quantitative trait into genetic *(G)*, environmental *(E)*, and $G \times E$ components is the first step toward understanding resemblance among relatives and population response to selection. The second step is partitioning the genetic variance into its underlying components:

$$V_G = V_A + V_D + V_I, \tag{12.2}$$

where V_A = the additive component; V_D = the dominance component; and V_I = the epistatic, or interaction, component.

Each component of genetic variance is related to a particular mode of gene expression, which relates to how alleles interact within and between loci to determine expression of the quantitative trait.

The **additive variance,** V_A, is the fraction of genetic variance attributable to loci for which phenotypic expression of the quantitative trait is determined by the summation of individual allelic effects; that is, the phenotype is *not* affected by interactions of alleles. It is important to know the magnitude of the additive variance because it is the chief cause of resemblance among relatives and, therefore, the chief determinant of the observable genetic properties of the population and of the response of the population to selection (Falconer and Mackay 1996:125). It is the only component of genetic variance that can be readily estimated from observations made on the population. Hence, in practice, quantitative geneticists partition genetic variance into two components, additive (σ_A) and nonadditive (σ_{D+I}) components. The ratio V_A/V_P expresses the extent to which progeny phenotypes are determined by the alleles transmitted from the parents. This ratio is called the **narrow-sense heritability,** or sometimes simply heritability, often denoted h^2_n.

The **dominance variance,** V_D, corresponds to the fraction of genetic variance attributable to loci for which phenotypic expression of a quantitative trait is affected by interactions of alleles *at a locus,* that is, by dominance allelic interactions. Dominance variance represents the effect of putting alleles together in pairs to make genotypes, the effect of which is not accounted for by the effects of the two alleles taken singly. The relationships between observed phenotypes, underlying genotypes, and dominance-mediated allelic interaction at a single locus are shown in Figure 12.1. The dominance variance is the sum of all such interactions across loci. The magnitude of dominance variance depends on allele frequencies in the population and, hence, is partly a population property and not simply a reflection of the degree of dominance (Falconer and Mackay 1996:116).

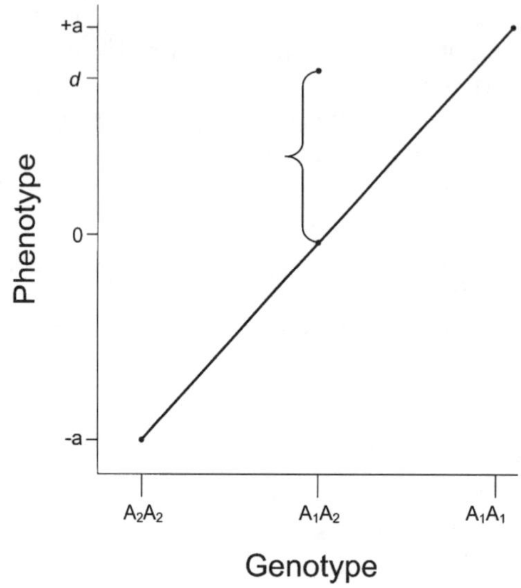

Figure 12.1 Relationships between observed phenotypes, underlying genotypes, and dominance-mediated allelic interaction at a single locus. The alleles at the locus are A_1 and A_2 and have frequencies p and q, respectively. Phenotypes corresponding to the A_1A_1 and A_2A_2 genotypes are assigned the values $+a$ and $-a$, respectively. The dominance effect, d, is defined as the difference between the phenotypic value of the heterozygote from the mid-point, 0, between phenotypic values $+a$ and $-a$. In this example, $d = \frac{3}{4}a$. Figure adapted from Falconer and Mackay (1996:117).

When a phenotype is determined by the interaction of two or more loci, genetic variance will include a second nonadditive component, the **epistatic or interaction variance,** V_I. Epistatic variance corresponds to the fraction of genetic variance attributable to loci for which phenotypic expression of the quantitative trait *is* affected by interactions of alleles *among loci.* If G_A is the genotypic value of an individual attributable to one locus, G_B that attributable to a second locus, and G the aggregate genotypic value attributable to both loci together, then

$$G = G_A + G_B + I_{AB}, \tag{12.3}$$

where I_{AB} is the deviation from additive combination of these genotypic values (Falconer and Mackay 1996:119). If I is nonzero for any combination of alleles at different loci, the alleles exhibit interaction, or **epistasis.** The term epistasis has a broader meaning in quantitative genetics than it has in Mendelian genetics. Loci may interact in pairs, threes, or higher numbers, with all interactions treated together as a single variance component. Although interaction between loci controlling quantitative characters is a frequent occurrence, it is difficult to estimate the amount of variance generated, and little is known about the relative importance of epistatic interactions as a source of quantitative variation (Crow and Kimura 1970; Barker 1979).

12.2.4 *Estimating the Components of Genetic Variance*

An understanding of quantitative genetics provides the basis for genetic improvement programs for agriculturally important organisms. In particular, a reliable estimate of additive genetic variation, or narrow-sense heritability, allows breeders to anticipate the efficacy of alternative selective breeding approaches. The theoretical bases for heritability estimation are well developed (Mather and Jinks 1949; Tave 1993; Falconer and Mackay 1996; Lynch and Walsh 1998). Population and evolutionary geneticists apply heritability estimation techniques to anticipate the responsiveness of a population to selection. Heritability can be estimated using several experimental designs.

12.2.4.1 **Estimates of Heritability.** Heritability can be estimated from measurements of phenotypic resemblance among relatives. The first method uses the regression or correlation among observations between parent and offspring groups to estimate h^2. Several choices of group relationships are possible (Table 12.2). To use the offspring–parent regression method, data are collected on the trait of interest on parents—one or the mean of both (midparent)—and the mean of the offspring. The regression coefficient for a one-parent design is multiplied by 2 or taken directly for the midparent design as the estimated heritability. An example of application of this design to estimate heritability for a meristic trait is presented in Figure 12.2.

The choice of a resemblance-among-relatives design depends in part upon circumstances. In addition to the practical matter of which relatives are available for measurement, issues of precision and bias also will affect choice of experimental design (Falconer and Mackay 1996:163–164). In general, the closer the degree of relatedness, the more precise the estimate. This is because the observed regression or correlation must be multiplied by a factor $1/r$ (where r is the coefficient associated with V_A in the covariance term in Table 12.2), which is larger with more distant relatives. Similarly, the standard error of the regression or correlation must be multiplied by the same factor to obtain the standard error of the estimated heritability. Bias is often more important than precision. Bias is introduced by environmental sources of covariance and, in the case of full sibs, by dominance. Generally, half-sib correlation and regression of offspring on father are the most reliable estimates. Regression of offspring on mother can give heritability estimates that are too high because of maternal effects, especially for traits measured early in life. Full-sib correlation is subject to a large environmental covariance due to shared environment and dominance variance and often provides an estimate of heritability that is higher than the true value.

The second method for estimating heritability based on resemblance among relatives involves estimation of variance from sib analysis. The data analysis is relatively complicated and will be only outlined here; for a more complete description, see Falconer and Mackay (1996). In one form of this experimental design, a number of males, or **sires,** are

Table 12.2 Expressions for estimating heritability (h^2) by measuring resemblance among relatives. Genetic variance components are additive (V_A), dominance (V_D), and common environment (V_{Ec}). Table from Falconer and Mackay (1996:163), with permission from Longman, Harlow, Essex, UK.

Relatives	Regression (*b*) or correlation (*t*)	Covariance[a]
Offspring and one parent	$b = \frac{1}{2}$	$\frac{1}{2}V_A$
Offspring and midparent	$b = h^2$	$\frac{1}{2}V_A$
Half sibs	$t = \frac{1}{4}h^2$	$\frac{1}{4}V_A$
Full sibs	$t \geq \frac{1}{2}h^2$	$\frac{1}{2}V_A + \frac{1}{4}V_D + V_{Ec}$

[a] The contributions of epistatic interactions are ignored, as well as possible environmental contributions to relatives other than full sibs.

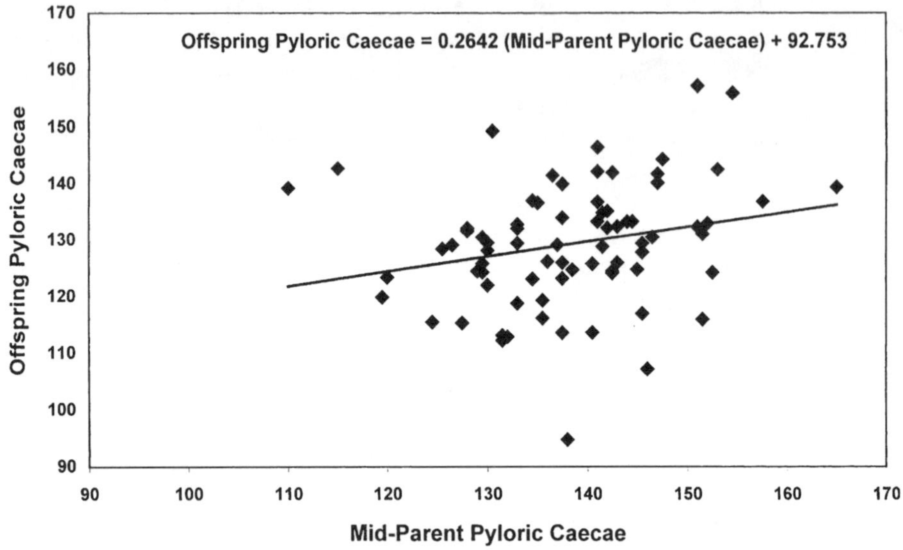

Figure 12.2 Estimation of heritability of pyloric caecae number in pink salmon *Oncorhynchus gorbuscha* using the regression of offspring on parent values. Estimated $h^2 = 0.26 \pm 0.13$, $r^2 = 0.044$, and $P = 0.038$. Figure provided courtesy of W. Smoker, University of Alaska-Juneau.

each mated to several females, or **dams.** Individuals are randomly chosen and paired for mating (Figure 12.3). Phenotypes from a number of offspring from each female are measured to provide the raw data. The progeny so measured represent a collection of full-sib families **nested** within half-sib families. Analysis of variance (ANOVA) (Table 12.3, top panel) is used to partition the phenotypic variance into components attributable to differences between the progenies of sires (the between-sire component, σ^2_S), between progenies of dams mated to the same sire (the between-dam, within-sire component, σ^2_D), and among individual offspring of the same dam (the within-progenies component, σ^2_W). The family design described is termed a **hierarchichal mating of dams nested within sires;** a corresponding design and analysis for **sires nested within dams** also is possible. The next step in the data analysis is to use the relationships among the observed variance components (σ^2_S, σ^2_D, and σ^2_W) to estimate the causal components of phenotypic variance, V_A, V_D, and V_E, as shown in Table 12.3 (bottom panel). A half-sib analysis estimating heritability of pyloric caecae number in pink salmon *Oncorhynchus gorbuscha* is presented in Table 12.4.

A third method for heritability estimation is based on measuring a selection differential *(S)* and a population's response to selection *(R)* to determine what heritability would relate the two parameters. Key parameters are μ, the mean phenotypic value of the trait for the population, and μ_s, the mean phenotypic value for selected individuals. Figure 12.4 presents a diagrammatic representation of the heritability estimation experiment, showing (a) the **selection differential,** $S = \mu_s - \mu$, imposed in the parental generation, and (b) the **response to selection,** $R = \mu' - \mu$, observed in the progeny generation, where μ' is the mean phenotypic value for selected individuals in the progeny generation. The **realized heritability** is estimated as $h^2 = R/S$. Purposeful application of a realized heritability design for estimating h^2 in natural populations generally would prove impractical, although some studies (e.g., Smith and Dhondt 1980) have used naturally imposed selection episodes to observe response and thereby to estimate realized heri-

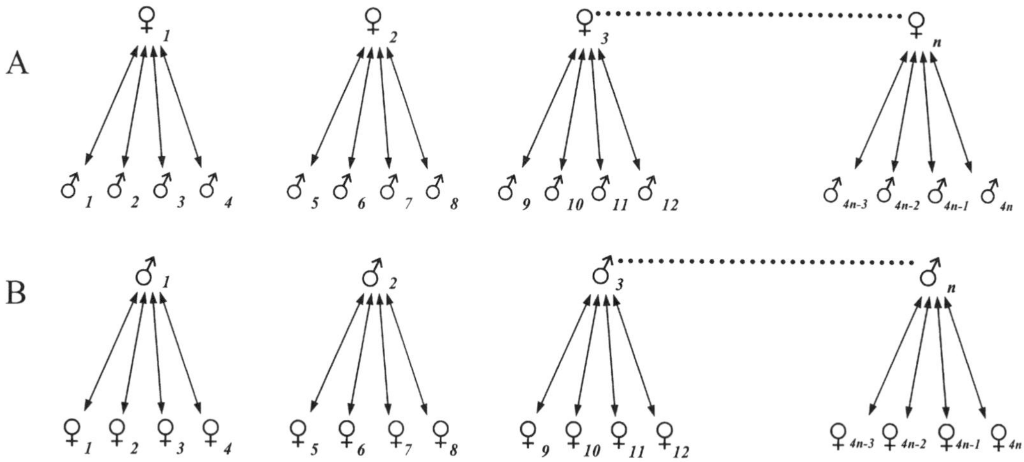

Figure 12.3 Diagrammatic representation of nested full- and half-sib families in hierarchical breeding designs: (A) sires nested within dams and (B) dams nested within sires. ANOVA for a trait within and between such families can be used for heritability estimation.

Table 12.3 Structure of analyses of variance for estimating heritability by measuring resemblance among full- and half-sib families in a dams-within-sires hierarchical breeding design. Analysis follows Falconer and Mackay (1996:167). Abbreviation are number of sires (*s*), number of dams per sire (*d*), number of offspring per dam (*k*), half sibs (HS), full sibs (FS), common environment (E_c), within full-sib family environment (E_w), variance among progeny (V_P), within-progeny component of variance (σ^2_W), between-dam within-sire component of variance (σ^2_D), between-sire component of variance (σ^2_S), and total variance (σ^2_T). Other abbreviations are as in Table 12.2.

Step 1: Analysis of observational variance components

Source	df	Mean square	Composition of mean square
Between sires	$s-1$	MS_S	$=\sigma^2_W + k\sigma^2_D + dk\sigma^2_S$
Between dams (within sires)	$s(d-1)$	MS_D	$=\sigma^2_W + k\sigma^2_D$
Within progenies	$sd(k-1)$	MS_W	$=\sigma^2_W$

Step 2: Analysis of causal variance components

Observational component	Covariance components estimated	Causal components estimated
Sires	$\sigma^2_S = \mathrm{cov}_{HS}$	$= \frac{1}{4}V_A$
Dams	$\sigma^2_D = \mathrm{cov}_{FS} - \mathrm{cov}_{HS}$	$= \frac{1}{4}V_A + \frac{1}{4}V_D + V_{Ec}$
Progeny	$\sigma^2_W = V_P - \mathrm{cov}_{FS}$	$= \frac{1}{2}V_A + \frac{3}{4}V_D + V_{Ew}$
Total	$\sigma^2_T = \sigma^2_S + \sigma^2_D + \sigma^2_W = V_P$	$= V_A + V_D + V_{Ec} + V_{Ew}$
Sires + dams	$\sigma^2_S + \sigma^2_D = \mathrm{cov}_{FS}$	$= \frac{1}{2}V_A + \frac{1}{4}V_D + V_{Ec}$

Table 12.4 Estimation of heritability for pyloric caecae number in pink salmon using analysis of variance within progeny groups in a hierarchical breeding design. Data and analysis courtesy of W. Smoker, University of Alaska-Juneau.

Estimation of total variance accounted for by genetic variance

Source	df	Sum of squares (SS)	Mean square (MS)	*F*-value	*P* > *F*	R^2	CV	√MS for error	PYLNR mean[a]
Model	110	95400.0116	867.2728	4.77	0.0001	0.346801	10.44033	13.4790	129.105
Error	989	179685.7557	181.6843						
Corrected total	1099	275085.7673							

Estimation of sire and dam-within-sire variance

Source	df	Type III SS	MS	*F*-value	*P* > *F*	Type III expected MS
Sire	59	62203.7739	1054.3013	5.80	0.0001	Var(error) + 8.4775var[dam(sire)] + 16.13var(sire)
Dam(sire)	51	25450.7059	499.0334	2.75	0.0001	Var(error) + 9.0695var[dam(sire)]

Estimation of heritability based on dams-within-sires variance as an error term

Source	df	Type III SS	MS	*F*-value	*P* > *F*
Sire	59	62203.7739	1054.3013	2.11	0.0035

Estimated contributions to variance and heritability

Var(environment) = 181.6843
Var(sire) = 24.1137
Var(dam) = 34.9908
$h^2 = 0.400$

[a] Mean number of pyloric caecae.

tability. Falconer and Mackay (1996:198) identify several reasons why realized heritability values may not provide valid estimates of the true heritability for the trait in the base population. If the population is subject to selection by environmental trends, or to inbreeding or random genetic drift, these effects will be confounded with the response to selection on the measured trait. Comparison of selection response relative to a control line can help remove these trends from the results, although suitable control lines are not generally available in studies of wild populations. Selection responses of characters with high heritabilities are expected to decline after the first generation of selection, so realized heritability after the first generation will underestimate heritability in the base population. Unbiased estimates of heritability can be obtained from response to selection if information on relationships of individuals within and across generations is available, although these generally are not known in natural populations.

12.2.4.2 **Properties of Heritability Estimates.** Heritability estimates have been made for a variety of quantitative traits in a wide range of species. Not surprisingly, most heritability estimates for fishes (compiled by Tave 1993) are for economically important traits of aquaculturally important species. Table 12.5 presents heritability estimates for a range of traits in rainbow trout, a relatively well-studied aquaculture species. The selected values shown in the table illustrate several key points about heritability estimation and interpretation (Falconer and Mackay 1996:161).

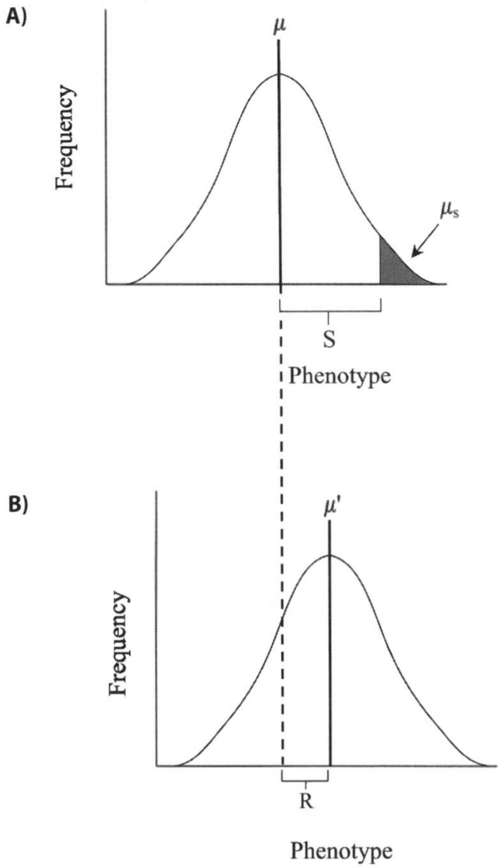

Figure 12.4 Diagrammatic representation of heritability estimation by the realized heritability method: **(A)** selection differential, $S = \mu_s - \mu$, in the parental generation, and **(B)** response to selection, $R = \mu' - \mu$, in the progeny generation. Realized heritability is estimated as $h^2 = R/S$.

—Heritability estimates for a given trait (e.g., weights at ~150 d and 1 year) differ among experiments. These differences reflect differences among populations or the conditions under which they were studied. It is important to realize that heritability is a property not only of a quantitative trait but also of the population studied, the environmental circumstances under which the experiment went forward, and the method by which the phenotype was measured. Examples of environmental effects on heritability estimates include the effects of temperature on number of lateral plates in threespine stickleback *Gasterosteus aculeatus* (Hagen 1973) and on hatching time, survival, weight, and length for fry of pink salmon and chum salmon *O. keta* (Beacham 1988).

—Heritability estimates are expected to be lower in small populations than in large populations. This difference occurs because the relative magnitudes of the genetic components of variance are influenced by allele frequency differences among populations, and small populations are subject to greater random genetic drift, which causes a higher rate of allelic fixation than that occurring in larger populations.

—Heritabilities may differ for the same trait measured in opposite directions. For example, an asymmetrical response to selection was observed for tolerance to high and low

Table 12.5 Estimated heritabilities for chosen traits in rainbow trout. Data were compiled by Tave (1993).

Trait	$h^2 \pm$ SE	Source
	Continuous traits	
147-d weight	0.26	Kincaid et al. (1977)
150-d weight	0.09 ± 0.10	Austad et al. (1972)
	0.50 ± 0.07	von Limbach (1970)
334-d weight	0.82 ± 0.38	Klupp (1979)
1-year weight	0.38 ± 0.25	Linder et al. (1983)
1-year weight	0.20 ± 0.11	Gall and Huang (1988a)
Tolerance to high temperature	0.48	Ihssen (1986)
Tolerance to low temperature	0.03	Ihssen (1986)
Age at sexual maturity	0.21 + 0.14	McKay et al. (1986)
Sexual precocity	0.3	Burger and Chevassus (1987)
Precocious spawning	0.1	Tofteberg and Hansen (1987)
Maturity at 2 years	–0.05 ± 0.12 to 0.28 ± 0.12	Gjerde and Gjedrem (1984)
Egg number	0.19 ± 0.06	Gall (1975)
	0.16 ± 0.10 to 0.67 + 0.15	Gall and Gross (1978)
	0.32 ± 0.14	Gall and Huang (1988b)
	0.33 ± 0.20	Haus (1984)
	Meristic traits	
Number of		
pyloric caecae	0.75 ± 0.34	Bergot et al. (1976)
	0.46 ± 0.10 to 0.53 ± 0.07	Chevassus et al. (1979)
	0.68 ± 0.08	Bergot et al. (1976)
anal fin rays	0.93 + 0.50	Leary et al. (1985)
dorsal fin rays	0.90 + 0.27	Leary et al. (1985)
gill rakers in upper arch	0.37 + 0.21	Leary et al. (1985)
gill rakers in lower arch	0.67 + 0.11	Leary et al. (1985)

temperatures in rainbow trout (Ihssen 1986) and for high and low growth rate in common carp (Moav and Wohlfarth 1976).

— Heritabilities cannot be estimated with great precision, and most estimates have relatively large standard errors.

— Because an estimate of heritability depends on the relative magnitudes of all the components of variance, a change in any one of them will affect the estimate.

— Environmental variance is dependent on experimental conditions. Hence, more variable conditions reduce heritability and more uniform conditions increase it.

It should be clear that these points are not an invalidation of the underlying theory or of the experiments themselves. Rather, a reported heritability value should be regarded as an estimate made for a given population, place, and time.

Another key observation is that traits exhibiting low heritability values frequently are connected with traits associated with fitness (e.g., life history traits; Price and Schluter 1991; Table 12.5). On the other hand, traits exhibiting high heritabilities often might be judged to have little bearing on fitness. This general relationship has been supported by surveys of heritabilities for various characters in fruit flies *Drosophila* spp. (Roff and Mousseau 1987) and in wild populations of a variety of other species (Mousseau and Roff 1987). For example, heritabilities for meristic traits in fishes (e.g., numbers of pyloric caecae, fin rays, and gill rakers) are generally rather high (Table 12.5). Expression of meristic traits is established early in development and, thus, is not affected by such factors as indeterminate growth (Allendorf et al. 1987). For example, the number of fin rays may be affected by the temperature of embryonic incubation. Individuals that developed at relatively low temperatures will, on average, exhibit a higher number of fin rays than will individuals that developed at relatively high temperatures. Meristic traits have little bearing on fitness, hence directional selection generally will not act to reduce additive genetic variance for meristic traits.

The high phenotypic variation observed in fish species is not necessarily associated with high levels of underlying genetic variability (Allendorf et al. 1987). Heritabilities for similar traits, such as body weight, generally are lower in fish populations than they are in populations of other vertebrates. High levels of phenotypic variability with relatively low heritabilities often indicate a strong effect of environmental factors determining expression of a quantitative trait. This is not particularly surprising given the unique physiology of fishes compared with other vertebrates. The indeterminate growth patterns of most fishes permit greater phenotypic accommodation to environmental factors, such as food availability and crowding (Purdom 1979). As poikilotherms, the rate of metabolism of fishes is highly sensitive to the effect of environmental temperature, which is not the case for homeothermic birds and mammals. Both age and size are related to sexual maturity in fishes, permitting flexibility in achieving reproductive success (Alm 1959; Jonsson et al. 1984; Gross 1985).

12.2.4.3 **Heritability Estimation in Wild Populations.** Most heritability estimates have been made for captive populations, most often in the context of genetic improvement of agricultural stocks or for laboratory model systems (e.g., Falconer and Mackay 1996:162). These systems present the best conditions for heritability estimation because geneticists know the histories of the (inbred) stocks, control matings, and control the environment in which the young are reared. Hence, reliable heritability estimates with relatively narrow standard errors often are achieved.

Relatively few heritability estimates have been made for wild populations. In contrast to captive stocks, for wild populations we generally do not know the breeding history, cannot control matings, and cannot control environmental conditions. A number of factors, mostly beyond the investigator's control, can affect heritability estimates in wild populations (Hailman 1986). Assortative mating (Chapter 1) raises h^2 estimates above their true values (see also Falconer and Mackay 1996:174). Treating each individual of a clutch (i.e., of a group reared together) as an independent observation inflates h^2 because the effects of shared environment are included in V_A, and it deflates the standard error of the h^2 estimate relative to the true value. A related environmental correlation involves the shared "parental environment" for species with parental care, which can raise h^2 when sibs are treated as independent observations. Within the control of the investigator, measuring the phenotype of offspring before they reach the same stage at which the trait was measured in the parents may inflate h^2 based on parent–offspring correlation. Improper data manipulation, for example "correcting" for environmental variation, can decrease V_E and inflate V_A. Mathematical transformations also can improve parameter estimations—consultation with a qualified statistical geneticist will help determine what constitutes appropriate data transformation. Considering all these factors shows the difficulty involved in estimating heritability for traits in wild populations. These factors also suggest that heritabilities for traits in wild populations are more often overestimated than underestimated. This inference should not be taken as a disqualification of heritability estimates on wild populations but should introduce a cautionary note.

Rather few studies have estimated heritabilities for fitness-related traits in wild populations of fishes. Heritabilities for reproductive (Snyder 1991) and meristic (Hagen 1973; Hagen and Blouw 1983) traits have been estimated for threespine stickleback. Heritabilities for meristic traits of viviparous eelpout *Zoarces viviparus* were reported by Kirpichnikov (1981). Examples of estimating genetic components for life history traits in wild pink salmon are discussed below in the context of conserving quantitative genetic variation.

Having considered how quantitative traits are expressed and how geneticists infer the magnitude of genetic variation, we are ready to consider the evolutionary dynamics and conservation importance of quantitative traits.

12.3 Quantitative Evolutionary Genetics

The goal of most genetic conservation programs is to preserve evolutionarily significant gene pools. Quantitative evolutionary genetics is the tool of choice for analyzing variation in quantitative traits, many of which have direct bearing on fitness traits relevant to conservation objectives (Barker and Thomas 1987; Lande 1988). This is particularly true because quantitative genetic theory is suitable for making predictions of a population's response to selective forces, at least through the short term. Anticipation of such responses is relevant for developing strategies for conserving genetic diversity (Box 12.2).

12.3.1 *Predicting the Response to Selection*

One way to apply quantitative genetics is to anticipate the response of a population to selection for a trait whose heritability has been estimated. Knowing h^2 and S, we can use the relationship $R = h^2S$ prospectively to predict the response to selection.

An alternative means of predicting response to selection is available for the case of **truncation selection,** in which selection favors all individuals with phenotypes above the truncation point, T. Assuming that the phenotype is normally distributed, the area of the normal curve beyond T equals B, and the height of the curve at point T equals Z (Hartl 1988:232–233). The properties of the normal distribution are such that

$$\frac{\mu_s - \mu}{\sigma^2} = \frac{Z}{B}, \qquad (12.4)$$

where, as above, μ_s and μ are the mean phenotypes for selected individuals and for the population, respectively, and σ^2 is the variance for the trait. Recognizing that $S = (\mu_s - \mu)$, we can define the parameter **selection intensity** as

$$i = \frac{S}{\sigma} = \frac{\sigma Z}{B}, \qquad (12.5)$$

where σ is the phenotypic standard deviation for the trait (Hartl 1988:243). Then R can be estimated as

■ Box 12.2
Molecular Variation, Quantitative Traits, and Conservation

Conservation genetics has grown rapidly over the past 2 decades, but its development largely has been independent of relevant progress in evolutionary genetics (Hard 1995). Conservationists have tended to rely heavily on assessments of presumably neutral molecular genetic variability to identify units for conservation, while relatively little attention has been paid to the genetic basis for phenotypic polymorphisms, despite their adaptive significance. Genetic and phenotypic variation in quantitative traits reflects adaptive differences among populations, and maintenance of genetic and ecological diversity within a species depends on recognition and preservation of these differences. Assessment of quantitative genetic variation in life history, physiological, and behavioral traits should be integrated into conservation strategies.

Differences in frequencies of genetic markers have an unambiguous genetic basis, and their generally selectively neutral qualities render them useful for ascertaining pedigrees, reconstructing phylogenies, identifying genetic stocks, and estimating migration frequencies, among other purposes. However, some studies in conservation biology have used molecular genetic data for additional purposes, for example, to infer adaptive features of population genetic structure (Lynch 1996). For example, lack of molecular variation in the cheetah (O'Brien et al. 1985) was taken to imply an absence of genetic variation for adaptive quantitative characters and, therefore, an enhanced risk of extinction due to genetic homogeneity. The confidence that can be attached to such extrapolations is limited (Caughley 1994). Lynch (1996) cites several reasons to doubt whether a strong connection normally should be expected among levels of molecular and quantitative genetic diversity within populations.

— Variability is introduced into a population at the per-locus mutation rate for the molecular marker typically on the order of 10^{-8} to 10^{-5} per year (Kimura 1983). However, variation for quantitative traits is introduced at a higher rate, approximately 10^{-3} to 10^{-2} per generation (Lynch 1988). Consequently, a population that loses considerable genetic variation due to a genetic bottleneck event (Chapter 9) will exhibit reduced molecular genetic variability for perhaps thousands of years while having ample time to recover normal levels of quantitative genetic variation (Lande and Barrowclough 1987).

— The expected level of heterozygosity at neutral molecular markers declines linearly with the inbreeding coefficient (Chapter 10). However, when there is significant nonadditive genetic variance for quantitative traits, that is, when there are considerable dominance or epistatic effects, fitness may not decrease in accordance with this expectation. Because allele frequencies change through random genetic drift, and because allele frequencies affect genetic variance for a trait, it is possible for additive genetic variance for a trait actually to increase with a population bottleneck (e.g., Bryant and Meffert 1993). However, inflation of genetic variance may be accompanied by a reduction in mean fitness.

— Even for a quantitative trait with a purely additive genetic basis, large variation among quantitative genetic parameters can arise among small populations (Avery and Hill 1977).

continued on next page

■ Box 12.2 *continued*

- —Screening molecular genetic markers will provide little insight into segregation at fitness-related loci unless many marker and fitness loci are tightly linked on the chromosomes.
- —Significant molecular divergence provides evidence that the opportunity for adaptive divergence also existed. However, lack of molecular divergence is uninformative.
- —Few studies have assessed the relationship between molecular and quantitative genetic variation, and only two have considered the genetic component of phenotypic variance. Briscoe et al. (1992) found a positive correlation between allozyme heterozygosity and heritability of bristle number in *Drosophila melanogaster*. Lynch and Spitze (data cited in Lynch 1996) found no association between allozyme heterozygosity and genetic variance for fitness characters in several populations of water fleas *Daphnia* spp.

The uncertain relationship between molecular genetic variability and quantitative and phenotypic variability has led several authors to suggest that conservation decisions based on molecular genetic information alone may not identify appropriate units for conservation or minimize risks to them (Hedrick et al. 1986; Hard 1995; Lynch 1996).

Given the desire to evaluate and conserve quantitative genetic variation better, there is a clear need for practical methods for more rapidly assessing quantitative variation in natural populations. A major stumbling block is the issue of what characters should be measured. In most cases, it is a judgement call as to which traits are most critical to survival and reproduction, even more so as to which traits are likely to be relevant to future selective challenges (Lynch 1996).

$$R = i\sigma h^2. \tag{12.6}$$

Because the parameter i is defined in terms of the variance, it is useful for comparing selective intensities among experiments or studies.

12.3.2 *Predicting the Response of Fitness to Natural Selection*

Noted geneticist R. A. Fisher recognized that fitness itself is a quantitative trait and proposed that natural selection can act as truncation selection. How, then, would fitness respond to natural selection? The solution put forward by Fisher (1930) has been termed the fundamental theorem of natural selection. Assuming that a given population has nonoverlapping generations, the increase in average fitness in one generation of natural selection, R_W, can be estimated as

$$R_W = \frac{\sigma_a^2}{\overline{W}}, \tag{12.7}$$

where σ^2_a is the additive genetic variance of fitness, and $\overline{W}$ is the average relative fitness in the parental generation (i.e., $\overline{W}$ is scaled to 1).

Hence, the increase in fitness at any time is equal to the additive genetic variance of fitness at that time. So long as additive genetic variance, σ^2_a, is nonzero, the average fitness of the population will increase.

The model's prediction that a population's mean fitness will always increase clearly is not borne out. Wright (1931) pointed out that not only contemporary additive genetic variance but also mutation, migration, random genetic drift, and environmental change affect the fitness of a population. Still, the fundamental theorem gave rise to a great deal of discussion about its validity and generality (Frank and Slatkin 1992; Edwards 1994; Falconer and Mackay 1996:339). A proof of the theorem and explanation of some of its difficulties are presented by Price (1972). Falconer and Mackay (1996:340) noted that any increase in fitness predicted by the theorem will not necessarily lead to an increase in population number because the population cannot exceed the carrying capacity of the habitat. Gene frequencies will change if there is additive variance for fitness, which can cause changes in characters correlated with fitness. These changes in the components of fitness often are of greater interest than is the change of fitness itself, and their prediction is the subject of a body of theory.

12.3.3 Not Responding to Selection: Selection Plateau

Sustained directional selection on quantitative traits in agricultural or laboratory populations sometimes leads to genetic progress for the trait for dozens of generations, often changing the mean phenotype by as much as 5 to 10 standard deviations (Falconer and Mackay 1996:215; Lynch 1996). After many generations of selection, such populations often exhibit a **selection plateau,** from which they respond only weakly or not at all to further selection. This is because of fixation of genes affecting expression of the trait and because of negative pleiotropies among genes affecting the trait.

In small populations, random genetic drift (Chapter 9) can prove a strong evolutionary force. Alleles affecting expression of a trait tend, over time, toward fixation or loss. With the loss of genetic variability, additive genetic variance can become exhausted. Only mutation or migration can renew the population's additive genetic variability and potential for response to selection. However, if the effective size of the population, N_e, is on the order of at least a few dozen individuals, it should possess sufficient quantitative genetic variation that, when confronted with a selective pressure, it would prove capable of expressing mean phenotypes outside the current range of variation (Lynch 1996).

12.3.4 Maintaining Adaptive Quantitative Variation

A population in the wild is subjected to a changing selective regime. Successful response to such changes is essential for the population to persist. Its reservoir of adaptive potential is its genetic variation, notably its quantitative genetic variation for morphological, physiological, and behavioral traits. What factors influence the maintenance of this adaptive variation? Because, for conservation purposes, we often focus on small populations, how does population size affect the dynamics of adaptive variation?

In small, isolated populations, that is, those not affected by migration, the level of genetic variation is influenced by a dynamic equilibrium among selection, random genetic drift, and mutation. Selection tends to reduce variation by eliminating extreme phenotypes, although balancing or diversifying selection may be at issue in some situations. In the case of directional selection, the degree of reduction of genetic variation is a function of the intensity of selection and the inheritance of fitness characters. Small populations will lose approximately $1/(2N_e)$ of the genetic variance each generation through random genetic drift (Chapter 9). At the same time, genetic variation is added to a population by mutation (Chapter 6). For a population at constant size under constant selection pressure, a quasi-equilibrium level of genetic variance will be reached at which losses to selection and drift are just balanced by mutational input (Lynch 1996).

For populations with an N_e smaller than a few hundred individuals, the expected genetic variance for a quantitative character is nearly independent of the strength of selection and is largely determined by the balance between mutation and drift (Keightley and Hill 1988; Barton 1989). For polygenic characters, the selection pressure is distributed across many loci, and that on any one locus will be small enough to be overwhelmed by drift. In such situations, characters with an additive genetic basis will have an expected genetic variance of $2N_e\sigma^2{}_m$, where $\sigma^2{}_m$ is the mutation rate. The implication for small populations is that a doubling in population size effectively doubles the evolutionary potential of the population (Lynch 1996).

As the N_e increases, random genetic drift becomes a less powerful evolutionary force. In large populations, the equilibrium level of genetic variance is determined by the balance between selection and mutation. The average genetic variance is essentially independent of population size once N_e exceeds 1,000 individuals (Lynch 1996), although this is not to say that a population of 1,000 is a genetically infinite population. The level of adaptive genetic variation is an important consideration in determining genetically viable population size (Chapter 18).

12.3.5 *Identifying Fitness-Related Loci*

A major gap in our knowledge of molecular, quantitative, and population genetics is identification of genes directly determining fitness. In other words, we do not generally measure genetic variation directly at fitness-related loci. Experimental desxigns have been developed by plant and animal breeders to detect segregation at loci directly influencing yield or other performance characters. These experimental designs also can be applied to detect segregation at fitness-related loci in wild populations (Box 12.3). Pioneering studies in this area are underway. Much important work characterizing fitness-related genes is yet to be done.

12.4 Application of Quantitative Genetic Theory to Fisheries Management

Quantitative genetic evolutionary theory and laboratory studies on model systems provide interesting and useful scientific foundations, but what are the implications for fisheries management? Can we detect a genetic component of fitness in managed fish populations, and, if so, what impact should that finding have on fisheries management actions? To answer such questions, it is useful to examine case studies.

Much of the quantitative phenotypic variation observed in fishes has a genetic basis, a finding best demonstrated in studies of salmonids (e.g., Ricker 1972; Beacham et al. 1988). Quantitative genetic variation in expression of life-history-related traits has been studied in populations of pink salmon in the Auke Creek watershed in southeast Alaska. Five subpopulations of spawning adults have been identified in both even and odd years based on time and place of spawning: early-run intertidal, early-run upstream, late-run intertidal, late-run upstream, and Lake Creek. A genetic marker has been bred into the

■ Box 12.3
Detection of Fitness-Related Loci

A major gap between theoretical and empirical evolutionary genetics concerns the general inability to identify and measure variation directly at fitness-related gene loci. Identification of such loci would in itself illuminate our concept of adaptation. The ability to track the dynamics of allele frequencies at such loci in natural populations as they adapt to ecological change is a goal of evolutionary genetics.

Similar problems—the inability to identify performance-related loci and to track allele frequencies—also exist in plant and animal breeding studies. Over the past 2 decades, applied population geneticists in the breeding community have developed experimental designs to detect genes affecting quantitative traits, or **quantitative trait loci (QTLs),** in agricultural populations, including fishes. Just as genetic markers can be mapped in relation to one another to generate a genetic map, genes affecting agricultural performance can be mapped in relation to genetic markers (Beckmann and Soller 1983; Soller and Beckmann 1983). Knowledge of linkages between marker and performance genes enhances our knowledge of the trait's **genetic architecture,** that is, the number, relative strengths, and modes of expression for genes affecting the trait. This knowledge is useful for designing practical breeding strategies. In the paradigmatic experimental design for animal systems (Soller 1978), a statistical test is carried out for each genetic marker to determine the significance of differences in mean performance of the two groups of progeny expressing alternative alleles clearly inherited from a particular parent. If a statistically significant difference is found, it is inferred that marker alleles are genetically linked to those at the QTLs. The difference between the means provides an estimate of the phenotypic effect of substituting one marked QTL allele for the other. Because of their high reproductive capability and the possibility of raising large groups of fish in common environments, fish may provide an attractive system for detecting QTLs and executing marker-assisted selection (Hallerman and Beckmann 1988; Lie et al. 1994; Poompuang and Hallerman 1997). Liu (1998) provides a thorough review of genetic mapping and QTL analysis for the interested reader.

Recognizing that fitness itself is a quantitative trait, experimental designs for QTL detection can be adapted and applied to wild populations for purposes of detecting fitness-related loci. Such experimental designs are being used for study of fitness in wild populations of fruit flies (A. Korol, Institute of Evolution, Haifa University (Israel), personal communication). Such experimental designs could be applied in studies of adaptation in fishes, perhaps most conveniently within the context of stock supplementation or ocean ranching programs. Knowledge of the dynamics of allele frequencies at fitness-related loci will prove interesting not only to evolutionary geneticists but also to fisheries managers.

late-run upstream subpopulation in order to study the dynamics of the subpopulations in the watershed (see Chapter 16, section 16.3.1).

Timing of the spawning run in anadromous salmonids is important to individual survival and population persistence. If the adults return too early or too late, the streams often will prove unsuitable for successful spawning or embryo development. A series of screenings of allozyme allele frequencies in the five runs of Auke Creek pink salmon showed little straying among subpopulations and a genetic component to run timing (Gharrett and Smoker 1993; McGregor et al. 1998). Observations of genetic marker frequencies suggested that even finer genetic structure existed within the subpopulations. To avoid the possibility of inadvertently including early-run fish in the marked population, screening of adults for the genetic marking experiment did not begin until the middle of the late run. Consequently, late-returning late-run fish were genetically marked. The inheritance of run timing within the late run was reflected in individuals returning two generations later. To estimate the genetic component of time of return, Smoker et al. (1998) conducted a breeding study using a two dams nested within sires design. Because two temporally and genetically distinct populations were at issue, two experiments were conducted, one involving fish taken randomly near the midpoint of the early run, the other at the midpoint of the late run. All families of early-run ancestry returned during the early run, and all families of late-run ancestry during the late run. The existence of within-subpopulation genetic variability for run timing also was demonstrated. Even after removing early-run or late-run effects, heritabilities of return within a run were 0.4 for females and 0.2 for males (Smoker et al. 1998). Directional selection experiments for time of return showed responses after one generation, which suggested high heritability for run timing (unpublished data cited in Gharrett and Smoker 1993). Other evidence for a genetic component of run timing is provided by the practice of egg collection in many hatcheries and its often unintended consequence. Egg collection often commences with the beginning of the run and ceases when the incubation capacity of the hatchery is reached. The consequence is selection for early return, resulting in temporal advancement of the run. For example, selection for early spawning steelhead (anadromous rainbow trout) at the Skamania Hatchery in Washington State changed the timing of the early run by 2 months over a 13-year time span (Millenbach 1973).

Timing of fry migration is another important life history character in anadromous salmonids. If fry migrate too early or too late, they miss access to the seasonably abundant foods important to early survival and growth. Gharrett and Smoker (1993) sampled 10 to 20 fry each day during the emigration period of the marked, odd-year subpopulation of pink salmon and screened frequencies of allozyme marker alleles. Fry produced from the genetically marked late-run subpopulation emigrated late among all pink salmon migrating seaward (Gharrett and Smoker 1993). The pattern of fry migration echoed that of the adult return, suggesting a genetic correlation.

The remaining portion of the anadromous life cycle is the period of growth and maturation at sea. Geiger et al. (1997) observed the number of pink salmon offspring, marked and tagged by family, returning to their natal site in southeast Alaska and uncovered evidence for a genetic component for marine survival. Matings were made using factorial (two dams × three sires) or hierarchical (two dams nested within each sire) designs, and the young were released after emergence. Return rates were tested using ANOVA to quantify the sire component for survival and family size. If half sibs with the same sire and different dams exhibited survival and family size more similar than those in the stock as a whole, it would be inferred that these traits are genetically influenced.

To the degree that survival was more similar among groups of half sibs with the same sire than among family groups with different parents, it would be taken as evidence of genetic influence on survival and average family size. One of the five groups, the 1983 late run, exhibited statistically significant sire effects, thereby showing that survival and family size are heritable traits. Return rates showed that stock size increased largely with increasing variance in family size and not by a uniform increase in family size across the stock. What, then, are the implications of these findings for fisheries management? Given that there is an interfamily genetic variance component underlying marine survival, why doesn't selection increase the productivity of the stock by favoring fitter families? Geiger et al. (1997) suggested that favored phenotypes must change frequently among generations, tracking a dynamic selective regime. They noted similar findings by McIntyre et al. (1988), who were unsuccessful in selecting for increased marine survival in coho salmon *O. kisutch*. If marine survival of salmon cannot be increased by selection, then any surplus production exploitable in a fishery must come from a changing and unpredictable portion of the population. The genetic component of marine survival and the changing environment suggest the importance of maintaining the variance in quantitative genetic characters in exploited or recovering salmon populations. Predictive models of fisheries production that incorporate only abundance parameters will not predict future decreases in stock productivity due to loss of genetic variability.

To generalize, it is important that variability of fitness-related quantitative traits is maintained in order to preserve the long-term fitness of the population. Fish populations live in variable environments. Genetic variation increases the likelihood that at least some of the individuals in the population will express a phenotype that will support survival and successful reproduction through the range of environmental challenges that the population inevitably will face.

12.5 Perspective

Relatively few studies have addressed detection and quantification of the genetic component of fitness in managed fish populations, but the knowledge gained provides insight into the management of wild populations. More work is needed in this important area of applied genetics.

Quantitative evolutionary genetics is a broad and actively growing area of population genetics, so much so that a chapter of this scope can only begin to introduce the topic. Additional material on evolutionary aspects of quantitative genetics can be found in Chapter 20 of Falconer and MacKay (1996). Readers interested in gaining a thorough grounding in the area, especially including its mathematical aspects, are referred to Lynch and Walsh (1998). Those interested in applications of quantitative genetics as applied in genetic improvement of aquaculture stocks might turn to Tave (1993) or Lutz (2001).

12.6 References

Allendorf, F. W., N. Ryman, and F. M. Utter. 1987. Genetics and fishery management: past, present, and future. Pages 1–19 *in* N. Ryman and F. M. Utter, editors. Population genetics and fishery management. University of Washington Press, Seattle.

Alm, G. 1939. Investigations of growth of different forms of trout. Reports of the Institute for Freshwater Research, Drottningholm 18:6–145. (In Swedish, English summary.)

Alm, G. 1946. Reasons for the occurrence of stunted fish populations. Meddelanden fran Statens Undersoknings och Foroksanstalt for Sot vattensfisket Stockholm 25:1–146.

Alm, G. 1959. The connection between maturity, size, and age in fishes. Reports of the Institute for Freshwater Research, Drottningholm 40:6–145.

Austad, D., T. Gjedrem, and H. Skjervold. 1972. Genetic and environmental sources of variation in length and weight of rainbow trout *(Salmo gairdneri)*. Journal of the Fisheries Research Board of Canada 29:237–241.

Avery, P. J., and W. G. Hill. 1977. Variability on genetic parameters among small populations. Genetical Research 29:193–213.

Barker, J. S. F. 1979. Inter-locus interactions: a review of experimental evidence. Theoretical Population Biology 16:323–346.

Barker, J. S. F., and R. H. Thomas. 1987. A quantitative genetic perspective on adaptive evolution. Pages 3–23 *in* V. Loeschcke, editor. Genetic constraints on adaptive evolution. Springer-Verlag, Berlin.

Barton, N.H. 1989. Divergence of a polygenic system subject to stabilizing selection, mutationm, and drift. Genetical Research 54:59-77.

Beacham, T. D. 1988. A genetic analysis of early development in pink *(Oncorhynchus gorbuscha)* and chum salmon *(Oncorhynchus keta)* at three different temperatures. Genome 30:31–35.

Beacham, T. D., R. E. Withler, C. B. Murray, and L. W. Barner. 1988. Variation in body size, morphology, egg size, and biochemical genetics of pink salmon in British Columbia. Transactions of the American Fisheries Society 117:109–126.

Beckmann, J. S., and M. Soller. 1983. Restriction fragment length polymorphisms in genetic improvement: methodologies, mapping, and costs. Theoretical and Applied Genetics 67:35–43.

Bergot, P., B. Chevassus, and J. M. Blanc. 1976. [Genetic analysis of the number of pyloric caecae in brow trout (*Salmo trutta* Linneaus) and rainbow trout (*Salmo gairdneri* Richardson). I. Character distribution and phenotypic variability within and between families.] Annals de Hydrobiologie 7:105–114. (In French.)

Briscoe, D. A., and six coauthors. 1992. Rapid loss of genetic variation in large captive populations of *Drosophila* flies: implications for the genetic management of captive populations. Conservation Biology 6:416–425.

Bryant, E. H., and L. M. Meffert. 1993. The effect of serial founder-flush cycles on quantitative genetic variation in the housefly. Heredity 70:122–129.

Burger, G., and B. Chevassus. 1987. [Potential of selection for age at first sexual maturation in rainbow trout (*Salmo gairdneri* R.).] Bulletin Francaise Peche Pisciculture 307:102–117. (In French.)

Caughley, G. 1994. Directions in conservation biology. Journal of Animal Ecology 63:215–244.

Chevassus, B., J. M. Blanc, and P. Bergot. 1979. [Genetic analysis of the number of pyloric caecae in brown trout (*Salmo trutta* Linneaus) and rainbow trout (*Salmo gairdneri* Richardson). II. Effect of the genotype, rearing environment, and feeding on the realization of the character in rainbow trout.] Annals Genetique et Selection de Animales 11:79–92. (In French.)

Crow, J. F., and M. Kimura. 1970. An introduction to population genetics theory. Harper and Row, New York.

Edwards, A. W. F. 1994. The fundamental theorem of natural selection. Biological Review 69:443–474.

Falconer, D. S., and T. F. C. Mackay. 1996. Metric characters under natural selection. Pages 335–355 *in* Introduction to quantitative genetics, 4th edition. Longman, Harlow, Essex, UK.

Fisher, R. F. 1930. The genetical theory of natural selection. Clarendon Press, Oxford, UK.

Frank, S. A., and M. Slatkin. 1992. Fisher's fundamental theorem of natúral selection. Trends in Ecology and Evolution 7:92–95.

Gall, G. A. E. 1975. Genetics of reproduction in domesticated rainbow trout. Journal of Animal Science 40:19–28.

Gall, G. A. E., and S. J. Gross. 1978. A genetic analysis of the performance of three rainbow trout stocks. Aquaculture 15:113–127.

Gall, G. A. E., and N. Huang. 1988a. Heritability and selection schemes for rainbow trout: body weight. Aquaculture 73:43–56.

Gall, G. A. E., and N. Huang. 1988b. Heritability and selection schemes for rainbow trout: female reproductive performance. Aquaculture 73:57–66.

Geiger, H. J., W. W. Smoker, L. A. Zhivitovsky, and A. J. Gharrett. 1997. Variability of family size and marine survival in pink salmon *(Oncorhynchus gorbuscha)* has implications for conservation biology and human use. Canadian Journal of Fisheries and Aquatic Sciences 54:2684–2690.

Gharrett, A. J., and W. W. Smoker. 1993. Genetic components of life history traits contribute to population structure. Pages 197–202 *in* J. G. Cloud and G. H. Thorgaard, editors. Genetic conservation of salmonid fishes. Plenum, New York.

Gjedrem, T. 1983. Genetic variation in quantitative traits and selective breeding in fish and shellfish. Aquaculture 33:51–72.

Gjerde, B., and T. Gjedrem. 1984. Estimates of phenotypic and genetic parameters for carcass traits in Atlantic salmon and rainbow trout. Aquaculture 36:97–110.

Gross, M. 1985. Disruptive selection for alternative life histories in salmon. Nature (London) 315:47–48.

Hagen, D. W. 1973. Inheritance of numbers of lateral plates and gill rakers in *Gasterosteus aculeatus.* Heredity 30:303–312.

Hagen, D. W., and D. Blouw. 1983. Heritability of dorsal spines in the fourspine stickleback *(Apeltes quadracus).* Heredity 50:275–281.

Hailman, J. P. 1986. The heritability concept applied to wild birds. Pages 71–95 *in* R. F. Johnston, editor. Current ornithology, volume 4. Plenum, New York.

Hallerman, E. M., and J. S. Beckmann. 1988. DNA-level polymorphism as a tool in fisheries science. Canadian Journal of Fisheries and Aquatic Sciences 45:1075–1087.

Hard. J. J. 1995. A quantitative genetic perspective on the conservation of intraspecific diversity. Pages 304–326 *in* J. L. Nielsen, editor. Evolution and the aquatic ecosystem: defining unique units in population conservation. American Fisheries Society, Symposium 17, Bethesda, Maryland.

Hartl, D. 1988. A primer of population genetics, 2nd edition. Sinauer Associates, Sunderland, Massachusetts.

Haus, E. O. 1984. [A genetic and phenotypic analysis of egg size, egg volume, and number in rainbow trout.] Doctoral dissertation, Agricultural University of Norway, As, Norway. (In Norwegian, cited by Refstie 1987.)

Hedrick, P. W., P. F. Brussard, F. W. Allendorf, J. A. Beardmore, and S. Orzack. 1986. Protein variation, fitness, and captive propagation. Zoo Biology 5:91–99.

Ihssen, P. E. 1986. Selection of fingerling rainbow trout for high and low tolerance to high temperature. Aquaculture 57:370.

Johnson, L. 1980. The Arctic charr, *Salvelinus alpinus*. Pages 15–98 *in* E. K. Balon, editor. Charrs, salmonid fishes of the genus *Salvelinus*. Dr. W. Junk, The Hague, The Netherlands.

Jonsson, B., K. Hindar, and T. G. Northcote. 1984. Optimal age at sexual maturity of sympatric and experimentally allopatric cutthroat trout and Dolly Varden char. Oecologia 61:319–325.

Keightley, P. D., and W. G. Hill. 1988. Quantitative genetic variability maintained by mutation–stabilizing selection balance in finite populations. Genetical Research 52:33–43.

Kimura, M. 1983. The neutral theory of molecular evolution. Cambridge University Press, Cambridge, UK.

Kincaid, H. L., W. R. Bridges, and B. von Limbach. 1977. Three generations of selection for growth rate in fall-spawning rainbow trout. Transactions of the American Fisheries Society 106:621–628.

Kirpichnikov, V. S. 1981. Genetic bases of fish selection. Springer-Verlag, New York.

Klupp, R. 1979. Genetic variance for growth in rainbow trout *(Salmo gairdneri).* Aquaculture 18:123–134.

Lande, R. 1981. The minimum number of genes contributing to quantitative variation between and within populations. Genetics 99:541–553.

Lande, R. 1988. Quantitative genetics and evolutionary theory. Pages 71–84 *in* B. S. Weir, E. J. Eisen, M. M. Goodman, and G. Namkoong, editors. Proceedings of the second international conference on quantitative genetics. Sinauer Associates, Sunderland, Massachusetts.

Lande, R., and G. F. Barrowclough. 1987. Effective populaton size, genetic variation, and their use in population management. Pages 87–123 *in* M. E. Soulé, editor. Viable populations for conservation. Cambridge University Press, Cambridge, U.K.

Leary, R. F, F. W. Allendorf, and K. L. Knudsen. 1985. Inheritance of meristic variation and the evolution of developmental stability in rainbow trout. Evolution 39:308–314.

Lie, O., A. Slettan, V. Grimholt, M. Lundin, M. Syed, and O. Olsaker. 1994. Fish gene maps and their implications for aquaculture. Animal Biotechnology 5:209–218.

Linder, D., O. Sumari, K. Nyholm, and S. Sirkkomaa. 1983. Genetic and phenotypic variation in production traits in rainbow trout strains and strain crosses in Finland. Aquaculture 33:129–134.

Lindsey, C. C. 1981. Stocks are chameleons: plasticity in gill rakers of coregonid fishes. Canadian Journal of Fisheries and Aquatic Scinces 38:1497–1506.

Liu, B. H. 1998. Statistical genomics: linkage, mapping, and QTL analysis. CRC Press, Boca Raton, Florida.

Lutz, C. G. 2001. Practical genetics for aquaculture. Fishing News Books, Oxford, U.K.

Lynch, M. 1988. The rate of polygenic mutation. Genetical Research 51:137–148.

Lynch, M. 1996. A quantitative genetic perspective on conservation issues. Pages 471–501 *in* J. C. Avise and J. L Hamrick, editors. Conservation genetics: case histories from nature. Chapman and Hall, New York.

Lynch, M., and B. Walsh. 1998. Genetics and analysis of quantitative traits. Sinauer Associates, Sunderland, Massachusetts.

Mather, K., and J. L. Jinks. 1949. Biometrical genetics. Chapman and Hall, London.

Mayr, E. 1969. Principles of systematic zoology. McGraw-Hill, New York.

McGregor, A. J., S. Lane, M. A. Thomason, L. A. Zhivitovsky, W. W. Smoker, and A. J. Gharrett. 1998. Migration timing, a life history trait important in the genetic structure of pink salmon. North Pacific Anadromous Fish Commission Bulletin 1:262–273.

McIntyre, J. D., A. R. Hemmingsen, and R. C. Simon. 1988. Selection to increase survival of smolts in four successive broods of coho salmon. Transactions of the American Fisheries Society 117:90–92.

McKay, L. R., P. E. Ihssen, and G. W. Friars. 1986. Genetic parameters of growth in rainbow trout, *Salmo gairdneri,* as a function of age and maturity. Aquaculture 58:241–254.

Millenbach, C. 1973. Genetic selection of steelhead trout for management purposes. International Atlantic Salmon Journal 4:253–257.

Moav, R., and G. Wohlfarth. 1976. Two-way selection for growth rate in the common carp (*Cyprinus carpio* L.). Genetics 82:83–101.

Mousseau, T. A., and D. A. Roff. 1987. Natural selection and the heritability of fitness components. Heredity 59:181–197.

O'Brien, S. J., and nine coauthors. 1985. Genetic basis for species vulnerability in the cheetah. Science 227:1428–1434.

Poompuang, S., and E. M. Hallerman. 1997. Toward detection of quantitative trait loci and marker-assisted selection in fish. Reviews in Fisheries Science 5:253–277.

Price, G. R. 1972. Fisher's fundamental theorem made clear. Annals of Human Genetics 36:129–140.

Price, T., and D. Schluter. 1991. On the low heritability of life-history traits. Evolution 45:853–861.

Purdom, C. E. 1979. Genetics of growth and reproduction in teleosts. Symposia of the Royal Society of London 44:207–217.

Refstie, T. 1987. Selective breeding and intraspecific hybridization of coldwater finfish. In K. Thiews, editor. Selection, hybridization, and genetic engineering in aquaculture, volume 1. Heenemann, Berlin.

Ricker, W. E. 1972. Hereditary and environmental factors affecting certain salmonid populations. Pages 27–160 *in* R. C. Simon and P. A. Larkin, editors. The stock concept in Pacific salmon. H. R. MacMillan Lectures in Fisheries. Institute of Fisheries, University of British Columbia, Vancouver.

Roff, D. A., and T. A. Mousseau. 1987. Quantitative genetics and fitness: lessons from *Drosophila*. Heredity 58:103–118.

Ryman, N. 1972. An attempt to investigate the magnitude of additive genetic variation of body size in the guppy-fish *Lebistes reticulatus*. Hereditas 71:237–244.

Smith, J. N. M., and A. A. Dhondt. 1980. Experimental confirmation of heritable morphological variation in a natural population of song sparrows. Evolution 34:1155–1160.

Smoker, W. W., A. J. Gharrett, and M. S. Stekoll. 1998. Genetic variation of return date in a population of pink salmon: a consequence of fluctuating environment and dispersive selection? Alaska Fishery Research Bulletin 5:46–54.

Snyder, R. J. 1991. Quantitative genetic analysis of life histories in two freshwater populations of the threespine stickleback. Copeia 1991:526–529.

Soller, M. 1978. The use of loci associated with quantitative traits in dairy cattle improvement. Animal Production 27:133–139.

Soller, M., and J. S. Beckmann. 1983. Genetic polymorphism in varietal identification and genetic improvement. Theoretical and Applied Genetics 47:179–190.

Tave, D. 1993. Genetics for fish hatchery managers, 2nd edition. Van Nostrand Reinhold, New York.

Totfteberg, P., and T. Hansen. 1987. Relationship between age at maturity and growth rate in farmed rainbow trout *Salmo gairdneri*. *In* K. Thiews, editor. Selection, hybridization, and genetic engineering in aquaculture, volume 1. Heenemann, Berlin.

von Limbach, B. 1970. Fish genetics laboratory. Progress in Sport Fishery Research 1970. U.S. Fish and Wildlife Service Research Publication 106:153–160.

Wilkens, H. 1971. Genetic interpretation of regressive evolutionary processes: studies on hybrid eyes of two *Astyanax* cave populations (Characidae, Pisces). Evolution 25:530–544.

Wohlfarth, G., R. Moav, and G. Hulata. 1975. Genetic differences between the Chinese and European races of the common carp. 2. Multicharacter variation—a response to the diverse methods of fish cultivation in Europe and China. Heredity 34:341–350.

Wright, S. 1931. Evolution in Mendelian populations. Genetics 16:97–159.

Practical Applications of Population Genetics

13

Genetic Stock Identification and Risk Assessment

James B. Shaklee and Kenneth P. Currens

13.1 Introduction

Fisheries scientists and managers long have been concerned about population structure and the implications of subcomponents of fishery harvests (multiple species, different stocks, or both) for sustainable, long-term management of the resource(s). They have used data on fish abundance, geographic and temporal distributions, catch statistics, meristic and morphometric variation, parasite burdens, immunological differences, population dynamics, variations in life history and other biological characteristics, and artificial tags to gain insight into patterns of population structure and mixed-stock harvests. In the last 10 to 15 years, fisheries biologists increasingly have used genetic data to investigate stock structure and stock contributions to mixed-stock fisheries. Similarly, biologists responsible for resource management and conservation have become increasingly concerned with the identification and assessment of the genetic risks associated with various actions and inactions.

Despite the fact that species are fundamental taxonomic and biological units, they are seldom entirely panmictic. Rather, sexually reproducing species typically are composed of subpopulations that are at least partially reproductively isolated and differentiated from one another. Thus, it is the individual demes and subpopulations that make up species that are the randomly interbreeding population units. It is now broadly recognized that fish and shellfish species are typically subdivided into more-or-less discrete subpopulations or stocks (note that although the terms subpopulation and stock are often used synonymously, the latter is sometimes reserved for components of species that are exploited in fisheries or actively managed). Coupled with such partial reproductive isolation, the typically restricted distributions of subpopulations in time and space provide the basis for local adaptation via selection (see Chapter 8). Given this background, it is easy to see that the overall productivity and evolutionary potential of a species is dependent upon maintaining the abundance and diversity of its component stocks. This perspective (paradigm), known as the **stock concept,** was initially developed for Pacific salmon *Oncorhynchus* spp. by fisheries scientists such as Ricker, Larkin, and Simon but is now a central theme in the management and study of nearly all fish and shellfish species (Berst and Simon 1981; Dizon et al. 1992).

Genetic stock identification (GSI) is a general term used in the fisheries science literature to refer to any of an array of genetically based methods used for species or population identification. These approaches are all based on the presumption that because genetic differences underlie species and population differentiation, genetic data and analyses can be used to identify and characterize these entities. Allozyme (allelic isozyme) and DNA analyses can be used at the species level to identify distinct species and interspecific hybrids, identify the species of origin of fishery products in forensic applications (Chapter 17), and investigate systematic interrelationships among taxa. In this chapter, we focus on (1) applications of genetic data for identifying stocks and investigating stock interrelationships, (2) applications of GSI for managing mixed-stock fisheries, and (3) the need for and approaches to genetic risk assessment and management. Other applications of GSI at the stock level include genetic marking (Chapter 16), forensic identification (Chapter 17), and investigation of hatchery–wild stock interactions (Chapter 15).

A strength of GSI is that it allows direct statistical testing of hypotheses regarding stock structure. The null hypothesis in all studies of population structure is **panmixia;** that is, all individuals are randomly interbreeding. Given this hypothesis, we expect multiple (random) collections of the species to differ genetically no more than would be expected due to sampling error. This expectation is based on the presumption of selective neutrality of the alternative character states (alleles) being screened. We reject the null hypothesis and conclude that significant population subdivision occurs when the probability that the collections were drawn from a single, panmictic gene pool falls below a predefined criterion. Like other tests of stock structure, failure to disprove the null hypothesis by use of GSI data does not establish the existence of a single, unit stock, although it is consistent with this conclusion.

13.2 Genetic Stock Identification Data Sets

13.2.1 Allozymes

Electrophoretic studies of nearly all genetic variation at protein-encoding loci have been the predominant approach to GSI for the past 25 years. Specifically, starch-gel and cellulose-acetate electrophoresis coupled with histochemical staining to reveal variation of allelic isozymes (allozymes) at various enzyme-coding loci and to estimate allele frequencies have provided extensive and informative data sets for identifying subpopulations of many fish and shellfish species (Richardson et al. 1986). The primary advantages of electrophoretic studies include

- the codominant expression of allozymes allows individual genotypes at specific, known loci to be reliably inferred from the banding patterns on gels;
- allozymes are inherited as simple Mendelian traits;
- the expression of allozymes is largely independent of the environmental history, age, and physiological state of the individual;
- large numbers of individuals can be screened electrophoretically at a relatively large number of different gene loci with simple equipment and at relatively modest cost; and
- the approach can be quite powerful for elucidating stock structure because many species exhibit considerable genetic variation at many enzyme-encoding loci.

Disadvantages include the need for relatively large fresh or frozen tissue samples, usually requiring lethal sampling, limited detectable variation at allozyme loci in some species (e.g., many decapod crustaceans), and possible differential selection upon allozyme variants.

Allozyme data sets have been useful in elucidating population subdivision in a wide variety of freshwater, estuarine, marine, and diadromous fish and shellfish species. Examples include centrarchids such as largemouth bass *Micropterus salmoides* in North America (Philipp et al. 1983), Japanese medaka *Oryzias latipes* (Sakaizumi et al. 1983), mummichog *Fundulus heteroclitus* (Ropson et al. 1990), the eelpout *Zoarces viviparus* (Christiansen et al. 1976), cobbler *Cnidoglanis macrocephalus* (Ayvazian et al. 1994), tarakihi *Nedadactylus macropterus* (Elliott and Ward 1994; Smith et al. 1996), pearl oyster *Pinctada maxima* (Johnson and Joll 1993), giant clam *Tridacna gigas* (Benzie and Williams 1995), barramundi perch *Lates calcarifer* (Shaklee et al. 1993), and Pacific salmon (Utter et al. 1989; Phelps et al. 1994, Wood et al. 1994). See Chapter 2 for more information about the collection and uses of allozyme data.

13.2.2 *Mitochondrial DNA*

There have been a large number of studies of population structure based on mitochondrial DNA (mtDNA) data in the past 10 years. Mitochondrial DNA consists of a small circular molecule (roughly 16,000–20,000 base pairs) that usually is inherited clonally from mother to offspring without recombination; it is functionally haploid. The mtDNA genome contains over 30 genes (encoding enzymes and RNAs) and noncoding sequences (D-loop), but it is treated as a single locus in population studies because of the absence of recombination. Whereas a growing number of mtDNA studies involve sequence analysis, others use restriction fragment length polymorphism (RFLP) analysis. This latter analysis allows the detection of base sequence changes to be revealed after digestion of the mtDNA with specific restriction endonucleases. These restriction enzymes recognize specific 4–6 base pair sequences and cleave the DNA at such sites. Advantages of this approach include the following:

- levels of detectable variation are often quite high;
- mtDNA typically evolves at a higher rate than do most allozyme loci;
- because mtDNA experiences effective population sizes (see Chapter 9) that are one-fourth the size of nuclear genes in the same species, genetic drift is higher and gene flow via dispersal is lower than they are for nuclear genes, including allozyme loci;
- there is the possibility of reconstructing the phylogenetic history of mtDNA and thereby the phylogenetic history of the populations (or species) being studied (because mtDNA does not recombine); and
- selective amplification of mtDNA using polymerase chain reaction (PCR) methods allows the analysis of very small tissue samples.

The maternal inheritance of mtDNA in most species, however, means that such markers usually cannot be used to investigate male contributions to population structure. Additionally, some species exhibit such low levels of mtDNA variation as to preclude informative analyses of population structure.

Mitochondrial DNA data sets also have been useful in elucidating population structure in a wide variety of freshwater, estuarine, marine, and diadromous organisms. These include mummichog (González-Villaseñor and Powers 1990), Eastern oyster *Crassostrea virginica* (Reeb and Avise 1990), sea turtles (Karl et al. 1992; Bowen et al. 1994), sea urchins *Heliocidaris* spp. (McMillan et al. 1992), tarakihi (Grewe et al. 1994), coconut crab *Birgus latro* (Lavery et al. 1996), rainbow smelt *Osmerus mordax* (Bernatchez and Martin 1996), Australian bass *Macquaria novemaculeata* (Chenoweth and Hughes 1997), carps (Cyprinidae) (Lu et al. 1997), billfishes (Istiophoridae) (Graves and McDowell 1995; Chow et al. 1997), southern bluefin tuna *Thunnus maccoyi* (Grewe et al. 1997), white sucker *Catastomus commersoni* (Lafontaine and Dodson 1997), brown trout *Salmo trutta* (Bernatchez 2001), and many other species of fishes (see Stepien 1995 and Shulman and Bermingham 1995). Avise (1994) provides an excellent overview of mtDNA methodologies and data analysis and a comprehensive review of earlier studies. See Chapter 4 for more information about the collection and uses of mtDNA data.

13.2.3 Mini- and Microsatellite Nuclear DNA

Two subclasses of variable number tandem repeat (VNTR) nuclear DNA have emerged in the past few years as highly informative and useful markers for stock identification studies—mini- and microsatellites. Variation in **mini- and microsatellite DNA** consists of alleles that are arrays of different numbers of copies of a short, basic repeating unit of base pairs (bp; 2–5 bp for microsatellites and typically 9–65 bp for minisatellites). The polymerase chain reaction (PCR) typically is used to amplify microsatellites by use of specific primers. The amplification products then are separated by means of electrophoresis as alleles (or allelic classes) on the basis of their size. Minisatellites typically are analyzed using Southern blot methods, whereas microsatellites typically are fluorescently labeled and detected using an automated sequencer (or radioactively labeled and visualized by means of autoradiography). Advantages of mini- and microsatellite DNA approaches include (1) the often very high levels of variation (heterozygosity and numbers of alleles) detectable, (2) the need for only very small tissue samples that can be obtained nonlethally and stored at room temperature, and (3) the prospects for efficient, semiautomated analysis of large numbers of samples. Disadvantages include the current high cost of analysis, limitations on the rate of sample processing, lack of established markers for many species of management interest, the general absence of existing large data sets for comparison and perspective, and the difficulty and expense of standardizing allele identifications among different laboratories.

Some examples of the utility of mini- and microsatellite DNA data for elucidating stock structure include Bentzen et al. (1996) and Ruzzante et al. (1996, 1998) for Atlantic cod *Gadus morhua;* Angers et al. (1995) for brook trout *Salvelinus fontinalis;* Beacham et al. (1999) for steelhead *Oncorhynchus mykiss;* Taylor et al. (1994) and Beacham (1996) for chum salmon *O. keta;* Beacham et al. (1995), Wood (1995), and Taylor et al. (1996) for sockeye salmon *O. nerka;* Wenburg et al. (1998) for cutthroat trout *O. clarki;* Small et al. (1998) for coho salmon *O. kisutch;* and Banks et al. (2000) for chinook salmon *O. tshawytscha.* See Chapter 5 for more information about the collection and uses of nuclear DNA data.

13.3 Statistical Analyses

The primary goal of quantitative analyses for GSI is usually to provide objective, stable classifications of individuals or samples into groups. **Objective** means that anyone repeating the analysis based on the same data and the same sequence of methods will produce the same classification. A **stable** classification is one that does not change if more individuals or more descriptive traits are added to the analysis.

Another goal can be to make inferences about processes affecting relationships among groups. Depending on the processes that are important to understand, methods for classification may be phylogenetic or phenetic. **Phylogenetic** classifications specifically attempt to show relationships based on reconstructing the evolutionary history of groups or unique genomic lineages, whereas **phenetic** analyses are based on similarity. Phenetic analyses may or may not reflect phylogenetic relationships. Phylogenetic analyses are based on implicit or explicit models for evolutionary change. A critical assumption for phylogenetic analyses is that gene flow among lineages has been rare. When gene flow is more common, phenetic analyses that focus on geographical structuring among related populations or lineages are more appropriate (Swofford et al. 1996).

Producing objective, stable classifications is not always easy because the results inevitably also depend on the skill and insight of the researcher. Most data sets for GSI are complex. They usually consist of measurements of many traits on many individuals from different samples. In addition, they may include traits that have different kinds of genetic control, such as codominant diploid traits from allozyme electrophoresis, haploid traits from mtDNA, and quantitative genetic traits. Consequently, the first challenge for an investigator is to select analyses that reduce large data sets to a manageable number of dimensions without losing important information about the differences among samples. In addition, the investigator must choose between the strengths and weakness of different kinds of analyses. As we shall see, not all numerical analyses are equally efficient at providing objective, stable classifications. Finally, investigators must find objective ways of comparing analyses from different kinds of genetic data.

This section focuses on analysis of single-locus traits. Quantitative traits (Chapter 12), many of which are important ecologically, usually are controlled by multiple loci and the environment. They require a very different kind of analysis that rarely is done on large geographic scales.

Analysis of single-locus traits for stock identification usually proceeds through three stages. The first stage involves testing the validity of using allele frequencies to describe the genotypic variation within samples or populations (Chapter 1). These are **genotypic analyses.** The second stage usually involves exploring the data for patterns. During this stage, scientists employ a variety of **taxonomic analyses,** which can provide quantitative, objective descriptions of the pattern(s) of differences among groups. The third stage focuses on **hypothesis-based analyses,** whereby scientists hypothesize that genetic differences are structured a certain way and then try to quantify or statistically test for those differences.

13.3.1 Genotypic Analyses

Genetic data for stock identification may be diploid for traits controlled by nuclear DNA or haploid in the case of traits encoded in mtDNA. For codominant diploid traits, such as those commonly obtained in allozyme or microsatellite DNA studies, data first are collected as genotypes. It is much simpler, however, to describe populations by the frequency of n alleles at a locus rather than the $n(n + 1)/2$ different genotypes at a locus, especially when you have examined a large number of loci. Consequently, the first step in analyzing codominant marker data is to determine whether allelic frequencies calculated from genotypes accurately characterize the samples.

Allelic frequencies characterize a population if the frequencies of alleles in one generation are expected to remain the same in the next generation. Recall from Chapter 1 that this occurs if a population is in Hardy–Weinberg equilibrium and has a relatively large effective size (Chapter 9). We can test for Hardy–Weinberg proportions among the genotypes by comparing how well the observed distribution of genotypes fits the expected multinomial distribution based on statistical goodness-of-fit tests, such as a χ^2 test or a log likelihood ratio test (G-test). Both are valid tests, but many statisticians favor the G-test because its mathematical properties allow more complex analyses. The G-test statistic is calculated from the observed and expected number of genotypes based on the expression

$$G = 2\Sigma(\text{observed})\log_e\left(\frac{\text{observed}}{\text{expected}}\right), \tag{13.1}$$

with degrees of freedom equal to the number of classes minus one, minus the number of parameters estimated from the data. If the expected number of genotypes is large (>5), the test statistic can then be compared with a table of χ^2 values found in most standard statistics books to determine the probability (P-values) that genotype frequencies are different from those expected. In many cases, however, the expected number of genotypes may be less than 5, and specific permutational P-values must be calculated by other methods. These methods are available in many recent statistical computer programs. If the fit of the observed genotype frequencies to the expected genotype frequencies is poor, at least one of the assumptions of Hardy–Weinberg equilibrium has been violated, and the investigator needs to examine the samples for possible causes to determine whether the data can be used for stock identification (Chapter 1). Sampling or data collection errors leading to departures from Hardy–Weinberg equilibrium, such as inadvertent mixing of samples that were genetically different or unequal resolution of different genotypes, need to be resolved before biological explanations can be explored. Common biological explanations for departures from Hardy–Weinberg equilibrium include migration (Chapter 7), nonrandom mating (Chapter 1), small population size (Chapter 9), and selection (Chapter 8).

It is not always possible to test for Hardy–Weinberg proportions. At some loci, for example, heterozygotes cannot be distinguished from one of the homozygote classes because of recessive alleles, incomplete dominance, or co-migration of allozyme products encoded by isoloci. In these cases, the investigator must assume Hardy–Weinberg proportions of genotypes in order to derive the expected allele frequencies. In contrast, for haploid data the number of haplotypes is equivalent to the number of alleles, and the haplotype frequencies completely describe the variation within a sample.

13.3.2 *Taxonomic Analyses*

13.3.2.1 **Genetic Distances.** Many taxonomic analyses use a matrix of similarities or dissimilarities among all the samples in the analysis. Consequently, a second step commonly employed in analyzing population genetic data is to reduce the allelic frequency differences at all loci among all the samples to a single measure of similarity or dissimilarity for every pair of samples. This usually involves calculating a genetic distance (Nei 1987; Swofford et al. 1996).

Numerous genetic distance measures have been proposed, and most computer programs for analyzing population genetic data include several measures. Genetic distance measures differ in terms of (1) their underlying theoretical assumptions regarding the predominant cause(s) of variation, (2) how well they represent biological properties, and (3) the geometric nature of distances between samples. Consequently, choice of genetic distance measure may influence how stable a classification may be.

One important difference to consider when choosing among different genetic analyses, such as genetic distance measures, is the assumptions they make about the evolution of alleles and whether they are appropriate for the data. Most analyses are based on one of two different assumptions. The **infinite allele model** of mutation (Chapter 6) assumes that most new mutations occurring in a population lead to new alleles. In contrast, the **stepwise mutation model** assumes that new alleles arise in a stepwise fashion by the gain or loss of repeated units. The practical consequence of this difference is that analyses based on the infinite allele model assume that alleles distinguished by their electrophoretic mobility or size represent mostly unique alleles (i.e., no homoplasy—indistinguishable alleles arising from reversals or parallel evolution). If homoplasies occur, genetic distances would tend to underestimate the true genetic differences among populations. In contrast, analyses based on the stepwise mutation model assume large amounts of homoplasy. Because of concern that some classes of microsatellite DNA may evolve more in accordance with the stepwise mutation model than the infinite allele model, geneticists have developed other genetic distance and genetic differentiation measures for this kind of data.

Popular genetic distance metrics for allelic frequency data consistent with the infinite allele model are Nei's genetic distance (Nei 1972, 1978) and Cavalli-Sforza and Edwards (1967) chord distance. Slatkin's average sum of squares distances (Slatkin 1995) are examples of genetic distance measures based on the stepwise mutation model.

Nei's standard genetic distance is

$$D_{\mathrm{NEI}} = -\log_e\left(\frac{J_{XY}}{\sqrt{J_X J_Y}}\right), \tag{13.2}$$

where x_i and y_i are the frequencies of the ith allele at a locus in samples X and Y, respectively, and J_X, J_Y, and J_{XY} are the arithmetic means across all loci of Σx_i, Σy_i, and $\Sigma x_i y_i$, respectively. When sample sizes are small, however, D_{NEI} gives a biased estimate. Much of this bias can be corrected by substituting unbiased estimates of J_X, J_Y, and J_{XY} into the above formula by use of the means of $j_x = (2n\Sigma x_i^2 - 1)/(2n - 1)$; $j_y = (2n\Sigma y_i^2 - 1)/(2n - 1)$; and $j_{xy} = \Sigma x_i y_i$, respectively, over all loci examined.

This popular distance measure emphasizes the role of mutations in the divergence of populations. It was proposed to estimate the number of codon substitutions per locus that have occurred since two populations diverged, although it is a valid estimate only if the rate of substitution is the same across both loci and populations. Another characteristic of Nei's genetic distance is that it is very sensitive to the level of heterozygosity within samples, which can be affected by different patterns of genetic drift (Hillis 1984).

In contrast, the Cavalli-Sforza and Edwards chord distance measures the distance between two populations as though the populations were on the surface of a multidimensional sphere. This distance is calculated by

$$D_{\mathrm{CHORD}} = \frac{2\sqrt{2}}{\pi}\sqrt{1 - \cos\theta}\ , \tag{13.3}$$

where $\theta = \Sigma\sqrt{x_i y_i}$ over all loci. The angular transformation standardizes the genetic distance with respect to possible effects of random genetic drift. Consequently, it is an example of a distance measure that may be more appropriate where random genetic drift rather than mutation is the primary force of divergence.

Slatkin's genetic distances calculate an average within-population distance, S_W, an average between-population distance, S_B, and an average distance for the entire population, $\overline{S}$, based on the average sums of squares of the difference in allele sizes from microsatellite data. These genetic distance measures take into account the consequences of the stepwise mutation model. Average within-population distance is calculated by

$$S_W = \frac{1}{d_s}\sum_{j=1}^{d_s}\frac{2}{2n(2n-1)}\sum_{i<i'}(a_{ij} - a_{i'j})^2, \tag{13.4}$$

where d_s is the number of populations (j) examined, n is the sample size of each population, a_{ij} is the number of repeats of the ith allele in the jth population, and i' and j' indicate unique elements in a symmetric matrix of all a_{ij}. Average between-population distance is

$$S_B = \frac{2}{(2n)^2 d_s(d_s-1)}\sum_{j<j'}\sum_{i<i'}(a_{ij} - a_{i'j'})^2. \tag{13.5}$$

Distances S_B and S_W are equivalent to D_0 and D_1, two genetic distances developed by Goldstein et al. (1995) to use for microsatellite DNA analysis. Average squared difference for the entire population then is

$$\overline{S} = \frac{2n-1}{2nd_s - 1}S_W + \frac{2n(d_s-1)}{2nd_s-1}S_B. \tag{13.6}$$

13.3.2.2 **Ordination.** It is difficult to visualize the relationships among samples based only on a matrix of genetic distances when there are more than three samples or populations in the matrix. **Ordination** and **multidimensional scaling** techniques provide means by which the many pairwise genetic distances between the samples are represented by points along a line, in a plane, or in three dimensions such that the Euclidean distance between the points reflects as closely as possible the same relationship as the distance in the original matrix. These techniques do not explicitly classify individuals or samples into groups, but they do let the investigator visually identify two kinds of patterns: discontinuities in the distributions of objects or trends in the variation. Although the results of different ordination techniques are displayed in much the same way (Figure 13.1A–C), the mathematical approaches to reducing multidimensional data to three or fewer dimensions can be very different. Two popular techniques that we will consider here are principal coordinates analysis and nonmetric multidimensional scaling. The mathematical details of principal coordinates analysis and nonmetric multidimensional scaling are available in many multivariate statistics or numerical taxonomy texts (e.g., Dunn and Everitt 1982).

The goal of **principal coordinates analysis** (PCO) is to produce a pattern of points that represents the original dissimilarities or distances in the genetic distance matrix as

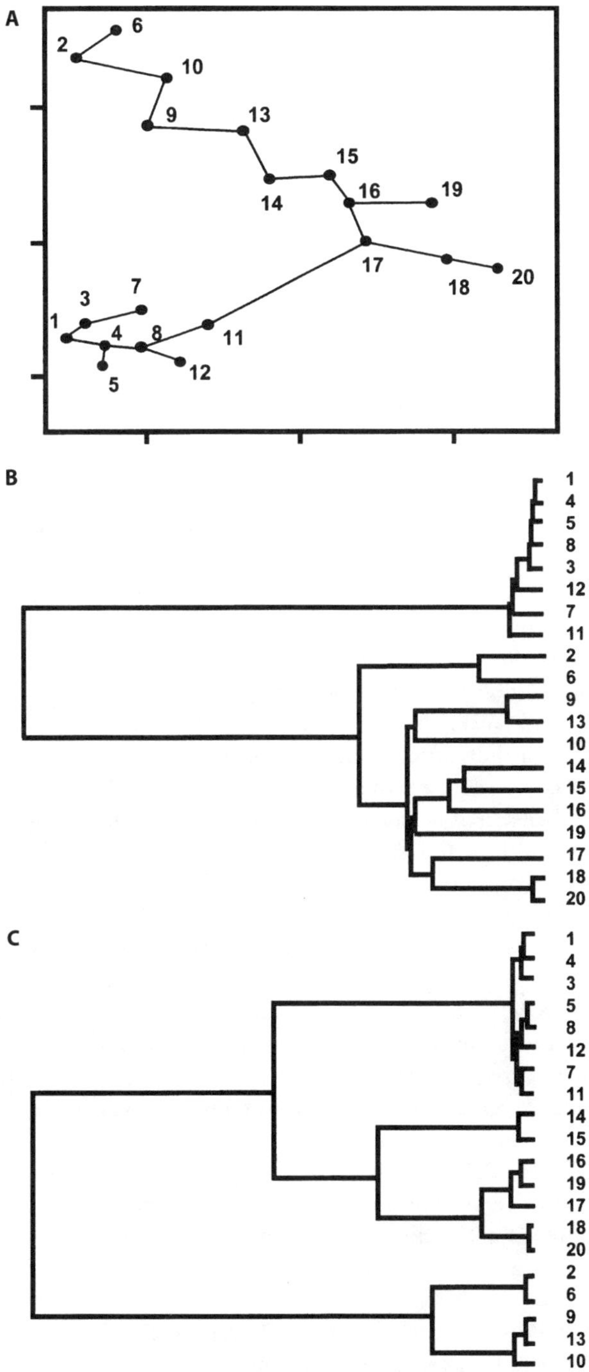

Figure 13.1 What is a cluster? **(A)** Ordination of genetic differences among fish from 20 sampling locations and a minimum spanning tree superimposed. **(B)** Dendrogram of genetic similarities among 20 samples in (A) based on the single-linkage algorithm; the tree shows two major clusters. **(C)** Dendrogram of genetic similarities among among 20 samples in (A) based on the complete-linkage algorithm; the tree shows three distinct clusters. Note that the cluster made up of samples 2, 6, 9, 10, and 13 in (C) is not readily apparent in (A) or (B).

closely as possible. The PCO is a generalized form of principal components analysis. The two are easily confused. **Principal components analysis** (PCA) uses the covariance matrix of allele frequencies to ordinate populations. As noted earlier, not all genetic distances are Euclidean, however, and the PCO allows investigators to use other distance or similarity matrices. If the original distances were Euclidean, then the results of the PCO would be equivalent to the results from the PCA. In both these analyses, if the different samples are considered points in multidimensional space, the computations project the samples' positions in a way that maximizes the spread or variance among them. The first axis is computed to explain the largest share of the variance, the second axis has the next largest share and sits at right angles to the first axis, and each additional axis contributes in lesser amounts until 100% of the variance has been described. By analogy, if we were to use PCO to reduce the position of points in three dimensions to a plane, we could imagine shining a light through the points and onto a sheet of paper. Mathematically, the computations in PCO are projections that involve rotating the light and the paper until we have found the greatest spread among the points as they appear on the paper.

Unlike PCO, which attempts to maintain a correspondence between distance and dissimilarity, the goal of **nonmetric multidimensional scaling** (MDS) is to maintain the original pattern in the genetic distance matrix by using the rank order of the genetic distance values. If, for example, we imagine the points in a three-dimensional space as a structure being joined by elastic bands that equal the genetic distance between samples, the computations in MDS attempt to find the minimum amount of distortion (bending, stretching, and squeezing of the rubber bands) required to flatten the structure to three or two dimensions while maintaining the rank order of distances between the samples that existed in the original distance matrix. Computationally, MDS is not as objective as PCO because most computer programs start from an arbitrary but reasonable arrangement of points and then search for the best solution. Different starting arrangements can yield slightly different results. Consequently, the calculation may be repeated multiple times to arrive at a final solution.

It is important to have a way of evaluating how much information about differences between samples is retained when multidimensional data are reduced to a few dimensions. In PCO, the greater the proportion of the original variation is explained by the first principle axis, the better the fit to the original data. Obviously, if the two axes in a two-dimensional plot of points from PCO together explain only 40% of the variation, we may not be able to detect important patterns in the data. In MDS, the fit of the ordination to the original data is estimated by a stress statistic, S, which calculates departure from the original rank order. An S of 0, for example, indicates a perfect fit (no distortion). In contrast, an S of 0.4 or more indicates poor fit. Another means of evaluating how well ordination preserves the similarities in the original data is to use minimum spanning trees. This technique will be discussed in the next section.

13.3.2.3 **Tree Analyses.** The results of a variety of numerical analyses can be displayed as trees. In graph theory, a tree is a minimally connected graph such that only one path exists between any two objects. It consists of points joined by lines that fork at nodes and single branches but does not include closed loops. Trees may be rooted, that is, there is a particular node that is the origin and a unique path to each other node or vertex within the tree, or they may be unrooted. In this section, we will discuss a variety of tree analyses that superficially may seem similar because the results are displayed as trees. However, to interpret the trees properly, it is important to understand the differences among these analyses.

Ordination and minimum spanning trees are useful for exploring patterns in genetic data, but they do not explicitly produce a classification of individuals or samples into groups. To produce objective classifications, investigators rely on a cluster analysis, which uses an explicit set of rules to classify individuals into groups. Many different algorithms are available. However, the most commonly used ones for GSI are agglomerative, hierarchical techniques that sequentially join individual samples into similar groups—or clusters—based the differences (or similarities) in the genetic distance matrix. These results are displayed as **dendrograms,** which are trees that show hierarchical structure in the data (Figure 13.1B, C).

Three basic agglomerative, hierarchical algorithms that are included in many computer packages are the single-linkage method, the complete-linkage method, and the unweighted pair-group method using arithmetic averages (UPGMA) (Sneath and Sokal 1973). The basic procedure is the same for all of them. The main computational difference is how the distance between two clusters is calculated. In the **single-linkage method,** the similarity between two clusters is defined as the distance between the two most similar pairs of samples; in the **complete-linkage method,** the distance between two clusters is the distance between the two least similar pairs of samples; and in the **group-average method,** the distance is defined by an average of the distances between samples. To show how these analyses work, let us consider the single-linkage method. The first step in the cluster analysis is to find the pair of samples in the genetic distance matrix that are most similar. This becomes the first cluster. It is displayed in the dendrogram as a path between the two samples linked at a node or vertex, where the length of the paths from the samples to the node indicates the distance between the samples. We can now reduce the pairwise genetic distance matrix by one fewer sample by calculating new genetic distances between this cluster and each of the other samples, using the criterion of the most similar pair from the original genetic distance matrix. Once again we look for the most similar pair of samples, but this time we are considering the previously identified cluster as a single unit. Now we can identify a new cluster. It might be a cluster consisting of our previous cluster and a new sample, or it might be a pair consisting of two other samples. Once again, we calculate a new genetic distance matrix with one fewer sample based on our distance criterion. We continue this iterative process until all the original samples are included in one cluster.

Although cluster analysis is objective, it is important to realize its limitations. First, different algorithms can produce different classifications depending on the relationships among the samples, which are reflected as differently shaped clusters in multidimensional space. For example, points in a scatter plot (two or three dimensions) may form dense clouds, they may be scattered linearly like a chain (Figure 13.1A), or they may have the shape of a kite with a tail. Different putative groups, such as samples from different geographical areas, may have different shapes, which influences what kind of clusters will be detected. Second, ties (equal distances) between two or more pairs of objects in the distance matrix can lead to different dendrograms. Finally, there is no objective way of determining beforehand which of the possible dendrograms may be best. Investigators may reject one clustering strategy over another based on the relationships seen in the ordination analyses. Clustering methods may produce artifactual clusters (Figure 13.1C), for example, when compared with ordination of data that show a linear distribution with no discrete clusters. Investigators also may choose dendrograms that fit a preconceived model of how evolution works. For example, if geological history suggests that some water bodies have been connected but others have not, we might expect genetic similarity to reflect the potential gene flow among some groups but not among others. An investigator may choose the dendrogram that is consistent with his or her knowledge of geological history and reject others.

Another class of tree analyses includes those that produce additive, unrooted trees. These include minimum spanning trees (Kruskal 1956) and additive phylogenetic methods, such as the neighbor-joining method (Saitou and Nei 1987) and the distance Wagner method (Farris 1972). The computational goal of these methods is to minimize the total length of a tree. Once constructed, unrooted additive trees can be rooted to a common vertex or evolutionary origin by including one or more outgroups—taxa that are not part of the group of interest but that share a close common ancestor with that group—or by making other extrinsic assumptions about the rate of genetic divergence. However, because many of the evolutionary assumptions underlying these methods are especially precarious when dealing with intraspecific divergence, investigators should use extreme caution in rooting trees or making phylogenetic inferences about the divergence of different stocks from these analyses.

A **minimum spanning tree** is a network that shows the shortest path among all the objects or samples. Very simply, the computer algorithm uses the pairwise genetic distances from the genetic distance matrix to find the tree that connects all the different samples and has the shortest total length. Because they are based on the same pairwise distance matrices as ordinations, minimum spanning trees can be superimposed on scatter plots from ordination to detect whether there are local distortions in the ordinations (Figure 13.1A). For example, two points may appear close together in a scatter plot when in another dimension they are separated. When a minimum spanning tree is superimposed on the plot, each of the two points that appear similar in the scatter plot may be joined independently by the tree to a third point farther away, which indicates that they are more similar to that point than to each other. This can be misleading, however, when an object is related to two or more other objects by distances of nearly equal magnitude because the minimum spanning tree will show only the single shortest branch for each population to its "nearest" neighbor. This can be a problem with genetic data sets that include large numbers of intraspecific populations.

The **neighbor-joining method** was first proposed as a method for finding a tree that is close to representing the true phylogeny of the groups of interest. Although it may be inappropriate to use it to infer phylogenies at the level of population differentiation, it can be used as an exploratory tool or a classification technique for GSI. In this case, there is no a priori criterion for choosing this method over other clustering strategies. Computationally, the process begins with a matrix of genetic distances and a completely unresolved, unrooted phylogenetic tree that has the shape of a star with a single internal node and branches to each of the terminal nodes or objects to be classified. New pairwise distances are calculated by adjusting the original genetic distance values based on their average divergence from all other nodes. The algorithm then searches for the closest pair of objects to group as a new internal node. Unlike the single-linkage method, the closest pair is the pair that would result in greatest reduction in the length of the branches. A new distance is calculated between the new node and all the other objects. The original pair of objects is then deleted, and the process is repeated to generate new internal nodes until only two nodes remain, which are separated by a single branch. This process defines the nodes and branches of the tree.

As we have noted already, it can be difficult to decide how to choose between different kinds of tree analyses. Bootstrapping and maximum likelihood methods are tools that can be used for evaluating different trees. **Bootstrapping** refers to a set of statistical methods for creating pseudoreplicate data sets in cases in which it is impossible to get true replicate data to evaluate whether the results of analysis can be explained by chance alone. If we have no a priori reason for choosing a tree, we can use bootstrapping to derive a measure of how reliable the different branches of a particular tree are compared with chance. Computationally, bootstrapping generates a large number of pseudoreplicate data sets by random sampling (with replacement) of the original data set. Each of these data sets is analyzed, and the results are used to calculate the frequency of occurrence of a given branch. In contrast, we may want to evaluate a particular tree topology because it represents a hypothesis about evolutionary relationships. **Maximum likelihood methods** can be used to evaluate the evolutionary history of groups, as represented by a particular tree topology, in terms of a proposed model of evolution. In this case, the investigator proposes a specific model for genetic change and then calculates the probability that the observed genetic data would have generated a particular tree. The probabilities for different trees are compared, and the best tree is the one with the highest probability. Bootstrapping and maximum likelihood methods are being increasingly used in GSI studies to assess tree robustness and alternative topologies. There are several useful computer programs available for this purpose, some of which are available in the public domain.

13.3.3 *Model-Based Analyses*

Ordination and most tree analyses do not assume an underlying pattern—or structure—to intraspecific genetic differences. In these analyses, the structure is derived from examining the data. In this sense, they are exploratory. In contrast, population geneticists sometimes find it useful to test specific hypotheses about the relationships among samples. These hypotheses might have been the basis for the study design or they may have been suggested by exploratory analyses. In this section, we review techniques for examining how well the observed data fit a genetic structure that is specified by the investigator.

The simplest genetic structure that an investigator may want to test is the null hypothesis of no allelic frequency differences among samples. Unplanned comparisons, however, usually have only limited value. Often, a more useful approach is to ask how much of the total observed genetic variation can be explained by partitioning the variation among alternative sources of differences. Where specimens of a freshwater fish have been collected from several drainages, for example, we might partition the variation in a nested scheme to examine potential differences among subpopulations in different river systems, differences among subpopulations within river systems, and differences among individuals within populations. This nested analysis may be extended to other sources of variation, such as differences in life history forms or temporal differences, if the sampling were appropriate (Box 13.1).

13.3.3.1 **Statistical Tests.** The most direct way to test for genetic differences is to use contingency table G-tests or χ^2 tests. Although investigators sometimes conduct unplanned or pairwise G-tests to explore their data, explicit hypothesis testing is preferable. A useful characteristic of G-tests that many investigators exploit is that the scores are additive. Consequently, individual G-tests can be nested within larger tests. Using this kind of nested contingency table analysis, for example, it is possible to test for differences at the various levels of geographical subdivision, such as in the example in Box 13.1, in a manner analogous to analysis of variance.

■ Box 13.1
An Example of Hierarchical Genetic Analysis

Consider a freshwater fish that spawns in both an inlet river flowing into a lake and the outlet river flowing from the lake. Within both the inlet and outlet rivers, the fish spawn in warm, thermal springs and cool, upland streams. We are curious whether the fish, which mix in the lake when they are feeding and rearing, are part of one big, panmictic population. In addition, we wonder whether fish spawning in thermal springs versus upland streams might be different populations because they face environmental differences that could have selected for different genotypes. We decide to sample 50 fish from each of two springs and two upland streams in both the inlet and outlet rivers and to collect allele frequency data for each of the eight samples. Because we have specific geographical hypotheses, we decide to examine allele frequency differences (using *G*-tests) and differences in level of heterozygosity (using gene diversity analysis) by partitioning the total variation into differences between inlet and outlet fish, between spring-spawning and stream-spawning fish, and between subpopulations within different spawning habitats. The figure below shows the hierarchy for how these differences can be partitioned. For simplicity, we will use data from a single locus with two alleles to illustrate the statistical analysis. From the frequency of one allele *(p)* and the number of genotypes in each sample, we can calculate the frequency of the alternate allele, Hardy–Weinberg expected heterozygosity *(H),* and all the statistics in the figure.

Differences in Allele Frequencies

The basic strategy for nested, hierarchical *G*-tests is to partition the test statistic so that $G_{\text{TOTAL}} = G_{\text{AMONG GROUPS}} + G_{\text{WITHIN GROUPS}}$ and then treat the within-group test statistic as a new total and partition it further. In our example, all our allele frequency differences can be summarized in an 8 × 2 contingency table in which each row is a separate sample and each column contains the numbers of allele *a* and allele *b* for that sample. A *G*-test on these data gives us our total *G*-test statistic, $G_T = 87.56$, with 7 df, which indicates that there are significant allele frequency differences in the data. We can find out more about where the important differences are by breaking these data down into contingency tables that reflect the groups in our hierarchy and calculating separate *G*-test statistics. For example, to test for differences between rivers (inlet versus outlet fish), we collapse the original table into a 2 × 2 table by summing all the allele counts for the four inlet populations as one row and all the allele counts for the outlet populations as the other row. This *G*-test gives us $G_{RT} = 43.78$ with 1 df. That means that an additional $G = 43.78$ must be explained by differences within rivers (87.56 = 43.78 + 43.78). This latter value, of course, represents the differences within both the inlet and the outlet groups of fish. We can calculate the test statistic for each of these separately by using the two 4 × 2 contingency tables that have the allele counts for the four inlet samples and the four outlet samples. For the inlet group, this is $G = 22,32$, with 3 df. It is also the sum of the differences between fish spawning in springs or streams and differences between different subpopulations of spring-spawning or stream-spawning fish (22.32 = 16.18 + 1.99 + 4.20). For the

continued on next page

■ Box 13.1 *continued*

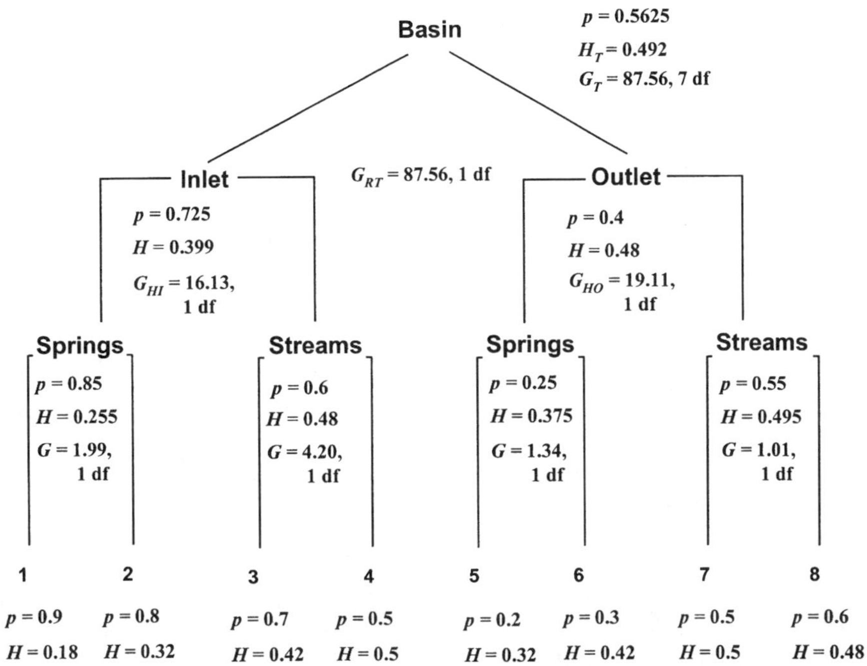

One possible hierarchy for genetic analyses. Allele frequency is indicated by *p,* heterozygosity by *H, G*-test statistics by *G,* and subscripts are T (total), R (river), H (habitat), I (inlet), and O (outlet).

outlet group, this test statistic is $G = 21.46$, with 3 df. Notice that the degrees of freedom are also additive. The table on the next page shows results from our example.

Differences in Heterozygosity

Based on equation (13.8), heterozygosity may also be partitioned into different groups. In our example (see figure), we want to partition heterozygosity according to the following model:

$$H_T = D_{\text{Rivers within basin}} + D_{\text{Habitats within rivers}} + D_{\text{Subpopulations within habitats}} + H_{\text{Subpopulations}}.$$

We begin by calculating total heterozygosity, H_T, using Hardy–Weinberg expectations (Chapter 1) from the allele frequency of p and $1 - p$ over all samples, as though they were a single sample. From the figure, p is $(0.9 + 0.8 + 0.7 + 0.5 + 0.2 + 0.3 + 0.5 + 0.6)/8$, or 0.563. We also calculate the average heterozygosity within subpopulations, $\bar{H}_S$. For the eight subpopulations, $\bar{H}_S$ is $(0.18 + 0.32 + 0.42 + 0.5 + 0.32 + 0.42 + 0.5 + 0.48)/8$, or 0.393. The absolute gene diversity, D_{ST}, then becomes $H_T - \bar{H}_S$, or $0.492 - 0.393 = 0.099$. Dividing D_{ST} by H_T (equation [13.9]) gives the relative magnitude of differentiation between subpopulations, or a G_{ST} of 0.201. This means that diversity among subpopulations explains 20.1% of the observed genetic variation.

continued on next page

■ Box 13.1 *continued*

An example of hierarchical *G*-tests. Groups nested within larger groups are indented.

Source of variation	G-test statistic	Degrees of freedom
Total	87.56	7
Between rivers (inlet versus outlet)	43.78	1
Within rivers	43.78	6
Inlet	22.32	3
Between habitats (spring versus stream)	16.13	1
Within habitats	6.19	2
Between spring subpopulations	1.99	1
Between stream subpopulations	4.20	1
Outlet	21.46	3
Between habitats (spring versus stream)	19.11	1
Within habitats	2.35	2
Between spring subpopulations	1.34	1
Between stream subpopulations	1.01	1

As our model indicates, we would like to partition the diversity among subpopulations into diversity explained by groups of fish in different rivers within the basin, different habitats within the rivers, and different subpopulations within habitats. At each level of the hierarchy we calculate a new p and $1 - p$ from the samples that make up that group (see figure), the expected heterozygosity based on those allele frequencies, and the average heterozygosity over all the groups at that level. For example, at the level of differences between inlet and outlet rivers, p for inlet fish and outlet fish is 0.725 and 0.400, respectively, and expected heterozygosity is 0.399 and 0.480, respectively. Average heterozygosity of the two groups is 0.439. Hence, the component of relative gene diversity explained by differences among fish from different rivers is (0.492 – 0.439)/0.492, or 0.108 (about 11% of the total diversity). Consequently, of the 20.1% of the total diversity that is due to differences among subpopulations, about half occurs because of the differences between fish from inlet and outlet streams. Similarly, the average heterozygosity among fish spawning in different habitats is (0.255 + 0.48 + 0.375 + 0.495)/4, or 0.401. We calculate the relative gene diversity by (0.439 – 0.401)/0.492 = 0.077. Likewise, we calculate the relative gene diversity among fish from different subpopulations within spawning habitats by (0.401 – 0.393)/0.492 = 0.016. For another example, see Chakraborty et al. (1982).

Interpretation

This example illustrates the complementary nature of these analyses. For example, in both analyses, the most important difference occurs between fish from inlet and outlet streams. From the gene diversity analysis, we estimate that this difference explains about 10% of the total gene diversity. From the *G*-tests, we know this is statistically

continued on next page

■ Box 13.1 *continued*

highly significant. Likewise, the differences among spring- and stream-spawning fish suggests that they are different populations and future studies to investigate possible adaptive differences might be worthwhile. In contrast, the differences between subpopulations within different spawning habitats account for less than 2% of the total gene diversity. This was statistically significant in only the inlet populations.

An alternative method of statistically testing hypotheses of population structure, which we have not illustrated, is to use nested analysis of variance of allele frequencies to estimate measures of divergence based on relatedness of either alleles or genotypes (F_{ST} and coancestry, θ). A complete description is available in Weir (1996), and a variety of computer programs implement the analysis.

13.3.3.2 Gene Diversity Analysis. Gene diversity analysis is an extension of a classic population genetic statistic for population differentiation known as Wright's F_{ST}, or the fixation index. This statistic takes advantage of the Wahlund principle (Chapters 1, 7), which is that the observed average heterozygosity in a pool of populations with different allele frequencies is reduced compared with the heterozygosity in the pooled population if all those individuals were randomly mating. Consequently, the **fixation index** can be calculated by

$$F_{ST} = \frac{H_T - \bar{H}_S}{H_T}, \qquad (13.7)$$

where H_T is the expected heterozygosity in an equivalent randomly mating total population and $\bar{H}_S$ is the average of the expected heterozygosity in randomly mating subpopulations. Gene diversity analysis uses a similar statistic, G_{ST}, which is calculated based on allele frequencies at several multiallelic loci. In this case, we can partition the total heterozygosity in a population, H_T, into the average absolute gene diversity between subpopulations, D_{ST}, and the average heterozygosity within subpopulations, $\bar{H}_S$, such that

$$H_T = D_{ST} + \bar{H}_S. \qquad (13.8)$$

The H_S in the s subpopulation is calculated by $1 - \sum_k x_{ik}^2$, where x_{ik} is the frequency of the kth allele in the ith subpopulation, and $H_T = 1 - \sum_k \bar{x}_k^2$. With more than one locus, H_S and H_T are averaged over all loci. Consequently, the relative magnitude of differentiation between subpopulations is

$$G_{ST} = \frac{\bar{H}_T - \bar{H}_S}{\bar{H}_T}. \qquad (13.9)$$

This partitioning of genetic variation can be extended to other levels. For example, using the scenario in Box 13.1, we might partition the genetic variation according to

$$\bar{H}_T = D_{RT} + D_{SR} + \bar{H}_S,$$

where D_{RT} is the absolute gene diversity among rivers and D_{SR} is the absolute gene diversity of subpopulations within rivers. From these, it is easy to calculate the corresponding components of relative gene diversity, G_{RT} and G_{SR}.

Analyses using F_{ST} generally assume that mutation follows an infinite allele model or that mutation is negligible and can be disregarded. Because evolution of microsatellites may be too rapid to be disregarded in some taxonomic studies and mutation of microsatellites may be more likely to follow a stepwise model, *R*-statistics have been developed. The *R*-statistics are analogous to *F*-statistics and account for loci undergoing stepwise mutation (Slatkin 1995). The R_{ST}, which is analogous to F_{ST}, is calculated by

$$R_{ST} = \frac{\bar{S} - S_W}{\bar{S}}, \tag{13.10}$$

where S_W and $\bar{S}$ are the average sum of squares of the differences in allele size within a subpopulation and the entire population, respectively, as described earlier.

13.3.4 Summary

Although numerical analyses attempt to provide objective, stable means of classifying individuals or samples into groups, in practice the results inevitably depend on the insights and abilities of the researcher. Numerical analyses for GSI usually proceed through three or four stages: analysis of genotypic distributions, taxonomic analyses, analyses of different models of genetic population structure, and analysis of mixture samples in many cases (covered in section 13.4.2). A fundamental genotypic analysis is testing for Hardy–Weinberg equilibrium. Taxonomic analyses are largely exploratory. Often, they begin with choosing and calculating pairwise genetic distances between samples. Ordination techniques, such as PCO and MDS, attempt to reduce the multidimensional data in a genetic distance matrix to a plot of points in one, two, or three dimensions that preserves the original relationships among the samples as much as possible. Tree analyses provide means of summarizing the relationships among samples as unrooted trees or dendrograms. These techniques include cluster analyses, minimum spanning trees, and neighbor-joining methods. Bootstrapping and maximum likelihood tree analyses can be used to evaluate different trees. In contrast, hypotheses about genetic structure specified by the investigator can be tested by contingency table analyses using χ^2 or *G*-tests. Similarly, gene diversity analysis allows investigators to partition the genetic variation into absolute and relative measures of diversity among nested subpopulations according to a model of genetic structure defined by the investigator.

13.4 Applications

Genetic theory and analyses are relevant to many aspects of fish biology and fisheries management. Genetic data can be used to (1) elucidate and characterize stocks and understand their evolutionary interrelationships and dynamics, (2) analyze unknown mixtures to estimate stock contributions to fishery catches or assign individuals to their most likely stock of origin, and (3) monitor and evaluate hatchery programs and analyze pedigrees and parentage. Because the fitness and productivity of fish populations and species are largely controlled by genetic traits that are affected by a multitude of natural and anthropogenic factors, genetic risk assessment and management are also important components of effective fishery management. Below we discuss some of these applications and present several specific examples.

13.4.1 *Stock Management and Conservation*

13.4.1.1 **Genetic Inventories.** A detailed genetic inventory and description of stocks is an essential prerequisite for meaningful, genetically based stock management or conservation programs. Optimal long-term management and conservation depend upon knowing the number, distribution, and characteristics of all component stocks and maintaining their integrity, diversity, and abundance. The more subdivided and diverse a species is, the more its overall productivity and evolutionary potential are likely to depend on maintaining its various components. Given a goal of maintaining natural systems (populations, species, communities, and ecosystems) in their natural state, populations and human activities should be managed to prevent or minimize anthropogenic genetic changes in stock structure (extinction, loss of among-stock diversity, loss of within-stock variability, or change in population characteristics) (see Ryman 1991).

Genetic stock identification methods can and should be used when developing a stock inventory because they provide the most direct measures we have of reproductive isolation and genetic differentiation (see Chapter 7). However, despite the emphasis we place on collecting and analyzing direct genetic data, there are three important caveats for such inventories: (1) direct biochemical or molecular genetic data may fail to detect all existing population subdivision, (2) such data will seldom if ever identify or even represent characteristics of direct importance to fitness (performance and productivity), and (3) other biological characteristics (with known or even unknown genetic bases), environmental conditions, and zoogeographic factors and history can provide useful information for elucidating stock structure.

Reproductive isolation usually is initiated by spatial, physical, temporal, and biological restrictions to gene flow. Discontinuous distributions of suitable habitat for any life history stage can preclude panmixia and thereby promote population subdivision within a species. Sockeye salmon in North America, for example, spawn in hundreds of separate rivers, streams, and lakes. Impassable physical barriers in otherwise continuous habitat (such as a high waterfall in a stream or a land mass separating an otherwise continuous marine environment) have the same isolating effect. Species distributions that exceed the dispersal and migrational capacities or behaviors of individuals [e.g., the supposed circumglobal distributions of bonefish (*Albula "vulpes"*) and striped mullet (*Mugil cephalus*)] can contribute to stock heterogeneity and even speciation. Similarly, differences in the timing of spawning can be effective barriers to gene flow and result in multiple subpopulations. In Washington State, for example, the tendency of local chum salmon to spawn consistently in summer, fall, or winter has resulted in multiple subpopulations, often occurring in the same locations. Behaviors such as the remarkable homing of various salmon species and differences in the season of spawning also can lead to reduced gene flow and increased genetic divergence among populations. Additionally, past geological events and conditions, such as continental drift, glaciation, and stream capture (the joining of two formerly independent streams) can have profound effects on past and present patterns of stock structure.

The sampling design for conducting a genetic stock inventory should be based on the known biology and distribution of the target species, information about the physical environment, and an understanding of the geological history of the region. Biological and environmental information and samples should be obtained from throughout the geographic range covered by the inventory, taking care to include collections from areas likely to be isolated (by distance, physical barriers, or other factors) and from populations that may be reproductively isolated by spawning season or year. The individuals sampled for each collection should be chosen at random (or at least without regard to sex, size, appearance, behavior, or other such factors). Additionally, collections should be made in such a way that possible movements of individuals among sites will not confound or prevent meaningful interpretation. For example, if juveniles move from separate spawning sites to communal juvenile rearing areas, samples of juveniles from such rearing areas would be unlikely to reveal an accurate picture of population subdivision, whereas sampling of adults on the spawning grounds would.

Next, one must identify and screen a sufficient number of genetic traits known to be variable in the species and likely to be informative regarding patterns and levels of gene flow. For example, investigating stock structure based on only 1 or 2 rarely variable loci is likely to be uninformative, whereas screening 20 or more loci polymorphic at the $P_{0.95}$ level will provide a very powerful test of population subdivision. As a general rule of thumb, we recommend using at least six polymorphic loci ($P_{0.95}$) in such studies to provide reasonable statistical power.

Once the genetic data have been collected, statistical analyses to test for significant heterogeneity are usually straightforward (see section 13.3.3). Although statistically significant differences may indicate biologically important subpopulation differentiation, other explanations are possible. These should be ruled out by the experimental design. For example, one obvious cause of differences is short-term temporal instability. Sampling error or random genetic drift in allele frequencies in populations with small effective sizes could lead to year-to-year differences in the genetic profiles of individual populations that might be mistakenly interpreted as evidence of population divergence. Analysis of temporally repeated collections and specific analyses of the data for evidence of small effective population size should identify such situations. Consequently, the sampling design should include some repeat collections in different seasons or years either to establish the stability and repeatability of the observed pattern of stock structure or to identify temporal variation as a possible complicating factor in understanding stock structure.

Although the direct analysis of genetic traits such as allozyme frequencies may provide an informative view of population structure, the inclusion of all available relevant information (other biological characteristics, distributions, and geological history of the region) likely will improve the understanding of population structure.

13.4.1.2 **Diversity Assessment.** Executing a genetic inventory of stocks is only part of the process for describing and managing genetic diversity. The second phase is to define and, if possible, to understand the interrelationships of stocks and quantify how genetic diversity is distributed among stocks within a species. This is a critical step if one is to manage and conserve diversity in the most effective and efficient manner.

Genetic variation in most species is hierarchically distributed. While all individual populations may contain much of the variation found in a species, some variation may occur in only one or a subset of stocks. Similarly, some stocks are usually genetically more similar to each other than they are to other stocks or stock groups. Any of the factors discussed above that promote population subdivision can be viewed as different levels in the hierarchical organization of genetic diversity within a species. Thus, geographic region, watershed, and river system might be major factors in explaining diversity in one species, whereas spawn timing, migratory behavior, and a major physical barrier might be the major factors responsible for the organization of genetic diversity in another species (Box 13.2).

There are a number of different statistical approaches for investigating the organization of genetic diversity within species. As discussed in sections 13.3.2 and 13.3.3, these include F_{ST} calculations, hierarchical gene diversity analysis, and dendrogram and multidimensional scaling analyses of genetic distances.

Understanding how genetic diversity is partitioned among stocks within a species should be the basis for both management and conservation when maintaining diversity is the goal. In most cases, such an approach will lead to the identification of the stocks that contribute most to overall within-species genetic diversity and priority can then be placed on these stocks when conflicts arise or when resources for management and conservation are limiting.

The above approaches of constructing a stock inventory and then determining how diversity is partitioned among stocks are the cornerstones for the ongoing wild stock initiative in Washington State and for the current National Marine Fisheries Service (NMFS) coastwide stock status reviews for anadromous salmonids under the U.S. Endangered Species Act of 1973 (ESA; 16 U.S.C. §§ 1531 to 1544). The Washington Department of Fish and Wildlife (WDFW) process is detailed in Box 13.2.

■ Box 13.2
Pacific Salmon and Steelhead Stock Inventory and Genetic Diversity Unit Determination

In 1992, the then-separate Washington departments of Fisheries and Wildlife and the Western Washington Treaty Indian Tribes initiated a statewide inventory of stocks of Pacific salmon and steelhead (SASSI) as part of an ambitious management plan to restore and conserve wild stocks (Washington Department of Fisheries et al. 1993). For this process, stocks were defined as "the fish spawning in a particular lake or stream(s) at a particular season, which fish to a substantial degree do not interbreed with any group spawning in a different place, or in the same place at a different season." Three criteria were used to recognize stocks: (1) distinct spawning distributions, (2) distinct temporal distributions (run timing, spawn timing, or both), and (3) distinct biological characteristics (primarily genetic stock identification [GSI] data, when available). Differences in any of these three types of information were deemed sufficient to designate separate stocks. Whereas both the distribution of spawners and run timing were evaluated subjectively or qualitatively, genetic data were analyzed statistically by means of *G*-tests or heterogeneity χ^2 tests and evaluated by dendrogram analysis. In addition to identifying stocks, the inventory process also evaluated the status of each stock based on estimates of abundance (e.g., escapement or run-size estimates) for the most recent 25-year period. In total, 435 stocks were identified statewide, distributed among species and stock status categories as summarized below.

continued on next page

■ Box 13.2 *continued*

Summary of Washington salmon and steelhead stock inventory.

Species	Total number of stocks	Stock status category: Healthy	Depressed	Critical	Unknown	Extinct[a]
Chinook salmon	108	54	35	5	14	
Chum salmon	72	48	3	2	18	1
Coho salmon	90	37	34	1	18	
Pink salmon	15	9	2	2	2	
Sockeye salmon	9	3	4	1	1	
Steelhead	141	36	44	1	60	
Total	435	187	122	12	113	1

[a] This category was applied in the inventory only to stocks that had been tracked since the mid-1960s in the agency's escapement and fisheries management databases and had become extirpated in their native range during this period. It did not include any of the large number of stocks believed to have become extinct prior to the mid-1960s, when the Washington Department of Fisheries initiated a salmon stock-monitoring program.

The second major phase of this process has been to determine how genetic diversity is distributed among stocks for each of these species. To date, this has been accomplished for chinook, chum, and sockeye salmon, pink salmon *O. gorbuscha,* and steelhead. We have sought to identify two primary hierarchical levels above stocks: genetic diversity units (GDUs), which are groups of one or more genetically similar stocks, and major ancestral lineages (MALs), which are assemblages of one or more genetically similar GDUs.

As an example, diversity assessment for pink salmon began with the list of 15 stocks defined in the SASSI inventory process and the analysis of allozyme data for 28 variable loci in 41 collections of Washington pink salmon. Genetic distances (Cavalli-Sforza and Edwards, 1967; modified Rogers [Rogers 1972; see Wright 1978]; and Nei 1978) were calculated and used for both dendrogram and multidimensional scaling (MDS) analysis by means of the BIOSYS-1 and NTSYS-pc computer programs of Swofford and Selander (1981) and Rohlf (1994), respectively. Both types of graphical analysis clearly showed that the even-year stock from the Snohomish River was highly diverged from all odd-year stocks. The pronounced divergence of the even-year stock caused us to place it in a separate MAL (and GDU) from the odd-year stocks. Multidimensional scaling analysis of the odd-year stocks by themselves revealed that the Hood Canal Hatchery stock was considerably different from the wild stocks. The hatchery stock, therefore, was placed in a GDU by itself. Further MDS analyses of the 12 naturally spawning odd-year stocks for which there were genetic data revealed that they tended to group into six basic assemblages—upper Dungeness, lower Dungeness, Nooksack, Hood Canal, Nisqually, and north + central Puget Sound (see below). Subsequent analyses that included nine stocks (16 collections) of southern British Columbia odd-year pink salmon were conducted to determine whether or not some of the Washington stocks belonged in a larger assemblage with some Canadian stocks and to assess whether or not the pattern of stock interrelationships resulting from the analysis of only Washington stocks would

continued on next page

■ Box 13.2 *continued*

be affected by inclusion of nearby stocks from southern British Columbia. These analyses resulted in little change in Washington stock assemblages and demonstrated that the southern British Columbia stocks were most similar to the north + central Puget Sound stock group (analysis not shown). Independent evaluation of information about spawner distribution and timing, adult size, freshwater habitat characteristics, and spawning maturation and coloration of Washington stocks generally supported the stock associations revealed by MDS but also led to the separation of the Puyallup River stock (in central Puget Sound) from the north Puget Sound stocks on the basis of distribution and freshwater habitat characteristics. In summary, genetic diversity in naturally spawning Washington pink salmon is hierarchically organized in seven odd-year GDUs in one MAL (odd-year), as shown in the figure below, and one even-year GDU in a second MAL (even-year). The Washington Department of Fish and Wildlife (WDFW) is using this information to guide its future management and conservation actions for Washington pink salmon.

In its recent coastwide status review of pink salmon, based on these and other data, the National Marine Fisheries Service (NMFS) determined that the odd- and even-year MALs identified here are equivalent to evolutionarily significant units under the U.S. Endangered Species Act (ESA) (Hard et al. 1996; see also Waples 1991; Nielsen 1995; Box 18.1).

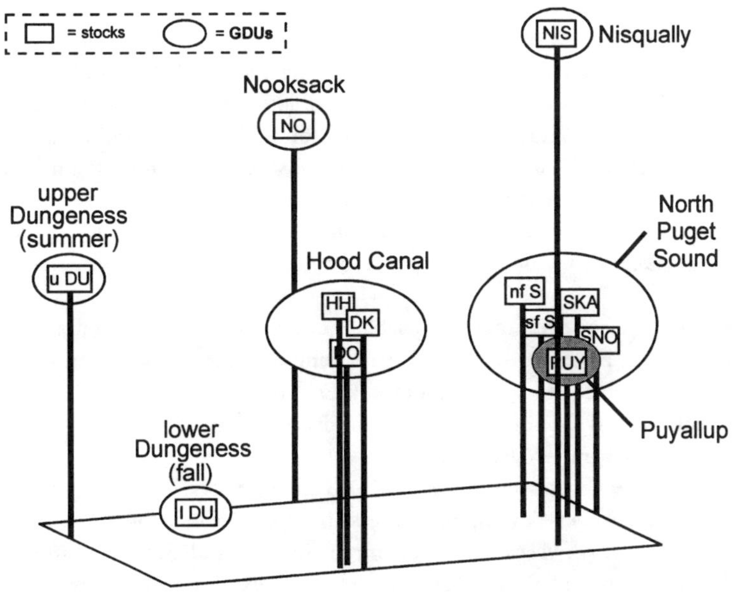

Genetic diversity of naturally spawning odd-year Washington pink salmon. Genetic diversity units were revealed by multidimensional scaling analysis of Cavalli-Sforza and Edwards (1967) chord distances based on 28 variable loci (stress2 = 0.147; good fit). Stocks analyzed are upper Dungeness (uDU), lower Dungeness (lDU), Nooksack (NO), Hamma Hamma (HH), Duckabush (DK), Dosewallips (DO), Nisqually (NIS), Skagit (SKA), north fork Stillaguamish (nf S), south fork Stillaguamish (sf S), Snohomish (SNO), and Puyallup (PUY).

13.4.2 Harvest Management: Mixed-Stock Fishery Analysis

Because the overall abundance, productivity, and genetic diversity of a species depend upon the persistence of all component stocks, successful management of mixed-stock fisheries relies upon obtaining adequate escapements of all contributing individual stocks. This goal, in turn, requires that stock contributions to mixed-stock fisheries be determined (estimated). Although there are many approaches to mixed-stock analysis (MSA), GSI methods are among the most informative and cost effective (Shaklee et al. 1990; Utter and Ryman 1993). The basis of the GSI approach is that distinct stocks exhibit detectable genetic divergence (see section 13.2 and Chapter 2) and that this divergence is large enough to provide a basis for the reliable estimation of stock contributions in a mixture of stocks.

To date, allozyme data have been used most extensively for genetic MSA, although mtDNA or nuclear DNA markers can and have been used. Maximum likelihood estimation (MLE) procedures are the statistical method of choice for such studies (Millar 1990). The basic approach is to elucidate the array of multilocus genotypes present in a fishery sample and then to use statistical procedures to find the proportions of potentially contributing stocks that best approximate the distribution of genotypes seen in the fishery (based on probabilities) (Ryman and Utter 1987; Wood et al. 1987). This is done by first calculating the probabilities (frequencies) of all possible multilocus genotypes in each baseline stock. Hardy–Weinberg equilibrium is assumed in the computer programs used by some agencies (e.g., Alaska Department of Fish and Game, NMFS, and WDFW), but observed genotype proportions are used in the programs employed by others (e.g., Canada Department of Fisheries and Oceans and Pacific Salmon Commission). Initial estimates of the proportion of each stock in the fishery sample then are assigned (typically these initial values are $1/N$ for each of N stocks). The initial estimates then are refined iteratively based on the probabilities of each multilocus genotype in the fishery, given the currently estimated proportion of each stock and the observed (or calculated) genotype frequencies in each baseline stock. The process continues until a set number of iterations has been completed or a given tolerance level is achieved. The resulting estimate represents the proportion of baseline stocks that provides the statistically most likely fit to the distribution of multilocus genotypes observed in the fishery sample.

The use of a large number of variable loci (and alleles) is generally key to successful genetic MSA. The reason for this is that genetic MSAs such as described above gain more power from each additional locus; this contrasts with many approaches to diversity assessment, such as those based on genetic distances, for which the discrimination is an average of the separation due to all loci.

Although the genetic approach to MSA can, in theory, be applied to any species and any fishery, it has been developed and used most extensively for various salmon fisheries in the Pacific Northwest (see, for example, Shaklee and Bentzen 1998; Shaklee et al. 1999). The discussion below is drawn largely from experiences and insights derived from these applications.

The basic experimental design for a genetic MSA begins with a clear formulation of the specific management question(s) to be answered, including expectations for sensitivity, accuracy, and precision. For example, which stocks or stock groups must be estimated, what levels of contribution are expected or important to estimate, and what accuracy and precision is acceptable? Additional considerations and planning involve the specific fishery strata (e.g., areas, times, and gear types) to be analyzed, the genetic methodology to be employed, the baseline data set (e.g., stocks, loci, and alleles) needed to support the planned analysis, the statistical algorithm to be used, and the fishery sampling design (e.g., sample size, tissues, and locations and timing of sampling) needed to answer the question(s) being asked. In practice, this design phase requires that several of the above issues be pursued more-or-less simultaneously because they interact with one another.

The creation of a baseline data set characterizing all stocks that are potential contributors to a mixed-stock fishery is an important but challenging process. Almost by definition, mixed-stock fisheries involve stocks from a wide geographic area. In many cases, the range of stocks that must be included in such a baseline exceeds the resources, jurisdiction, or both, of any single agency or laboratory. This simple fact means that success is dependent upon coordination, collaboration, and data standardization among participating researchers (Shaklee and Phelps 1990). For any given fishery analysis

—spawning ground (or other appropriate) samples from all major, potentially contributing stocks throughout the range must be obtained;
—information about specific data collection methodologies should be shared;
—standardized data sets (loci and alleles) with consistent nomenclature must be generated for each stock and made available to users; and
—procedures to assure that appropriate within- and among-laboratory quality control standards are being met should be implemented.

An informal GSI consortium of federal, state, and other participants serves as a forum to promote these goals for genetic analyses of Pacific salmon mixed-stock fisheries in North America.

Computer simulation analysis is used to evaluate and refine aspects of the MSA such as fishery sample size, loci and alleles to be analyzed, and reporting groups (stocks or groups of stocks for which contribution estimates are considered reliable and accurate) for each analysis (Shaklee et al. 1999). The starting point for such simulations consists of genetic characterizations of all potentially contributing stocks (i.e., baseline data). Baseline data usually consist of multilocus genotypes for 100 or more individuals from each spawning stock. Simulations involve creating hypothetical stock mixtures of varied proportions (based on random sampling from the allele frequencies for each stock in the baseline data, usually assuming Hardy–Weinberg proportions, and then analyzing these known, simulated mixtures by means of a suitable computer algorithm [e.g., Fournier et al. 1984; Millar 1987]). Such simulations typically are run for 50 or more replications to provide estimates of accuracy and precision. Jackknifing or bootstrapping can be used to estimate the variance of the estimates. Simulations provide insight regarding the sensitivity, accuracy, and precision that can be expected for a given baseline data set, a given stock mixture in the fishery, and a given sample size from the fishery. As such, they are invaluable for defining appropriate sample sizes to be used.

Simulations also provide information regarding the reliability of estimates of individual stocks. In many (perhaps most) allozyme-based applications, among-stock differences are not large enough to provide reliable contribution estimates for individual stocks, so it is necessary to combine estimates for genetically similar stocks and report contribution estimates for groups of stocks instead of for individual stocks.

One consistent artifact of the MLE process is that it tends to overestimate the contribution of minor or noncontributing stocks and to underestimate the contributions of abundant stocks. A possible refinement of typical genetic MSA is to correct stock composition estimates for such bias provided one has an estimate of the bias (as can be obtained from appropriate simulation analyses).

After simulation analyses have been completed and the study design adjusted to take their results into account, samples from the various fishery strata must be obtained. It is imperative that the fishery sampling be as unbiased and representative of the fishery as possible. Samples should be proportional to harvests with regard to, for example, areas and times fished, sizes and sexes harvested, and gear types used. Although the genetic and statistical methods can be refined to provide maximally accurate and precise estimates, if the fishery sample does not represent the fishery, the resulting estimates will not accurately describe the fishery and will be of little or no value.

Genetic analysis of the fishery samples and subsequent MLE analysis of the baseline and fishery data are usually straightforward. The resulting estimates of stock or stock-group contributions to various fishery strata then can be used to determine harvest impacts or to shape subsequent fisheries or for other purposes, such as run-size estimation or analysis of migratory patterns and timing. Although it is common for such determinations to be done after a fishing season and the results applied to future years' fisheries, in-season genetic analyses have been conducted for several Pacific salmon fisheries and the results used immediately (see Box 13.3).

■ Box 13.3
Mixed-Stock Fishery Analysis of a Lower Columbia River Chinook Fishery

Several resource management agencies on the west coast of North America have used genetic methods based on allozyme data to analyze mixed-stock fisheries of various Pacific salmon species since the mid-1980s. One of the most visible of such applications has been that concerning the winter gill-net fishery for spring-run chinook salmon in the lower Columbia River. The experimental basis for this application was developed by NMFS-Seattle personnel. The fishery, which is jointly managed by Washington and Oregon, occurs in the lower 225 km of the river and is targeted on lower river hatchery stocks (predominantly those from the Willamette River). Nevertheless, because of the location and timing of the fishery, upper river spring-run stocks (from various Columbia and Snake river tributaries) also are intercepted. Conservation concerns due to the declining status of upper river runs have led to increasing restrictions on the fishery. Beginning in 1988, a federal court order established a ceiling for interceptions of upper river stocks at 5% of the run.

Staff at the Washington Department of Fisheries began using maximum likelihood estimation methods to determine stock-group contributions to this fishery in 1986. From 1988 through 1994, the duration of the fishery was adjusted based on in-season GSI estimates of upper river stock interceptions. The basis for this genetic application is summarized below.

A dendrogram summarizing the genetic interrelationships of stocks contributing to this fishery based on the electrophoretic analysis of 21 variable loci is shown on the next page. The large genetic distance between lower and upper river stocks revealed by this analysis provides the statistical power for estimating the contributions

continued on next page

■ Box 13.3 *continued*

of these two major stock groups of primary management concern. Much smaller genetic distances distinguish Snake River spring chinook salmon stocks (now listed as "threatened" under the ESA) from the upper Columbia River stocks and provide some basis for estimating their contributions (as a separate stock group) to fishery harvests. (See figure below.)

An example of the accuracy and precision expected for genetic estimates of stock-group contributions derived from simulations is presented in the table on the next page. The results of this analysis indicate high accuracy and precision of estimates based on fishery samples of only 180 fish.

The estimated stock-group contributions to the lower Columbia River chinook salmon gill-net fishery in three different years are summarized in the figure on the next page.

continued on next page

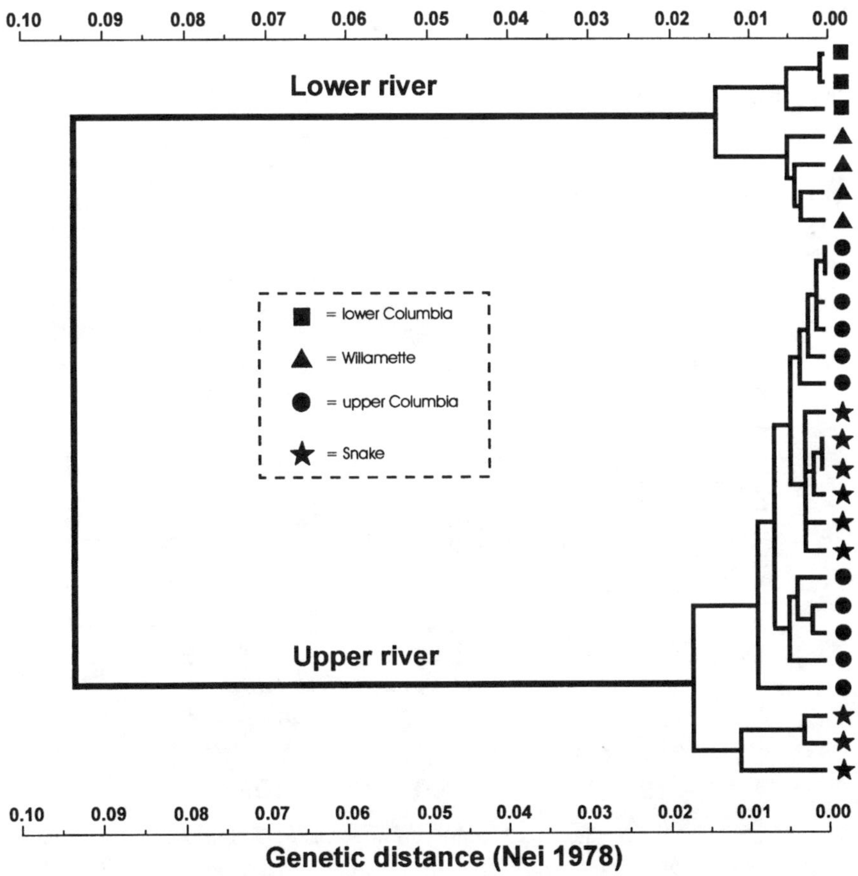

Dendrogram showing genetic interrelationships of lower and upper Columbia River spring chinook salmon stocks. Baseline data are from Waples et al. (1993); Waples, NMFS-Seattle, unpublished; and Washington Department of Fisheries, unpublished.

■ Box 13.3 *continued*

Expected accuracy and precision of GSI estimates of stock-group contributions to the lower Columbia River chinook salmon fishery. These estimates are based on simulation analysis for three different hypothetical stock mixtures and two sample sizes. Estimates are expressed as percent contribution (– 1 SD), *N* is the sample size from the fishery, and *T* is the true (simulated) proportions in each mixture.

	Mixture 1			Mixture 2			Mixture 3		
		Estimated contribution			Estimated contribution			Estimated contribution	
	T	*N* = 180	*N* = 540	T	*N* = 180	*N* = 540	T	*N* = 180	*N* = 540
Lower river	71	71 (3.2)	72 (2.4)	82	83 (2.9)	82 (2.0)	96	96 (1.3)	96 (0.8)
Lower Columbia River	8	9 (3.1)	9 (2.9)	0	3 (2.6)	3 (1.8)	8	10 (3.9)	10 (2.4)
Willamette River	63	62 (3.9)	63 (3.6)	82	80 (3.7)	79 (2.4)	88	86 (4.1)	86 (2.8)
Upper river	29	29 (3.2)	28 (2.4)	18	17 (2.9)	18 (2.0)	4	4 (1.3)	4 (0.8)
Upper Columbia River	20	20 (4.7)	20 (2.8)	11	10 (3.3)	11 (2.6)	3	2 (1.4)	3 (0.9)
Snake River	9	10 (3.8)	9 (2.5)	7	7 (2.9)	7 (2.0)	1	2 (1.2)	1 (0.7)

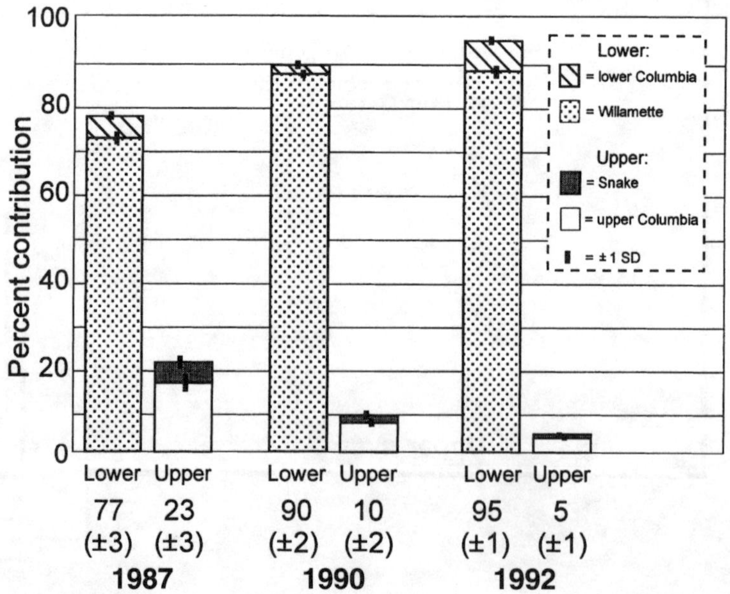

Actual in-season stock group contribution estimates (±SD) for lower Columbia River chinook salmon gill-net fishery harvests.

Recently, microsatellite DNA data have been used to refine mixture analysis of salmonid fishes from the level of estimating stock group proportions to the assignment of individuals in a mixture to their specific stock of origin with considerable power and accuracy (Banks et al. 2000). The very high levels of genetic variation (high heterozygosities and large numbers of alleles) and significant divergence among populations exhibited by many microsatellite loci allow such assignment-based analyses of stock mixtures. This approach has been used to assign individuals from admixed baseline collections, mixed-stock fishery samples, and other mixed samples to their stock of origin (Banks et al. 2000; Olsen et al. 2000). It has also been used to assign parentage to individual offspring in hatchery populations (Estoup et al. 1998). The power of microsatellite-DNA-based analyses to determine parentage of individual offspring is also currently being used by several research groups to estimate the relative reproductive success of hatchery and wild fish.

13.4.3 *Approaches to Uncertainty: Risk Assessment and Management*

One important use of GSI is to identify risk where genetic diversity might be lost, for example, in ESA considerations (National Research Council 1995). However, genetic analyses rarely provide unambiguous answers to such conservation and fishery problems. Uncertainty arises from problems in measuring natural variability, in expanding from the scale at which the measurements were made to the scale at which management will occur, in using inadequate models to predict the results of making specific decisions and from the inherent unpredictability of biological, environmental, and other natural events (Lemons 1996). Consequently, it is important for decision makers to know the sources, magnitude, and implications of uncertainties. Genetic risk assessment is the process of identifying and describing vulnerability to genetic changes, such as losses of genetic diversity and fitness. Genetic risk management means reducing genetically based threats while maximizing any associated benefits.

13.4.3.1 **Components of Risk Assessment.** Risk assessments usually have four steps. The first step is the identification of events that are likely to lead to losses of genetic diversity or fitness. In managed fish populations, these may be classified into four general kinds of hazards: (1) extinction, (2) loss of genetic diversity within populations, (3) loss of genetic diversity among populations, and (4) domestication, or loss of fitness in the wild from artificial propagation (Busack and Currens 1995). The second step in risk assessment is estimation of the likelihood, or probability, of each event. The third step is an evaluation of the consequences, in terms of losses, given the likelihood of a hazardous event. This is often called the vulnerability or risk, although "risk" can also be confused with the probability of an event. Finally, the fourth step is an audit of what kind of protective actions could be implemented as risk management. Risk assessment in the context of hatchery management is discussed in Chapter 14.

Based on these steps, the most basic heuristic relationship between vulnerability (V), probability (p), and loss (L) can be expressed as (Smith 1992):

$$V = p \times L. \tag{13.11}$$

In actual application, formulas are more complex because they take into account the distributions of possible p and L values.

When the losses of interest are always the same, as in the case of extinction, then the risk assessment may be based on only the frequency of occurrences. This is one of the basic approaches to population viability analysis, which is a special kind of risk assessment (Chapter 18). By simulating population growth for a specified period of time under a variety of conditions, the frequency of extinctions can be tallied, and risk of extinction can be expressed as a probability of extinction in a specific period of time (Burgman et al. 1993).

13.4.3.2 **Types of Risk Assessment.** Not all situations lend themselves to statistical treatments of vulnerability. In many cases, biologists do not have a historical database from which they can associate the probability of occurrence of an event with the losses that could occur. In some cases, they may know too little about the dynamics of the populations to construct realistic models. In other cases, decision makers may not need elaborate analyses to make a decision. In general, then, the possible kinds of risk assessment depend on the amount of uncertainty that is present for a system and the level of precision that is needed to make a decision (Figure 13.2). When uncertainty is high and precision is low, qualitative descriptions may be all that is appropriate. In contrast, when uncertainty is low because biologists have adequate data and models to understand cause and effect relationships, models can be used deterministically to predict the consequences with high precision. Warnings are useful when uncertainty is low but great precision is not necessary. The warnings on cigarette packages are an example of a situation in which scientific uncertainty is low but greater precision is not required.

For genetic hazards other than extinction, predetermined thresholds for judging losses may not exist because too little is known about the biological system. This poses a problem for risk assessment. Recognizing the uncertainty, decision makers may want more precision than descriptions or warnings provide; at the same time, however, risk assessments are limited by biological uncertainty. A variety of approaches can provide descriptive risk assessments with confidence bounds. These include expert panels, event tree analyses, and probabilistic modeling.

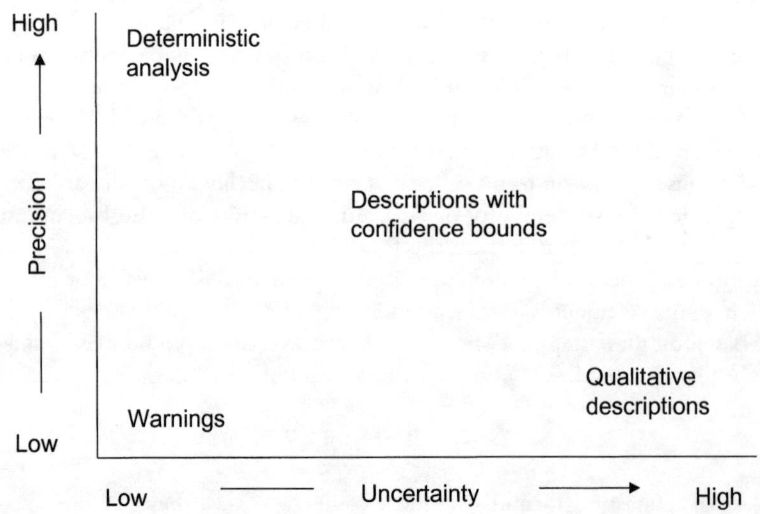

Figure 13.2 Types of risk assessment appropriate to varying levels of management uncertainty and precision (modified from Fiksel and Covello (1986)).

Expert panels use a group of experts on a subject to arrive at risk assessments. The selection of the experts and the process by which they arrive at estimates of risk are two important factors in successfully using expert panels. Experts may be chosen from outside an agency when the agency does not have the expertise to conduct its own risk assessment or when the issue is controversial and it is important to obtain public confidence regarding scientific objectivity and rigor. Because many natural resource issues are complex, it also can be important to include a variety of experts that have been carefully chosen to represent all the areas of uncertainty. For genetic issues in fisheries, for example, this might include population geneticists, quantitative geneticists, fisheries managers, fish culturists, and fish ecologists. Likewise, the process by which expert panels arrive at a risk assessment also is carefully structured. One that is widely used is the Delphi process (Linstone and Turoff 1975). In this process, experts are asked to provide answers to carefully structured questions in an anonymous, iterative format. The experts' answers then can be used to estimate central tendencies and dispersion associated with risk.

Event tree analysis can be useful when a known chain of events must occur to produce a disaster. This is easy to imagine for mechanical or electrical systems, such as might exist in hatcheries (Figure 13.3). If a pump fails in a fish hatchery, for example, the probability that fish may die depends on whether the warning system detects the failure, the hatchery technician sees the warning and responds appropriately, and the back-up pump starts in time and functions to maintain oxygen levels in the tanks. Loss of genetic diversity in natural systems associated with different management strategies is much more complex, but it usually consists of a failure somewhere in a series of four types of events: (1) specific management direction or guidelines for how to maintain genetic diversity, (2) equipment or logistical planning, (3) natural variability that confounds the intended management action, and (4) ability of technicians, biologists, or managers to complete the tasks or to recognize a failure in the preceding events and take the appropriate action. Reducing potential disasters to a series of events helps biologists and decision makers understand vulnerability. In addition, when it is possible to use historical data or experts to estimate probabilities for each independent event, the probability of success is

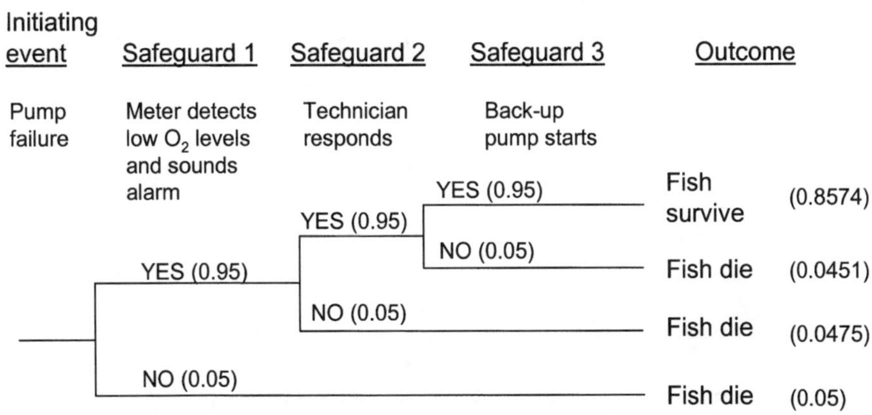

Figure 13.3 Hypothetical event tree analysis for pump failure in a trout hatchery. Probabilities of each outcome are in parentheses. Probabilities of the final outcome are the product of the probabilities of each independent event. Normally, we would round the final probabilities to the lowest level of precision used in the analysis.

a product of the probabilities of each independent event. When not all the probabilities can be estimated, ranking the events by likelihood of occurrence and attempting to estimate the probability of success of the one event that was ranked least successful can provide an upper estimate of the success of the project as a whole.

Probabilistic modeling is another approach for describing vulnerability. Genetic theory provides many mathematical models for understanding loss of genetic diversity and fitness due to genetic drift, gene flow, or natural selection and mutation under a set of specific conditions (Chapters 6–9). When these conditions hold and the values of the parameters are known, these models provide precise predictions. However, one or more of the conditions for prediction may not exist in natural populations, and natural variability makes it difficult to assign a single value to a parameter. These conditions appear to occur randomly because we are unable to predict them from the available information. Consequently, the actual outcome is more likely to fall within a certain set of values. Probabilistic modeling uses computer simulations to estimate the distribution of possible outcomes. **Monte Carlo simulations** are a class of techniques that involve using random numbers generated by a computer to simulate stochastic processes. This is done by solving the mathematical equation that describes the potential genetic changes many times with randomly chosen values each time and keeping track of the outcomes.

13.4.3.3 Reliability and Resilience in Risk Management. Risk assessment often means describing the status of a population or species. However, decision makers may also want to know the risks associated with different kinds of management strategies, such as translocating fish, starting a captive broodstock program, or increasing harvest. These decisions involve risk management as well as risk assessment. Equation (13.11) indicates that vulnerability can be reduced by lowering the probability that an event happens and by reducing the losses that occur from the event. Consequently, risk management usually involves components of *reliability* and components of *resilience* (Currens and Busack 1995). Safeguards that reduce the probability of an event promote reliability, whereas safeguards that reduce potential losses promote resilience. Genetic stock identification, for example, is a safeguard that can be used to reduce the probability that different stocks will be unknowingly mixed (in a hatchery broodstock) or that fisheries will inadvertently drive a small, unidentified stock to extinction when it occurs in the same time and place as a large, productive stock. In contrast, genetic monitoring programs and reserves are examples of safeguards that promote resilience.

13.5 Summary

Detailed genetic information is an essential prerequisite for meaningful, genetically based fishery management and conservation programs. Methods of GSI are used when developing stock inventories and descriptions because they provide the most direct measures we have of reproductive isolation and genetic differentiation. These methods have four important steps: (1) sampling programs are designed based on the biology of the species, the physical environment, and the biogeographical history of the region; (2) sufficient numbers and kinds of genetic traits that are likely to be informative are identified; (3) data are collected; and (4) data are analyzed to detect interrelationships of stocks and to quantify how genetic diversity is distributed.

For fishery management, an important use of genetic data is to identify what stocks are being caught. If appropriate baseline genetic data have been collected, MLE procedures can be used to estimate statistically the proportions of potentially contributing stocks that best approximate the distribution of genotypes seen in the fishery.

Another important use of GSI is to identify where genetic diversity might be lost. Genetic risk assessment is the process of identifying and describing vulnerability to extinction, loss of genetic diversity within populations, loss of genetic diversity among populations, and loss of fitness in the wild from artificial propagation or other human intervention. Risk or vulnerability to loss of genetic diversity and fitness is the product of the probability of a hazardous event and the losses that would be incurred. Both the amount of uncertainty that is present and the level of precision that is needed to make management decisions help determine the kind of risk assessment that is appropriate. Risk assessments may use expert panels, event tree analysis, or probabilistic modeling to estimate uncertainty and arrive at useful levels of precision. Vulnerability then can be reduced by reducing the probability that an event happens and by reducing the losses that occur from the event.

13.6 Acknowledgments

We thank our colleagues, especially those at WDFW and the Northwest Indian Fisheries Commission, for their numerous intellectual contributions to ideas about stock inventories, genetic diversity assessment, and genetic risk issues expressed in this chapter. We also acknowledge the substantial contributions of Anne Marshall, Marc Miller, Paul Hirose, and others involved in the GSI analysis of the lower Columbia River chinook salmon fishery described herein. We thank NMFS-Seattle staff, especially Robin Waples and David Teel, for allowing us to use baseline data for seven Snake River stocks and one Columbia River stock in this fishery analysis. We thank Jim Ames and Don Hendrick for their contributions to the pink salmon GDU analysis, Sewall Young and his staff for field collections, and all those who contributed to the SASSI process and documents. Funds from the U.S. Anadromous Fish Act, the Pacific Salmon Treaty Act, the Pacific Salmon Commission, and the State of Washington General Fund supported the fishery investigations and the genetic inventory (SASSI) and diversity assessment of pink salmon described herein.

13.7 References

Angers, B., L. Bernatchez, A. Angers, and L. Desgroseillers. 1995. Specific microsatellite loci for brook trout reveal strong population subdivision on a microgeographic scale. Journal of Fish Biology 47 (Supplement A):177–185.

Avise, J. C. 1994. Molecular markers, natural history and evolution. Chapman and Hall. New York.

Ayvazian, S. G., M. S. Johnson, and D. J. McGlashan. 1994. High levels of genetic subdivision of marine and estuarine populations of the estuarine catfish *Cnidoglanis macrocephalus* (Plotosidae) in southwestern Australia. Marine Biology 118:25–31.

Banks, M. A., V. K. Rashbrook, M. J. Calavetta, C. A. Dean, and D. Hedgecock. 2000. Analysis of microsatellite DNA resolves genetic structure and diversity of chinook salmon *(Oncorhynchus tshawytscha)* in California's Central Valley. Canadian Journal of Fisheries and Aquatic Sciences 57:915–927.

Beacham, T. D. 1996. The use of minisatellite DNA variation for stock identification of chum salmon, *Oncorhynchus keta*. U.S. National Marine Fisheries Service Fishery Bulletin 94:611–626.

Beacham, T. D., S. Pollard, and K. D. Le. 1999. Population structure and stock identification of steelhead in southern British Columbia, Washington, and the Columbia River based on microsatellite DNA variation. Transactions of the American Fisheries Society 128:1068–1084.

Beacham, T. D., R. E. Withler, and C. C. Wood. 1995. Stock identification of sockeye salmon by means of minisatellite DNA variation. North American Journal of Fishery Management 15:249–265.

Bentzen, P., C. T. Taggart, D. E. Ruzzante, and D. Cook. 1996. Microsatellite polymorphism and the population structure of Atlantic cod *(Gadus morhua)* in the northwest Atlantic. Canadian Journal of Fisheries and Aquatic Sciences 53:2706–2721.

Benzie, J. A. H., and S. T. Williams. 1995. Gene flow among giant clam *(Tridacna gigas)* populations in Pacific does not parallel ocean circulation. Marine Biology 123:781–787.

Bernatchez, L. 2001. The evolutionary history of brown trout *(Salmo trutta)* inferred from phylogeographic, nested clade, and mismatch analysis of mitochondrial DNA variation. Evolution 55:351–379.

Bernatchez, L., and S. Martin. 1996. Mitochondrial DNA diversity in anadromous rainbow smelt, *Osmerus mordax* Mitchell: a genetic assessment of the member-vagrant hypothesis. Canadian Journal of Fisheries and Aquatic Sciences 53:424–433.

Berst, A. H., and R. C. Simon, editors. 1981. Proceedings of the stock concept international symposium (STOCS). Canadian Journal of Fisheries and Aquatic Sciences 38:1457–1923.

Bowen, B. W., N. Kamezaki, C. J. Limpus, G. R. Hughes, A. B. Meylan, and J. C. Avise. 1994. Global phylogeography of the loggerhead turtle *(Caretta caretta)* as indicated by mitochondrial DNA haplotypes. Evolution 48:1820–1828.

Burgman, M. A., S. Ferson, and H. R. Akçakaya. 1993. Risk assessment in conservation biology. Chapman and Hall, London.

Busack, C. A., and K. Currens. 1995. Genetic risks and hazards in hatchery operations: fundamental concepts and issues. Pages 71–80 *in* H. L. Schramm, Jr., and R. G. Piper, editors. Uses and effects of cultured fishes in aquatic ecosystems. American Fisheries Society, Symposium 15, Bethesda, Maryland.

Cavalli-Sforza, L. L., and A. W. F. Edwards. 1967. Phylogenetic analysis: models and estimation procedures. American Journal of Human Genetics 19:233–257.

Chakraborty, R., M. Haag, N. Ryman, and G. Stahl. 1982. Hierarchical gene diversity analysis and its application to brown trout population data. Hereditas 97:17–21.

Chenoweth, S. F., and J. M. Hughes. 1997. Genetic population structure of the catadromous perciform: *Macquaria novemaculeata* (Percichthyidae). Journal of Fish Biology 50:721–733.

Chow, S., H. Okamoto, Y. Uozumi, Y. Takeuchi, and H. Takeyama. 1997. Genetic stock structure of the swordfish *(Xiphias gladius)* inferred by PCR-RFLP analysis of the mitochondrial DNA control region. Marine Biology 127:359–367.

Christiansen, F. B., O. Frydenberg, J. P. Hjorth, and V. Simonsen. 1976. Genetics of *Zoarces* populations. IX. Geographic variation at the three phosphoglucomutase loci. Hereditas 83:245–256.

Currens, K. P., and C. A. Busack. 1995. A framework for assessing genetic vulnerability. Fisheries 20(12):24–31.

Dizon, A. E., C. Lockyer, W. F. Perrin, D. P. DeMaster, and J. Sisson. 1992. Rethinking the stock concept: a phylogenetic approach. Conservation Biology 6:24–36.

Dunn, G., and B. S. Everitt. 1982. An introduction to mathematical taxonomy. Cambridge Studies in Mathematical Biology 5. Cambridge University Press, Cambridge, UK.

Elliott, N. G., and R. D. Ward. 1994. Enzyme variation in jackass morwong, *Nemadactylus macropterus* (Schneider, 1801) (Teleostei: Cheilodactylidae), from Australian and New Zealand waters. Australian Journal of Marine and Freshwater Research 45:51–67.

Estoup, A., K. Gharbi, M. SanCristobal, C. Chevalet, P. Haffray, and R. Guyomard. 1998. Parentage assignment using microsatellites in turbot *(Scophthalmus maximus)* and rainbow trout *(Oncorhynchus mykiss)* hatchery populations. Canadian Journal of Fisheries and Aquatic Sciences 55:715–725.

Farris, J. S. 1972. Estimating phylogenetic trees from distance matrices. American Naturalist 106:645–668.

Fiksel, J., and V. T. Covello. 1986. The suitability and applicability of risk assessment methods for environmental applications of biotechnology. Pages 1–34*in* J. Fiksel and V. T. Covello, editors. Biotechnology risk assessment: issues and methods for environmental introductions. Pergamon Press, New York.

Fournier, D. A., T. D. Beacham, B. E. Riddell, and C. A. Busack. 1984. Estimating stock composition in mixed stock fisheries using morphometric, meristic, and electrophoretic characteristics. Canadian Journal of Fisheries and Aquatic Sciences 41:400–408.

Goldstein, D. B., A. R. Linares, M. W. Feldman, and L. L. Cavalli-Sforza. 1995. An evaluation of genetic distances for use with microsatellite loci. Genetics 139:463–471.

González-Villaseñor, L. I., and D. A. Powers. 1990. Mitochondrial-DNA restriction-site polymorphisms in the teleost *Fundulus heteroclitus* support secondary intergradation. Evolution 44:27–37.

Graves, J. E., and J. R. McDowell. 1995. Interocean genetic divergence of istiophorid billfishes. Marine Biology 122:193–204.

Grewe, P. M., N. G. Elliott, B. H. Innes, and R. D. Ward. 1997. Genetic population structure of southern bluefin tuna *(Thunnus maccoyii)*. Marine Biology 127:555–561.

Grewe, P. M., A. J. Smolenski, and R. D. Ward. 1994. Mitochondrial DNA diversity in jackass morwong (*Nemadactylus macropterus*: Teleostei) from Australian and New Zealand waters. Canadian Journal of Fisheries and Aquatic Sciences 51:1101–1109.

Hard, J. J., R. G. Kope, W. S. Grant, F. W. Waknitz, L. T. Parker, and R. S. Waples. 1996. Status review of pink salmon from Washington, Oregon, and California. NOAA (National Oceanic and Atmospheric Adminsitration) Technical Memorandum NMFS (National Marine Fisheries Service) NWFSC-25, Northwest Fisheries Science Center, Seattle.

Hillis, D. M. 1984. Misuse and modification of Nei's genetic distance. Systematic Zoology 33:238–240.

Johnson, M. S., and L. M. Joll. 1993. Genetic subdivision of the pearl oyster *Pinctada maxima* (Jameson, 1901) (Mollusca: Pteriidae) in northern Australia. Australian Journal of Marine and Freshwater Research 44:519–526.

Karl, S. A., B. W. Bowen, and J. C. Avise. 1992. Global population genetic structure and male-mediated gene flow in the green turtle *(Chelonia mydas)*: RFLP analyses of anonymous nuclear loci. Genetics 131:163–173.

Kruskal, J. B., Jr. 1956. On the shortest spanning subtree of a graph and the traveling salesman problem. Proceedings of the American Mathematical Society 7:48–50.

Lafontaine, P., and J. J. Dodson. 1997. Intraspecific genetic structure of white sucker *(Catostomus commersoni)* in northeastern North America as revealed by mitochondrial DNA polymorphism. Canadian Journal of Fisheries and Aquatic Sciences 54:555–565.

Lavery, S., C. Moritz, and D. R. Fielder. 1996. Indo-Pacific population structure and evolutionary history of the coconut crab *Birgus latro*. Molecular Ecology 5:557–570.

Lemons, J. (editor). 1996. Scientific uncertainty and environmental problem solving. Blackwell Scientific Publications, Cambridge, Massachusetts.

Linstone, H. A. and M. Turoff. 1975. The Delphi method: techniques and applications. Addison-Wesley, Reading, Massachusetts.

Lu, G.-Q., S. Li, and L. Bernatchez. 1997. Mitochondrial DNA diversity, population structure, and conservation genetics of four native carps within the Yangtze River, China. Canadian Journal of Fisheries and Aquatic Sciences 54:47–58.

McMillan, W. O., R. A. Raff, and S. R. Palumbi. 1992. Population genetic consequences of developmental evolution in sea urchins (Genus *Heliocidaris*). Evolution 46:1299–1312.

Millar, R. B., 1987. Maximum likelihood estimation of mixed stock fishery composition. Canadian Journal of Fisheries and Aquatic Sciences 44:583–590.

Millar, R. B. 1990. Comparison of methods for estimating mixed stock fishery composition. Canadian Journal of Fisheries and Aquatic Sciences 47:2235–2241.

National Research Council. 1995. Estimating risk. Chapter 7 *in* Science and the Endangered Species Act. National Academy Press, Washington, D.C.

Nei, M. 1972. Genetic distance between populations. American Naturalist 106:283–292.

Nei, M. 1978. Estimation of average heterozygosity and genetic distance from a small number of individuals. Genetics 89:583–590.

Nei, M. 1987. Genetic distance and molecular phylogeny. Pages 193–223 *in* N. Ryman and F. Utter, editors. Population genetics and fishery management. University of Washington Press, Seattle.

Nielsen, J. L., editor. 1995. Evolution and the aquatic ecosystem: defining unique units in population conservation. American Fisheries Society, Symposium 17, Bethesda, Maryland.

Olsen, J. B., P. Bentzen, M. A. Banks, J. B. Shaklee, and S. Young. 2000. Microsatellites reveal population identity of individual pink salmon to allow supportive breeding of a population at risk of extinction. Transactions of the American Fisheries Society 129:232–242.

Phelps, S. R., L. L. LeClair, S. Young, and H. L. Blankenship. 1994. Genetic diversity patterns of chum salmon in the Pacific Northwest. Canadian Journal of Fisheries and Aquatic Sciences 51 (Supplement 1):65–83.

Philipp, D. P., W. F. Childers, and G. S. Whitt. 1983. A biochemical genetic evaluation of the northern and Florida subspecies of largemouth bass. Transactions of the American Fisheries Society 112:1–20.

Reeb, C. A., and J. C. Avise. 1990. A genetic discontinuity in a continuously distributed species: mitochondrial DNA in the American oyster, *Crassostrea virginica*. Genetics 124:397–406.

Richardson, B. J., P. R. Baverstock, and M. Adams. 1986. Allozyme electrophoresis. Academic Press, Sydney.

Rogers, J. S. 1972. Measures of genetic similarity and genetic distance. University of Texas Publication 7213:145–153.

Rohlf, F. J. 1994. NTSYS-pc. Numerical taxonomy and multivariate analysis system, version 1.80. Exeter Software, Setauket, New York.

Ropson, I. J., D. C. Brown, and D. A. Powers. 1990. Biochemical genetics of *Fundulus heteroclitus* (L.). VI. Geographical variation in the gene frequencies of 15 loci. Evolution 44:16–26.

Ruzzante, D. E., C. T. Taggart, D. Cook, and S. Goddard. 1996. Genetic differentiation between inshore and offshore Atlantic cod (*Gadus morhua* L.) off Newfoundland: microsatellite DNA variation and antifreeze level. Canadian Journal of Fisheries and Aquatic Sciences 53:634–645.

Ruzzante, D. E., C. T. Taggart, and D. Cook. 1998. A nuclear DNA basis for shelf- and band-scale population structure in northwest Atlantic cod (*Gadus morhua*): Labrador to Georges Bank. Molecular Ecology 7:1663–1680.

Ryman. N. 1991. Conservation genetics considerations in fishery management. Journal of Fish Biology 39 (Supplement A):211–224.

Ryman, N., and F. Utter, editors. 1987. Population genetics and fishery management. University of Washington Press, Seattle.

Saitou, N., and M. Nei. 1987. The neighbor-joining method: a new method for reconstructing phylogenetic trees. Molecular Biology and Evolution 4:406–425.

Sakaizumi, M., K. Moriwaki, and N. Egami. 1983. Allozyme variation and regional differentiation in wild populations of the fish *Oryzias latipes*. Copeia 1983:311–318.

Shaklee, J. B., T. Beacham, L. Seeb, and B. A. White. 1999. Managing fisheries using genetic data: case studies from four species of Pacific salmon. Fisheries Research 43:45–78.

Shaklee, J. B., and P. Bentzen. 1998. Genetic identification of stocks of marine fish and shellfish. Bulletin of Marine Science 62:589–621.

Shaklee, J. B., C. Busack, A. Marshall, M. Miller, and S. R. Phelps. 1990. The electrophoretic analysis of mixed-stock fisheries of Pacific salmon, Pages 235–265 *in* Z.-I. Ogita and C. L. Markert, editors. Isozymes: structure, function, and use in biology and medicine. Progress in Clinical and Biological Research 344. Wiley-Liss, New York.

Shaklee, J. B., and S. R. Phelps. 1990. Operation of a large-scale, multiagency genetic stock identification program. Pages 817–830 *in* N. C. Parker, A. E. Giorgi, R. C. Heidinger, D. B. Jester, Jr., E. D. Prince, and G. A. Winans, editors. Fish-marking techniques. American Fisheries Society, Symposium 7, Bethesda, Maryland.

Shaklee, J. B., J. Salini, and R. N. Garrett. 1993. Electrophoretic characterization of multiple genetic stocks of barramundi perch in Queensland, Australia. Transactions of the American Fisheries Society 122:685–701.

Shulman, M. J., and E. Bermingham. 1995. Early life histories, ocean currents, and the population genetics of Caribbean reef fishes. Evolution 49:897–910.

Slatkin, M. 1995. A measure of population subdivision based on microsatellite allele frequencies. Genetics 139:457–462.

Small, M. P., T. D. Beacham, R. E. Withler, and R. J. Nelson. 1998. Discriminating coho salmon *(Oncorhynchus kisutch)* populations within the Fraser River, British Columbia, using microsatellite DNA markers. Molecular Ecology 7:141–155.

Smith, K. 1992. Environmental hazards: assessing risk and reducing disaster. Routledge, Chapman and Hall, New York.

Smith, P. J., C. D. Roberts, S. M. McVeagh, and P. G. Benson. 1996. Genetic evidence for two species of tarikihi (Teleostei: Cheilodactylidae: *Nemadactylus*) in New Zealand waters. New Zealand Journal of Marine and Freshwater Research 30:209–220.

Sneath, P. H. A., and R. R. Sokal. 1973. Numerical taxonomy. Freeman, London.

Stepien, C. A. 1995. Population genetic divergence and geographic patterns from DNA sequences: examples from marine and freshwater fishes. Pages 263–287 *in* J. L. Nielsen, editor. Evolution and the aquatic ecosystem: defining unique units in population conservation. American Fisheries Society, Symposium 17, Bethesda, Maryland.

Swofford, D. L., G. J. Olsen, P. J. Waddell, and D. M. Hillis. 1996. Phylogenetic inferences. Pages 407–514 *in* D. M. Hillis, C. Moritz, and B. K. Mable, editors. Molecular systematics. 2nd edition. Sinauer Associates, Sunderland, Massachusetts.

Swofford, D. L., and R. B. Selander. 1981. BIOSYS-1: a FORTRAN program for the comprehensive analysis of electrophoretic data in population genetics and systematics. Journal of Heredity 72:281–283.

Taylor, E. B., T. D. Beacham, and M. Kaeriyama. 1994. Population structure and identification of north Pacific Ocean chum salmon *(Oncorhynchus keta)* revealed by an analysis of minisatellite DNA variation. Canadian Journal of Fisheries and Aquatic Sciences 51:1430–1442.

Taylor, E. B., C. J. Foote, and C. C. Wood. 1996. Molecular genetic evidence for parallel life-history evolution within a Pacific salmon (sockeye salmon and kokanee, *Oncorhynchus nerka*). Evolution 50:401–416.

Utter, F., G. Milner, G. Stahl, and D. Teel. 1989. Genetic population structure of chinook salmon, *Oncorhynchus tshawytscha,* in the Pacific Northwest. U.S. National Marine Fisheries Service Fishery Bulletin 87:239–264.

Utter, F., and N. Ryman. 1993. Genetic markers and mixed stock fisheries. Fisheries 18(8):11–21.

Waples, R. S. 1991. Definition of a "species" under the Endangered Species Act: application to Pacific salmon. NOAA (National Oceanic and Atmospheric Administration) Technical Memorandum NMFS (National Marine Fisheries Service) F/NWC-194, Northwest Fisheries Science Center, Seattle.

Waples, R. S., and six coauthors. 1993. A genetic monitoring and evaluation program for supplemented populations of salmon and steelhead in the Snake River Basin. Bonneville Power Administration Annual Report 1992 (project 89-096), Portland, Oregon.

Washington Department of Fisheries, Washington Department of Wildlife, and Western Washington Treaty Indian Tribes. 1993. 1992 Washington State salmon and steelhead stock inventory (SASSI). Washington Department of Fisheries, Olympia.

Weir, B. S. 1996. Genetic data analysis II. Sinauer Associates, Sunderland, Massachusetts.

Wenburg, J. K., P. Bentzen, and C. J. Foote. 1998. Microsatellite analysis of genetic population structure in an endangered salmonid: the coastal cutthroat trout *(Oncorhynchus clarki clarki).* Molecular Ecology 7:733–749.

Wood, C. C. 1995. Life history variation and population structure in sockeye salmon. Pages 195–216 *in* J. L. Nielsen, editor. Evolution and the aquatic ecosystem: defining unique units in population conservation. American Fisheries Society, Symposium 17, Bethesda, Maryland.

Wood, C. C., S. McKinnell, T. J. Mulligan, and D. A. Fournier. 1987. Stock identification with the maximum likelihood mixture model: sensitivity analysis and application to complex problems. Canadian Journal of Fisheries and Aquatic Sciences 44:866–881.

Wood, C. C., B. E. Riddell, D. T. Rutherford, and R. E. Withler. 1994. Biochemical genetic survey of sockeye salmon *(Oncorhynchus nerka)* in Canada. Canadian Journal of Fisheries and Aquatic Sciences 51 (Supplement 1):114–131.

Wright, S. 1978. Evolution and the genetics of populations. Volume 4. Variability within and among natural populations. University of Chicago Press, Chicago.

14

Genetic Guidelines for Hatchery Supplementation Programs

Loren M. Miller and Anne R. Kapuscinski

14.1 Introduction

Stocking of hatchery-reared fishes is an important component of fisheries management programs (Schramm and Piper 1995). Ecologists and population geneticists, however, have raised concerns about the possible effects of such programs on native fish populations, leading to great debate over the appropriate role of hatcheries in modern fisheries management (Incerpi 1996; Grimes 1998; Waples 1999). In the past, hatcheries typically were used simply to produce fish, either to create or support a fishery or to mitigate habitat loss. Now hatcheries increasingly are called upon to perform a new role—to rehabilitate depressed natural populations. This new role is often called **supplementation,** which we define as the use of artificial propagation, while conserving genetic resources, for the goal of restoring or augmenting self-sustaining populations (Kapuscinski et al. 1991). Other definitions have been given (e.g., Cuenco et al. 1993), but all have two common themes: increasing population size and avoiding unnatural alteration of genetic resources. The two alternative goals of supplementation, restoration or augmentation, reflect the status of the population. **Restoration** requires the re-establishment of self-sustaining populations in habitat no longer occupied by that species. **Augmentation** requires the rehabilitation of a depressed population by integrating naturally and hatchery-produced fish to maximize the seeding of appropriate habitat by naturally reproducing adults.

A fundamental premise underlying this chapter is that hatchery supplementation programs, when properly designed and implemented, provide an important tool for the rehabilitation of depleted fish populations. Three caveats, however, must be kept in mind. First, without proper adherence to genetic, evolutionary, and ecological principles, intended integration of hatchery and naturally reproducing fish may lead to a number of adverse consequences for naturally reproducing fish, eventually damaging restoration efforts and jeopardizing sustainable use of the resource over the long term. Many past attempts to enhance wild populations via artificial propagation programs that did not adhere to important genetic and ecological principles have produced few benefits; in some cases, these attempts may have contributed to the decline of wild populations (Nickelson et al. 1986; Miller et al. 1990; Reisenbichler and Rubin 1999). Second, hatchery supplementation is still an unproven technology and should be conducted with a commitment to the concept of adaptive management. **Adaptive management** requires explicit design and implementation of management actions or programs as experiments. Furthermore, it requires regular monitoring for obtaining valid experimental data and for tracking progress toward program goals and objectives. Adaptive management also requires systematic evaluation of actions and, most importantly, implementation of adaptive changes (midcourse corrections) based on conclusions drawn from such evaluations (Walters 1986). Finally, hatchery supplementation must be viewed as only one part of a comprehensive strategy to remedy factors causing decline or impeding recovery. A comprehensive strategy also must address habitat availability and quality, impediments to fish movements, and harvest.

This chapter examines the genetic hazards imposed by supplementation hatchery programs and provides genetic guidelines to reduce the impact of these hazards. First we describe the four genetic hazards imposed by hatchery programs as defined by Busack and Currens (1995). We then divide hatchery programs into their four main components and for each component discuss the potential for hatchery activities to cause one of the four genetic hazards. In addition, we provide guidelines to reduce the genetic risks imposed by hatchery activities. These guidelines are designed specifically for supplementation programs, but many of the principles discussed will also apply to more traditional hatchery programs in which hatchery fish are temporally or spatially separated from natural fish.

Many of the examples we provide are from studies of salmonids. This is because literature on hatchery uses and effects is more extensive for salmonids than for most other taxa, especially literature relating to hatchery supplementation. The genetic principles and guidelines, however, apply to any cultured species, but distinct life history characteristics must be considered (Utter 1998). The best advice to any manager is to understand basic genetic principles that apply to hatchery operations and collaborate with a geneticist to determine an appropriate approach for your species of interest.

This chapter contains genetic guidelines only; costs, technical feasibility, harvest demands, and sociopolitical dynamics are among other factors that will affect decision making in hatchery supplementation programs. If the long-term sustainability of managed populations is to be restored and maintained, however, genetic conservation must be given high priority. This priority is imperative because long-term sustainable use of fish populations depends greatly on maintaining populations' capacity to evolve in natural environments, and such evolution depends on maintaining their genetic diversity (Reisenbichler 1997).

14.2 Genetic Concepts Relevant to Hatchery Programs

The goal of conserving genetic resources in fish populations requires maintenance of similar genetic resources and life history patterns in hatchery fish and wild or natural populations with which hatchery fish may genetically interact (Kapuscinski 1991; Hard 1995). Throughout this chapter, we make a distinction between wild and natural fish. **Wild fish** are those with no hatchery-produced ancestors. **Natural fish** are those with hatchery ancestors that are now spawning in the natural environment. Fisheries managers should be aware of hatchery-related genetic hazards that are responsible for loss or alteration of genetic variation in populations (Chapter 13). Busack and Currens (1995) presented four types of genetic hazards in relation to hatchery activities: (1) extinction, (2) loss of within-population genetic variation, (3) loss of between-population genetic variation (**population identity**), and (4) domestication selection. They further distinguished genetic risk from hazards. **Genetic hazards** are the adverse consequences listed above and **genetic risk** is the probability that these hazards might occur. Here we explain each of the four hazards and the genetic processes that affect them.

14.2.1 *Extinction*

Extinction is ultimately a demographic process. It is the least likely immediate hazard in hatchery programs but is possible under certain conditions. An extreme example is a captive broodstock comprising an entire population. A catastrophe during hatchery rearing could cause extinction, that is, demographic loss of all individuals in a small population. Genetic processes do not directly lead to extinction, but genetic impacts are relevant. In this example, the impact was the loss of an entire species or one component of the genetic diversity of a species. Genetic processes can contribute to extinction risk in more subtle ways. Recent experimental (Newman and Pilson 1997) and empirical (Saccheri et al. 1998) studies have attributed increased rates of extinction to reduced levels of genetic variation.

14.2.2 *Loss of Within-Population Genetic Variation*

Loss of within-population genetic variation has several causes, the most important being genetic drift due to sampling gametes from finite populations (Chapter 9). Loss of genetic variation due to drift occurs at a rate inversely proportional to the genetically effective population size. In a process known as the extinction vortex (Gilpin and Soulé 1986), inbreeding (Chapter 10) and loss of variation due to genetic drift (Chapter 9) can lead to reduced fitness. This loss of fitness may reduce N_e, thereby leading to greater inbreeding and loss of variation, which reduces fitness further. The continued reduction in population size exposes the population to increased demographic risk of extinction. Loss of within-population genetic variation is the most common hazard associated with a number of decisions made in the hatchery, particularly the numbers of broodstock used and how broodstock are mated.

Uncertainty remains about the effect of losing a specific amount of genetic variation on the long-term fitness of a specific fish population and about the relative sensitivity of different species and populations to reductions in genetic variation. Because such uncertainty may be virtually impossible to resolve, the guideline most consistent with evolutionary principles is to minimize human-caused losses of genetic variation (Waples et al. 1990).

14.2.3 *Loss of Between-Population Variation*

Crosses made among fish from multiple populations results in loss of genetic distinctness of each individual population, that is, population identity. Such crossing has the same effect as the genetic process of migration (Chapter 7), which is neither unnatural nor necessarily detrimental. Many fish species are composed of **metapopulations,** in which relatively isolated subpopulations are connected by various amounts of gene flow via migrants. Isolation allows subpopulations to adapt to local environmental conditions (Chapters 8, 11). Low amounts of migration can counter the loss of genetic variation within isolated populations without overwhelming the local forces of natural selection. However, many past hatchery programs (and stock transfers) have led to increased migration rates among populations or to mixing of populations that would not otherwise have occurred. One outcome of mixing distinct populations is **outbreeding depression** (Chapter 11), a loss of fitness due to disruption of local adaptation or coadapted gene complexes.

14.2.4 *Domestication Selection*

Waples (1999) defines **domestication selection** as any change in the selection regime of a cultured population relative to that experienced by the natural population. The result is that the genetic composition of a population within a hatchery program will likely differ from what it would have been in the absence of hatchery propagation. A premise of our application of genetic principles to hatchery management is that natural selection can best produce distributions of genetically determined traits and that any significant alterations of these distributions will reduce a population's fitness in the wild (Reisenbichler 1997).

Alteration of selection regimes can occur in many ways (Busack and Currens 1995; Campton 1995; Waples 1999). An obvious example is intentional selection on traits such as size or spawning time. Another potential source of this genetic hazard is nonrandom collection of hatchery broodstock from a spawning population. The distinction between the two is that in the second case there is no intention to alter traits. For example, gametes may be collected as soon as a spawning period begins and end when sufficient numbers are taken, even though only early maturing fish have been sampled. The parents have been unintentionally selected for early spawning time, a trait with a heritable component of variance in some fish species (Siitonen and Gall 1989). A third form of domestication selection is the unintentional selection that occurs in the hatchery environment, for example, selection imposed by rearing conditions or feeding methods that lead to heritable changes in agonistic behavior (e.g., Berejikian et al. 1996). These second and third forms of domestication selection may be referred to as **inadvertent selection.** A fourth, less recognized, form of this hazard is the release of juvenile fish from the pattern of natural selection that would have been imposed on them had they been in the natural environment (Waples 1999). When hatchery-produced fish eventually spawn in the wild, their offspring may express early life stage traits that are maladaptive in the natural environment.

■ 14.3 Genetic Guidelines for Hatchery Programs

There are four major components of hatchery production: collecting broodstock, spawning adults, rearing offspring, and releasing offspring into the wild. Each of these components may induce one or more of the four genetic hazards, and how they are conducted will affect the ultimate level of genetic risk imposed on the populations involved. For the remainder of this chapter, we discuss the potential for particular hatchery activities to cause one or more of these genetic hazards (see examples in Table 14.1). We then provide guidelines for reducing genetic risk during each of the major phases of hatchery production.

14.3.1 *Broodstock Collection*

14.3.1.1 **Choice of Donor Population.** The choice of a donor population to use in hatchery popagation is the first decision to be made regarding broodstock collection. It is a criti-

Table 14.1 Genetic hazards of hatchery programs, specific hatchery activities causing such hazards, and genetic processes causing these hazards (expressed as changes from natural patterns or amounts).

Type of genetic hazard	Examples of hatchery activity	Genetic process
Extinction[a]	Excessive numbers of adults removed from donor stock to supply hatchery broodfish	Increase in genetic drift; increase in inbreeding
Loss of within-population variation	Low numbers of parents mated; inappropriate mating scheme; hatchery matings inflate natural amounts of family size variation	Increase in genetic drift; increase in inbreeding
Loss of between-population variation (population identity)	Hatchery broodstock collected from population(s) genetically unrelated to supplemented and other interbreeding populations	Migration (changes in patterns and increases in quantity)
Domestication selection	Hatchery broodstock not collected from all portions of a spawning run; unnatural hatchery environment; release from natural selection in the wild	Artificial selection: disruption of patterns or amounts (or both) of natural selection

[a] Demographic forces are most important in causing this hazard, but drift and inbreeding depression can exacerbate it (Lande 1988).

cally important decision for achieving genetic conservation and population sustainability because it affects the risk level of two genetic hazards. First, if either the donor or target populations are severely depleted, risk of extinction may be high. Second, donor choice has great influence on the risk of losing population identity because distinct lineages may become mixed.

■ **Guideline: Choose the donor population that, compared with the population targeted for management, shows the greatest possible similarity in terms of genetic lineage, life history patterns, and ecology of originating environment.**

In an attempt to minimize these two genetic risks, we developed similarity criteria for choice of donor populations (Box 14.1). It is evident from these criteria that the first choice for a broodstock is the target population itself (the population into which fish will be stocked). This imposes no risk of population identity loss but may affect extinction risk in a number of ways. If a catastrophe occurs in the hatchery (e.g., disease outbreak or dissolved oxygen depletion), then the entire reproductive output of the broodstock may be lost. If the survival and subsequent reproductive success of hatchery fish are lower than they are for naturally reproducing fish, the hatchery program will effectively "mine" natural fish as the population fitness is lowered.

The use of a nontarget population in the choice of broodstock also poses the hazard of loss of between-population genetic variation. Mixing or interbreeding of populations should be avoided because it incurs loss of population identity, which is a particularly serious concern in cases in which a critically small population may be all that remains of a native genetic lineage. Additionally, the outcome of intentional interbreeding of distinct populations is unpredictable before the fact: the mixed population might exhibit reduced fitness due to outbreeding depression (Chapter 11), temporarily increased fit-

■ Box 14.1 Criteria for Broodstock Selection

When establishing a new hatchery stock for supplementation purposes, choose the donor population that, compared with the population targeted for management, shows greatest possible similarity in terms of

— genetic lineage,
— life history patterns, and
— ecology of originating environment.

Taken together, adherence to these three similarity criteria will increase the probability that hatchery offspring will have an evolutionary history as similar as possible to that of the extirpated or depleted population. In turn, this similar history increases the probability that hatchery offspring and their descendants will have high fitness in the target population's environment.

Similarity in Genetic Lineage

Similarity in genetic lineage, as deduced from genetic population structure analysis (Chapter 13) of existing populations in the surrounding region, is necessary to avoid accidental mixing of ancestrally divergent populations that may reside in adjacent drainages. For instance, chinook salmon *Oncorhynchus tshawytscha* in California's Klamath River appear to be descended from a lineage distinct from that of adjacent coastal populations (Utter et al. 1989; Bartley and Gall 1990). In restoration, if it is impossible to meet this similarity criterion due to lack of data on the extirpated population, estimation of the most closely related donor will be necessary. In augmentation, determination of genetic relationships between the target and candidate donor populations is feasible in all cases and can supply information on differences in genetic makeup (e.g., from electrophoretic analyses of protein or DNA variation).

continued on next page

ness due to hybrid vigor in the F_1 generation with possible decreased fitness in subsequent generations, or neither (Templeton 1986).

14.3.1.2 **Collection of Broodstock.** After a donor source population has been chosen, the collection of individuals to make up the hatchery broodstock imposes additional genetic hazards on fish populations. By means of their effect on sampling bias, the number of broodstock collected and the procedures of collection will determine the extent of the risk. The number collected determines the extent to which random processes may change gene frequencies, leading to loss of genetic variation. Collection procedures, on the other hand, may selectively change gene frequencies, leading to domestication selection of the first two forms: intentional selection and unintentional selection due to biased sampling.

Genetic and demographic factors interact to determine an appropriate number of fish to collect for broodstock. If hatchery offspring and their descendants return at high

■ Box 14.1*continued*

Similarity in Life History Patterns

Similarity in life history patterns is important because it partly reflects similarity in genetic makeup for these evolutionarily important traits (Ricker 1972) and increases chances that life history patterns of hatchery progeny will be adaptive in the supplementation environment. Sole reliance on data from analyses of proteins or DNA variation is inadvisable because these data may not reflect important, genetically based differences in life history traits (Chapter 12).

In addition to reflecting similar patterns in fitness-related traits, similar life history patterns are particularly important in augmentation because they minimize risks of depleting naturally reproducing fish due to behavioral, physiological, or demographic factors. Inattention to this principle of similar life history patterns probably contributed to estimated declines in naturally produced juvenile coho salmon *Oncorhynchus kisutch* in streams populated with hatchery fish (Nickelson et al. 1986) and inferior reproductive success of naturally spawning hatchery steelhead *Oncorhynchus mykiss* (Chilcote et al. 1986). Ignoring this principle over consecutive years of supplementation could deplete populations in the wild to levels incurring significant risks of extinction and loss of within-population genetic variation.

Similarity in Ecology of Originating Environment

A similar originating environment is indicative of similarity in evolutionary history and increases the chances that the donor population will be highly adaptive in the environment targeted for supplementation. A population's evolutionary history through numerous events of natural selection and random genetic change determines its current genetic makeup (shifting balance theory, Chapter 8). This evolutionary history is greatly influenced by the suite of past ecological conditions it has encountered. In cases when the donor population is different from the target population, evaluation of ecological similarity should take into account anthropogenic alterations of the donor's originating set of freshwater and marine environments and of the environments that will receive hatchery fish.

rates, the population may increase rapidly. If these hatchery fish are not genetically similar to wild fish, however, the population will be swamped with hatchery fish of an altered genetic composition. This may affect the population's sustainability in the face of changing environmental conditions, leading to eventual decline.

The number of spawners collected for broodstock will affect genetic composition at two stages of a hatchery program. First, based on the proportion of spawners sampled, allele frequencies in the hatchery fish and the remaining wild fish will differ from those of the original population. Second, based on the number sampled, each spawning subpopulation—the hatchery subpopulation and remaining wild subpopulation—may lose genetic heterozygosity and rare alleles at a greater rate than would the original whole population because these losses are inversely proportional to effective population size (N_e). Reduced genetic variation in hatchery stocks has been documented for various species (e.g., black porgy *Acanthopagrus schlegeli* [Taniguchi et al. 1983]; ayu *Plecoglossus altivelis* [Iguchi et al. 1999]; giant clam *Tridacna gigas* [Benzie and Williams 1996]; Dover sole *Microstomus pacificus* [Exadactylos et al. 1999]; and salmonids [Allendorf and Ryman 1987, Bartley et al. 1992]). Small broodstock sizes likely contributed to the reductions in many of these cases, along with other factors discussed below. Simon et al. (1986) suggested that low N_e in a hatchery-augmented population of coho salmon was caused in part by mating only a small fraction (<4%) of returning adults. In a review of federal and state hatchery spawning practices for trout, Kincaid (1995) found that broodstock sizes ranged from only 3 to as many as 2,000 adults per year, with many examples of broodstock sizes below 100.

When considering numbers to collect, remember that offspring of fish mated in the hatchery and of those spawning in the natural environment will eventually interact. Therefore, the population of interest includes both wild (or natural) and hatchery fish, and this combined group's N_e is of importance. In 1991, Ryman and Laikre called attention to a critical trade-off inherent to supplementation programs. They showed theoretically that the augmentation of populations using the target population as a broodstock source can reduce N_e, and thus genetic variation, below levels expected if no hatchery production took place. This reduction is caused by an increased variance in family size in the overall population due to the enhancement of family size for those fish spawned in the hatchery. For a population subdivided into captive (hatchery) and wild spawners with effective numbers of parents N_C and N_W, respectively, the total N_e is

$$\frac{1}{N_e} = \frac{x_C^2}{N_C} + \frac{x_W^2}{N_W}, \tag{14.1}$$

where x_C and x_W are the relative contributions of offspring from the hatchery and wild parents, respectively. It can be shown that $N_e = N_C + N_W$ only when the contributions (x) from wild and hatchery spawners are proportional to their effective sizes (N); in all other cases, N_e is smaller than $N_C + N_W$. For a hatchery program to be successful, the portion of the population brought into the hatchery must produce more offspring per individual than do the naturally spawning fish. The more successful the stocking, the more the genetic composition of the population will resemble that of the hatchery subset. This highlights a potential trade-off: gains in offspring numbers can lead to decreases in N_e and thus increased losses of genetic variation (see Boxes 14.2 and 15.1 for further examination of the Ryman–Laikre effect).

How fish are collected will have an important effect on the extent of domestication selection, which will alter the variability of genes influencing life history traits—the key to long-term fitness (i.e., sustainability) of populations in the wild. Although geneticists are unlikely to identify the majority of these genes, this essential genetic diversity can be conserved by ensuring that, in each breeding season, the hatchery broodstock constitutes a representative sample of the existing patterns in life history traits of the donor population. The genetic consequences of biased broodstock collection are well documented. For example, Steward and Bjornn (1990) reviewed numerous observations of hatchery stocks of salmon shifting to early run timing following consecutive collections of hatchery gametes primarily from early returning adults. This shift occurs because run timing has a relatively high heritability (Smoker et al. 1998). Other examples of relevant traits include age and size structure of spawning adults, fecundity, and diversity in habitat preferences.

As a hatchery program continues, an important decision regarding broodstock collection presents itself—whether or not to choose broodstock from among wild and natural fish or hatchery returns. Exclusive collection of wild returns, one generation after the next, is appropriate only if hatchery returns show sufficiently high fitness in the natural environment to meet the hatchery's specified objective for the rate of rebuilding per generation. That is, sufficient numbers of hatchery returns must move into the natural spawning grounds that they were intended to seed and then successfully reproduce in these habitats. If, instead, hatchery returns exhibit low fitness, continued collection of wild fish will cause "reproductive mining" of the remaining donor population, greatly increasing risks of loss of within-population genetic variation and extinction.

Researchers have shown reduced reproductive success of hatchery fish in natural environments (reviewed for Pacific salmon *Oncorhynchus* spp. by Reisenbichler and Rubin 1999). Unfortunately, most past studies have confounding factors that make their relevance to supplementation programs uncertain. For one hatchery program involving steelhead (Chilcote et al. 1986), success of naturally spawning hatchery returns in producing smolt offspring was only 28% of that for wild spawners. When followed to adulthood, the survival of offspring from naturally spawning hatchery returns declined to 10% of that for offspring of wild spawners (Leider et al. 1990). These data, however, may not be relevant to all hatchery planning because (1) the hatchery population was not derived from the wild population for which enhancement was intended (loss of population identity and violation of similarity criteria) and (2) this nonnative hatchery population had undergone selection for early run timing since 1958 (domestication selection), resulting in advanced spawning time and a probable impact on genetic variation (loss of within-population genetic variation). Skaala et al. (1996) found survival to be three times higher in wild brown trout *Salmo trutta* than in F_1 offspring of wild and hatchery fish. But here also, the hatchery fish were derived from a nonnative population, one found at a higher elevation and thus a different ecological environment (violation of similarity criteria). In a third study, Unwin (1996) compared survival of naturally and hatchery-produced chinook salmon from the same system, Glenariffe Stream in New Zealand. He found that fry-to-adult survival of hatchery-reared chinook salmon was much lower than expected given the size advantage of hatchery over naturally produced juveniles. However, chinook salmon are not native to New Zealand, and the hatchery release strategy used there is atypical of practices used in the native chinook salmon range of the Pacific Northwest. It is, therefore, uncertain if this case of lower survival of hatchery fish is applicable to salmon hatchery programs in North America.

■ **Guideline:** Collect an appropriate number of fish to minimize genetic differences between the hatchery- and wild-spawning subpopulations and, ultimately, to minimize genetic alterations of the overall population.

There is no one, universally appropriate minimum number of fish to collect as hatchery broodstock. The goal and scope of the hatchery program and the status and life history of the populations involved influence the appropriate course of action. For example, a minimum N_e of 50 is commonly recommended in the conservation literature to prevent detrimental effects of inbreeding depression (e.g., Frankel and Soulé 1981). This minimum might be appropriate for a restoration broodstock with which the population will be perpetuated from the original founders. In augmentation programs, however, rates of inbreeding and losses of variation will be determined by both the hatchery stock and the naturally spawning target population (see the discussion of the Ryman–Laikre

■ Box 14.2
Examination of the Ryman Laikre Effect: Does Supplementation Lower Effective Population Size?

Ryman and Laikre (1991) suggested that hatchery augmentation can produce an unavoidable trade-off between increased numbers and reduced genetic variation. They considered, however, only one generation of supplementation. Waples and Do (1994) considered multiple generations and age structure in simulations of inbreeding resulting from augmentation using captive broodstock (i.e., potential for extreme family size differences between captive and wild fish). They found the critical factor that affected levels of inbreeding was the success of a population after supplementation ended. If the population crashed, inbreeding increased (and thus genetic diversity declined) at a greater rate than if no supplementation had occurred. This was because most of the remaining individuals were related members of the reproductively successful families raised in the hatchery. If the population sustained its increased size, inbreeding declined after several generations. Simulations revealed one scenario in which a captive broodstock program would not considerably affect inbreeding levels despite a population crash: when the ratio of effective population size (N_e) to census size (N) in the captive program was greater than that ratio in the wild. Hedrick et al. (1995) showed that this scenario might be realistic in empirical situations.

Hedrick et al. (1995) used Ryman and Laikre's equations to estimate N_e for the hatchery- augmented winter-run chinook salmon of the Sacramento River in California. This population declined from over 85,000 spawners in the 1960s to about 500 in the 1980s, leading managers to initiate a hatchery program in 1989 in hopes of rebuilding this federally endangered population. Using various assumptions about the ratio of N_e to N, they found that the hatchery program likely resulted in a 15% increase to a 5% decrease in N_e. Potential increases were attributed to a higher N_e/N ratio in their hatchery broodstock than that in the wild. Their results showed, however, that for higher N_e/N ratios in the wild, even slight increases in hatchery production, that is, captive proportion x_C, would have led to much greater reductions in N_e (see figure on next page). Because of this risk, Ryman and Laikre (1991) suggested capturing all adults when populations are small to avoid differential reproductive rates for wild and hatchery spawners.

Continued on next page

effect earlier in this section). An appropriate number to collect will be determined in part by the survival and reproductive success of hatchery fish relative to fish in the natural environment. It is important to have the genetic diversity of the hatchery broodstock represent that in the natural population if hatchery descendants will predominate in future generations.

The most conservative approach would be to assume that the entire population may derive from the current hatchery broodstock. Assuming this, one should sample enough fish to minimize genetic differences between the hatchery and wild subpopulations. When assessing genetic population structure, researchers find that only 50 (or fewer) individuals allow a reasonable estimate of most allele frequencies. The expected change

■ Box 14.2 *continued*

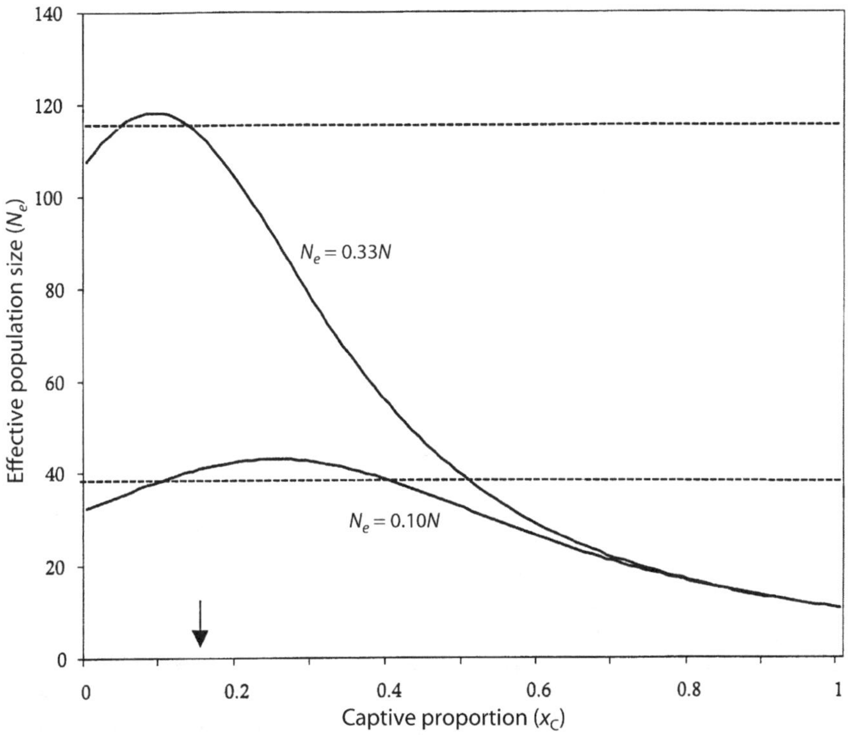

This figure illustrates the impact of hatchery augmentation on the estimated N_e for the augmented 1993 winter-run chinook salmon in the Sacramento River (Hedrick et al. 1995). The effective number of hatchery parents (N_C) was estimated from spawning records and family survival data. The effective number of wild parents (N_W) was estimated as 0.33 or 0.10 of the census number of spawners in the wild. The arrow pointing to the *x*-axis indicates the actual captive proportion for this run. The horizontal dashed lines indicate the N_e that would have existed had no hatchery production taken place, that is, all adults spawned in the wild (adapted from Hedrick et al. 1995: Figure 1). The segments of the curves below the dashed lines represent captive proportions for which this hatchery augmentation would reduce N_e.

in allelic frequencies due to sampling will be determined by the allele frequency in the population and the proportion of fish sampled for broodstock. Lower frequency alleles require that greater numbers be sampled to limit frequency changes for a given population size (Table 14.2, compare across rows), and greater proportions must be sampled as population size decreases to obtain the same probability of change in allele frequencies (Table 14.2, compare down columns). When multiple low-frequency alleles are present, larger sample sizes are needed to capture most alleles in the population. Kincaid (1983) recommended a minimum of 100 parents (equal numbers of males and females), and Allendorf and Ryman (1987) recommended a minimum of 200.

Table 14.2 Effects of sampling on allele frequencies. Shown are the proportions *(s)* and numbers *(n)* collected so that, given a population of size *N*, a sample allele frequency will not differ proportionally from the true population allele frequency *(p)* by more than *d* with a probability of α.

	Population allele frequency *(p)*					
	0.5		0.3		0.2	
N	*s*[a]	*n*	*s*[a]	*n*	*s*[a]	*n*
5,000	0.01	34	0.02	78	0.03	132
1,000	0.03	33	0.07	73	0.12	119
500	0.06	32	0.14	68	0.21	106
100	0.25	25	0.44	44	0.58	58
50	0.40	20	0.61	31	0.73	37

[a] The equation used to determine *s* is based on an equation for sample size requirements in binomial sampling:

$$s = \frac{1}{\left[2N \cdot \left(\frac{d}{z(\alpha)}\right)^2 \left(\frac{p}{1-p}\right)\right] + 1}.$$

For this example, $d = 0.2$ and $\alpha = 0.9$, $[z(\alpha)] = 1.645$.

Less conservative numbers may be needed when the biology of the species and the duration of the hatchery program are taken into account. Allelic changes and losses should be considered over generations and for the entire population. For iteroparous species, the spawning population from one year to the next may be essentially the same, plus recruitment and minus mortality. Changes in allele frequencies or alleles lost in one hatchery cohort may be ameliorated by another year's augmentation. For the semelparous life history of the Pacific salmon, each year's spawners represent only a portion of the genetic diversity of a population. Natural differences in allele frequencies will exist depending on the intermixing of age-classes in a spawning run. Waples (1990) showed that N_e for an entire generation of Pacific salmon is equal to the effective number of breeders (N_b) per year times generation length. Hatchery broodstock collected over 4 or 5 years, a typical generation length for salmon, might all be combined when considering a recommendation based on N_e.

■ **Guideline:** Collect a minimum of 50 spawners, preferably with equal numbers of males and females, to reduce loss of rare alleles.

Allelic frequencies also will change upon mating of the broodstock. Considering the expected loss of rare alleles in small populations as a limiting factor, Allendorf and Ryman (1987) recommended a minimum of 50 for collecting broodstock (also Ryman and Stahl 1980; Shaklee 1983). The expected loss of rare alleles can be calculated from binomial sampling probabilities using the formula $(1 - p)^{2N_e}$, where p is frequency of the allele and N_e is effective population size of the broodstock. For example, when N_e equals N (where N is the number of mating adults), alleles at frequencies of 0.05 will be retained in the offspring generation with 95% probability if 30 spawners are mated. In the most likely event that N_e is less than N, a conservative approach of collecting 50 spawners will allow some leeway before exceeding a 5% probability of losing the allele.

Loss of alleles from an entire population is especially detrimental because lost alleles can be recovered only by migration (Chapter 7) or mutation (Chapter 6). The adaptive importance of any given allele, however, is generally unknown. In the case of augmentation using the target population for broodstock, the probability of losing an allele may increase in one or another subpopulation but not in the overall population (i.e., not in both hatchery and naturally spawning subpopulations simultaneously). Therefore, the critical uncertainty is the importance of an allele for survival in the hatchery and in the natural environment.

■ **Guideline:** Collect less than 50% of the wild spawners if minimum size requirements can be met easily.

The proportion of a population to use for hatchery production is an area of great uncertainty. Taking too many fish into the hatchery will quickly deplete a population if hatchery descendants have reduced reproductive success, and it will quickly alter genetic composition if genetic hazards are imposed. Taking too few will slow the rate of rebuilding, causing increased demographic risks if the population is small or reducing opportunities for increased harvest at larger population sizes. In populations for which minimum size requirements easily can be met, we recommend collecting less than 50% of the wild spawners and adjusting collection percentages in future generations based on monitoring information. Important and generally unknown factors (research needs) are (1) relative return rates of hatchery- and wild-spawned fish and (2) reproductive success of matings involving hatchery descendants. Taking 50% or less gives reasonable assurance that wild fish will persist for sufficient generations to monitor return rates and population size changes. If adverse effects of the hatchery program are indicated, some wild fish will remain with which to try alternative rebuilding measures. An exception is when populations are critically small (in the 10s to a few 100s). Then consider using all adults for broodstock, perhaps in a captive broodstock program (discussed below).

■ **Guideline:** Generally exclude hatchery returns from broodstock collections but reevaluate this practice and the entire supplementation program if hatchery returns exhibit low fitness on spawning grounds.

Advantages of excluding hatchery returns are twofold. First, the repeated enhancement of the same families and their descendants will reduce overall genetic variation by overrepresenting their genes in the population, leading to decreased N_e. Excluding hatchery returns guarantees that no family will be enhanced for two generations in a row. Second, the potential impacts of domestication selection also will be reduced if families are not exposed to the hatchery environment for two subsequent generations.

Rather than blindly following this guideline, it is imperative to monitor the fitness of hatchery returns in their intended spawning habitat. Indication of low fitness in a given generation is cause for alarm, particularly if it is part of an ongoing downward trend in population size. Low fitness of hatchery returns should trigger a thorough reexamination of the entire augmentation program and of all related efforts to remedy anthropogenic causes of decline. Before continuation of the program, corrective measures must be identified. If corrective measures still include a hatchery component, it may be useful to develop a ratio of wild to hatchery returns in collected broodstock that would strike an appropriate balance between reducing genetic risks of hatchery propagation and reducing extinction risk for the depleted wild or natural stock.

■ **Guideline:** Collect adults for broodstock so that they constitute an unbiased sample of the naturally spawning donor population with respect to spawning timing, size, age, sex ratio, and any other traits identified as important for long-term fitness.

Assuming adequate numbers of spawners have been collected, unbiased sampling of spawning populations can best be attained by random sampling throughout a spawning season and, where appropriate, throughout spawning grounds. Unbiased sampling, however, requires careful planning that incorporates known characteristics of the population. An estimation of expected timing and duration of spawning is needed so that the required total number of fish can be collected proportionally over the duration of the spawning season. For example, in order to obtain a previously specified number of adults, collection of too few fish early in the season could lead to overrepresentation of late-season fish. If, on the other hand, too many fish are collected early, in spite of careful attempts to avoid doing so, managers should avoid the temptation to underrepresent the later portions of the run. Unfortunately, this situation may call for culling excessive numbers of early progeny. In turn, culling may impose unnecessary genetic and demographic hazards on the naturally spawning population because it leaves a smaller number of naturally reproducing adults than would have been available if excessive collection of early spawners had been avoided. Therefore, it is important to estimate accurately spawn timing so that random sampling can be proportionally distributed throughout the season. Historical records and counts at downstream locations (for migratory species) can be used to estimate numbers and run timing.

Stratified random sampling within a day's collection may be considered if fish are known to segregate by some observable or measurable trait that results in some characteristics being over- or underrepresented in a sample. **Stratified sampling** involves separating fish according to a trait and sampling an equal proportion from each subgroup. For example, if a certain size-group tends to be collected too often, spawners could be separated into large, medium, and small males and females. Equal proportions would then be taken from each of these six groups. This method ensures that each individual has an equal opportunity to be collected. Sampling all parts of a spawning run is actually one form of stratification. This is likely to be the most important and most practical stratification hatchery operators can perform. If spawners tend to have different traits at different times of the day, collecting several times throughout a day is also recommended. Other stratifications may require collecting excess fish, handling them, and releasing those not kept for broodstock. However, the extra labor involved and the stress imposed on the fish may not make this worthwhile.

■ **Guideline:** Consider a captive broodstock program if the target population is critically small and extinction is probable.

When population size is critically low, subdivision of a population creates two smaller subpopulations and puts each at very high risk of genetic alteration and demographic-based extinction. Based on demographic concerns alone, drastic procedures, such as a captive broodstock program, are recommended to increase population size to a level that reduces demographic risk of extinction. In captive broodstock programs, gametes (from ripe adults) or juveniles are collected, then fish are raised to maturity and spawned, and their offspring are released. Such programs differ from the usual augmentation programs in that fish are held in captivity for an entire life cycle. The benefit of such a program is the greatly increased survival to adulthood, allowing production of large numbers of offspring in a single generation. Captive broodstock programs may be especially useful for giving critically small populations an initial boost in numbers. However, Waples and Do (1994) caution that if captive-born fish survive and reproduce effectively, the poststocking population can become greatly inbred unless population increases can be sustained after stocking ends (see Box 14.2). One way to avoid this increased inbreeding is to develop the captive broodstock from all of the adults.

Adult-based captive broodstock programs should be considered when the logistical constraints of collection make it is easier to collect all adults than to assure an unbiased sample of juveniles. For example, this option was chosen for Snake River sockeye salmon *Oncorhynchus nerka* in Redfish Lake, Idaho, which had only zero to eight adult returns yearly from 1991 to 1996 (Flagg et al. 1995). All adults were collected to maximize juvenile production and avoid the genetic hazard of reduced within-population variation due to preferentially enhancing the reproductive success of only a portion of all families (Ryman and Laikre 1991). For this same reason, up to 25% of juveniles were later collected for captive broodstock to assure that they would represent all families in the population. The risk of loss of within-population variation may be greatest with the use of captive broodstock because of the potential for high mean family size in captive families (Waples and Do 1994).

Captive broodstock programs are largely unproven as a rehabilitation measure and should be regarded as experimental (Hard et al. 1992). Therefore, they should be avoided for larger populations. Because these larger populations more easily permit adherence to guidelines aimed at minimizing unnatural genetic alterations, the benefits of increased spawner numbers generally do not outweigh the possible genetic and demographic risks and dollar costs associated with maintaining a population in a hatchery for an entire life cycle.

Protecting captive broodstock against catastrophic loss or high mortality is of utmost concern (Hard et al. 1992). In the Redfish Lake sockeye salmon recovery program, fish were divided among federal and state hatcheries to guard against a facility-wide failure. Survival from fry stage to adulthood in the hatchery currently averages over 50%. Fish are being released at various life stages (egg plants, smolts, and prespawn adults) and by various methods (direct release and holding pens) to determine which release methods result in the best survival. This experimental aspect exemplifies adaptive management discussed above.

■ Maintain captive broodstock for only one generation.

To reduce genetic risks, captive broodstock should be held for only one generation. The magnitude of genetic risk due to captive rearing is largely unknown. Evidence consistent with occurrence of domestication selection has been reported for hatchery broodstock of brook trout *Salvelinus fontinalis,* which were bred in captivity over many generations (Vincent 1960; Flick and Webster 1964; Moyle 1969). The minimum number of generations at which domestication selection may occur at intensities sufficient to reduce fitness in the wild will vary across different populations, hatchery practices, and facilities.

14.3.2 Spawning

Spawning practices can impose a high risk of loss of within-population genetic variation. Population genetics theory shows that within-population genetic variation will decrease more quickly as N_e decreases (Chapter 9). As discussed previously, Ryman and Laikre (1991) have shown how supplemental stocking may inherently reduce N_e. However, simulations by Waples and Do (1994) and the empirical studies of Hedrick et al. (1995) both describe a situation that prevents the decrease in N_e—when the ratio of effective size to census size *(N_e/N)* is considerably higher in the hatchery population than it is in the wild (see Box 14.2). Using an appropriate mating scheme is the first step toward increasing N_e/N in the hatchery. For a given number of broodstock, the mating scheme affects N_e in two ways—by determining the sex ratio and by affecting the resulting variance in family size. An obvious example of a mating practice that skews sex ratios is the use of a single male to fertilize multiple females' eggs. Such a scheme may be used because of its convenience but directly creates a skewed sex ratio in the adults. An indirect way in which sex ratios may be skewed is the use of pooled sperm from multiple males. Withler (1988) found that pooling sperm may cause unequal contribution of gametes from some males due to differences in potency of their sperm; consequently, N_e is reduced because fewer males contribute gametes than were actually crossed or because variation in number of progeny per male (variance in family size) increases.

Past hatchery practices often have led to an N_e much smaller than the census population size, as low as 0.03 in a coho salmon hatchery (Simon et al. 1986), 0. 0001 and 0.04 in two chinook salmon hatcheries, and 0.4 in a white seabass *Atractoscion nobilis* hatchery (Bartley et al. 1992). Furthermore, reduced within-population genetic variation resulting from improper genetic management of hatchery broodstocks has been documented (e.g., Taniguchi et al. 1983; Allendorf and Ryman 1987; Benzie and Williams 1996; Exadactylos et al. 1999; Iguchi et al. 1999). Kincaid (1995), in a review of breeding records for 100 trout broodstocks, found that 21 had an N_e less than 100 due to low numbers of breeders and unequal sex ratios. Simon et al. (1986) attributed decreases in N_e to crosses done with unequal sex ratios, pooling of eggs and sperm from multiple fish, greater than random variation in family size, and exportation of entire families to other streams. Bartley et al. (1992) also suggested that pooling sperm from multiple males may have contributed to the decreases in N_e in a chinook salmon hatchery population. In some hatchery broodstocks, deleterious changes in traits, such as increased incidence of abnormal vertebral columns and missing fins, have been associated with low genetic variation (e.g., Allendorf and Leary 1988; see also Chapter 10, Box 10.4).

The examples above refer to losses of genetic variation within hatchery populations, not subsequent losses in the natural populations being stocked. In contrast, Campton (1995) stated that he was not aware of any documented losses of genetic variation due to genetic swamping by hatchery fish in wild Pacific salmon populations, a group among the most stocked and well studied of all fishes. He made it clear, though, that this was not because such losses were unlikely but rather because it is very difficult to demonstrate them empirically.

■ **Guideline:** Within each group of adults that are ripe for spawning on a given day, randomize matings with respect to size and other phenotypic traits.

The foremost reason for this recommendation is that mating behaviors are so complex, for the few cases in which anything in known about them (e.g., Fleming 1996; Fleming and Gross 1994), that it is unlikely that hatchery operators could mimic them. Nonrandom matings that do not accurately mimic natural patterns may impose the genetic hazards of loss of within-population genetic variation and domestication selection without the potential benefits of maintaining natural patterns.

Another rationale for this recommendation is that the adaptive value of previously evolved natural mating patterns may be altered when demographic risk of extinction becomes high (i.e., when the census size of the naturally reproducing population has substantially declined). Little is known about the nature of possible interactions between demographic extinction risk and the adaptive value of certain mating behaviors and patterns. A simple example would be sex ratio at spawning, which has the effect of reducing N_e if it is not 1:1. A skewed sex ratio, if it had occurred naturally in a past abundant population, may have had some adaptive value. It would have reduced N_e, but in a large population most genetic variation would still be conserved. If, on the other hand, the population was drastically reduced in number due to harvest or habitat degradation, this same mating pattern may cause a reduction in N_e below a critical value for the population. In this situation, a "natural" pattern may not be the best in the face of new exploitation and environmental pressures.

■ **Guideline:** For all fish collected as hatchery broodstock, choose the mating scheme that will maximize N_e.

■ One-by-One Mating Scheme Guidelines: Within each group of males and females that are ripe for spawning on a given day, cross one female with one male. If one sex is in excess, divide gametes from individuals of the less numerous sex to allow crosses with all adults of the excess sex.

■ For cases in which male infertility is frequent, pool semen from overlapping pairs of males and immediately use each pooled pair to fertilize one lot of eggs.

A baseline for making crosses, assuming equal reproductive potential of each individual, would be by pairs of a single male and a single female (Figure 14.1A). However, random collection of broodstock on most days will result in unequal numbers of each sex, and the potential for gamete inviability should be taken into account. If one sex is in excess, divide gametes from individuals of the less numerous sex to allow crosses with all of the excess sex (Figure 14.1B). Semen from a single male is preferred for each cross because the alternative of using pooled semen may cause unequal contribution of gametes from some males due to differences in potency of their sperm (Withler 1988). An exception is when it is suspected that many males are infertile. In this case, pool semen from overlapping pairs of males (e.g.,1 and 2, 2 and 3, 3 and 4) and immediately use each pooled pair to fertilize eggs of one female (Figure 14.1C). Pooling pairs of males will reduce the incidence of unsuccessful fertilization because at least one of the pair is likely to be fertile (Gharrett and Shirley 1985). The degree of sperm potency differences versus numbers of completely infertile males will determine whether or not overlapping males is advantageous. A disadvantage of pooling semen is the loss of parentage information, although the advent of highly variable genetic markers (e.g., microsatellite DNA) may still allow determination of paternity when semen is pooled.

Bartley et al. (1992) estimated N_e for a number of hatchery populations (see Box 9.1). One population of chinook salmon had a drastically reduced N_e, which they attributed in part to mating practices. A hatchery population of white seabass, which mass spawned in groups of approximately 25 fish, had an N_e less than one-half the number of spawners. In contrast, N_e approximated the census size in a hatchery stock of rainbow trout (nonanadromous *Oncorhynchus mykiss*) in which a one-to-one mating scheme was applied.

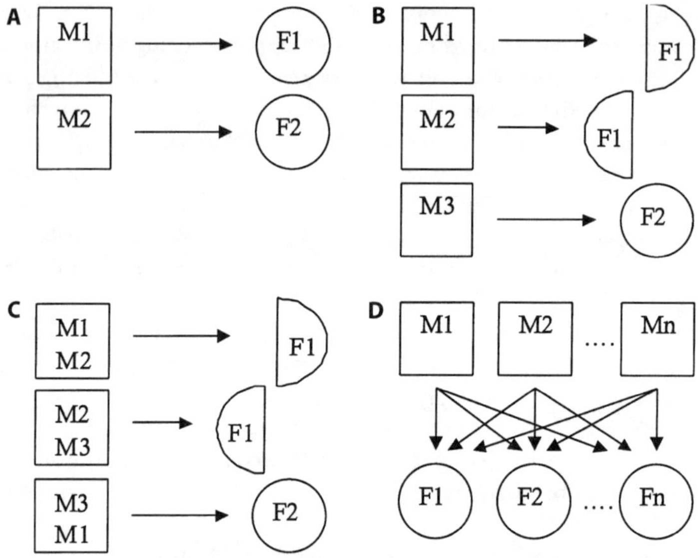

Figure 14.1 Mating schemes for hatchery crosses. **(A)** One-to-one: assuming equal reproductive potential, make baseline crosses in which each male fertilizes one female. **(B)** Division of gametes: when one sex is in excess, divide gametes to allow spawning of all ripe adults. **(C)** Overlapping males: when male infertility is common, mix semen from overlapping pairs of males to reduce the loss of a given female's entire egg production. **(D)** Factorial crosses: to increase diversity of genotypes, cross all males to all females (full factorial). Alternatively, make partial factorial crosses within subsets of adults (e.g., four 5 × 5 crosses among 20 males and 20 females).

■ Factorial Mating Scheme Guidelines: Apply a factorial cross to as many matings as possible.

■ For critically small populations, consider a full factorial cross on each spawning date.

Time and available labor permitting, apply a factorial cross to as many matings as possible. **Factorial matings** involve dividing gametes so that a portion of each female's eggs is fertilized by each male. Simulations have shown that partial or full factorial matings can increase N_e relative to single-pair matings (C. Busack, Washington Department of Fisheries and Wildlife, personal communication). In a full factorial mating design, gametes of both sexes are divided so that all males are mated to all females (Fig 14.1D). Partial factorials involve making factorial crosses within subsets of available adults (e.g., 20 fish of each sex divided into four 5 × 5 factorial crosses). Larger factorial crosses (i.e., closer to full factorial) lead to greater increases in N_e relative to single-pair matings. Full factorials, however, demand more handling and require meticulous tracking of different lots of gametes. They may be feasible when hatchery operators are handling only small numbers of broodstock and when the objective is to maximize diversity of genotypes produced.

Factorial matings increase N_e through a reduction in variance in family size. By providing multiple mates for each fish, extremes in reproductive capacity are moderated, especially when infertile fish are involved. For example, if pairs are mated singly and one male is infertile, the entire reproductive output for that male and its mate would be zero.

If a factorial cross is used, the infertile male's output would still be zero, but each female would contribute offspring, although with a reduced total output. Busack (personal communication) found that the greatest gains in N_e occurred when the coefficient of variance for family size (variance/mean) was high.

A possible negative consequence of using factorial matings (which also applies to using overlapping males) is the increase of crosses that produce relatives among the offspring, thus potentially increasing levels of inbreeding. The relative advantages of factorial versus one-by-one crosses will likely depend on the variance in reproductive success among the hatchery parents. Further work is needed to determine when factorial matings should be applied and what size of factorial mating provides the most benefit and least possible negative consequences.

14.3.3 Rearing

Rearing practices mostly affect the potential for hatchery environments to impose domestication selection on behavioral and other performance traits, thereby leading to poor performance in natural environments. Effects that artificial conditions might have on the subsequent performance and survival of hatchery juveniles released into the wild are poorly understood. Although domestication is very likely to occur if hatchery broodstocks are propagated in captivity over consecutive generations, the risk of domestication is uncertain for programs in which fish usually experience the hatchery environment for only part of a generation (i.e., adults are gathered from the wild each generation). However, genetic theory for populations experiencing different selective pressures but connected by gene flow (e.g., Lythgoe 1997), as is the case for hatchery and natural populations, predicts that the overall population could become better adapted to the hatchery environment as easily as to the natural environment (Waples 1999).

Numerous studies have shown behavioral differences between hatchery and naturally reproduced fish (reviewed in Maynard et al. 1995). However, these studies were not designed to determine if differences were due to genetic causes (changes in the makeup of genes controlling expression of behavioral traits) or to environmental causes (inappropriate environmental conditions in the hatchery for fish to develop natural behaviors). Some studies, however, implicate genetic factors in the alteration of agonistic behavior. Swain and Riddell (1990) concluded that genetic factors were responsible for increased aggressiveness of juvenile coho salmon of hatchery parents compared with those of parents produced in the wild. Swain and Riddell used two pairs of hatchery and wild populations, each pair taken from the same region so that the ecological and physical environments for both populations were similar. In both comparisons, in a common, novel environment—laboratory aquaria—hatchery juveniles challenged their mirror image more often than did wild juveniles. Berejikian et al. (1996) compared behavior of a wild population of steelhead and a hatchery population derived from it seven generations earlier. This design eliminated any confounding effects of initial population differences with hatchery propagation differences. In both natural and hatchery environments, the two groups displayed differences in agonistic behavior (swimming against a mirror) that were attributed to domestication selection over this relatively short time.

■ **Guideline:** Produce fish that are qualitatively similar to wild fish of the targeted population in size, morphology, behavior, physiological status, health, and other ecological attributes important for fitness in the wild.

Rearing guidelines are based on the principle that adaptive traits important for high fitness in the wild are quantitative traits controlled by a combination of factors including genotype (genetic makeup), experienced environment, and genotype × environment interaction (Chapter 12). Rearing practices, therefore, must aim to reduce unnatural changes in genetic and environmental components of these adaptive traits. Culture practices also must reduce within-hatchery mortalities that are nonrandom with respect to important life history traits because nonrandom mortalities incur the risk of domestication selection.

- **Provide for volitional movement of swim-up fry to feeding areas in rearing troughs, raceways, or ponds.**
- **Simulate the natural environment in incubation for incubation densities, substrate, light, temperature, and oxygen.**
- **Simulate natural juvenile rearing conditions for rearing densities, container hydraulics, habitat complexity, feeding conditions, feeding and predator avoidance behaviors, and fish size and nutritional status at release.**

Alternative rearing practices may produce quasi-natural physiological, behavioral, and morphological characteristics in hatchery fish, resulting in improved survival upon release. The effects of rearing environment on Pacific salmon are being researched at the Northwest Fisheries Science Center of the National Marine Fisheries Service in Seattle, Washington. Researchers there call their experimental rearing conditions the natural rearing enhancement system (NATURES) (Maynard et al. 1996). Salmon are reared in raceways with overhead cover, instream structure (aquatic plants), substrate (sand or gravel), and nonintrusive feeding systems. In addition, fish are exercised, trained to avoid predators, fed more natural foods, and reared at lower densities than those used in typical production hatcheries. Initial studies (Maynard et al. 1996) indicated that chinook salmon reared in and released from a NATURES environment had significantly higher postrelease instream survival than did controls reared under traditional hatchery conditions. The fish reared in the NATURES environment had 25–50% higher survival to weir collection points 0.2–21 km downstream than did their conventionally reared counterparts. Researchers attributed this difference to the more natural camouflage coloration and fright responses of fish reared in the NATURES environment. Ongoing studies are examining whether or not survival advantages remain to adulthood. Further study should determine which natural elements lead to the greatest survival when fish are released into the natural environment. Similar studies are needed for other cultured species. Critical elements then can be incorporated into the design and operation of traditional hatcheries.

- **Unless conducting experiments on the effects of different rearing methods, rear all fish of a given population under the same conditions.**
- **Mix families randomly as soon as possible, taking into consideration plans to mark families.**

In addition to providing natural conditions that may increase subsequent survival of all individuals, care should be taken to provide similar conditions for all fish so that no subset is either favored or disadvantaged. Exposing all fish of a population to the same environmental conditions will reduce the risk of some forms of domestication selection and hatchery-induced increases in variance in family size. Mixing of families may be especially important because some incidents may affect only a portion of the hatchery (e.g., mortalities in one rearing unit due to low oxygen or high water temperature). If the affected rearing unit contains one or more entire families, then these families lose their entire genetic contribution.

Marking of hatchery fish is important. At a minimum, bulk marking (e.g., fin clip or thermal otolith mark) allows comparisons of survival between naturally and hatchery-produced fish and avoidance of repeated enhancement of hatchery fish genomes. Waples and Do (1994) showed that essentially 100% marking of hatchery fish is necessary to avoid significant increases in inbreeding during captive broodstock programs. Family-specific marks, including physical tags or genotype profiles for parentage assignment, allow evaluation of factors that contribute to differential family survival. This information will address many of the questions concerning the importance of domestication selection in hatcheries.

14.3.4 Release

Methods for releasing hatchery fish into the wild have the potential to impose all four types of genetic hazards. Improper release practices could lead to poor survival, resulting in greater losses of within-population variation and increased probability of extinction. Hatchery releases may affect population mixing, leading to losses of population identity. Finally, the life stage at release may significantly affect the likelihood that domestication selection occurs.

Release practices primarily impose the hazards of extinction and loss of within-population genetic variation by indirectly affecting adult return rates of hatchery-produced fish, a major component of the short-term fitness. Survival may be reduced if the release site does not match the type of site where natural fish of the same life stage can be found. For example, consider the spring chinook salmon population of the Yakima River in Washington. Juveniles begin moving downstream soon after emergence and many make a second winter migration to areas farther downstream (Fast et al. 1991). Young salmon released at the wrong site (e.g., too far upstream) may be at an inappropriate stage of development to complete their seaward migration successfully.

Release practices may also impose the hazard of loss of population identity above and beyond the hazard imposed by using a broodstock other than the target stock. For homing species, the site and timing of release will affect imprinting to the stocked locale. In a review of straying in salmonids, Quinn (1993) concluded that, beyond the use of nonnative populations, moving fish from their rearing place and releasing them outside the normal migration period can lead to increased straying. Unusually high levels of straying may lead to a breakdown of the genetic identity of individual populations.

The choice of when to release fish (i.e., life stage at release) involves a trade-off that is inevitable in hatchery operations. The longer that fish are held in the hatchery, the greater the potential for enhancing survival relative to their natural counterparts. However, reducing exposure to the hatchery environment likely will reduce the extent of domestication selection for two reasons. First, there is less time for inadvertent selection due to hatchery practices and conditions. Second, the concern that hatchery fish have reduced exposure to natural selection patterns goes hand-in-hand with the primary rationale for using a hatchery, which is to increase survival at early life stages. Hatchery-propagated fish are not exposed to the same combination of random and natural selection forces that may have otherwise eliminated certain individuals. Released from this early mix of natural selective patterns and random mortality, some individuals may survive to reproduce that would not naturally have done so. Two uncertainties, which may never be resolved, make it difficult to assess the degree of genetic risk incurred by this lack of exposure to natural environmental patterns: (1) the relative importance of natural selection versus random mortality is unknown for natural environments and may vary widely across years and populations; and (2) the ability of natural selection to compensate for relaxed natural selection by causing mortality of unfit hatchery fish after they are released. The potential for impairment of genetically based adaptation in hatchery fish populations that did not experience natural environmental conditions is therefore a major uncertainty in determining release practices.

Release guidelines are designed to maximize short-term fitness of released hatchery fish in the natural environment and to minimize any negative genetic or ecological interactions between released hatchery fish and the naturally produced fish with which they may breed or interact. These objectives may be met by giving attention to four areas of concern: (1) how many to release, (2) what life stage to release, (3) where to release, and (4) how to conduct the release.

■ **Guideline:** The sum of the numbers of hatchery fish released and naturally reproduced fish (with none displaced) should be equal to or less than the estimated carrying capacity of the environment for juvenile fish.

Release all hatchery fish up to the point at which the number of hatchery and naturally produced fish is equal to the carrying capacity for the supplemented species. The carrying capacity applies to all components of the aquatic environment through which released fish may disperse or migrate. If the number of fish available for release exceeds the carrying capacity, reduce the number of broodstock collected in the following season to avoid this problem.

It is important to consider the disease and disease carrier status of hatchery-propagated fish. Fish showing disease symptoms should be culled because their survival chances after release are greatly reduced, and their shedding of pathogens or parasites increases the exposure of naturally reproduced fish. Some past releases of nonnative hatchery fish have introduced pathogens into the natural environment (examples given in Currens et al. 1997; Waples 1999). In these cases, it is uncertain if hatchery propagation itself, beyond being a conduit for population translocation, increased the likelihood of disease transfer to natural populations.

■ **Guideline:** Release at the earliest life stage possible considering the trade-off between adult return rate and genetic risks.

The condition of fish at release and timing of release are probably the two most important hatchery-related factors that affect the adult return rate of hatchery-produced fish, a major component of short-term fitness. These release factors also influence some types of genetic hazards and, therefore, long-term fitness of hatchery-produced fish. There may be an optimal life stage that yields high adult return rates within a range of acceptable genetic risk.

The life stage that achieves this optimal balance of return rate and minimal risk is uncertain; it undoubtedly differs between species and probably between populations and even years. Surely, there also is an interaction effect between life or physiological stage and quality of freshwater habitat. The best approach may be to determine the life stage at release that consistently yields high adult return rates and to evaluate levels of associated genetic risks. Then it should be determined if release at any life stage yielding lower return rates decreases genetic risk sufficiently enough to warrant the trade-off of lower adult return rate. Such determinations will require an adaptive management approach of experimentation and monitoring followed by appropriate revision of release strategies. Certainly, if survival is equal across a range of life stages, fish should be released at the earliest of such life stages.

■ **Guideline:** Release fish in locations where the target population s naturally produced fish occur at the same life stage.

■ **Guideline:** Disperse release sites.

Once efforts have been made to maintain the development of hatchery fish in parallel with wild fish, it is important to release fish in habitat types where the life stage naturally occurs. This will ensure that the proper habitat for that life stage is available. For instance, many lake-dwelling species spawn in shallow nearshore environments, and the young may stay in the protection of the weeds or gravel for days to weeks. As they grow, their needs for protection and prey change so that they shift to deeper, midlake locations. For migratory fish, it is important that released fish are in the appropriate location to complete their movement patterns in coordination with the natural population.

A massive influx of stocked fish into a single region of a stream may overwhelm available resources (e.g., food and cover). This could result in decreased survival of planted fish, of natural fish of the targeted population, and of other species in the local area. Close and Anderson (1992) found that stocked steelhead fry did not disperse evenly throughout the available stream habitat. This resulted in locally high densities of juveniles near release sites. At high densities, growth was reduced and fewer fish reached the critical size necessary to survive the first winter in these streams. Close and Anderson recommended stocking based on a threshold-weight carrying capacity (i.e., the maximum number of fish that will grow large enough to reach the threshold size for winter survival) and dispersing release sites so that this threshold could be achieved throughout the available habitat.

■ **Guideline:** Acclimate juveniles by exposing them to water from the target stream.

■ **Guideline:** Allow volitional movement of juveniles to the target stream.

The remarkable homing precision of some species (e.g., anadromous salmonids) is clearly important for their reproductive success and the maintenance of locally adapted traits. Salmonids imprint on freshwater olfactory and visual cues before and during smoltification (physiological transformation in preparation for entering the marine environment); therefore, hatchery fish should be exposed to the imprinting stimuli of the location to which they are expected to return as adults. Proper imprinting is required to reduce excessive straying, which may reduce between-population differences.

In many cases, the simplest way to ensure imprinting is to build small-scale hatcheries near the locations to be supplemented. Fish can be reared with water from the receiving site and allowed volitional out-migration. Volitional migration also should reduce the potential for swamping a site with a large influx of stocked fish. If compelling reasons require fish to be released away from the hatchery site, they should be moved before they are physiologically ready to imprint and allowed to acclimate in ponds adjacent to and supplied with water from the receiving location.

14.4 Conclusion

Although the role of hatcheries in fisheries management is highly controversial (Incerpi 1996; Grimes 1998; Waples 1999), their use for rebuilding populations likely will expand as fish populations decline. By recognizing how each stage of hatchery production can impose genetic hazards—extinction, loss of within-population variation, loss of between-population variation, and domestication selection—managers can reduce the genetic risks of hatchery programs. Because of the many uncertainties and unmeasurable risks that remain, hatchery programs should be treated as adaptive management experiments, with careful attention to monitoring and re-evaluation of goals and protocols. Hatchery managers should keep up to date with new research findings because new information may help further reduce risks or point out unavoidable problems that cannot be circumvented. Finally, the preceding points suggest that these guidelines should be continuously re-examined and revised to incorporate future research findings and lessons learned by ongoing hatchery program evaluation.

14.5 References

Allendorf, F. W., and R. F. Leary. 1988. Conservation and distribution of genetic variation in a polytypic species, the cutthroat trout. Conservation Biology 2:170–184.

Allendorf, F. W., and N. Ryman. 1987. Genetic management of hatchery stocks. Pages 141–159 *in* N. Ryman and F. Utter, editors. Population genetics and fishery management. University of Washington Press, Seattle.

Bartley, D. M., and G. A. E. Gall. 1990. Genetic structure of chinook and gene flow in chinook salmon populations of California. Transactions of the American Fisheries Society 119:55–71.

Bartley, D., M. Bagley, G. Gall, and B. Bently. 1992. Use of linkage disequilibrium data to estimate effective size of hatchery and natural fish populations. Conservation Biology 6:365–375.

Benzie, J. A. H., and S. T. Williams. 1996. Limitations in the genetic variation of hatchery produced batches of the giant clam, *Tridacna gigas*. Aquaculture 139:225–241.

Berejikian, B. A., S. B. Mathews, and T. P. Quinn. 1996. Effects of hatchery and wild ancestry and rearing environments on the development of agonistic behavior in steelhead trout *(Oncorhynchus mykiss)* fry. Canadian Journal of Fisheries and Aquatic Sciences 53:2004–2014.

Busack, C. A., and K. P. Currens. 1995. Genetic risks and hazards in hatchery operations: fundamental concepts and issues. Pages 71–80 *in* Schramm and Piper (1995).

Campton, D. E. 1995. Genetic effects of hatchery fish on wild populations of Pacific salmon and steelhead: what do we really know? Pages 337–353 *in* Schramm and Piper (1995).

Chilcote, M. W., S. A. Leider, and J. J. Loch. 1986. Differential reproductive success of hatchery and wild summer-run steelhead under natural conditions. Transactions of the American Fisheries Society 115:726–735.

Close, T. L., and C. S. Anderson. 1992. Dispersal, density-dependent growth, and survival of stocked steelhead fry in Lake Superior tributaries. North American Journal of Fisheries Management 12:728–735.

Cuenco, M. L., T. W. H. Backman, and P. R. Mundy. 1993. The use of supplementation to aid in natural stock restoration. Pages 269–294 *in* J. G. Cloud and G. H. Thorgaard, editors. Genetic conservation of salmonid fishes. Plenum Publishing, New York.

Currens, K. P., and five coauthors. 1997. Introgression and susceptibility to disease in a wild population of rainbow trout. North American Journal of Fisheries Management 17:1065–1078.

Exadactylos, A., A. J. Geffen, and J. P. Thorpe. 1999. Growth and genetic variation in hatchery-reared larval and juvenile Dover sole, *Solea solea* (L.). Aquaculture 176:209–226.

Fast, D. E., J. Hubble, M. Kohn, and B. Watson. 1991. Yakima spring chinook enhancement study, final report. Bonneville Power Administration, Division of Fish and Wildlife, Project 82-16, U.S. Department of Energy, Portland, Oregon.

Flagg, T. A., C. V. W. Mahnken, and K. A. Johnson. 1995. Captive broodstocks for recovery of Snake River sockeye salmon. Pages 81–90 *in* Schramm and Piper (1995).

Fleming, I. A. 1996. Reproductive strategies of Atlantic salmon: ecology and evolution. Reviews in Fish Biology and Fisheries 6:379–416.

Fleming, I. A., and M. R. Gross. 1994. Breeding competition in a Pacific salmon (coho: *Oncorhynchus kisutch*): measures of natural and sexual selection. Evolution 48:637–657.

Flick, W. A., and D. A. Webster. 1964. Comparative first year survival and production in wild and domestic strains of brook trout, *Salvelinus fontinalis*. Transactions of the American Fisheries Society 93:58–69.

Frankel, O. H., and M. E. Soulé. 1981. Conservation and evolution. Cambridge University Press, Cambridge, UK.

Gharrett, A. J., and S. M. Shirley. 1985. A genetic examination of spawning methodology in a salmon hatchery. Aquaculture 47:245–256.

Gilpin M. E., and M. E. Soulé. 1986. Minimum viable populations: the processes of extinction. Pages 13–34 *in* M. E. Soulé, editor. Conservation biology: the science of scarcity and diversity. Sinauer Associates, Sunderland, Massachusetts.

Grimes, C. B. 1998. Marine stock enhancement: sound management or techno-arrogance? Fisheries 23(9):18–23.

Hard, J. J. 1995. Genetic monitoring of life-history characters in salmon supplementation: problems and opportunities. Pages 212–225 *in* Schramm and Piper (1995).

Hard, J. J., R. P. Jones, Jr., M. R. Delarm, and R. S. Waples. 1992. Pacific salmon and artificial propagation under the Endangered Species Act. NOAA (National Oceanic and Atmospheric Administration) Technical Memorandum NMFS (National Marine Fisheries Service)-NWFSC-2.

Hedrick, P. W., D. Hedgecock, and S. Hamelberg. 1995. Effective population size in winter-run chinook salmon. Conservation Biology 9:615–624.

Iguchi, K., K. Watanbe, and M. Nishida. 1999. Reduced mitochondrial DNA variation in hatchery populations of ayu *(Plecoglossus altivelis)* cultured for multiple generations. Aquaculture 178:235–243.

Incerpi, A. 1996. Hatchery-bashing: a useless pastime. Fisheries 21(5):28.

Kapuscinski, A. R. 1991. Genetic analysis of policies and guidelines for salmon and steelhead hatchery production in the Columbia River Basin. Prepared for the Northwest Power Planning Council, Agreement 90-037, Portland, Oregon.

Kincaid, H. L. 1983. Inbreeding in fish populations used in aquaculture. Aquaculture 33:215–227.

Kincaid, H. L. 1995. An evaluation of inbreeding and effective population size in salmonid broodstocks in federal and state hatcheries. Pages 193–204 *in* Schramm and Piper (1995).

Lande, R. 1988. Genetics and demography in biological conservation. Science 241:1455–1460.

Leider, S. A., P. L. Hulett, J. J. Loch, and M. W. Chilcote. 1990. Electrophoretic comparison of the reproductive success of naturally spawning transplanted and wild steelhead trout through the returning adult stage. Aquaculture 88:239–252.

Lythgoe, K. A. 1997. Consequences of gene flow in spatially structured populations. Genetical Research 69:49–60.

Maynard, D. J., T. A. Flagg, and C. V. W. Mahnken. 1995. A review of seminatural culture strategies for enhancing the postrelease survival of anadromous salmonids. Pages 307–314 *in* Schramm and Piper (1995).

Maynard, D. J., T. A. Flagg, and C. V. W. Mahnken, editors. 1996. Development of a natural rearing system to improve supplemental fish quality. 1991–1995 progress report to the Bonneville Power Administration, Contract DE-A179-91-BP20651, Portland, Oregon.

Miller, W. H., T. C. Coley, H. L. Burge, and T. T. Kisanuki. 1990. Analysis of salmon and steelhead supplementation: emphasis on unpublished reports and present programs. Part 1 *in* W. H. Miller, editor. Analysis of salmon and steelhead supplementation, parts 1–3. Bonneville Power Administration, Technical Report 88–100, U.S. Department of Energy, Portland, Oregon.

Moyle, P. B. 1969. Comparative behavior of young brook trout of domestic and wild origin. The Progressive Fish-Culturist 31:51–56.

Newman, D., and D. Pilson. 1997. Increased probability of extinction due to decreased genetic effective population size: experimental populations of *Clarkia pulchella*. Evolution 51:354–362.

Nickelson, T. E., M. F. Solazzi, and S. L. Johnson. 1986. Use of hatchery coho salmon *(Oncorhynchus kisutch)* presmolts to rebuild wild populations in Oregon coastal streams. Canadian Journal of Fisheries and Aquatic Sciences 43:2443–2449.

Quinn, T. P. 1993. A review of homing and straying of wild and hatchery-produced salmon. Fisheries Research 18:29–44.

Reisenbichler, R. R. 1997. Genetic factors contributing to declines of anadromous salmonids in the Pacific Northwest. Pages 223–244 *in* D. J. Stouder, P. A. Bisson, and R. J. Naiman, editors. Pacific salmon and their ecosystems: status and future opportunities. Chapman and Hall, New York.

Reisenbichler, R. R., and S. P. Rubin. 1999. Genetic changes from artificial propagation of Pacific salmon affect the productivity and viability of supplemented populations. ICES Journal of Marine Sciences 56:459–466.

Ricker, W. E. 1972. Hereditary and environmental factors affecting certain salmonid populations. Pages 19–160 *in* R. C. Simon and P. A. Larkin, editors. The stock concept in Pacific salmon. H. R. MacMillan Lectures in Fisheries, University of British Columbia, Vancouver.

Ryman, N., and L. Laikre. 1991. Effects of supportive breeding on the genetically effective population size. Conservation Biology 5:325–329.

Ryman, N., and G. Stahl. 1980. Genetic changes in hatchery stocks of brown trout *(Salmo trutta)*. Canadian Journal of Fisheries and Aquatic Sciences 37:82–87.

Saccheri, I., and five coauthors. 1998. Inbreeding and extinction in a butterfly metapopulation. Nature (London) 392:491–494.

Schramm, H. L., Jr., and R. G. Piper, editors. 1995. Uses and effects of cultured fishes in aquatic ecosystems. American Fisheries Society, Symposium 15, Bethesda, Maryland.

Shaklee, J. B. 1983. The utilization of isozymes as gene markers in fisheries management and conservation. Pages 213–247 *in* Isozymes: current topics in biological and medical research, vol 11. Alan R. Liss, New York.

Siitonen, L., and G. A. E. Gall. 1989. Response to selection for early spawn date in rainbow trout, *Salmo gairdneri*. Aquaculture 78:153–161.

Simon, R. C., J. D. McIntyre, and R. A. Hemmingsen. 1986. Family size and effective population size in a hatchery stock of coho salmon *(Oncorhynchus kisutch)*. Canadian Journal of Fisheries and Aquatic Sciences 43:2434–2442.

Skaala, O., K. E. Jorstad, and R. Borgstrom. 1996. Genetic impact of two wild brown trout *(Salmo trutta)* populations after release of non-indigenous hatchery spawners. Canadian Journal of Fisheries and Aquatic Sciences 53:2027–2035.

Smoker, W. W., A. J. Gharrett, and M. S. Stekoll. 1998. Genetic variation of return date in a population of pink salmon: a consequence of fluctuating environment and dispersive selection? Alaska Fishery Research Bulletin 5:46–54.

Steward, C. R., and T. C. Bjornn. 1990. Supplementation of salmon and steelhead stocks with hatchery fish: a synthesis of published literature. Part 2 *in* W. H. Miller, editor. Analysis of salmon and steelhead supplementation, parts 1–3. Technical Report 90–1, Bonneville Power Administration, U.S. Department of Energy, Portland, Oregon.

Swain, D. P., and B. E. Riddell. 1990. Variation in agnostic behavior between newly emerged juveniles from hatchery and wild populations of coho salmon, *Oncorhynchus kisutch*. Canadian Journal of Fisheries and Aquatic Sciences 47:566–571.

Taniguchi, N., K Sumantadinata, and S. Iyama. 1983. Genetic change in the first and second generations of hatchery stock of black seabream. Aquaculture 35:309–320.

Templeton, A. R. 1986. Coadaptation and outbreeding depression. Pages 105–116 *in* M. E. Soulé, editor. Conservation biology: the science of scarcity and diversity. Sinauer Associates, Sunderland, Massachusetts.

Unwin, M. J. 1996. Fry-to-adult survival of natural and hatchery-produced chinook salmon *(Oncorhynchus tshawytscha)* from a common origin. Canadian Journal of Fisheries and Aquatic Sciences 54:1246–1254.

Utter, F. M. 1998. Genetic problems of hatchery-reared progeny released into the wild, and how to deal with them. Bulletin of Marine Science 62:623–640.

Utter, F. M., G. Miller, G. Stahl, and D. Teel. 1989. Genetic population structure of chinook salmon, *Oncorhynchus tshawytscha*, in the Pacific Northwest. U.S. National Marine Fisheries Service Fishery Bulletin 87:239–264.

Vincent, R. E. 1960. Some influences of domestication upon three stocks of brook trout (*Salvelinus fontinalis* Mitchell). Transactions of the American Fisheries Society 89:35–52.

Walters, C. 1986. Adaptive management of renewable resources. Macmillan Publishing Company, New York.

Waples, R. S. 1990. Conservation genetics of Pacific salmon. III. Estimating effective population size. Journal of Heredity 81:277–289.

Waples, R. S. 1999. Dispelling some myths about hatcheries. Fisheries 24(2):12–21.

Waples, R. S., and C. Do. 1994. Genetic risk associated with supplementation of Pacific salmonids: captive broodstock programs. Canadian Journal of Fisheries and Aquatic Sciences 51 (Supplement 1):310–329.

Waples, R. S., G. A. Winans, F. M. Utter, and C. Mahnken. 1990. Genetic approaches to the management of Pacific salmon. Fisheries 15(5):19–25.

Withler, R. E. 1988. Genetic consequences of fertilizing chinook salmon *(Oncorhynchus tshawytscha)* eggs with pooled milt. Aquaculture 68:15–25.

15

Genetic Impacts of Fish Introductions

Fred Utter

15.1 Introduction

Purposeful introductions of species and populations of fishes that are not native to a particular locality long have been practiced. The ubiquitous presence of species such as common carp *Cyprinus carpio*, rainbow trout *Oncorhynchus mykiss*, and western mosquitofish *Gambusia affinis* in suitable habitats far beyond their native ranges attests to the worldwide and long-term practice of primarily intentional introductions. There also have been numerous inadvertant introductions, typified in the North American upper Great Lakes by the sea lamprey *Petromyzon marinus* and the round goby *Neogobius melanostomus* (Fuller et al. 1999), and human-facilitated range extensions, such as those arising from the opening of the Suez Canal.

Concerns about introduced fishes reflect a growing awareness of humanity's responsibility to understand and manage natural ecosystems properly (Vitousek et al. 1997). Effects of introductions contribute to and are compounded by habitat disruptions (Courtenay 1995) and inevitably intensify as our human population exceeds 6 billion. Indeed, introduced fishes were factors in 70% of 69 listings under the U.S. Endangered Species Act of 1973 (16 U.S.C. §§ 1531 to 1544) (ESA) reported by Lassuy (1995).

This chapter's focus on the genetic impacts of introductions limits its content to those taxonomic levels at which the potential for genetic interactions exist, either within species or among closely related species. This focus complements information relative to introductions at or above the species level, as in the examples noted above. Recognition of the value of within-species genetic resources has risen with public awareness of the general need for conservation. The maintenance of genetic variability within and between populations has become an accepted goal for genetic conservation (Ryman 1991). Particularly noteworthy in this regard are the inclusion of "distinct population segments" for protection under such legislation as the ESA (reviewed in Waples 1991; Nielsen 1995; see also Chapter 18) and international agreements such as the Convention on International Trade in Endangered Species of Wild Fauna and Flora (1973). Relevant discussions also are found in the proceedings of a 1990 symposium (summarized in Allendorf 1991) that reviews the magnitude and ecological and genetic effects of fish introductions and in recent reviews by Hindar et al. (1991) and Ryman et al. (1995b).

This chapter focuses on the biological consequences of fish introductions within a taxonomic range in which actual or potential direct and indirect genetic effects can occur on local populations. Attention is given to direct effects of **introgressive hybridization,** defined here as the infiltration of the genome of one species or population into that of another, as well as to potential indirect effects (e.g., from competition, displacement, disease transfer, and nonintrogressive hybridization).

15.2 Detection of Genetic Changes

15.2.1 Traditional Markers

Past problems in detecting genetic impacts of fish introductions have been more practical than theoretical. The possibility of genetic consequences to populations brought about by human activities is not difficult to grasp by conceptualizing breakdowns of population and even species structures through hybridizations or losses of within-population genetic variability through reduced numbers of breeders (Chapter 9). Until fairly recently, the primary difficulty has been the absence of reliable genetic markers with which to measure these presumed alterations (see more detailed discussions in Chapters 2–5). The unknown genetic and environmental determinants of most traditional morphological markers have restricted their reliability largely to detection of first-generation interspecific hybrids. Breakdowns of within-species population structures traditionally have been inferred from records of transfers (e.g., Simon 1972). Losses of genetic variability within populations have been inferred from field data regarding fish and redd counts and timing and location of migration and spawning (e.g., MacLean and Evans 1981; Thorpe and Koonce 1981). The limited variability detected by karyotyping and the labor-intensive nature of the technique placed strong restrictions on studying chromosome sets as a source of genetic variation (Ihssen et al. 1981; Campton 1987; Thorgaard and Allen 1987; Chapter 3).

15.2.2 Molecular Markers

Since the late 1960s, molecular genetic procedures have revolutionized the use of genetic data for studying populations because of the sudden availability of abundant and readily detectable genetic variation (Utter 1991; see Chapters 2, 4, and 5 for history and applications in fishes). The ability to enumerate allelic differences at defined loci through these ubiquitous markers displaced much conjecture concerning genetic effects with sound empirical information. This molecular genetic information commonly has generated new and unexpected insights about distinct lineages both within and among species, as well as quantifying and clarifying many prevailing understandings about taxonomic relationships (Avise 1994; Kocher and Stepien 1997). In addition, allelic and genotypic information at the level of individual populations has proven extremely valuable for identifying mixtures of gene pools, monitoring genetic changes, and measuring genetic losses (e.g., Allendorf et al. 1987; Altukhov and Salmenkova 1987; Bugert et al. 1995; Utter and Ryman 1993).

The dramatic insights arising from applications of molecular genetic markers, however, have obscured the persisting and complementary value of traditional markers. Commonly influenced by large numbers of undefined genes coupled with sizeable and variable environmental influences, adaptive traits such as migration and spawning times, temperature tolerance, and disease resistance remain almost exclusively within the domain of traditional quantitative variation (Chapter 12). Hence, direct insights into adaptive traits have not been provided by new molecular markers. The validated assumption of more or less neutrality of most molecular variants (e.g., Eanes 1987; Nei 1987) has elevated their primary value as ancestral markers reflecting the evolutionary processes of drift and migration (and to a lesser degree, mutation, for examining within-species lineages). Reliable estimates of times since divergence permit, indeed mandate, inferences that differential adaptations have occurred but otherwise remain uninformative about the nature of such distinctions. Thus, traditional and molecular markers form complementary data sets for distinguishing populations because they reflect different evolutionary processes and time scales (Utter et al. 1993; Grant et al. 1999). Where not explicitly noted, the limitations of molecular data are implied throughout this chapter, although most examples relate to genetic impacts measured through molecular markers because of their ability to identify lineages and hybridizations.

■ 15.3 Insights from Introduced Salmonids

Salmonid fishes are a useful group for examining genetic effects of fish introductions on wild populations. Their commercial and sporting value and relative ease of culture have led to a long history of hatchery operations (e.g., Davis 1956). These attributes, coupled with a diversity of life history adaptations, have resulted in salmonid populations being distributed far beyond their native ranges in temperate freshwater and marine environments worldwide (e.g., MacCrimmon 1971; Welcomme 1988; Quinn et al. 1996; Fuller et al. 1999). Considerable literature has arisen on the genetic consequences of this extended culture and redistribution, much of which is generally pertinent to questions concerning genetic effects of fish introductions. Some of this information is explored in this section as a basis for subsequent considerations of pertinent generalities.

Evaluating the extent and manner by which native populations are genetically affected by releases of cultured fish commonly is complicated by the inability to separate the effects of exogenous (nonnative) origins from those of the cultured environment from which most such releases originate (Campton 1995; Utter 1998). In order to clarify this distinction, attention here primarily is directed toward concerns related to distinct lineages of native and introduced fishes, whereas Miller and Kapuscinski (Chapter 14) probe concerns relating to culture. An overview of some potential indirect and direct genetic effects arising from introductions of exogenous populations (Figure 15.1) covers a broad range of possibilities.

15.3.1 Indirect Genetic Effects

Indirect genetic effects (left, Figure 15.1) are those inducing genetic changes within or among populations without actual infiltration of exogenous genomes (i.e., introgressive hybridization) into native populations. Reductions in numbers within a population threaten its capabilities to adapt through lowered genetic variability (Box 15.1) and possibly inbreeding depression (Chapter 10); reduced numbers also increases a population's likelihood of extinction, the ultimate genetic loss (Chapter 18). Indirect genetic effects arise among subpopulations when isolation of formerly connected groups disrupt normal patterns of migration and gene exchange. As a consequence of diminished or eliminated reproductive, rearing, or migratory habitats, the reproductive and adaptive potential of the normally genetically interacting native subpopulations of a particular area

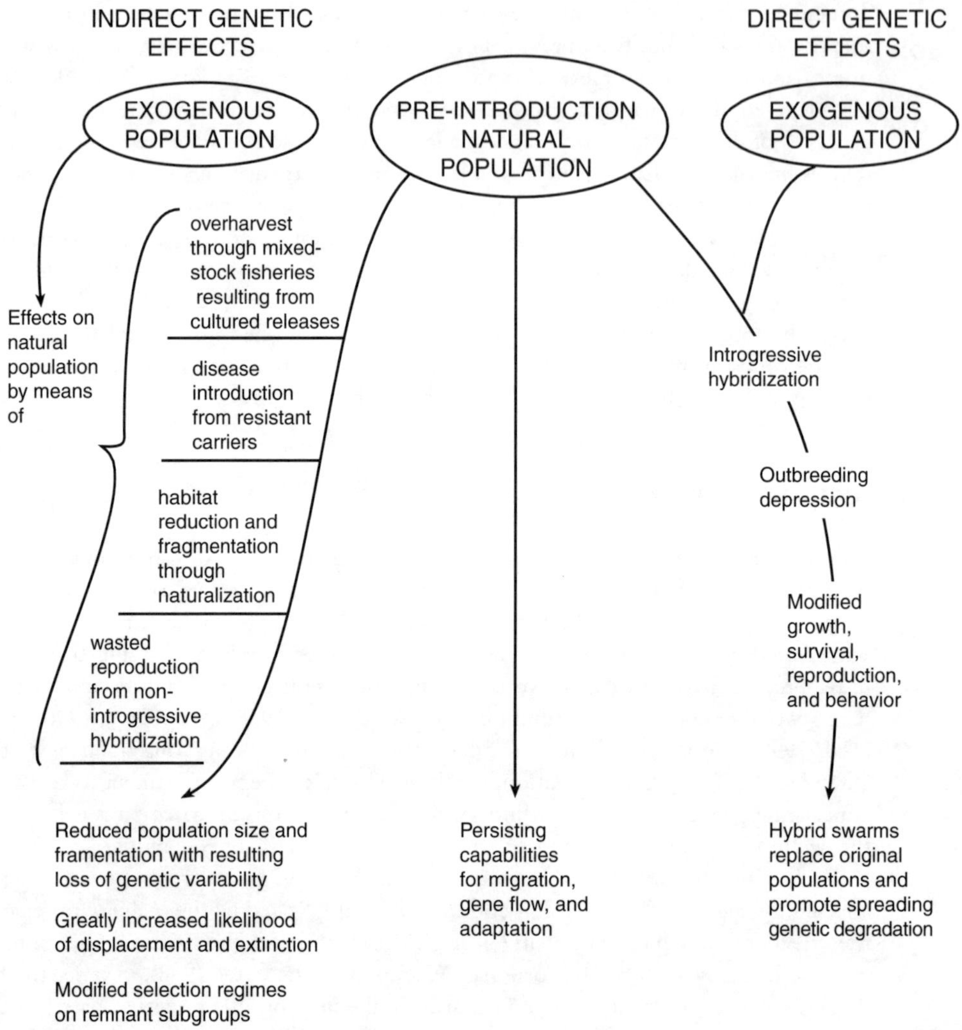

Figure 15.1 Some direct and indirect genetic effects of releases of exogenous, hatchery-reared fish on native populations. Figure adapted from Utter 1998.

become modified and threatened. A few examples in the context of those outlined in Figure 15.1 are discussed here (c.f. Hindar et al. 1991; Krueger and May 1991; Waples 1995 for more extended descriptions).

15.3.1.1 Mixed-Stock Fisheries. The presence of large numbers of cultured fish may stimulate excessive harvest pressure on wild populations. This topic has been extensively discussed and documented for anadromous salmon and trout populations of the genus *Oncorhynchus* in western North America (e.g., Wright 1981; Fraidenburg and Lincoln 1985; Lichatowich and McIntyre 1987). Recognition of excessive harvest pressure has prompted considerable activity among federal, state, tribal, and regional agencies, resulting in largely genetically based mixed-stock analysis (Utter and Ryman 1993). This analysis is currently being used in postseason and real-time harvest management applications aimed at protecting weaker wild populations of Pacific salmon (Shaklee et al. 1999;

■ Box 15.1 Effect of Supportive Breeding on Effective Size of Total Population

Eric Hallerman

The release of captive-bred fishes into the wild often is practiced to support demographically weak or threatened and endangered populations (Chapter 14). In many instances, the justification for a hatchery program is to boost the targeted population's numbers by increasing survival through the early life stages by means of captive rearing. Typically, a fraction of wild parents are captured and bred in captivity, and the offspring subsequently are released into the natural ecosystem where they mix with wild conspecifics. However, this process also implies that the reproductive success of one segment of the population may be favored, which will result in an increase in the total variance of family size, a parameter of critical importance in determining the genetically effective size (N_e) of the overall population (Chapter 9). The reduction of N_e that may result from breeding–release programs aimed at supporting natural populations was first described by Ryman and Laikre (1991). Such reduction in N_e through supportive breeding has been referred to as the **Ryman–Laikre effect** (see also Box 14.2).

The magnitude of the Ryman–Laikre effect can be estimated if a fisheries manager has a rough idea of the approximate effective sizes of the wild and captive populations and of their relative contributions to the overall population of offspring. Let's assume that there is a population of breeders divided into n groups, and let N_i be the effective number of breeders on the ith group (i = 1, 2, ..., n). Because the reproductive contribution may differ among groups, let x_i be the relative contribution of progeny from the ith group, so that the fraction x_i of individuals in the offspring generation is produced by the N_i effective breeders of that group. The effective sizes of the contributing groups and of the total population can be related as

$$\frac{1}{N_e} = \sum_{i=1}^{n}\left(\frac{x_i^2}{N_i}\right).$$

Note that *only* when contributions from the different groups are proportional to their effective sizes does N_e equal ΣN_i. Note also that in all *other* cases, when the offspring contributions from the respective groups are disproportional to their effective sizes, N_e is smaller than ΣN_i.

Let's go on, as did Ryman and Laikre (1991), to consider expressly a typical management scenario. A wild population of N_W effective parents reproduces in the wild; N_C effective parents are bred in captivity and their offspring are released into the wild to supplement natural recruitment. In this case, the total effective size of the population, N_e, is given by

$$\frac{1}{N_e} = \left(\frac{x^2}{N_C}\right) + \left(\frac{1-x^2}{N_W}\right),$$

continued on next page

■ Box 15.1 *continued*

where x is the relative contribution of offspring by captive parents and $(1 - x)$ that of wild parents. The figure below shows N_e as a function of the relative contributions of the wild and captive components of the overall population. It shows how supportive breeding may reduce the total N_e well below what it would have been without supportive breeding. For example, considering an example that would typify a salmonid stock enhancement program, suppose there was an N_W of 200 and an N_c of 20 that contributed 40% of the total number of offspring. Under these assumptions, the total N_e would be approximately 100, or only half that of the wild population alone. Studies using genetic markers have shown that the Ryman–Laikre effect is realized in the field. For example, Tessier et al. (1997) reported a 50% reduction in N_e in a wild Atlantic salmon population caused by the release of hatchery fish.

Supportive breeding, thus, poses a trade-off. Although there is a gain in total production of offspring, there likely will be a simultaneous reduction in the total N_e, resulting in loss of genetic variability. In many cases, the loss of genetic variability may be regarded as justified given the overall gain in production. However, when the absolute size of the wild population is small, supportive breeding can lead to serious depletion of genetic variability of the overall population. Hence, when extinction probabilities are high because of demographic processes and supportive breeding is most needed, supportive breeding actually may increase genetic stochasticity (Chapter 18). Ryman and Laikre (1991) concluded that even in a case in which only a one-generation boost in reproductive success appears justified to reduce the probability of extinction, the trade-off must be carefully evaluated. In such a situation, it may be appropriate to consider capture of the entire wild population for a single generation of captive propagation, thereby avoiding the effects of differential reproductive success of wild and captive parents (see also section 14.3.1.2).

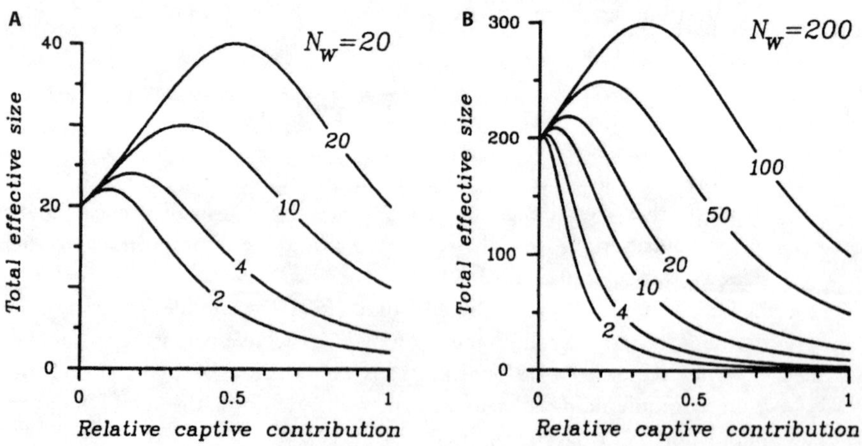

Total effective population size (wild + captive) when a natural population of (A) 20 and (B) 200 effective parents is supported by different numbers (labeled curves) of effective parents from a hatchery. The x-axis represents the proportion of offspring contributed by captive breeders. Figure adapted from Ryman and Laikre (1991).

Lincoln 1994; Utter and Ryman 1993; Chapter 13). Although such protective efforts are encouraging, mixed-stock fisheries persist as a major contributor to declines of natural populations, being associated in 100 of 214 naturally spawning groups of Pacific salmon recently identified as at risk of extinction (Nehlsen et al. 1991).

The problem of mixed-stock fisheries is not limited to human harvests. For instance, in Oregon predators attracted to large-scale releases of cultured fish have been implicated in the decline of co-migrating native populations of coho salmon *O. kisutch* (Anonymous 1987). The possibility of similar predation by northern pikeminnow *Ptychocheilus oregonensis* attracted by releases of juvenile salmonids in the Columbia River has been described (Steward and Bjornn 1990).

15.3.1.2 **Disease Introduction.** Reductions of native populations from diseases introduced by nonnative fishes are well documented. A particularly severe problem of current interest concerns the fluke *Gyrodactylus salaris*, presently infesting susceptible Norwegian populations of Atlantic salmon *Salmo salar* by means of introductions from resistant Baltic fish (Bakke et al. 1990). The problem has been compounded by furunculosis introduced to Norwegian cultured fish from cultured Scottish Atlantic salmon infected with the bacterium *Aeromonas salmonicida*. In addition to spreading furunculosis, escaped cultured fish have competed and possibly interbred with or displaced the decimated wild populations in spawning areas (e.g., Hindar 1992; Saegrov et al. 1997). The long-term viability of Norwegian wild Atlantic salmon has been questioned if these conditions persist (Fleming et al. 2000). Whirling disease, caused by the myxosporidean *Myxobolus cerebralis*, was introduced to North America from shipments of European brown trout *Salmo trutta* in 1955 (Hulbert 1996). Presently threatening many native and nonnative trout fisheries, the disease has become a major fish health issue (Nehring and Walker 1996; Bergersen and Anderson 1997; Allendorf et al., 2001).

15.3.1.3 **Wastage of Gametes.** **Nonintrogressive hybridization** (i.e., where sterility or reduced fitness precludes reproductive success of hybrids) between released hatchery and native fish promotes displacement of native fish through wastage of gametes and habitat. An increased incidence of first-generation hybridization between Atlantic salmon and brown trout (Verspoor and Hammar 1991) has occurred since escaped net-pen-reared Atlantic salmon have returned to spawn in Scottish rivers (Youngson et al. 1993), indicating a potential threat of such escapes to native populations of both species. This process also appears to be contributing to the displacement of endangered bull trout *Salvelinus confluentus* by introduced brook trout *Salvelinus fontinalis* in the western United States (Leary et al. 1993, 1995).

15.3.1.4 **Fragmentation of Habitat.** However mediated (e.g., mixed-stock fisheries, disease, competition, or nonintrogressive hybridization), displacements by naturalized nonnative populations or species have contributed to fragmentation of native populations in diverse marine and freshwater habitats (e.g., Ryman et al. 1995b). Such disruptions inevitably reduce the amount of contiguous habitat available for population maintenance. The resulting restrictions in gene flow and rearing and reproductive space reduce the numbers of breeders, promote short- and long-term reductions of adaptive capabilities through genetic degradation, and increase the likelihood of local extinctions. The threat to survival of species that are dependent on maintenance of substantial contiguous habitat, such as bull trout (Rieman and McIntyre 1995), grows as fragmentation increases.

Modified and restricted habitats inevitably impose different selective pressures on populations that survive disruptions inflicted by nonnative introductions, although measurement of such changes in terms of selective pressures and responses may be difficult (Waples 1991). For instance, the appearance of riverine sockeye salmon *O. nerka* (an atypical life history adaptation circumventing the normal need of a lake for juvenile rearing and adult maturation; see Gustafson and Winans 1999; Wood 1995) in tributaries of the upper Columbia River is correlated with major habitat disruptions and relocations occurring within this region over the past 60 years (Chapman et al. 1995). Characteristics of populations colonizing formerly uninhabited areas also provide insights into adaptive divergence, as discussed in section 15.4.

15.3.2 Direct Genetic Effects through Introgressive Hybridization

The extensive aquaculture and translocation involving *Oncorhynchus* spp. (e.g., Nielsen 1995; Schramm and Piper 1995) provide a useful basis for seeking possible generalities concerning the occurrence and effects of introgression among subgroups of a single major lineage. Divergence into two separate lineages about 5 million years ago (Behnke 1992) led to the trout of (predominantly) western North America and the Pacific salmon (Figure 15.2). Progressively finer branchings of subgroups represent hierarchical differences in degree of genetic and ecological divergence. Within species, the branching typically represents subgroups that retain genetic distinction through some continuing degree of reproductive isolation (e.g., Figure 15.3).

The potential for introgression within either of the major branches of *Oncorhynchus* more often relates to complexity of life history than to degree of genetic divergence (Utter 2000). Particularly notable are native anadromous salmon populations that commonly resist intrusions by nonnative conspecifics (Ricker 1972; Withler 1982; Hendry et al. 1996). Presumably, introduced fish lack the detailed local adaptations particular to native populations (e.g., maturation schedules, routes and timing of migrations, stamina, temperature tolerance, and disease resistance), differences that minimize introgression or displacement. The rarity of successful introductions of anadromous sockeye salmon within the species' native range contrasts with the relative ease with which nonmigratory kokanee (lacustrine sockeye salmon), with a much simpler life history, have become established (Allendorf and Waples 1995). Similarly, myriad worldwide introductions of rainbow trout (MacCrimmon 1971) contrast with steelhead (anadromous rainbow trout) populations being resistant to conspecific intrusions (Phelps et al. 1994a; Utter 1998). Native Atlantic salmon populations in Spain (Moran et al. 1994) and Denmark (Nielsen et al. 1999) also have resisted introgression or displacement despite extensive and long-term introductions of nonnative fish.

However, an anadromous life history does not release a population from concerns about intraspecific introgression or displacement. Although chinook salmon of the upper Columbia River basin have resisted introgression or displacement by introduced downstream populations (Utter et al. 1995), introgression has occurred from cultured upper Columbia River fall-run fish straying into a genetically distinct adjacent hatchery population on the Snake River (Bugert et al. 1995). The failure of the downstream translocations appears to reflect, in part, the absence of adaptation to the rigors of upstream migration. The introgression in the Snake River occurred between two groups of more recent divergence, both adapted to upstream migration and a common spawning time. The greater success of translocations of upstream populations to downstream or coastal areas (Utter et al. 1989) is consistent with this reasoning. Notably, these successful

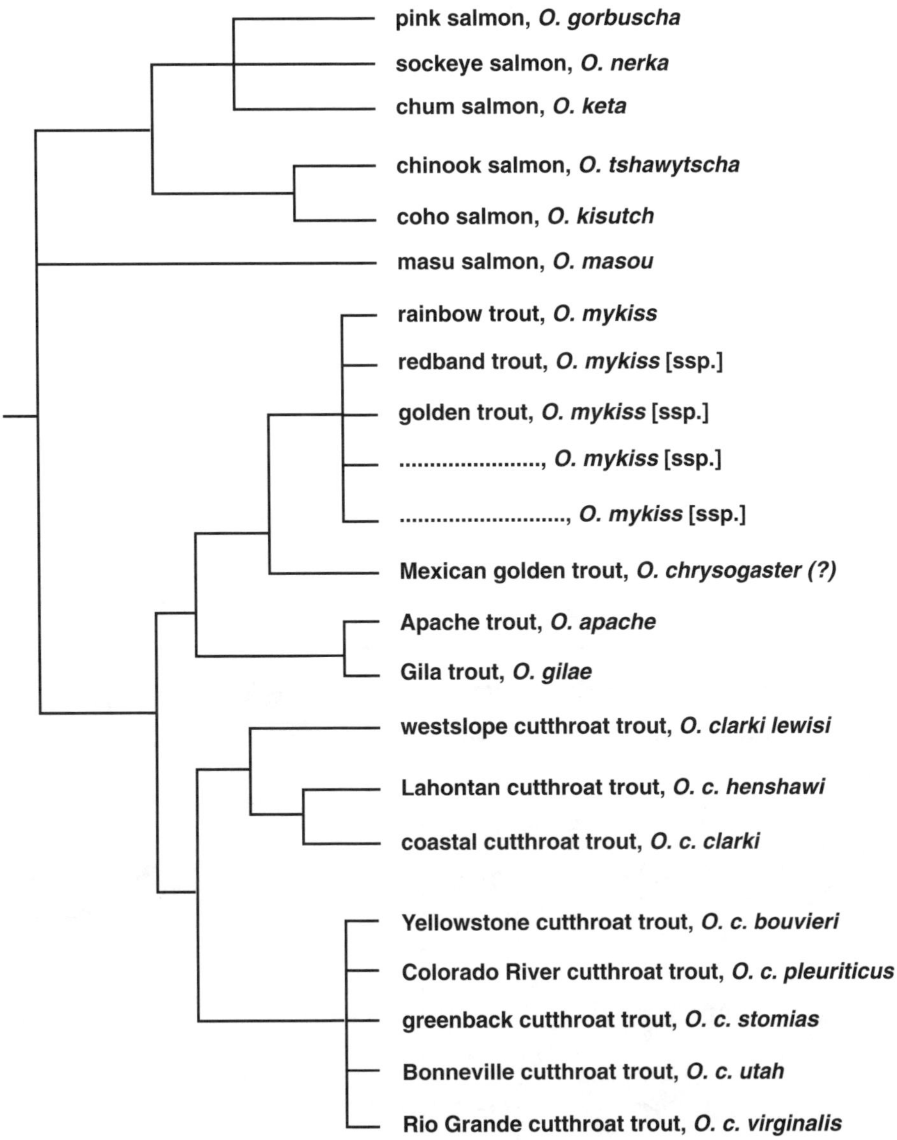

Figure 15.2 Consensus phylogenetic tree of *Oncorhynchus* based on a review of morphological and molecular data. Branch lengths approximate relative divergences among taxa. The question mark reflects uncertainty about the supposed Mexican golden trout collection. Dotted lines represent additional undescribed but diverged conspecific taxa of *O. mykiss* (e.g., Berg 1987). Figure adapted from Utter and Allendorf 1994.

translocations have not resulted in measurable introgressions with native populations (Utter 2000). These differences are considered later in conjunction with discussion on outbreeding depression (section 15.4).

In contrast to resistance to introgression between some conspecific anadromous lineages, introgressions between species and distinct subspecies of western trout (Figure 15.2) commonly involve nonanadromous groups that have diverged allopatrically and for a much longer time. The persisting capabilities of these allopatric taxa for naturally

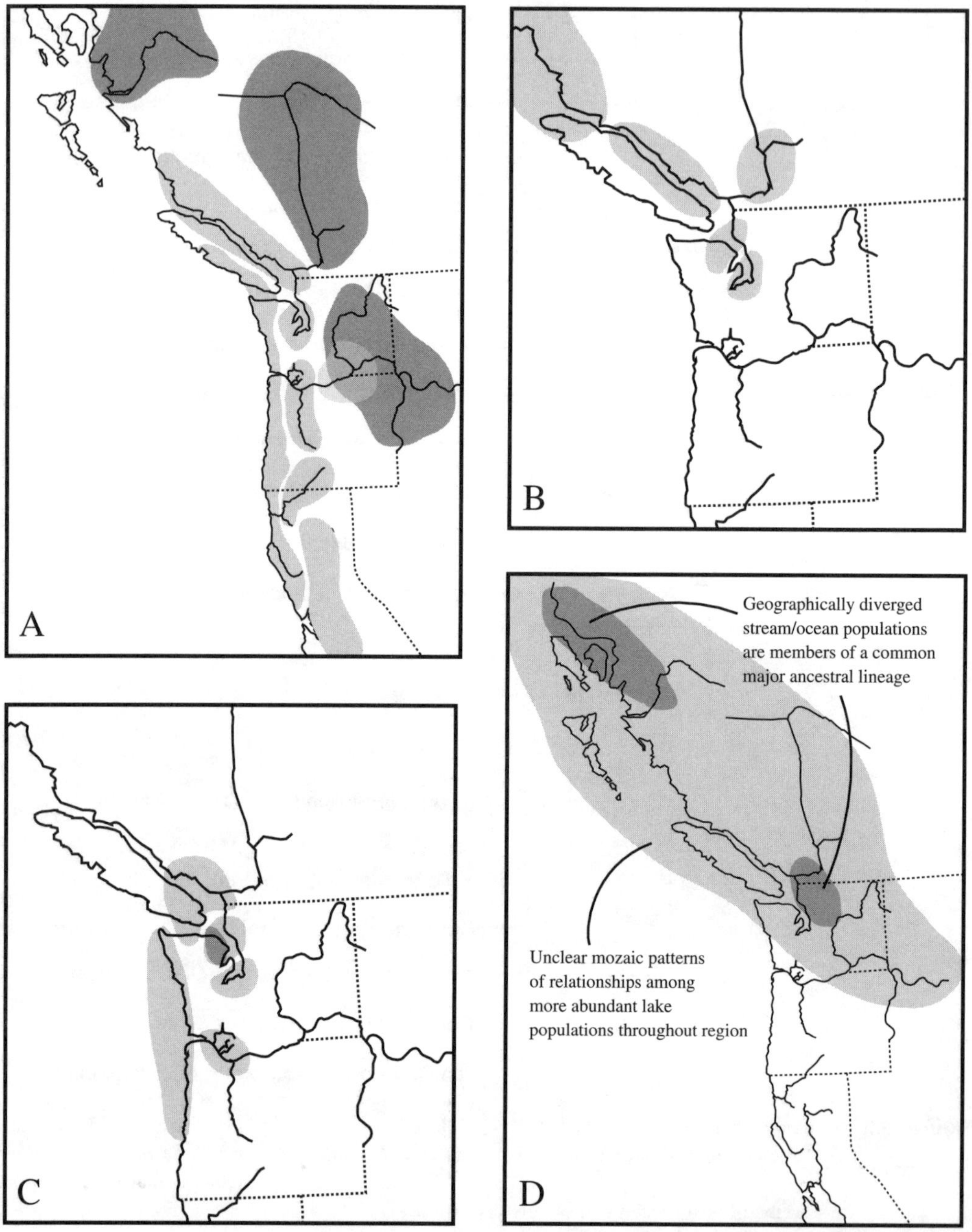

Figure 15.3 Major population groupings of four species of Pacific salmon in the Pacific Northwest. **(A)** Chinook salmon, contrasting major ancestral groupings putatively descended from Bering (dark groupings) or north Pacific (light groupings) refugia (Utter et al. 1989; 1992, 1995; Teel et al. 2000). Index of genetic differentiation among populations (F_{ST}) = 0.123. **(B)** Pink salmon, odd-year cycle (Shaklee et al. 1991). F_{ST} = 0.007. **(C)** Chum salmon, contrasting early- (light groupings) and late-run (dark grouping) lineages (Phelps et al. 1994b; Johnson et al. 1997). F_{ST} = 0.028. **(D)** Anadromous sockeye salmon, contrasting mosaic relationships of lake-type (light grouping) and close affinities of riverine (dark groupings) sockeye salmon populations (Wood 1995; Gustafson and Winans 1999). F_{ST} = 0.172 (lake type), 0.013 (riverine).

producing viable progeny contrast with congeneric species that have evolved sympatrically, such as coastal cutthroat trout and steelhead, which occasionally hybridize but appear to have resisted large-scale introgressions (e.g., Campton and Utter 1987). These allopatric groups apparently have retained the intrinsic capability for producing viable progeny in breedings among them in the absence of mechanisms that strengthen their reproductive isolation. Particularly notable are inland populations of native trout introgressed with introduced nonnative rainbow trout or introduced subspecies of cutthroat trout *O. clarki* (Carmichael et al. 1995; Leary et al. 1995).

The widespread occurrence of such introgressions masks the apparent resistance to interbreeding among similar introduced congeneric or conspecific fish inherent in some native populations. For instance, local environmental conditions coupled with innate migratory behavior of introduced Yellowstone cutthroat trout apparently prevented their establishment and introgression with native westslope cutthroat trout in Avalanche Lake in Glacier National Park (Verschuren and Marnell 1997). Similarly, due to adaptations to higher temperatures of native fish, Columbia River redband trout *O. mykiss gairdneri* have resisted introgression or displacement following widespread stocking of nonnative hatchery rainbow trout in Idaho (Wishard et al. 1984; Williams et al. 1996; but see also Leary et al. 1995).

Introgressive hybridizations following introductions of other nonanadromous populations in other salmonid and nonsalmonid genera are well documented. Recent reviews on this topic include those of Verspoor and Hammar (1991), Leary et al. (1995), and Sheridan (1995). Introduced hatchery populations of brown trout have widely introgressed with, or displaced, native Spanish (Cagigas et al. 1999; Garcia-Marin et al. 1999) and French (Guyomard 1989) populations. Of 38 brook trout populations surveyed in Great Smoky Mountains National Park (North Carolina), 8 were hybrid swarms with introduced northeastern hatchery populations, and 8 others were of pure hatchery origin (Hayes et al. 1996).

Among nonsalmonids, numerous introgressive hybridizations involving freshwater black bass *Micropterus* spp. have been reported. Introductions of the Florida largemouth bass *M. salmoides floridanus* have resulted in introgressions with native populations of northern largemouth bass *M. salmoides salmoides* in many locations (Philipp 1991). Introgression following introductions of smallmouth bass *M. dolomieu* have been recorded with Guadalupe bass *M. treculi* (Whitmore 1983) and spotted bass *M. punctulatus* (Avise et al. 1997; Pierce and Van Den Avyle 1997); in the latter instance, introgression was not detected in lakes where both species were naturally sympatric. Other reported nonsalmonid introgressions include threats to Pecos pupfish *Cyprinodon pecosensis* from introduced sheepshead minnow *C. variegatus* in New Mexico (Wilde and Echelle 1992), potential breakdowns of native walleye *Stizostedion vitreum* and sauger *Stizostedion canadense* populations through stockings of hybrid saugeye (sauger ♂ × walleye ♀) (White and Schell 1995), and erosions of native Monterrey platyfish *Xiphophorus couchianus* populations in Mexico from introduced congenerics (Contreras and Escalante 1984).

15.4 Effects of Hybridization and Translocation on Performance and Fitness

The displacements and hybridizations involving native and closely related introduced taxa described in the preceding sections have both long- and short-term implications. The long-term considerations relate to the need to identify and maintain biological diversity over a continuum of levels from individual organisms through local ecosystems to the totality of life. Such a legacy is the basis for the presently intensifying conservation orientation of individuals, groups, and governments (e.g., Ryman et al. 1995b). Continued erosions of native populations and higher taxa initiated and mediated through introductions of nonnative organisms reduce the overall biological diversity and promote the ultimate bankruptcy of the biological reserves required for the continued adaptation of organisms and ecosystems in a constantly changing biosphere.

However, the affected local populations are much more than reservoirs of biological diversity. Evidence for predominantly superior performance of native over introduced or introgressed populations continues to accumulate. Expectations that hybridized populations could provide new genetic vitality in conditions of environmental instability (e.g., Moav et al. 1978) are not strongly supported by empirical data. Rather, accumulating circumstantial evidence (e.g., Ricker 1972; Hindar et al. 1991; Utter 2000) and direct experimental data consistently document the superior performance of local populations over hybridized and purely nonnative ones in areas where habitats remain relatively unchanged, as indicated in examples in Table 15.1.

Outbreeding depression (e.g., Templeton 1986, Chapter 11) reflects breakdown of local adaptations through hybridizations and is a more realistic expectation of hybridization. Two types of outbreeding depression may affect hybridized progeny from parents of diverged lineages (Gharrett and Smoker 1991; Lynch 1991). The first category concerns local adaptations that result in native populations being better adapted than are either introduced or hybridized groups. The second category concerns the interactions of sets of genes affecting a specific adaptation that may differ between two diverged lineages. For instance, even- and odd-year pink salmon of Auke Creek, Alaska, are presumably adapted to a common environment, although they have been diverged for over 10,000 years (Zhivotovsky et al. 1994). Survival of second-generation between-lineage hybrids dropped markedly from that of first-generation crosses or parent populations (Gharrett and Smoker 1991; Gharrett et al. 1999; Chapter 11). Full sets of coadapted genes present in the first hybrid generation purportedly masked distinct coadaptations, which broke down during the recombination accompanying reproduction. Adapted to a common environment, each lineage apparently found "a different genetic solution to a common problem of local adaptation" (Gharrett and Smoker 1993b).

The distinction between these two types of outbreeding depression has important consequences with regard to intraspecific introgression. As with even- and odd-year pink salmon, major lineages of anadromous salmonids generally appear to resist introgression from other lineages because of both kinds of outbreeding depression. The pattern of characteristics that define distinct lineages differ among congeneric species that have overlapping geographic distributions (Figure 15.3) despite such common attributes as semelparity, strong homing tendencies, anadromy, and common access to deglaciated areas during the past 10,000 years (bounded just south of the U.S.–Canadian border). A clear and primarily geographic pattern of lineages is evident in three of the figured species (chinook, pink, and chum salmon). This configuration attests to introgressive resistance from other lineages, considering a long history of attempted translocations for these species throughout this range (Ricker 1972; Withler 1982; Utter 2000). A similar history supports a distinct lineage status for each lake-type sockeye salmon population (Allendorf and Waples 1995; Wood 1995). The higher indices of genetic differentiation among populations (F_{ST}) of chinook salmon and lake-type sockeye salmon (Figure 15.3) are consistent with stronger resistance to between-lineage introgressions in these groups.

However, gene flow among populations within major anadromous lineages appears to be much less constrained (e.g., see examples in Utter 2000). The existence of or capa-

Table 15.1 Comparisons of performance of native, introduced, and hybridized populations.

Observation	Reference
Superior survival and growth of parent species over hybrids in experimental crosses of western trout	Leary et al. 1995
Reduced survival and growth of pure and introgressed Florida largemouth bass relative to native northern largemouth bass in northern habitats	Philipp 1991
Reduced survival and growth among relocated northern largemouth bass populations relative to native populations	Philipp and Claussen 1995
Reduced survival or homing ability relative to native pink salmon in hybrids from distant populations or different year-classes	Bams 1976; Gharrett and Smoker 1991
Greater survival of rainbow trout fry of wild origin than those of hybrid or hatchery origin	Reisenbichler and McIntyre 1977
Reduced survival of naturally produced rainbow trout fry of hatchery parents relative to fry of wild parents	Chilcote et al. 1986; Campton et al. 1991
Reduced reproductive success (16%) of escaped farmed Atlantic salmon relative to native fish, while displacement and competition from invasions of escapees depressed native productivity by more than 30%	Fleming et al. 2000
Reduced resistance to endemic parasitic infections of wild rainbow trout populations following long-term introgression with nonnative hatchery fish	Currens et al. 1997.
Established populations of pink salmon in the Great Lakes and chinook salmon in New Zealand became differentiated from parent populations in allelic and life history characteristics	Withler 1982; Gharrett and Thomason 1987; Quinn and Unwin 1993; Quinn et al. 1996; Kinnison et al. 1998
Distinct colonization patterns of genotypically and phenotypically diverged brown trout populations that were simultaneously introduced to trout-free Swedish area	Palm and Ryman 1999

bility for contemporary gene flow within lineages prevents accumulation of strong coadaptations among gene loci, the reduction of which presumably reduces or removes the second major category of outbreeding depression. The resulting weakened hybrid penalties affecting progeny of within-lineage hybridizations increase the hybridized progeny's relative reproductive fitness and the potential for gene flow among subgroups where divergence (e.g., temporal or geographic) has occurred. This substructure appears to be a widespread but elusive and ephemeral reflection of constantly shifting microhabitat and climatic dynamics (e.g., Gharrett and Smoker 1993a,b; Sugg et al. 1996; Kinnison et al. 1998; Hendry et al. 1999). Because this divergence exists near or beyond the threshold of molecular genetic markers, subgroup identification and protection is particularly challenging (Utter et al. 1993; Grant et al. 1999; Whitlock and McCauley 1999). Two pertinent conclusions arise regarding introductions within anadromous salmonid lineages: (1) activities that promote unnaturally high levels of gene flow (e.g., outplantings or strayings from fish culture operations) inevitably break down existing substructure, and (2) continued excesses of such effective strayings prevent substructure reformation to the detriment of optimal natural productivity.

These proposed differential dynamics of outbreeding depression between and within major lineages of anadromous salmonids represent apparent tendencies rather than rigid principles. Nevertheless, the cumulative data point to important differences that provide stimuli for further studies to refine or modify the generality of the tendencies and to appreciate more fully the importance of outbreeding depression relative to all introductions of fishes.

Once established, introduced populations continue to evolve and diverge in their new habitats as indicated in the final examples of Table 15.1. Pink salmon were introduced into the Great Lakes through a single small and inadvertant release in 1955. The ensuing colonists diverged from the strict 2-year life history of all native populations to multiple spawning years for a given year-class in their new habitat (e.g., Withler 1982; Gharrett and Thomason 1987). Although the genetic and environmental components of the divergence remain to be determined, chinook salmon introduced into New Zealand have genetically diverged in different segments of their new habitat with regard to timings of return and maturation (Kinnison 1999). Useful insights also arise from studies of diverged conspecific groups simultaneously introduced into unoccupied areas, such as reported from a 9-year monitoring of brown trout in a Swedish area previously vacant of this species (Palm and Ryman 1999). Differential colonization, maturation schedules, and interbreeding of two genotypically and phenotypically distinct brown trout populations provided valuable guidelines for general consideration of translocations of these different conspecific lineages prior to the attempts. A word of caution must accompany such testimonies to the evolutionary flexibility of these examples of new inhabitants—the unmeasured but likely displacements resulting from such "successful" translocations (e.g., Sheridan 1995).

15.5 Perspective: General Relevance of Genetic Effects following Introductions

The examples of genetic impacts illustrated to this point outline general areas of concern for the protection of native populations directly or indirectly affected by introductions. The repeatedly expressed basic principles for appropriately dealing with these concerns relate to identifying and conserving genetic resources (e.g., FAO/UNEP 1981; Hindar et al. 1991; Ryman 1991; Allendorf and Waples 1995; Grant et al. 1999). These two topics are considered here more fully.

15.5.1 *Identification of Genetic Resources*

Understanding population structures underlies all other genetic concerns in considering interactions of introduced and native populations. This critical need has been outlined in a position statement of the Genetics Section of the American Fisheries Society (Utter 1995) presented at the symposium "Uses and Effects of Cultured Fishes in Aquatic Ecosystems" (Schramm and Piper 1995). The requirements for adequate understanding included (1) field studies to verify the existence and ecological parameters of natural populations, (2) laboratory studies to identify the geographic patterns of native ancestral groupings and the presence of nonnative or introgressed populations resulting from introductions, and (3) determinations of genetic and ecological relationships among hatchery, native, and other naturally reproducing populations. Mismanagement is inevitable without such information because relationships between introduced and native populations remain conjectural and detecting genetic changes due to introgression, displacement, or excessive inbreeding is precluded.

Mounting recognition of the importance of local adaptations has motivated the development of guidelines for management within biologically meaningful subspecific groupings. Concerns about defining such groupings intensified under need to define "distinct population segments" under the ESA. A particularly useful concept emerging from this need is to manage within **evolutionarily significant units** (ESU; Waples 1991; Nielsen 1995; Currens et al. 1998; Wainwright and Waples 1998; Box 18.1). To be considered an ESU, a population must be substantially reproductively isolated from other conspecific populations units (based on genetic distinctions) and must represent an important component in the evolutionary legacy of the species (based on unique habitats and adaptations and degree of loss to the species should this population become extinct).

15.5.2 Conservation of Genetic Resources

The above descriptions of indirect and direct genetic effects of introductions are necessarily based on information arising from introductions. Each introduction represents a unique set of interactions with native fish affected by complex ecological and genetic differences that defy accurate a priori predictions (Ryman et al. 1995a). Such generalities as "most introductions fail," "all aquatic systems are invasible," and "disturbed environment helps alien establishment" (Moyle and Light 1996; Bartley and Casal 1998) reflect this underlying uncertainty. The predominance of negative ecological effects from fish introductions recorded in FishBase (Bartley and Casal 1999) mandates caution. Any introduction having potential interactions of nonnative and native fish, then, puts at risk the genetic integrity of native resources. Ironically, the best possible outcome of the introduction often is a failure and thus a wasted effort (e.g., Utter et al. 1995). Retrospectively, the apparent short-term benefits of enhanced fisheries are more than counterbalanced by losses through displacement or introgression of irreplaceable gene pools, which themselves may have served better as sources for local culture and enhancement (e.g., Garcia-Marin et al. 1991).

This generally negative view toward use of introduced fishes is intended as a strong word of caution rather than a proscription against such activities. Under appropriate conditions (e.g., disease concerns), colonized nonnative populations represent potentially valuable gene banks for reintroductions to restored habitats of origin where ancestral populations have become extirpated (e.g., Stockwell et al. 1996). For instance, nonnative California populations of pure Colorado River cutthroat trout have been reintroduced to areas of Colorado where the native populations have been compromised by introgression with rainbow trout (Pister 1990; Behnke 1992). A valid place also exists for nonnative populations in areas where habitats have been drastically modified and native populations displaced (Utter 1995) in the absence of such expatriates. Furthermore, some introductions have proven beneficial in controlling others in already disturbed habitats (e.g., Shafland 1999).

Still, the pressure to use nonnative species or populations because "conservation of wild stocks is no longer an issue" due to presumed extinction (e.g., Isaakson 1988) must not be passively and prematurely accepted. For instance, the unexpected finding of native breeding populations of sockeye salmon (Hendry et al. 1996) and steelhead

(Phelps et al. 1994a) in the Lake Washington drainage near Seattle following decades of intensive nonnative introductions strongly attests to the need to test any assumption of extinction with reliable genetic information.

The need for such caution has been stated repeatedly. In summarizing a symposium on ecological and genetic effects of fish introductions, Allendorf (1991) noted, "Purposeful introductions rarely have achieved their objectives. Moreover, both intentional and unintentional introductions usually have been harmful to native fishes and other taxa through predation, competition, hybridization and the introduction of diseases." More specifically, Ryman et al. (1995b) observed, "The globally increasing aquaculture industry will result in increasing incidental and intentional releases of huge numbers of fish, although those fish will represent a fairly limited number of species and populations. Thus, the threat of hybridization from cultured fish to the genetic integrity of local populations must be considered a major concern." The short-term expedience of intentionally or accidentally releasing nonnative fish must be carefully balanced by the sobering knowledge of irreversible and unpredictable effects of past introductions.

15.6 References

Allendorf, F. W. 1991. Ecological and genetic effects of fish introductions: synthesis and recommendations. Canadian Journal of Fisheries and Aquatic Sciences 48 (Supplement 1):178–181.

Allendorf, F. W., N. Ryman, and F. Utter. 1987. Genetics and fishery management: past present, and future. Pages 1–19 *in* Ryman and Utter (1987).

Allendorf, F. W., P. Spruell, and F. M. Utter. 2001. Whirling disease and wild trout: Darwinian fisheries management. Fisheries 26(5):27–29.

Allendorf, F. W., and R. S. Waples. 1995. Conservation and genetics of salmonid fishes. Pages 238–280 *in* J. C. Avise and J. L. Hamrick, editors. Conservation genetics: case histories from nature. Chapman and Hall, New York.

Altukhov, Y. P., and E. A. Salmenkova. 1987. Stock transfer relative to natural organization, management and conservation of fish populations. Pages 333–344 *in* Ryman and Utter (1987).

Anonymous. 1987. The impact of Oregon Aqua-Foods operation on the wild coho salmon in the Yaquina River. Staff Report, Oregon Department of Fish and Wildlife, Fish Division, P. O. Box 59, Portland, Oregon 97207.

Avise, J. C. 1994. Molecular markers, natural history and evolution. Chapman and Hall. New York.

Avise, J. C., and five coauthors. 1997. Cytonuclear introgressive swamping and species turnover of bass after an introduction. Journal of Heredity 88:14–20.

Bakke, T. A., P. A. Jansen, and L. P. Hansen. 1990. Differences in host resistance of Atlantic salmon, *Salmo salar* L., stocks to the monogenean *Gyrodactylus salaris* Malmberg, 1957. Journal of Fish Biology 37:577–587.

Bams, R. A. 1976. Survival and propensity for homing as affected by presence or absence of locally adapted paternal genes in two transplanted populations of pink salmon *(Oncorhynchus gorbusha)*. Journal of the Fisheries Research Board of Canada 33:2716–2725.

Bartley, D., and C. V. Casal. 1998. Impacts of introductions on the conservation and sustainable use of aquatic biodiversity. FAO (Food and Agricultural Organization of the United Nations) Aquatic Newsletter (FAN) 20:15–20.

Behnke, R. J. 1992. Native trout of western North America. American Fisheries Society, Monograph 6, Bethesda, Maryland.

Berg, W. J. 1987. Evolutionary genetics of rainbow trout, *Parasalmo gairdnerii* (Richardson). Doctoral dissertation. University of California, Davis.

Bergersen, E. P., and D. E. Anderson. 1997. The distribution and spread of *Myxobolus cerebralis* in the United States. Fisheries 22(8):6–7.

Bugert, R., C.W. Hopley, C.A. Busack, and G.W. Mendel. 1995. Maintenance of stock integrity in Snake River fall chinook salmon. Pages 267–276 *in* Schramm and Piper (1995).

Cagigas, M. E., E.Vazquez, G. Blanco, and J. A. Sanchez. 1999. Genetic effects of introduced hatchery stocks on indigenous brown trout (*Salmo trutta* L.) populations in Spain. Ecology of Freshwater Fish 8:141–150.

Campton, D. E. 1987. Natural hybridization and introgression in fishes: methods of detection and genetic interpretations. Pages 161–192 *in* Ryman and Utter (1987).

Campton, D. E. 1995. Genetic effects of hatchery fish on wild populations of Pacific salmon and steelhead: what do we really know? Pages 337–353 *in* Schramm and Piper (1995).

Campton, D. E., and six coauthors. 1991. Reproductive success of hatchery and wild steelhead. Transactions of the American Fisheries Society 120:816–827.

Campton, D. E., and F. M. Utter. 1987. Genetic structure of anadromous cutthroat trout *(Salmo clarki clarki)* populations in the Puget Sound area: evidence for restricted gene flow. Canadian Journal of Fisheries and Aquatic Sciences 44:573–582.

Carmichael, G. J., J. N. Hanson, J. R. Novy, K. J. Meyer, and D. C. Morizot. 1995. Apache trout management: cultured fish, genetics, habitat improvements, and regulations. Pages 112–121 *in* Schramm and Piper (1995).

Chapman, D., and seven coauthors. 1995. Status of sockeye salmon in the mid-Columbia region. Don Chapman Consultants, Inc., Boise, Idaho.

Chilcote, M.W., S. A. Leider, and J. J. Loch. 1986. Differential reproductive success of hatchery and wild summer-run steelhead under natural conditions. Transactions of the American Fisheries Society 115:726–735.

Contreras, S., and M. A. Escalante. 1984. Distribution and known impacts of exotic fish in Mexico. Pages 102–130 *in* W. R. Courtenay, Jr., and J. R. Stauffer, Jr., editors. Distribution, biology, and management of exotic fishes. Johns Hopkins University Press, Baltimore, Maryland.

Convention on International Trade in Endangered Species of Wild Fauna and Flora (CITES). 1973. [Text of the Convention.] CITES Secretariat, Châtelaine-Genève, Switzerland. (For more information, go to *www.wcmc.org.uk/CITES.*)

Courtenay, W. R. 1995. The case for caution with fish introductions. Pages 413–424 *in* Schramm and Piper (1995).

Currens, K. P., A. R. Henmingsen, R. A. French, D.V. Buchanan, C. B. Schreck, and H.W. Li. 1997. Introgression and susceptibility to disease in a wild population of rainbow trout. North American Journal of Fisheries Management 17:1065–1078.

Currens, K. P., and eight coauthors. 1998. Conservation of Pacific salmon: response to Wainwright and Waples. Conservation Biology 12:1148–1149.

Davis, H. S. 1956. Culture and diseases of game fishes. University of California Press, Berkeley.

Eanes, W. 1987. Allozymes and fitness: evolution of a problem. Trends in Ecology and Evolution 2:44–47.

FAO/UNEP (Food and Agriculture Organization, United Nations, and United Nations Environmental Program). 1981. Conservation of the genetic resources of fish: problems and recommendations. Report of the Expert Consultation of the Genetic Resources of Fish. FAO Fisheries Technical Paper 217.

Fleming, I. A., and five coauthors. 2000. Lifetime success, and interactions of farmed salmon invading a native population. Proceedings of the Royal Society of London B 267:1517–1524.

Fraidenburg, M., and R. Lincoln. 1985. Wild chinook salmon management: an international conservation challenge. North American Journal of Fisheries Management 5:311–329.

Fuller, P. L., L. G. Nico, and J. D. Williams. 1999. Nonindigenous fishes introduced into inland waters of the United States. American Fisheries Society, Special Publication 27, Bethesda, Maryland.

Garcia-Marin, J. L., P. E. Jorde, N. Ryman, F. Utter, and C. Pla. 1991. Management implications of genetic differentiation between native and hatchery populations of brown trout *(Salmo trutta)* in Spain. Aquaculture 95:235–249.

Garcia-Marin, J. L., and C. Pla. 1996. Origins and relationships of native populations of *Salmo trutta* (brown trout) in Spain. Heredity 77:313–323.

Garcia-Marin, J. L., N. Sanz, and C. Pla. 1999. Erosion of the native genetic resources of brown trout in Spain. Ecology of Freshwater Fish 8:151–158.

Garcia-Marin, J. L., A. Vila, and C. Pla. 1990. Genetic variation in the Iberian toothcarp, *Aphanius iberus* (Cuvier & Valenciennes). Journal of Fish Biology 37 (Supplement A):233–234.

Gharrett, A. J., and W. W. Smoker. 1991. Two generations of hybrids between even- and odd-year pink salmon *(Oncorhynchus gorbuscha)*: a test for outbreeding depression? Canadian Journal of Fisheries and Aquatic Sciences 48:1744–1749.

Gharrett, A. J., and W. W. Smoker. 1993a. A perspective on the adaptive importance of genetic infrastructure in salmon populations to ocean ranching in Alaska. Fisheries Research 18:45–58.

Gharrett, A. J., and W. W. Smoker. 1993b. Genetic components in life history traits contribute to population structure. Pages 197–202 *in* J. G. Cloud and G. H. Thorgaard, editors. Genetic conservation of salmonid fishes. Plenum, New York.

Gharrett, A. J., W. W. Smoker, R. R. Reisenbichler, and S. G. Taylor. 1999. Outbreeding depression in hybrids between odd-and even-broodyear pink salmon. Aquaculture 173:117–129.

Gharrett A. J., and M. A. Thomason. 1987. Genetic changes in pink salmon *(Oncorhynchus gorbuscha)* following their introduction into the Great Lakes. Canadian Journal of Fisheries and Aquatic Sciences 44:787–792.

Grant, W. S., J. L. Garcia-Marin, and F. M. Utter. 1999. Defining population boundaries in fisheries management. Pages 27–72 *in* S. Mustafa, editor. Genetics in sustainable fisheries management. Blackwell Scientific Publications Ltd., Oxford, UK.

Gustafson, R. G., and G. A. Winans. 1999. Distribution and population genetic structure of river- and sea-type sockeye salmon in western North America. Ecology of Freshwater Fish 8:181–193.

Guyomard, R. 1989. Diversite genetique de la truite commune. Bulletin Francaís de la Peche et de la Pisciculture 314:118–135.

Hayes, J. P., S. Z. Guffey, F. J. Kriegler, G. F. McCracken, and C. R. Parker. 1996. The genetic diversity of native, stocked, and hybrid populations of brook trout in the southern Appalachians. Conservation Biology 10:1403–1412.

Hendry, A. P., O. K. Berg, and T. P. Quinn. 1999. Condition dependence, and adaptation-by-time: breeding date, life history, and energy allocation within a population of salmon. Oikos 85:499–514.

Hendry, A. P., T. P. Quinn, and F. M. Utter. 1996. Genetic evidence for the persistence and divergence of native and introduced sockeye salmon *(Oncorhynchus nerka)* within Lake Washington, Washington. Canadian Journal of Fisheries and Aquatic Sciences 53:823–832.

Hindar, K. 1992. Conservation and sustainable use of Atlantic salmon. Pages 168–186 *in* O. T. Sandlund, K. Hindar and A. H. D. Brown, editors. Conservation of biodiversity for sustainable development. Scandinavian University Press, Oslo.

Hindar, K., N. Ryman, and F. M. Utter. 1991. Genetic effects of cultured fish on natural fish populations. Canadian Journal of Fisheries and Aquatic Sciences 48:945–957.

Hulbert, P. J. 1996. Whirling disease: a resource stewardship challenge. Fisheries 21(6):26–27.

Ihssen, P. E., and five coauthors. 1981. Stock identification: materials and methods. Canadian Journal of Fisheries and Aquatic Sciences 38:1838–1855.

Isaksson, A. 1988. Salmon ranching: a world review. Aquaculture 75:1–33.

Johnson, O. W., and five coauthors. 1997. Status review of chum salmon from Washington, Oregon, and California. NOAA (National Oceanic and Atmospheric Administration) Technical Memorandum NMFS (National Marine Fisheries Service)-NWFSC (Northwest Fisheries Science Center)-32.

Kinnison, M. T. 1999. Life history divergence and population structure of New Zealand chinook salmon. Doctoral dissertation. University of Washington, Seattle.

Kinnison, M. T., M. Unwin, N. Boustead, and T. Q. Quinn. 1998. Population-specific variation in body dimensions of adult chinook salmon *(Oncorhynchus tshawytscha)* from New Zealand and their source population, 90 years after introduction. Canadian Journal of Fisheries and Aquatic Sciences 55:5554–5563.

Kocher, T. D., and C. A. Stepien, editors. 1997. Molecular systematics of fishes. Academic Press, London.

Krueger, C. C., and B. May. 1991. Ecological and genetic effects of salmonid introductions in North America. Canadian Journal of Fisheries and Aquatic Sciences 48 (Supplement 1):66–77.

Lassuy, D. R. 1995. Introduced species as a factor in extinction and endangerment of native fish species. Pages 391–396 *in* Schramm and Piper (1995).

Leary, R. F., F. W. Allendorf, and K. L. Knudsen. 1993. Conservation genetics of bull trout in the Columbia and Klamath River drainages. Conservation Biology 7:856–865.

Leary, R. F., F. W. Allendorf, and G. K. Sage. 1995. Hybridization and introgression between introduced and native fish. Pages 91–101 *in* Schramm and Piper (1995).

Lichatowich, J., and J. McIntyre. 1987. Use of hatcheries in the management of Pacific anadromous salmonids. Pages 131–136 *in* M. J. Dadswell, R. J. Klauda, C. M. Moffitt, R. L. Saunders, R. A. Rulifson, and J. E. Cooper, editors. Common strategies of anadromous and catadromous fishes. American Fisheries Society, Symposium 1, Bethesda, Maryland.

Lincoln, R. 1994. Molecular genetics applications in fisheries: snake oil or restorative? Reviews in Fish Biology and Fisheries 4:389–393.

Lynch, M. 1991. The genetic interpretation of inbreeding depression and outbreeding depression. Evolution 45:622–629.

MacCrimmon, H. 1971. World distribution of rainbow trout *(Salmo gairdneri)*. Journal of the Fisheries Research Board of Canada 28:663–704.

MacLean, J. A., and D. O. Evans. 1981. The stock concept, discreteness of fish stocks, and fisheries management. Canadian Journal of Fisheries and Aquatic Sciences 38:1889–1898.

Matthews, G., and R. S. Waples. 1991. Status review for Snake River spring and summer chinook salmon. NOAA (National Oceanic and Atmospheric Administration) Technical Memorandum NMFS (National Marine Fisheries Service) F/NWC-195.

Mills, E. L., J. H. Leach, J. T. Carlton, and C. L. Secor. 1991. Exotic species in the Great Lakes: a history of biotic crises and anthropogenic introductions. Great Lakes Fishery Commission Research Report, Ann Arbor, Michigan.

Moav, R., T. Brody, and G. Hulata. 1978. Genetic improvements of wild fish populations. Science 201:1090–1094.

Moran, P., A. M. Pendas, E. Garcia-Vazquez, and J. Izquierdo. 1991. Failure of a stocking policy, of hatchery reared brown trout, *Salmo trutta* L., in Asturias, Spain, detected using *LDH-5** as a genetic marker. Journal of Fish Biology 39 (Supplement A):117–122.

Moran, P., A. M. Pendas, E. Garcia-Vazquez, and J. T. Izquierdo. 1994. Electrophoretic assessment of the contribution of transplanted Scottish Atlantic salmon *(Salmo salar)* to the Esva River (northern Spain). Canadian Journal of Fisheries and Aquatic Sciences 51:248–252.

Moyle, P. B., and T. Light. 1996. Biological invasions of fresh water: empirical rules and assembly theory. Biological Conservation 78:149–161.

Nehlsen, W., J. E. Williams, and J. A. Lichatowich. 1991. Pacific salmon at the crossroads: stocks at risk from California, Oregon, Idaho, and Washington. Fisheries 16(2):4–21.

Nei, M. 1972. Genetic distance between populations. American Naturalist 106:283-292.

Nei, M. 1987. Molecular genetic evolution. Columbia University Press, New York.

Nielsen, E. E., M. M. Hansen, and V. Loeschcke. 1999. Genetic variation in time and space: microsatellite analysis of extinct and extant populations of Atlantic salmon. Evolution 53:261–268.

Nielsen, J. L., editor. 1995. Evolution and the aquatic ecosystem: defining unique units in population conservation. American Fisheries Society, Symposium 17, Bethesda, Maryland.

Palm, S., and N. Ryman. 1999. Genetic basis of phenotypic differences between transplanted stocks of brown trout. Ecology of Freshwater Fish 8:181–193.

Phelps, S. R., B. M. Baker, P. L. Hulett, and S. A. Leider. 1994a. Genetic analysis of Washington steelhead: initial electrophoretic analysis of wild and hatchery steelhead and rainbow trout.

Fisheries Management Program, Washington Department of Fish and Wildlife, Report 94–9, Olympia.

Phelps, S. R., L. L. LeClaire, S. Young, and H. L. Blankenship. 1994b. Genetic stock structure of western Alaska chum salmon and a comparison with Russian far east stocks. Canadian Journal of Fisheries and Aquatic Sciences 51 (Supplement 1):65–83.

Philipp, D. P. 1991. Genetic implications of introducing Florida largemouth bass, *Micropterus salmoides floridanus*. Canadian Journal of Fisheries and Aquatic Sciences 48 (Supplement 1):58–65.

Philipp, D. P., and J. E. Claussen. 1995. Fitness and performance differences between two stocks of largemouth bass from different river drainages within Illinois. Pages 236–243 *in* Schramm and Piper (1995).

Pierce, P. C., and M. J. Van Den Avyle. 1997. Hybridization between introduced spotted bass and smallmouth bass in reservoirs. Transactions of the American Fisheries Society 126:939–947.

Pister, E. P. 1990. Pure Colorado trout saved by California. Outdoor California 51:12–15.

Quinn, T., and six coauthors. 1996. Origin and genetic structure of chinook salmon *(Oncorhynchus tshawytscha)* transplanted from California to New Zealand: allozyme and mtDNA evidence. U.S. National Marine Fisheries Service Fishery Bulletin 94:506–521.

Quinn. T. P., and M. J. Unwin. 1993. Variation in life history patterns among New Zealand chinook salmon *(Oncorhynchus tshawytscha)* populations. Canadian Journal of Fisheries and Aquatic Sciences 50:1414–1421.

Reisenbichler, R. R., and J. McIntyre. 1977. Genetic differences in growth and survival of juvenile hatchery and wild steelhead trout, *Salmo gairdneri*. Journal of the Fisheries Research Board of Canada 34:123–128.

Ricker, W. E. 1972. Hereditary and environmental factors affecting certain salmonid populations. Pages 19–160 *in* R. Simon and P. Larkin, editors. The stock concept of Pacific salmon. H. R. MacMillan Lectures in Fisheries, University of British Columbia, Vancouver.

Rieman, B., and J. McIntyre. 1995. Occurrence of bull trout in naturally fragmented habitat patches of varied size. Transactions of the American Fisheries Society 124:285–296.

Ryman, N. 1991. Conservation genetic considerations in fishery management. Journal of Fish Biology 39 (Supplement A):211–224.

Ryman, N., and L. Laikre. 1991. Effects of supportive breeding on the genetically effective population size. Conservation Biology 5:325–329.

Ryman, N., and F. Utter, editors. 1987. Population genetics and fishery management. University of Washington Press, Seattle.

Ryman, N., F. Utter, and K. Hindar. 1995a. Introgression, supportive breeding, and genetic conservation. Pages 341–366 *in* J. K. Ballou, M. Gilpin, and T. J. Foose, editors. Population management for survival and recovery. Columbia University Press, New York.

Ryman, N., F. Utter, and L. Laikre. 1995b. Protection of intraspecific biodiversity of exploited fishes. Reviews in Fish Biology and Fisheries 5:417–446.

Saegrov, H., K. Hindar, S. Kalas, and H. Lura. 1997. Escaped farmed Atlantic salmon replace the original salmon stock in the River Vosso, western Norway. ICES Journal of Marine Science 54:1166–1172.

Schramm, H. L., Jr., and R. G. Piper, editors. 1995. Uses and effects of cultured fishes in aquatic ecosystems. American Fisheries Society, Symposium 15, Bethesda, Maryland.

Seeb, J. E., and five coauthors. 1999. Allozyme and mitochondrial DNA variation describe ecologically important genetic structure of even-year pink salmon inhabiting Prince William Sound, Alaska. Ecology of Freshwater Fish 8:122–140.

Shafland, P. L. 1999. The introduced butterfly peacock *(Cichla ocellaris)* in Florida. I. Fish community analyses. Reviews in Fisheries Science 7:71–94.

Shaklee, J., T. C. Beacham, L. Seeb, and B. A. White. 1999. Managing fisheries using genetic data: case studies from four species of Pacific salmon. Fisheries Research 43:45-78.

Shaklee, J. B., D. C. Klaybor, S. Young, and B. A. White. 1991. Genetic stock structure of odd-year pink salmon *Oncorhynchus gorbuscha* (Walbaum), from Washington and British Columbia and potential mixed-stock fisheries applications. Journal of Fish Biology 39 (Supplement A):21–34.

Sheridan, A. K. 1995. The genetic impacts of human activities on wild fish populations. Reviews in Fisheries Science 3:91–108.

Simon, R. C. 1972. Gene frequency and the stock problem. Pages 161–172 *in* R. C. Simon and P. A. Larkin, editors. The stock concept in Pacific salmon. H. R. MacMillan Lectures in Fisheries, University of British Columbia, Vancouver.

Steward, C. R., and T. C. Bjornn. 1990. Supplementation of salmon and steelhead stocks with hatchery fish: a synthesis of published literature. *In* Analysis of salmon and steelhead supplementation. Report To Bonneville Power Administration, Technical Report 90–1, Portland, Oregon.

Stockwell, C. A., M. Mulvey, and G. L. Vinyard. 1996. Translocations and the preservation of allelic diversity. Conservation Biology 10:1133–1141.

Sugg, D. W., Chesser, R. K., Dobson, F. S., and J. L. Hoogland. 1996. Population genetics meets behavioral ecology. Trends in Evolution and Ecology 11:338–342.

Teel, D. J., G. B. Milner, G. A. Winans, and W. S. Grant. 2000. Genetic population structure, and origin of life history types in chinook salmon from British Columbia, Canada. Transactions of the American Fisheries Society 129:194-209.

Templeton, A. R. 1986. Coadaptation and outbreeding depression. Pages 105–116 *in* M. E. Soulé, editor. Conservation biology. The science of scarcity and diversity. Sinauer, Sunderland, Massachusetts.

Tessier, N., L. Bernatchez, and J. M. Wright. 1997. Population structure and impact of supportive breeding inferred from mitochondrial and microsatellite DNA analyses in land-locked Atlantic salmon, *Salmo salar* L. Molecular Ecology 6:735–750.

Thorgaard, G., and S. Allen. 1987. Chromosome manipulation and markers in fishery management. Pages 319–332 *in* Ryman and Utter (1987).

Thorpe, J. E., and J. F. Koonce. 1981. Assessing and managing man's impact on fish genetic resources. Canadian Journal of Fisheries and Aquatic Sciences 38:1899–1907.

Utter, F. M. 1991. Biochemical genetics and fishery management: an historical perspective. Journal of Fish Biology (Supplement A):1–20.

Utter, F. M. 1994. Detrimental aspects of put-and-take trout stocking. Fisheries 19(8):8–9.

Utter, F. M. 1995. Position statement of Genetics Section. Page 588 *in* Schramm and Piper (1995).

Utter, F. M. 1998. Genetic problems of hatchery-reared progeny released into the wild, and how to deal with them. Bulletin of Marine Science 62:623–640.

Utter, F. M. 2000. Patterns of subspecific anthropogenic introgression in two salmonid genera. Reviews in Fish Biology and Fisheries 3:1–15.

Utter, F. M., and F. W. Allendorf. 1994. Phylogenetic relationships among species of *Oncorhynchus*: a consensus view. Conservation Biology 8:864–867.

Utter, F. M., D. Chapman, and A. Marshall. 1995. Genetic population structure and history of chinook salmon of the upper Columbia River. Pages 149–165 *in* J. L. Nielsen, editor. Evolution and the aquatic ecosystem: defining unique units in population conservation. American Fisheries Society, Symposium 17, Bethesda, Maryland.

Utter, F. M., G. B. Milner, G. Stahl, and D. Teel. 1989. Genetic population structure of chinook salmon, *Oncorhynchus tshawytscha*, in the Pacific Northwest. U.S. National Marine Fisheries Service Fishery Bulletin 87:239–264.

Utter, F. M., and N. Ryman. 1993. Genetic markers and mixed stock fisheries. Fisheries 18(8):11–21.

Utter, F. M., J. E. Seeb, and L. W. Seeb. 1993. Complementary uses of ecological and biochemical genetic data in identifying and conserving salmon populations. Fisheries Research 18:59–76.

Utter, F. M., R. S. Waples, and D. Teel. 1992. Genetic isolation of previously undistinguishable chinook salmon populations of the Snake and Klamath rivers: limitation of negative data. U.S. National Marine Fisheries Service Fishery Bulletin 90:770–777.

Verschuren, D., and L. Marnell. 1997. Fossil zooplankton and the historical status of westslope cutthroat trout in a headwater lake of Glacier National Park, Montana. Transactions of the American Fisheries Society 126:21–34.

Verspoor, E., and J. Hammar. 1991. Introgressive hybridization in fishes: the biochemical evidence. Journal of Fish Biology 39 (Supplement A):309–334.

Vitousek, P. M., H. A. Mooney, J. Lubchenco, and J. M. Melillo. 1997. Human domination of earth's ecosystems. Science 277:494–499.

Wainwright, T. C., and R. S. Waples. 1998. Prioritizing Pacific salmon stocks for conservation: response to Allendorf et al. Conservation Biology 12:1144–1147.

Waples, R. S. 1991. Pacific salmon and the definition of "species" under the Endangered Species Act. Marine Fisheries Review 53:11–22.

Waples, R. W. 1995. Genetic effects of stock transfers of fish. Pages 51–69 *in* D. Philipp, editor. Proceedings of the World Fisheries Congress, theme 3. Oxford and IBH, New Delhi.

Welcomme, R. L. 1988. International introductions of inland aquatic species. FAO (Food and Agricultural Organization of the United Nations) Fisheries Technical Paper 294.

White, M. M., and S. Schell. 1995. An evaluation of the genetic integrity of Ohio River walleye and sauger stocks. Pages 52–60 *in* Schramm and Piper (1995).

Whitlock, M. C., and D. E. McCauley. 1999. Indirect measures of gene flow and migration: F_{ST} is not equal to 1/(4Nm+1). Heredity 82:117–125.

Whitmore, D. H. 1983. Introgressive hybridization of smallmouth bass *(Micropterus dolomieu)* and Guadalupe bass *(M. treculi)*. Copeia 1983:672–769.

Wilde, G. R., and A. A. Echelle. 1992. Genetic status of Pecos pupfish populations after establishment of a hybrid swarm involving an introduced congener. Transactions of the American Fisheries Society 121:277–286.

Williams, R. N., D. K. Shiozawa, J. E. Carter, and R. F. Leary. 1996. Genetic detection of putative hybridization between native and introduced rainbow trout populations of the upper Snake River. Transactions of the American Fisheries Society 125:387–401.

Wishard, L. N., J. E. Seeb, F. M. Utter, and D. Stefan. 1984. A genetic investigation of suspected redband trout populations. Copeia 1984:120–132.

Withler, F. C. 1982. Transplanting Pacific salmon. Canadian Technical Report of Fisheries and Aquatic Sciences 1079.

Wood, C. C. 1995. Life history variation and population structure in sockeye salmon. Pages 195–216 *in* J. L. Nielsen, editor. Evolution and the aquatic ecosystem: defining unique units in population conservation. American Fisheries Society, Symposium 17, Bethesda, Maryland.

Wright, S. 1981. Contemporary Pacific salmon fisheries management. North American Journal of Fisheries Management 1:29–40.

Youngson A. F., J. H. Webb, C. E. Thompson, and D. Knox. 1993. Spawning of escaped farmed Atlantic salmon *(Salmo salar)*: hybridization of females with brown trout *(Salmo trutta)*. Canadian Journal of Fisheries and Aquatic Sciences 50:1986–1990.

Zhivotovsky, L. A., A. J. Gharrett, M. K. Glubokovsky, and M. W. Feldman. 1994. Gene differentiation in Pacific salmon (*Oncorhynchus* sp.): facts and models with reference to pink salmon *(O. gorbuscha)*. Canadian Journal of Fisheries and Aquatic Sciences 51 (Supplement 1):223–232.

16

Genetic Marking

Eric Hallerman

16.1 Introduction

Fisheries scientists and managers long have marked individual fish and shellfish for purposes of tracking individual movement, estimating population size, or estimating contributions from source populations to a mixed-stock fishery (Nielsen 1992). These physical marks often take the form of clipped fins, numbered plastic tags, binary-coded pieces of wire, or passive integrated transponders. Such marks are useful for distinguishing individuals but are not heritable; they are not transmitted from parents to progeny. Thus, for each generation a large marking effort would be required to mark members of a population.

There are many contexts in which fisheries scientists or managers want a heritable marker, one inherently linked to transmission of genetic information from generation to generation. Where the desire is to mark a cultured stock, these research or management needs could be addressed by deliberate **genetic marking** of the fish stock. That is, propagation of the cultured stock could be managed in order to produce an array of allele frequencies differing from that of the receiving population. As a consequence, we could detect any contribution of alleles from the cultured stock to the receiving population. There are a number of contexts in which deliberate genetic marking might be practiced.

Many fisheries management plans include hatchery-based supplementation of exploited stocks. Fisheries managers and administrators might want to know whether, and to what degree, a hatchery program contributes to the harvest. Many times, hatchery personnel insert coded wire or other physical tags into a sample of the fish released, subsequently yielding a direct estimate of fishery returns for a single generation. However, whether unharvested, released fish recruit to the spawning stock and contribute to future generations of the population cannot be ascertained with physical marks. Further, physical marks or insertion of physical tags can injure small fish and reduce their survival rate (Morrison and Zajac 1987), and physical tags can be lost.

In another common management context, fisheries managers might want to enhance a stock demographically by adding fry, juveniles, or spawners to a depleted population. These programs should be evaluated to determine whether stocked individuals contributed to any subsequent growth of the population.

Domesticated stocks may become differentiated from wild, locally adapted stocks, leading to concerns regarding possible genetic impacts of escaped or released domesticated stocks on receiving populations (Chapter 15). In such contexts, a means of identifying domesticated fish or their progeny would prove highly desirable. Only heritable markers permit identification of offspring.

Hence, there are a number of contexts in which fisheries scientists or managers might want to mark populations of cultured fish genetically. This chapter explores the theoretical underpinnings of genetic marking and presents several case studies demonstrating its application.

■ 16.2 Theoretical Background

Genetic marking is appropriate when discrimination among populations by means of existing physical, biological, or genetic characteristics is not possible or practical. Genetic marking uses selective breeding to alter allele frequencies intentionally in the marked population so that it can be distinguished from unmarked populations. When a genetic marking program is planned, several questions must be addressed, including how to choose a marker allele and how to evaluate its performance. Gharrett and Seeb (1990) provided quantitative guidelines for choosing appropriate marker alleles and described how to examine the selective neutrality of a marker allele. Their contribution was drawn upon heavily to produce the discussion below and should be consulted for additional detail.

16.2.1 Premarking Considerations

16.2.1.1 **Biological Factors.** An initial set of premarking considerations includes biological factors affecting the design of an effective genetic marking program. Researchers will need to know the distribution and abundance of the populations involved, as well as where and when they spawn. The population or stock to be marked must be large enough to be detectable at the time and place it will be sampled. The relative and absolute sizes of the populations are both important factors in deciding the potential utility of a genetic mark. Genetic marking also will require a means, such as a weir or hatchery, for producing and subsequently recovering fish with particular genotypes.

Genetic marking can occur where no individuals of a given species previously existed or where populations already exist. Where no population exists, a marked population can be established by transplanting previously screened and selected eggs, embryos, or juveniles. Where populations already exist, allele frequencies in one or more populations can be altered by selective breeding, as described below. This alteration of allele frequencies will require knowledge of how spawning groups are separated spatially or temporally. Effective genetic marking also will require accounting for alteration of the frequency of the marker allele by natural production in the system of populations (e.g., as in Lane et al. 1990).

Most species of fish have overlapping generations, which will require genetic marking in successive years throughout the life span of the longest-lived individuals. For a transplanted population, this marking must be done each year until marked fish contribute to reproduction. Marking an existing stock also requires marking in successive years. However, after the first returns of marked fish, only fish from year-classes that have not been marked need to be screened, provided that year-classes are distinguishable. The dilution of marks added to an existing population may require additional marking effort.

16.2.1.2 **Genetic Factors.** Other important premarking planning considerations include genetic factors, that is, baseline genetic information, the effects of altered allele frequencies on the population, and effective population sizes (Chapter 9).

Because genetic marking involves deliberate alteration of a stock's allele frequencies to distinguish it from other populations, planners will need baseline allele frequency data, both for the stock to be marked and for all populations with which the marked population may mix. This baseline information will be needed to choose marker alleles, as discussed below.

The array of allele frequencies in the genetically marked population will differ from its original array. It generally is presumed that the manipulation of frequencies of marker alleles will not affect the fitness of the marked stock. Still, natural selection could eliminate fish that express maladaptive marker alleles or that carry maladaptive alleles at closely linked loci. Hence, the marked stock should be monitored for several generations to determine whether fitness is being negatively affected by selection for the marker. Monitoring also might determine whether representation of marked fish decreases over time because of immigration of unmarked fish.

The process of selective breeding used to select the genetically marked broodstock greatly reduces the effective population size (N_e) because the intensive selection for the marker allele(s) results in a limited number of breeders. Also, if the marker allele(s) is (are) relatively rare in the population, the likelihood of selecting related individuals for broodstock is increased. Hence, inbreeding is a potential problem posed by a genetic marking program. Using a large number of spawners decreases the likelihood that deleterious, hitchhiking alleles will be increased in frequency in the marked population. To estimate the effect of inbreeding, we may use the equation

$$N_e = \frac{4N_m N_f}{N_m + N_f}, \tag{16.1}$$

where N_m is the number of males and N_f is the number of females spawned. The N_e can be related to the rate of inbreeding by $\delta F = 1/2N_e$, where δF is the one-generation increase in inbreeding. If, say, 200 spawners (100 males and 100 females) are used to found the marked population and in each succeeding generation to maintain it, then the change in the inbreeding coefficient will be 0.0025 per generation. More than 20 generations will be required before a moderate inbreeding level of an F of 0.05 is reached. For species with overlapping generations, N_e is approximately the number of broodstock in a given year times the average age of the breeders (Falconer 1981). Inbreeding resulting from the selection of related individuals will be reduced by increasing the number of breeders, which should increase the representation of more families in the broodstock.

16.2.2 *Genetic Marking*

After it has been determined that genetic marking is appropriate for a given application, and after baseline data on allele frequencies have been obtained for the population(s) in the system, a genetic marking strategy can be designed. Choice of strategy will depend on the genetic variability among populations as well as within the population to be marked. A potential marker allele cannot be abundant in populations other than the marked population.

Genetic marking is executed by breeding only individuals carrying the desired marker allele or alleles. Against the background of genetic baseline data and resources available for the program, detailed design of a genetic marking program involves consideration of marking effort and marking strategy. **Marking effort** is the ratio of the number of fish screened to the number of fish subsequently spawned. **Marking strategy** is the approach of selecting for or against alleles at one or more loci. Among potential marking strategies are (1) selecting for a single allele at one locus, (2) simultaneously selecting for two alleles, one at each of two loci, and (3) selecting for an allele at one locus and against one or several (in the case of microsatellite loci) alleles at another locus.

The rate change of allele frequencies achieved by genetic marking will depend on the premarking allele frequency, the marking effort, and the marking strategy (Figure 16.1). For strategies involving selection for marker alleles, increased marking effort will result in the progressive elimination of genotypes with the fewest marker alleles, until only those homozygous for markers remain. For example, if selecting for the single marker allele A' (Figure 16.1, top), the AA genotype is eliminated first, then AA'. If the initial frequency of the A' allele is greater than 0.30 or 0.40, then other alleles may be eliminated with relatively modest marking effort. If the initial frequency is less than 0.025, then the marking effort may entail removing only homozygous AA genotypes. For intermediate initial frequencies, progress in marking will exhibit an inflection point at some level of marking effort as selection against the AA genotypes is completed and removal of AA' genotypes begins. Selection for marker alleles at two loci is similar to selection for that at one locus, except that there are nine possible genotypes. Curves describing progress in elevating marker frequencies as a function of marking effort (Figure 16.1, bottom) will show a number of inflections as different genotypic classes are progressively eliminated from the broodstock. Selection for one allele at one locus and against alleles at a second locus is accomplished by first eliminating the undesired allele and then following the strategy for selecting a single allele.

16.2.2.1 **Estimation of Contribution of a Marked Stock.** The proportion of contribution of genetically marked fish to a mixture, x, is estimated (Lane et al. 1990) using the expression

$$x = \frac{p_m - p_2}{p_1 - p_2}, \tag{16.2}$$

where p_m is the frequency of the marker allele in the mixture, p_1 is the frequency of the marker allele in the marked population, and p_2 is the frequency of the marker allele in the unmarked population (Wallace 1981). The estimated contribution is subject to error, and hence has a variance, $V(x)$, which is calculated as

$$V(x) = \left[\frac{1}{p_1 - p_2}\right]^2 \cdot V(p_m) + \left[\frac{p_m - p_2}{(p_1 - p_2)^2}\right]^2 \cdot V(p_1) + \left[\frac{p_m - p_1}{(p_1 - p_2)^2}\right]^2 \cdot V(p_2), \tag{16.3}$$

where $V(p_i) = p_i(1 - p_i) / 2N_i$, and N_i is the size of the sample from which the p_i values were estimated.

16.2.2.2 **Detectability.** It is appropriate to assess the experimental power of a genetic marking experiment before it is undertaken. The minimum proportion of a genetically marked population that can be detected in a mixture with other populations depends on the

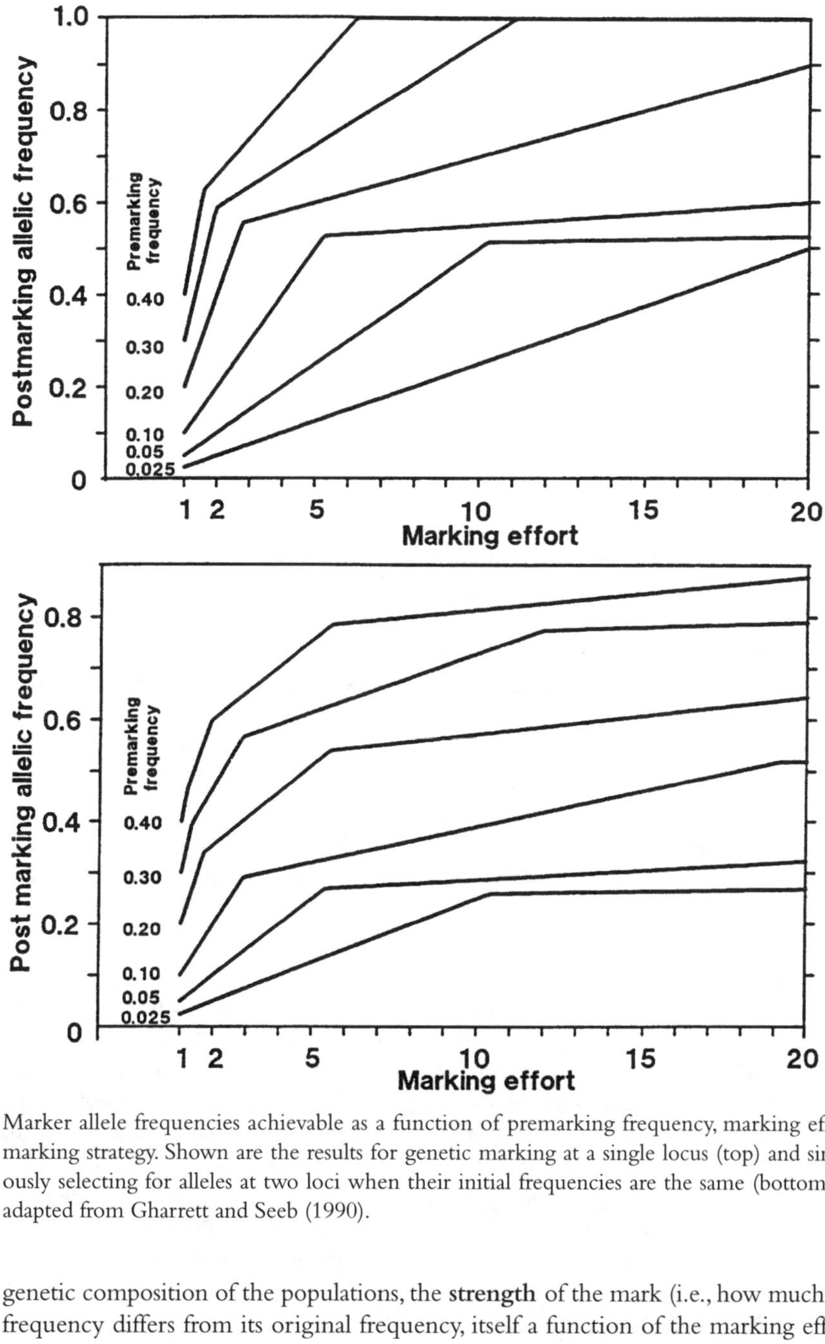

Figure 16.1 Marker allele frequencies achievable as a function of premarking frequency, marking effort, and marking strategy. Shown are the results for genetic marking at a single locus (top) and simultaneously selecting for alleles at two loci when their initial frequencies are the same (bottom). Figure adapted from Gharrett and Seeb (1990).

genetic composition of the populations, the **strength** of the mark (i.e., how much its new frequency differs from its original frequency, itself a function of the marking effort and the marking strategy), the size of the sample screened, and the critical value of the statistical test adopted for reaching a conclusion.

In a mixture comprising a genetically marked population and unmarked populations (for which genetic marker frequencies equal the premarking frequencies of the marked stock),

$$\mathrm{pmix}_i = m\,(\mathrm{pm}_i) + (1 - m)p_i = m\,(\mathrm{pm}_i - p_i) + p_i, \tag{16.4}$$

where $pmix_i$ are the frequencies (allelic or genotypic) observed in the mixture, m is the proportional contribution of the marked population to the mix of populations, i is the index for alleles, genotypes, or gamete types, pm_i is the frequency of a mark in the marked population, and p_i is the frequency of the mark in the unmarked contributors to the mixture (Gharrett and Seeb 1990). The power of the experiment is computed from the probability density function of the goodness-of-fit statistic when the alternative hypothesis is true; it is the area under the density function for values greater than the critical value of the test. The distribution of the alternative hypothesis is described by a noncentral chi-square distribution with a noncentrality parameter:

$$\lambda = m\sum_i \frac{(\mathrm{pm}_i - p_i)}{p_i} + m^2(N-1)\sum_i \frac{(\mathrm{pm}_i - p_i)^2}{p_i}, \quad (16.5)$$

where λ is the noncentrality parameter for the noncentral chi-square distribution and N is the number of individuals in the sample for marker genotypes or twice the number of individuals for marker alleles or haplotypes. To design an experiment with desired power (e.g., 90%), one would use a computational software program, enter a starting value for m, and solve for chi-square, iteratively modifying m until the chi-square value equals 0.90. For the set of genetic markers at issue, this procedure would reveal the minimum proportion for which detection of the marked population is 90% certain.

An example of the relationship between recovery effort, or sample size N, and detectability, m, is shown in Figure 16.2 for the case of selecting for a marker allele at one locus, given a marking effort of 10 fish screened for each spawned. Sample sizes of N between 300 and 1,000 approximate the minimum generally necessary to resolve a 5% contribution by marked fish within a relatively large recovery effort. An N of 300 fish per sampling week was used in Sumner Strait, Alaska, for composition estimates for wild populations of sockeye salmon *Oncorhynchus nerka* in southeastern Alaska (Pella et al. 1998).

Relationships between marking effort and detectability indicate that the premarking allele frequencies in the population, the marking effort, and the recovery effort all must be considered when choosing the optimum marking strategy (Figure 16.3). The best marking strategy for a marking effort of 5 fish screened per fish spawned and a recovery sample of 300 fish is quite different from the best strategy for a marking effort of 10 fish screened per fish spawned and a recovery of 1,000 fish. Selection for a relatively low frequency allele at a single locus is often the best strategy because selection at a single locus produces the largest allele frequency difference between the marked and unmarked populations. As marking effort is increased, progressively less frequent alleles become the best markers. With a marking effort of five fish screened per fish spawned, a marker allele with an initial frequency of 0.10 is useful. When marking effort exceeds 15 fish screened per fish spawned, an allele with an initial frequency of 0.025 is appropriate. Alleles with intermediate frequencies are appropriate for marking efforts between 5 and 15 screened fish. Practical considerations often will determine which marking strategy proves most appropriate.

16.2.2.3 **Evaluation.** After a population has been genetically marked, it should be monitored to ensure the persistence of the marker in the population. Decrease in the frequency of the marker allele over time diminishes its utility. The frequency of the marker allele is expected to decrease if it is not selectively neutral or if there is immigration of unmarked fish into the ecosystem. It is not simple to distinguish changes due to selec-

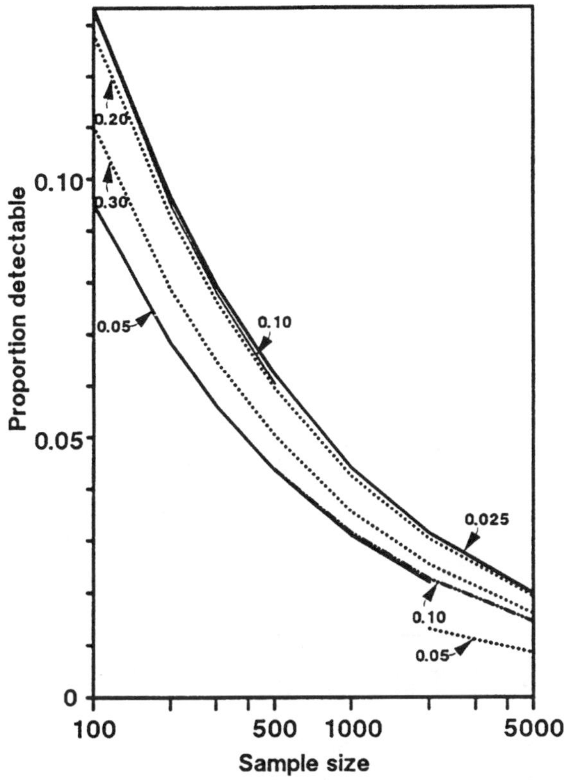

Figure 16.2 The minimum proportion of a genetically marked population that can be detected with a probability of 0.90 in a mixture of populations, assuming one locus was selected and 10 fish were screened per fish spawned. The minimum detectable proportion is shown as a function of premarking allele frequencies (numbers along the curves) and the size of the sample from a mixture. Shown are the lesser of values calculated from allele frequencies (solid lines) and genotype frequencies (broken lines). Figure adapted from Gharrett and Seeb (1990).

tion from those due to migration. Tests for selection against genetic markers are weak. Using a recovery sample of 1,000 fish, Gharrett and Seeb (1990) found that a selection coefficient of 0.15, which could result in a 10% decrease in fitness of a population, is only minimally detectable. Still, little if any decrease in fitness is expected due to genetic marking of a population. Any decrease most likely will be offset by an improved understanding of population dynamics and improved fishery management due to knowledge gained in the genetic marking effort.

16.3 Examples of Application

As suggested in the introduction to this chapter, uniquely identifiable stocks of fish are useful for scientific experimentation (Reisenbichler and McIntyre 1977; Gharrett and Shirley 1985) and for evaluation of the effectiveness of fish hatchery programs (Murphy et al. 1983; Taggart and Ferguson 1984; Seeb et al. 1986; Gharrett and Seeb 1990; Lane et al. 1990; Utter and Seeb 1990). Following are selected case studies illustrating diverse applications of deliberate genetic marking.

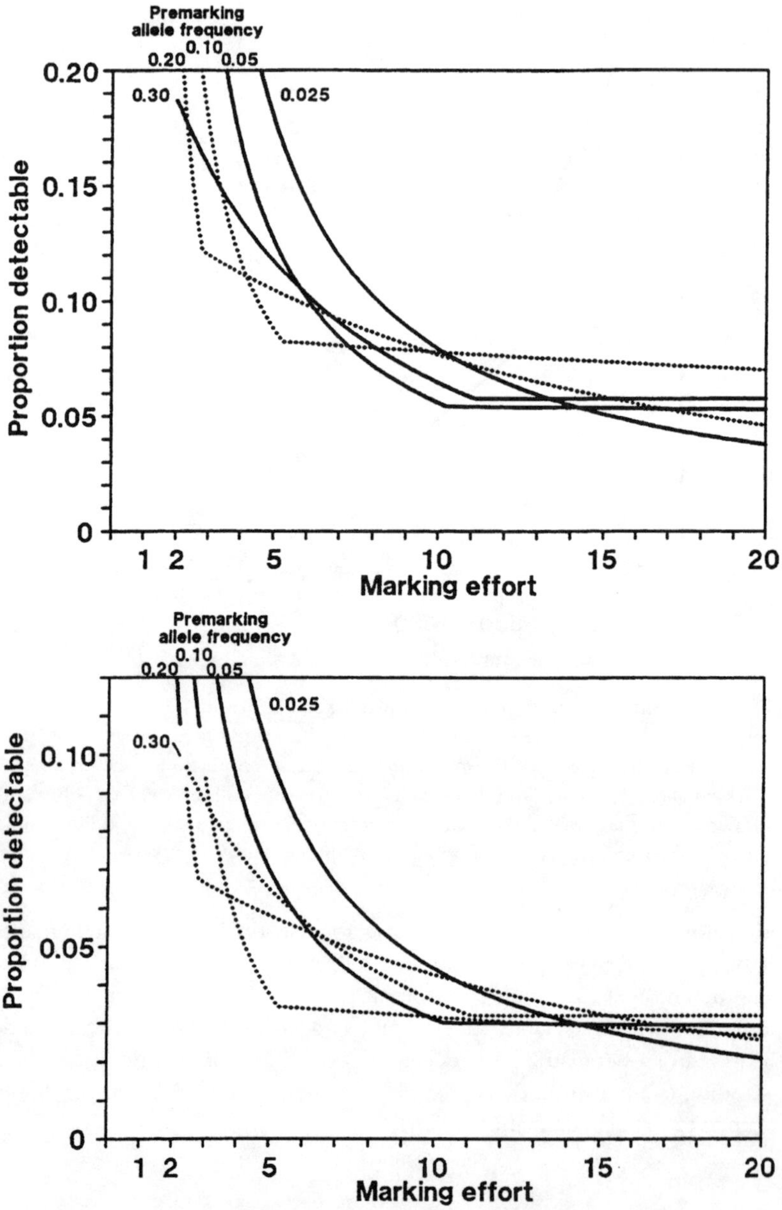

Figure 16.3 The minimum proportion of a genetically marked population that can be detected in a mixed sample of 300 (top) or 1,000 (bottom) fish with a probability of 0.90, shown for a number of premarking allele frequencies as a function of marking effort. Shown are the lesser of values calculated for allele frequencies (solid lines) and genotype frequencies (broken lines). Figure adapted from Gharrett and Seeb (1990).

16.3.1 Auke Creek Pink Salmon

To address questions regarding the practicality of genetic marking and persistence of marks, Lane et al. (1990) bred genetic markers into two populations of pink salmon *Oncorhynchus gorbuscha* in southeastern Alaska. In the life history of pink salmon, fry migrate to sea soon after resorbing their yolk sacs; they mature and return to spawn and die in their second year. As a result of their rigid 2-year life cycle, no genetic interchange occurs between odd- and even-year runs. Because the arrays of allele frequencies differ between odd- and even-year runs, the appropriate loci for genetic marking also differ. Lane et al. (1990) evaluated three aspects of their marking experiment.

1. Allele frequencies at the marker locus were monitored through the life cycle of the marked hatchery fish. If the marker allele was indeed selectively neutral, its frequency should not change during the life cycle.
2. Because many of the genetically marked fish would spawn naturally in the stream when they returned, the population of fish in the stream would become genetically marked. The frequency of the marker allele, then, should persist in the stream population, unless gene flow or selection altered it.
3. The selection required to mark the hatchery fish genetically could cause changes in allele frequencies at other loci due to chromosomal linkage or gametic disequilibrium. The likelihood of allele frequency changes caused by indirect means should be minimized by using a sufficient number of breeders. Hence, allele frequencies at unmarked loci were monitored, and changes were tested for gametic disequilibrium.

The two pink salmon populations marked were native to Auke Creek, just north of Juneau, Alaska. The National Marine Fisheries Service operates a salmon hatchery and weir for collecting juvenile and adult fish at the mouth of Auke Creek. The stream has natural runs of both odd- and even-year pink salmon. Within each year-class are five identifiable subpopulations whose spawning is separated spatially or temporally. Marker alleles were chosen for the odd- and even-year runs on the basis of existing baseline data and on the marking effort possible. Adults returning to the weir were screened for the appropriate marker allele. Protein electrophoresis was used to determine the genetic composition of biopsies of white skeletal muscle. Fish not used for breeding were released to the creek.

For marking the odd-year (1979) run, Lane et al. (1990) expected to obtain a broodstock of 400 males and females by screening 4,000 returning adults. With this marking effort, an allele present at an initial frequency of 0.05 could be increased to a frequency of about 0.50. The *sMDH-B1,2*70* allele (malate dehydrogenase, enzyme number 1.1.1.37; IUBMBNC 1992) was chosen as the marker allele because (1) its frequency had remained stable in nearby Fish Creek for six generations, (2) the frequency of the *sMDH-B1,2*100* allele was similar among populations in this and many nearby populations, and (3) *sMDH* activity is encoded at two pairs of loci (*sMDH-A1,2** and *sMDH-B1,2**), thereby possibly buffering any deleterious effects of increasing the frequency of the marker allele. In addition to increasing the frequency of the **70* allele, Lane et al. (1990) removed the **130* allele from the population. The marking effort involved fish returning in the latter part of the run, a procedural detail that subsequently proved important. Of 3,906 fish returning, 407 expressed the **70* allele but not the **130* allele and were selected as broodstock. In spring 1980, 178,219 genetically marked fry were released from Auke Creek Hatchery, 60,000 of which were fin-clipped to identify them later as returning, genetically marked spawners. A sample of the hatchery fry was taken for electrophoretic analysis. Additionally, 74,047 wild-produced fry emigrated past the Auke Creek weir. In 1981, 412 of the fin-clipped adult pink salmon returning were spawned in the hatchery. Their progeny were released in spring 1982. A sample was taken for electrophoretic analysis.

The *G3PDH-1*200* allele (glycerol-3-phosphate-dehydrogenase, 1.1.1.8) was chosen for marking the even-year run because (1) it is strongly expressed in muscle, (2) its frequencies were similar among nearby populations, and (3) its frequency was stable over five generations in nearby Fish Creek. In fall 1980, 7,710 adults were screened, and 396 were selected to produce the marked hatchery stock, a marking effort of 19.5. Hatchery fry were sampled for genetic analysis. Of 175,827 genetically marked fry released in spring 1981, 85,747 were fin-clipped. A sample was collected for electrophoretic analysis. Of 111,416 wild-produced fry passing over the weir, 27,026 were fin-clipped for future identification.

In the odd-year, genetically marked portion of Auke Creek pink salmon, the frequency of the *sMDH-B1,2*70* allele was increased from 0.055 to 0.508 and that of the **130* allele decreased from 0.046 to 0.000. Comparison of marker allele frequencies among 1979 hatchery spawners, fry sampled from the hatchery release in 1980, and fin-clipped returns from these in 1981 indicated no detectable changes during this generation of marked fish. Similarly, there were no changes in frequencies of the marker allele between the 1981 hatchery spawners and the resulting fry in 1982. These results suggest no direct negative effects of marking on fitness. Among other protein loci screened, no changes of allele frequencies through the life cycle were significant once adjustment was made for multiple testing. Only one of the 165 tests for gametic disequilibrium was significant at the 0.05 level, fewer than would be expected by chance alone. These results suggest that marking had no effect on survival through indirect effects of linkage and gametic disequilibrium. Adults without clipped fins that returned in 1981 exhibited an increase in the frequency of the marker allele. Lane et al. (1990) estimated that marked fish contributed 47.6 ± 3.5% of the late-run return. The frequency of the marker allele in late upstream spawners of the odd-year run was followed through four generations. No statistically significant changes were observed, although a trend of decreasing frequency was apparent.

In the even-year run, the frequency of the *G3PDH-1*200* allele was increased from 0.199 in the wild run to 0.876 in the hatchery broodstock. Comparisons of marker allele frequencies among 1980 hatchery spawners, hatchery fry released in 1981, and 1982 fin-clipped adult returns showed no significant differences during the life cycle. Similarly, no differences were observed through the life cycle of the 1982 year-class. No significant changes in allele frequencies were noted at other loci in either year-class. Genetically marked fish contributed 12.5 ± 2.4% to the Auke Creek run of September 1982, but virtually nothing, 0.0 ± 6.9%, to returns in 1984.

The results of Lane et al. (1990) show that if survival of marked hatchery fish in the wild is high, the mark will become strongly established; however, if survival is poor, the mark will be weak. Genetic marking often will be undertaken to estimate the contribution of a hatchery-based enhancement program on enhancement of a targeted population of fish. If a genetic mark is bred into a population by releasing a large component of hatchery-reared fish to spawn with an unmarked wild component, the success of both the enhancement and marking programs will depend on the survival to breeding age of the hatchery-produced fish.

Long-term monitoring of allele frequencies in the odd-year run and nearby populations led to inferences on the temporal and spatial substructure of pink salmon populations in southeastern Alaska. Over the next five generations, there was no detectable genetic exchange between the marked population and other temporally or spatially distinct spawning populations. No increases in frequency of the marker allele were detected, even in pink salmon spawning intertidally at Auke Creek or at Waydelich Creek, just 1 km away. This suggests that these populations are reproductively isolated, at least in the short term (Gharrett and Smoker 1993b).

Long-term monitoring results (McGregor et al. 1998) suggested that there is some degree of genetic influence on the time of return and the closely correlated time of spawning. Genetically marked fish returning in 1981 interbred with naturally produced fish, genetically marking the wild run. Through five generations, the genetic marker was detected only in the late part of the late (September) run, the segment of the adult return that actually had been screened and marked in 1979 (Gharrett and Smoker 1993a). Analysis of samples taken throughout the run showed that the marked component of the run remained in the late portion of the late upstream component for several generations (A. J. Gharrett, School of Fisheries and Ocean Sciences, University of Alaska–Fairbanks, unpublished). The marker frequency subsequently declined, either as a result of gene flow or an area-wide decrease in pink salmon returns in late September. Additional evidence that timing is an important component of population structure is that emigrant fry produced by the late-run component emigrated later, on average, than did fry produced by earlier returning parents (Gharrett and Smoker 1993a; Gharrett et al. 1999). The genetic marking experiment provided evidence that timing was an important component of population structure and probably had a genetic basis because the marked component produced offspring that had the same timing. A breeding experiment (Smoker et al. 1998) confirmed the existence of a heritable genetic component in time of return, even for timing within a particular run.

16.3.2 Kennedy Creek Chum Salmon

Seeb et al. (1986) genetically marked a population of chum salmon *Oncorhynchus keta* in Kennedy Creek, a tributary of Totten Inlet in southern Puget Sound, Washington. The selection process for genetic marking was carried out during 1976 to 1980. The Washington Department of Fisheries constructed a weir and trap across the creek, which intercepted all upstream migrants. All females were either passed upstream for natural spawning or used for the egg box take. Eggs were placed in streamside gravel incubation boxes in the hope of increasing the egg-to-adult survival rate. Males were selected for the purpose of genetic marking. Each day, males were taken from the trap, anaesthetized, and tagged, and their muscles were biopsied for electrophoretic analysis. Male spawners were selected for expression of the particular alleles, *sAAT-1,2*115* (aspartate aminotransferase, 2.6.1.1) and *PGDH*90* (phosphogluconate dehydrogenase, 1.1.1.44). The other males were passed upstream or discarded. The frequencies of the marker alleles were increased from approximately 0.12 and 0.01, respectively, in unselected fish to 0.25 and 0.11 among the progeny of selected males.

Seeb et al. (1990) used data from 5 years of marking and 6 years of monitoring the allozyme frequencies of returning adults to estimate the contribution of the gravel incubation boxes each brood year to the overall Kennedy Creek chum salmon run. Tissue samples were collected from freshly killed individuals immediately after spawning. Numbers were assigned to all samples so that the allozyme phenotypes and age could be matched for each individual; the age of each fish was determined from reading annuli on scales. All tissue samples were screened for variation at the *sAAT-1,2** and *PGDH** loci. Close to 17 million chum salmon eggs were genetically marked and placed in the streamside gravel incubation boxes. Phenotype frequencies of wild and egg box juveniles from each brood year (Seeb et al. 1986) were estimated and used to estimate the expected proportion of each group in the returning adult population. The proportion of adults from each source was estimated using maximum likelihood analyses of their allozyme frequencies. The success of the egg box program was assessed by comparing the rates of return among wild- and egg-box-produced adults.

Maximum likelihood estimates for five brood years of the proportions of chum salmon adults originating from the naturally spawned and egg box fry are shown in Table 16.1. The precision of the estimates of proportional contribution depended on the sample size. The numbers of fry produced in the stream and released from the egg boxes each year were known, as were the numbers of adults from each source that subsequently returned. Estimated indices of relative success ranged from 1.1 to 8.0, indicating that fry-to-adult survival rates of wild fry often were higher than those of egg box fry. The egg box fry performed poorly in the first year, 1976, when high egg-to-fry mortality rates occurred because of siltation in the egg boxes. In later years, settling boxes were installed to keep silt from entering the egg boxes. However, wild fry continued to return as adults at higher rates than did egg box fry. This finding contributed to the decision to discontinue the egg box program.

16.3.3 *Claytor Lake Walleye*

Walleye *Stizostedium vitreum* have been stocked widely throughout North America to establish new populations, maintain nonreproducing populations, and augment poor

Table 16.1 Maximum likelihood estimates of the proportion of chum salmon adults originating from naturally spawning (wild) and hatchery (egg box) fish. Data from Seeb et al. (1990).

Brood and year of return	Age	*N*	Estimated contribution		SD
			Wild	Egg box	
1976 brood					
1979	3	106	0.67	0.33	0.110
1980	4	26	[a]	[a]	
1977 brood					
1980	3	768	0.80	0.20	0.043
1981	4	462	0.74	0.26	0.052
1978 brood					
1981	3	48	0.50	0.50	0.176
1982	4	303	0.32	0.68	0.069
1979 brood					
1982	3	308	0.23	0.77	0.062
1983	4	379	0.09	0.91	0.056
1984	5	66	0.09	0.91	0.124
1980 brood					
1983	3	44	0.21	0.79	0.156
1984	4	260	0.39	0.61	0.069

[a] Sample size too small to make an estimate.

reproduction. The success of maintenance stocking is difficult to evaluate unless stocked fish can be identified and their contribution to harvest quantified. Schweigert et al. (1977) followed the frequencies of an sMDH* allozyme unique to stocked fish to show that the stocked juvenile walleye contributed 7% of the 1972 year-class in West Blue Lake, Manitoba. Murphy et al. (1983) assessed supplemental stocking effectiveness in Claytor Lake, Virginia, by quantifying shifts of *sMDH** alleles frequencies caused by stocking juvenile walleyes possessing allele frequencies different from those of naturally reproduced fish.

A small native population of walleye is believed to have inhabited the New River prior to the formation of Claytor Lake in 1939. Claytor Lake has a complex stocking history. Between 1939 and 1946, Pennsylvania (Lake Erie) walleye were stocked, and they established a self-sustaining population. Through the 1970s, however, the population declined, and four subsequent stockings planted fish from Kansas and Nebraska that had been derived from the Red River area of Minnesota. Hence, the lake might have contained as many as three geographic strains of walleye: a native Ohio River drainage strain, a Great Lakes strain, and a Hudson Bay strain.

Murphy et al. (1983) evaluated contributions from the various sources on recruitment of walleye in Claytor Lake by electrophoretic analysis of white muscle tissue collected from walleye from the lake and from original stocking sources. Frequencies of *sMDH-B** alleles varied significantly among cohorts. However, no heterogeneity was evident between cohorts that did not include stocked fish (1976 and 1978) or between cohorts that included stocked fish (1977 and 1979). Recruitment from the 1976 and 1978 cohorts originated entirely from natural reproduction, so the *sMDH-B2** allele frequency of 0.55 in the combined cohorts was taken as the best estimate of allele frequency among walleyes spawning in the lake during the entire study period, 1976 to 1979. Heterogeneity of allele frequencies was observed among the following groupings of walleyes: (1) Kansas and Pennsylvania, (2) the Claytor Lake group with stocked fish, and (3) Nebraska and the Claytor Lake group without stocked fish. Observed differences in allele frequencies were used to estimate the contribution of stocked age-0 Kansas fish to recruitment in an augmented cohort as

$$C = \left| \frac{(a-u)(100)}{s-u} \right|, \tag{16.6}$$

where C equals the percentage of the augmented cohort consisting of stocked fish, and a, u, and s are the observed *sMDH-B2** allele frequencies, respectively, of the augmented cohort in question, the combined unstocked cohorts, and the stocked age-0 Kansas fish. Confidence limits (CI 95%) for C were calculated as

$$C \pm t\left[\frac{C(100-C)}{n}\right]^{1/2}, \tag{16.7}$$

with t corresponding to a p of 0.05 for $(n-1)$ degrees of freedom, n being the number of possible alleles at the locus in question. Based on differences in allele frequencies between Claytor Lake cohorts with and without stocked fish, Murphy et al. (1983) estimated that stocked fish contributed 62% (CI, 52–72%) of the 1977 year-class at ages 2 and 3 and 72% (CI, 57–87%) of the 1979 year-class at ages 1 and 2. The genetic mark based estimate of recruitment of stocked fish, combined with estimates of the cost of hatchery production and return to the creel, showed that each stocked walleye harvested during 1976 to 1978 cost the state US$27.35. This study is an example of how through use of genetic marks it is possible to assess stock structure of a complex fishery and cost-effectiveness of a supplemental stocking program.

16.3.4 *Texas Red Drum*

In Texas, the population of red drum *Sciaenops ocellatus* began a dramatic decline in the 1970s, prompting the Texas Parks and Wildlife Department (TPWD) to implement an assessment and recovery program (McEachron et al. 1995). Management actions included monitoring of relative abundance of red drum, implementation of restrictive harvest regulations, and development and implementation of a hatchery-based enhancement program. The TPWD conducted extensive allozyme surveys on red drum populations. Findings suggested that the species should be regarded as a single genetic stock (King et al. 1995). The allozyme studies also estimated baseline allele frequencies and identified rare alleles that could be used as genetic markers for assessing the efficacy of the supplemental stocking program. Four loci, esterase D (*ESTD**, 3.1.1.56), tripeptide aminopeptidase (*PEPB**, 3.4.11.4), proline dipeptidase (*PEPD**, 3.4.13.9), and peptidase S (*PEPS**, 3.4.11.1), were identified as potential markers. The locus *ESTD** was chosen for the genetic marking primarily because stain ingredients are less toxic and costly, and its products are more easily resolved than those for peptidase stains. Further, this locus is resolvable using noninvasive sampling, that is, by screening enzyme activity in fin tissue. The *ESTD*95* allele averaged a 6% frequency along the Texas coast, a sufficiently low frequency that individuals homozygous for the marker would be rare in the population.

From fall 1991 through spring 1993, more than 700 red drum from the entire Texas coast were screened for rare alleles. Approximately 50 mature *ESTD*100/95* and 1 mature *ESTD*95/95* red drum were collected (King et al. 1993, 1995). Progeny of heterozygous broodfish ($N = 7$ spawns) were cultured to 30 d of age to detect fitness differences among genotypes and to confirm Mendelian inheritance of the markers in the F_1 fish. Results of this preliminary investigation suggested that *ESTD*95* is a biologically sound genetic marker because red drum possessing the rare allele appeared to perform as well as fish possessing the common allele.

A red drum gene-marking study based on these results was implemented in fall 1993. Approximately 250,000, 30-d-old fingerlings were stocked into East Matagordo Bay, Texas. Assessment of the success of the enhancement program consisted of monitoring changes in the frequency of the *ESTD*95* marker at the stocking site and adjacent bay systems. The recovery phase of monitoring, which began immediately, consisted of removal of fin tissue from all red drum collected in East Matagorda Bay bag-seine and gill-net samples by TPWD personnel, as well as from harvest surveys. Ward et al. (1995a) reported plans for stocking a variable number of fingerlings (250,000–1,500,000) in East Matagorda Bay each year for 4 years. The sampling and harvest survey would continue for 4 years after the last stocking of genetically marked fish. The stocking program would be considered successful if stocked fish contributed to the population, that is, if the frequency of the marker allele or the frequency of individuals homozygous for the marker allele increased above baseline values.

Over 200 million eggs, fry, and fingerlings of red drum have been stocked into Texas coastal waters since the inception of the enhancement program (Ward et al. 1995a). A preliminary assessment of program success involved regression analyses of the relationship between stocking rate and subsequent relative abundance as measured by gill-net harvest. Results suggested that stocking had contributed to abundance of red drum in the lower Laguna Madre, East Matagorda Bay, San Antonio Bay, and possibly Matagorda Bay (Ward et al. 1995b). Little or no evidence of stocking success was found in other bays. Results from screening of genetic markers would strengthen such inferences.

16.4 Perspective

These case studies demonstrate that genetic marking experiments can prove useful for investigating a range of questions relevant to both applied and basic fisheries research. Hatchery-based fisheries management practices have come under increasing scrutiny from view points of both cost-effectiveness and the possibility of ecological and genetic impacts of stocked fish on receiving populations (Schramm and Piper 1995; Chapter 15). Interest in genetic marking comes at a time when addition of new DNA-level markers to the geneticist's toolbox enhances technical capabilities for deliberate genetic marking of hatchery stocks. Hence, genetic marking may be incorporated more frequently into future hatchery and fisheries management plans.

16.5 References

Falconer, D. S. 1981. Introduction to quantitative genetics. Longman, New York.

Gharrett, A. J., and J. E. Seeb. 1990. Practical and theoretical guidelines for genetically marking fish populations. Pages 407–417 *in* Parker et al. (1990).

Gharrett, A. J, and S. M. Shirley. 1985. A genetic examination of spawning methodology in a salmon hatchery. Aquaculture 47:245–256.

Gharrett, A. J., and W. W. Smoker. 1993a. Genetic components in life history traits contribute to population structure. Pages 197–202 *in* J. G. Cloud and G. H. Thorgaard, editors. Genetic conservation of salmonid fishes. Plenum, New York.

Gharrett, A. J., and W. W. Smoker. 1993b. A perspective on the adaptive importance of genetic infrastructure in salmon populations to ocean ranching in Alaska. Fisheries Research 18:45–58.

Gharrett, A. J., and nine coauthors. 1999. Genetic structure in Auke Creek pink salmon and its role in productivity. Bulletin of the National Research Institute of Aquaculture (Japan), Supplement 1:19–26.

IUBMBNC (International Union of Biochemistry and Molecular Biology, Nomenclature Committee). 1992. Enzyme nomenclature, 1992. Academic Press, San Diego, California.

King, T. L., R. Ward, and I. R. Blandon. 1993. Gene marking: a viable assessment method. Fisheries 18(2):4–5.

King, T. L., R. Ward, I. R. Blandon, R. L. Colura, and J. R. Gold. 1995. Using genetics in the design of red drum and spotted seatrout stocking programs in Texas: a review. Pages 499–502 *in* Schramm and Piper (1995).

Lane, S., A. J. McGregor, S. G. Taylor, and A. J. Gharrett. 1990. Genetic marking of an Alaskan pink salmon population, with an evaluation of the mark and the marking process. Pages 395–406 *in* Parker et al. (1990).

McEachron, L. W., C. E. McCarty, and R. R. Vega. 1995. Beneficial uses of marine fish hatcheries: enhancement of red drum in Texas coastal waters. Pages 161–166 *in* Schramm and Piper (1995).

McGregor, A. J., and five coauthors. 1998. Migration timing, a life history trait important in the genetic structure of pink salmon. North Pacific Anadromous Fish Community Bulletin 1:262–273.

Morrison, J., and D. Zajac. 1987. Histologic effect of coded wire tagging in chum salmon. North American Journal of Fisheries Management 7:439–441.

Murphy. B. R., L. A. Nielsen, and B. J. Turner. 1983. Use of genetic tags to evaluate stocking success for reservoir walleyes. Transactions of the American Fisheries Society 112:457–463.

Nielsen, L. A. 1992. Genetic identification and marking. Pages 155–181 *in* L. A. Nielsen, editor. Methods of marking fish and shellfish. American Fisheries Society, Special Publication 23, Bethesda, Maryland.

Olito, C., and B. Davis. 1983. Genetic marking at Tutka Lagoon Hatchery, August 1981. Alaska Department of Fish and Game, Division of Fisheries Rehabilitation, Enhancement and Development, Juneau.

Parker, N. C., A. E. Giorgi, R. C. Heidinger, D. B. Jester, Jr., E. D. Prince, and G. A. Winans, editors. 1990. Fish-marking techniques. American Fisheries Society, Symposium 7, Bethesda, Maryland.

Pella, J., and six coauthors. 1998. Stock composition of some sockeye salmon, *Oncorhynchus nerka*, catches in southeast Alaska, based on incidence of allozyme variants, freshwater ages, and a brain tissue parasite. NOAA (National Oceanic and Atmospheric Administration) Technical Report NMFS (National Marine Fisheries Service) 132.

Reisenbichler, R. R., and J. D. McIntyre. 1977. Genetic differences in growth and survival of juvenile hatchery and wild steelhead, *Salmo gairdneri*. Journal of the Fisheries Research Board of Canada 34:123–128.

Schramm, H. L., and R. G. Piper, editors. 1995. Uses and effects of cultured fishes in aquatic ecosystems. American Fisheries Society, Symposium 15, Bethesda, Maryland.

Schweigert, J. F., F. J. Ward, and J. W. Clayton. 1977. Effects of fry and fingerling introductions on walleye *(Stizostedion vitreum vitreum)* production in West Blue Lake, Manitoba. Journal of the Fisheries Research Board of Canada 34:2142–2150.

Seeb, J. E., L. W. Seeb, and F. M. Utter. 1986. Use of genetic marks to assess stock dynamics and management programs for chum salmon. Transactions of the American Fisheries Society 115:448–454.

Seeb, L. W., J. E. Seeb, R. L. Allen, and W. K. Hershberger. 1990. Evaluation of adult returns of genetically marked chum salmon, with suggested future application. Pages 418–425 *in* Parker et al. (1990).

Smoker, W. W., A. J. Gharrett, and M. S. Stekoll. 1998. Genetic variation of return date in a population of pink salmon: a consequence of fluctuating environment and disruptive selection? Alaska Fishery Research Bulletin 5(1):46–54.

Taggart, J. B., and A. Ferguson. 1984. An electrophoretically detectable genetic tag for hatchery-reared brown trout (*Salmo trutta* L.) Aquaculture 41:119–130.

Utter, F. M., and J. E. Seeb. 1990. Genetic marking of fishes: overview focussing on protein variation. Pages 426–438 *in* Parker et al. (1990).

Wallace, B. 1981. Basic population genetics. Columbia University Press, New York.

Ward, R., T. L. King, I. R. Blandon, and L. W. McEachron. 1995a. Use of gene marking to assess stocking success of red drum in Texas bays. Page 551 *in* Schramm and Piper (1995).

Ward, R., L. W. McEachron, B. E. Fuls, and M. Muoenke. 1995b. Multivariate analysis of red drum stocking in Texas bays. Page 552 *in* Schramm and Piper (1995).

17

Forensics

Eric Hallerman

17.1 Introduction

Forensic science is the application of science for discussion, debate, argumentative, or legal purposes (Lapedes 1978). Within the scope of this chapter, forensics is the use of scientific methods to make inferences regarding past events, especially with regard to questions relevant to fisheries, wildlife, and conservation law enforcement. Forensics as a field focuses on determining and proving who perpetrated criminal acts. These acts include misappropriation of fisheries resources, misrepresentation of the content of fisheries products, and illegal commerce in fish or fisheries products. In this chapter, I focus on the use of genetic tools for forensic investigations involving fisheries or fisheries products.

In the classical situation, conservation law enforcement officials suspecting a crime will seize a quantity of fish or fisheries product. In order to establish or prove their case, they must demonstrate that the evidence constitutes a taking from a prohibited species or population, a taking made out of season, or a taking illegally introduced into commerce. The seized material is analyzed in a laboratory, where the question posed essentially is, what is this? Depending on the application, the question can be considered one of species, population, or individual identification, as detailed below. The theoretical underpinnings of hypothesis formulation, interpretation of DNA or other genetic marker profiles, and the means of expressing evidentiary strength in courts of law are covered by Evett and Weir (1998). Practical aspects of laboratory organization, analytical techniques, probability and statistical analyses, and quality control, as well as legal and ethical considerations and case studies, are presented by Kirby (1990).

Because forensics so often involves criminal prosecutions, case work rarely is reported in the scientific literature. I will use published examples of forensic studies as available but will have to use hypothetical situations to illustrate certain points.

17.2 Species-Specific Identification

For questions of species identification, forensic scientists need **fixed** (i.e., in this context, species-specific) diagnostic genetic markers. A number of molecular genetic markers have been used in various applications, including allozymes, mitochondrial DNA (mtDNA), ribosomal DNA, and other nuclear DNA markers. Two basic questions recur in the context of forensic species identification. In both cases, the forensic investigation must be based on baseline information about the species.

17.2.1 Is This a Legally Taken or Traded Species?

Many species are protected by legislation or regulations because their population levels have become low. Protection often will include a ban on taking (e.g., under the Endangered Species Act of 1973 [16 U.S.C. §§ 1531 to 1544] or the Convention on International Trade in Endangered Species of Wild Fauna and Flora [1973]), which may be year-round or seasonal. Takings, however, may continue illegally. Should law officers confiscate a collection of whole fish as evidence of suspected illegal activities, morphological identification of the fish can be made. However, if fillets or processed fisheries products are involved, species identification becomes more complex.

Whales provide a good case study of a forensic problem of species identification analogous to that for fisheries. Taking of most species of whales was prohibited or dramatically restricted by the International Whaling Commission (IWC) in 1986. Fishers of some IWC member nations, however, have continued to take whales, including protected species, under scientific permits, for subsistence or aboriginal use, or by means of poaching. In many cases, harvested whales are processed at sea in the absence of on-vessel fisheries observers. The resulting whale products are openly marketed. Discrimination between products of legally versus illegally taken species is difficult. Baker and Palumbi (1994) and Baker et al. (1996) purchased a variety of whale products through commercial channels in Japan and Korea. They extracted DNA and used the polymerase chain reaction to amplify and sequence 155 to 178 base pairs of the mtDNA control region. Sequences of DNA for 16 products of unknown species affinity were aligned and compared with reference sequences for 16 cetacean species, including geographic variants where known. Bootstrap simulations unambiguously grouped 14 of the samples of unknown affinity with a reference species sequence (Figure 17.1) with at least 90% confidence, providing statistical support for species identifications. Eight samples grouped with either southern or northern minke whales *Balaenoptera acutorostrata,* four with fin whales *B. physalus,* two with dolphins (Hector's dolphin *Cephalorhynchus hectori* and Commerson's dolphin *C. commersonii*), and two were not clearly classified. It was difficult to reconcile availability of some of the products commercially available in Japan and Korea with the legal catch records made available to the IWC. Relatively few minke whales were taken legally, and take of humpback whale *Megaptera novaeangliae* and fin whale had been prohibited since 1966 and 1986, respectively. The simplest explanation is that products available in these markets included species that had been hunted, processed, or marketed illegally.

A number of forensic species identification cases have been solved using evidence from allozyme analyses. In 1978, a Japanese trawler fishing within the 200-mi buffer claimed as U.S. coastal waters was suspected of illegally harvesting the rockfish, Pacific ocean perch *Sebastes alutus*. The trawler's log, however, bore contradicting claims. Fillets confiscated by enforcement officers were identified to species based on allozyme markers (Seeb 1986; Utter 1991), showing that the protected species had indeed been taken. A case in Texas involved a suspected illegal sale of frozen fillets from flathead catfish *Pylodictis olivaris*. Allozyme analyses confirmed the species identification and led to a fine being levied against the seller (Harvey 1990). In another case in Texas, allozyme analyses

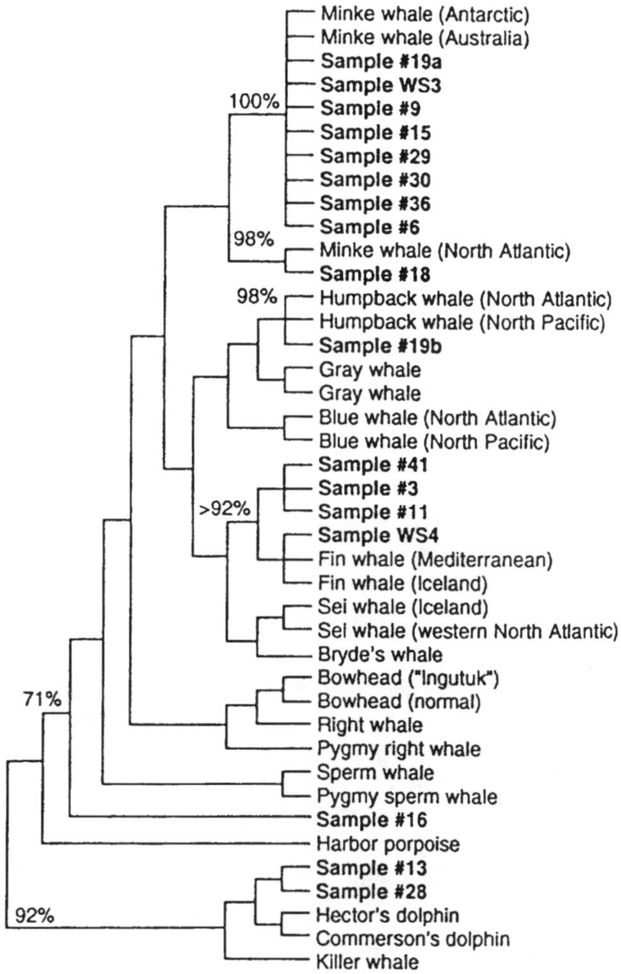

Figure 17.1 Phylogenetic relationships of mitochondrial DNA control region sequences from samples of whale products of unknown species affinity (shown in bold type) to reference samples of known species affinity. Reprinted with permission from C. S. Baker and S. R. Palumbi. Which whales are hunted? A molecular genetic approach to monitoring whaling. Science 265:1538–1539. Copyright 1994, American Association for the Advancement of Science.

were used to show that anglers were innocent of charges of illegal possession of fillets from red drum *Sciaenops ocellatus* and spotted seatrout *Cynoscion nebulosus* (Harvey 1990).

17.2.2 Is This Product What the Seller Claims?

Because some fisheries products command high prices in the marketplace, there is the temptation to market a less-valued product as the more expensive one. For example, fish processors openly market cylindrical sections stamped from the enlarged pectoral fins of skates as imitation, or mock, scallops. However, there are fewer instances when fin meat is misleadingly sold at a higher price as scallops. Similarly, while highly valued caviar is made from sturgeon roe, large eggs from other fishes can be marketed fraudulently as sturgeon caviar. Use of diagnostic genetic markers fixed for given species can demonstrate whether or not the product under question is what the seller claims.

Highly processed fisheries products, such as surimi (e.g., mock crab, often made from walleye pollack *Theragra chalcogramma*), are difficult to characterize regarding what species provided the starting material (O. Ramon, Department of Food Engineering and Biotechnology, Technion University, personal communication). Surimi products command different prices, depending on the species or mix of species used in their preparation. Further, certain countries demand certification of the content of fisheries products prior to import. There are a variety of reasons for this requirement, ranging from health to economic reasons to even cultural or social reasons, For example, Israel demands disclosure of content because not all fishes are regarded as fit for consumption under Jewish religious tradition. Regardless of the driving interest, certification is possible only if reliable identification of contents is possible, such as that provided by screening of genetic markers.

17.3 Population-Specific Identification

There are a number of situations in which exploitation or human-mediated transfer of a genetic stock of fish (Chapter 13) is restricted or prohibited as a conservation measure. In such cases, forensic scientists are faced with the need to develop population-specific identification methods. In such situations, fixed allelic differences among populations are unlikely. However, molecular genetics provides a variety of methods for developing and screening polymorphic genetic markers, and the mathematics of population genetics provides useful methods for assessing populations' different allele frequencies. The frequency or rate of occurrence of these characters serves as a marker not for the individual but for the population as a whole. Problems involving population-specific identification include the four following forensic questions.

17.3.1 From Where Did This Catch Come?

Whereas some stocks of a species may support harvest, others may not, and fisheries managers may ban taking from such stocks. For practical reasons, especially for marine fisheries, managers may not be able to monitor take comprehensively. They often have to rely on the understanding and voluntary compliance of fishers. Ultimately, the temptation to harvest from a protected stock proves too great for some fishers. By sampling a suspected illegal catch and determining its genetic stock affiliation, it is possible to support or refute a hypothesis regarding illegal taking.

As a case study involving population-specific identification, consider the sequence of events following the landing of a fishing boat loaded with unusually large Alaskan red king crabs *Paralithodes camtschaticus*. The size of the red king crabs aroused suspicion that they had been taken from Bristol Bay, a closed area. The captain denied this, saying the red king crabs had been caught near Adak Island. The Alaska Department of Fish and Game seized the catch and asked Seeb et al. (1990) to compare the array of allozyme frequencies in the catch against baseline data. The array of allozyme frequencies closely matched those of the red king crab population of the Bering Sea (Figure 17.2), an area closed at the time to fishing. Confronted with the findings, the captain confessed to fishing in the closed area. He agreed to pay a fine of US$565,000. The settlement money was used to support crab fishery research. In a study of similar design, Seeb and Seeb

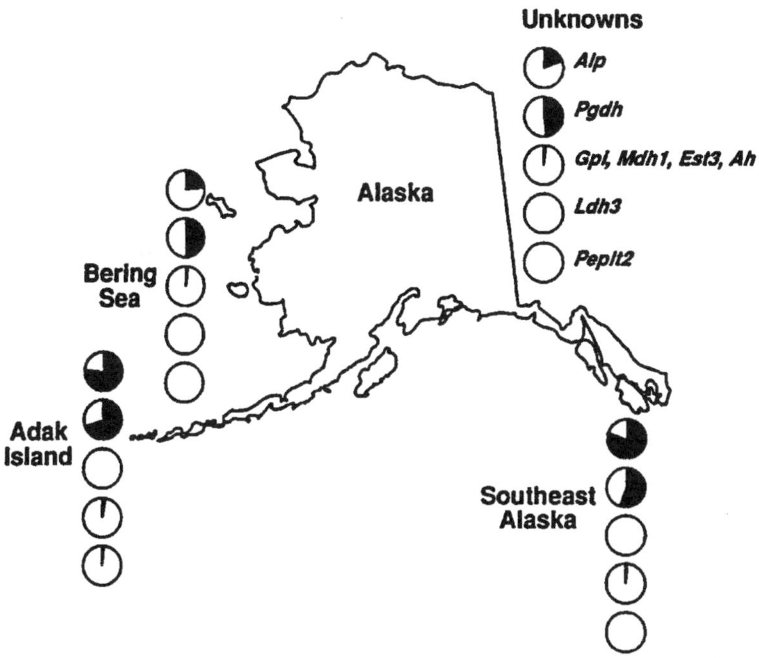

Figure 17.2 Pie charts showing arrays of allozyme frequencies at eight polymorphic loci for red king crab populations. Allozyme frequencies for a catch from an unknown location closely matched those from the population from the Bering Sea . Loci are *ALP** (alkaline phosphatase, enzyme number 3.1.3.1; IUBMBNC 1992), *PGDH** (phosphogluconate dehydrogenase, 1.1.1.44), *GPI** (glucose-6-phosphate isomerase, 5.3.1.9), *sMDH** (malate dehydrogenase, 1.1.1.37), *EST-3** (esterase, 3.1.1.-), *AH** (aconitate hydratase, 4.2.1.3), *LDH-3** (L-lactate dehydrogenase, 1.1.1.27), and *PEPlt-2** (peptidase leucyl–tyrosine, 3.4.-.-). Figure adapted from Seeb et al. (1990).

(1986) assigned the continent of origin to confiscated chum salmon *Oncorhynchus keta*, supporting the conservation officers' allegations that the fish were harvested illegally from closed areas.

17.3.2 Is This a Legally Transported Stock?

Many states have restrictions on moving particular fish stocks. Should the source and donor stocks have diagnostic genetic markers or sufficiently distinct allele frequencies, the nonnative character of the imported stock can be inferred. For example, certain northern states prohibit the importation of Florida largemouth bass *Micropterus salmoides floridanus*. However, because such stocks are sold in general commerce and because some people perceive Florida largemouth bass as desirable, individuals in northern states may attempt to import them. Should fisheries management authorities suspect an illegal importation, observation of alleles diagnostic of the Florida subspecies would support the conclusion that the collection of fish in question or their ancestors had been imported. For example, at a bass fishing tournament in Texas, the apparent winner was suspected of having smuggled in a large fish from outside the tournament site. Allozyme analysis of tissue samples from the fish showed that it expressed genes characteristic of the Florida subspecies, evidence that it had indeed been smuggled into the tournament site. The angler was charged and found guilty of fraud (Harvey 1990).

17.3.3 Did Someone Improperly Sell a Licensed Stock?

It is increasingly common that fish producers purchase fingerlings on a contract basis from a corporation, which then will purchase the entire stock at a fixed price following grow out. Other sorts of restrictive licensing agreements might become common, especially as biotechnologically manipulated fish stocks become more widely used in commercial aquaculture. In the licensed stock scenario, a fish farmer may be faced with the temptation to sell to a third party at a higher price or to reproduce and sell fingerlings from the licensed stock. In either case, should the corporation allege that the grower illegally sold some of the licensed stock, the forensic question is one of population-specific identification. Resolution of the issue would be straightforward if (1) the licensed stock had been deliberately genetically marked (Chapter 16), (2) the biotechnological manipulation itself produced a unique genetic mark, or (3) the stock displayed an otherwise unique mark.

17.3.4 Did Someone Illegally Release Fish?

Fishes often escape from aquaculture operations. In a subset of these instances, such escapes pose fisheries management concern for ecological, genetic, or fish health reasons (Chapter 15). Concern is particularly acute when a nonindigenous species is at issue. Escapement from an aquaculture operation may occur as a result of operator error, equipment failure, inadequate facility design, or, occasionally, intentional release. For example, producers of tropical fishes have been known to stock excess production into natural systems for later recovery. Producers of salmon in net-pens have been suspected of sabotaging their nets when market prices decreased below the insured value of their stock. In any case, fisheries management authorities may want to link recovered fish to their alleged source. Forensic scientists would attempt to show that genetic marker frequencies of the fish recovered from the natural system match those of fish remaining on the farm and that those marker frequencies differ from any other possible sources, notably the local native stock. Here again, unique marks (whether naturally occurring or deliberate marks) can provide information useful for determining the source of escapement. However, a mark does not need to be entirely unique or fixed. Rather, as long as the frequency of the mark is discernable among populations, it can be used to establish origin.

In a variation of this application, genetic markers could be used to show that fisheries products were *not* from an aquaculture stock, as a seller might claim. This would provide a means for investigating allegations that products from poached stocks were being laundered through commercial operations (Thomasen et al. 1989).

17.4 Individual-Specific Identification

In a common law enforcement case, authorities find the carcass of an illegally taken animal stripped of valued parts, such as tusks from elephants, gall bladders from bears, or fins from sharks. They may have a suspect who has in his possession products from the species at issue, but that suspect may claim the products were taken legally from another animal, in season. The enforcement question at issue in this context is, did the suspect take this illegally harvested individual? In such a context, a forensic scientist would need highly polymorphic, individual-specific genetic markers to tie the poached carcass directly and confidently to the product in the suspect's possession. In such cases, either multilocus DNA fingerprinting or a collection of single-locus VNTR (variable number of tandem repeat) markers (Chapter 5) is likely to prove the molecular genetic tool of choice.

Should enforcement officials have access to gram-quantity samples of seized fisheries product, they can prepare large-enough quantities of high-quality DNA that screening for multilocus DNA fingerprints on Southern blots is a viable technical possibility (Chapter 5). Should the banding patterns for one or more probe × restriction enzyme combinations for the poached carcass and the seized material match, successful prosecution is likely. A panel of DNA samples for other individual fish not related to the alleged crime should be included on the blot to demonstrate that complete matching of bands between the poached carcass and the seized material was not likely by chance alone.

In other, more challenging cases, the suspect may not have the products of the alleged poaching in his or her possession, but there may be materials from the alleged crime, such as blood or scales, on the suspect's tools, clothing, boat, or vehicle. Because little DNA can be prepared, a polymerase chain reaction based procedure (Chapter 5) may be useful to amplify marker loci for comparison between the poached carcass and collected samples. A variety of genetic markers might serve the intended purpose; the key point is screening a collection of markers with high numbers of alleles per locus, for example, microsatellite loci. Matching of alleles at a number of loci will be needed to support a successful prosecution. Screening of a panel of DNA from conspecific fish unrelated to the alleged crime will be needed to show that the collection of markers was sufficiently discriminatory to identify individual fish.

17.5 Perspective

As discussed in this chapter, a maturing area of focus in forensic science is applied population genetics. Fisheries managers and enforcement officers benefit by understanding the connection between population genetics and forensics. They need to recruit expert assistance in order to prosecute suspected cases of wrongdoing involving populations they manage, and they need to understand the genetic data collected. Relatively few fisheries and wildlife population geneticists are regularly involved in forensic work. Among the few notable exceptions, the U.S. Fish and Wildlife Service has a dedicated forensics laboratory in Ashland, Oregon, which largely is involved with cases concerning illegal importation of products made from internationally protected (i.e., threatened or endangered) species. More and more state fisheries management agencies are starting to use molecular tools and population genetic approaches for enforcing conservation laws.

17.6 References

Baker, C. S., F. Cipriano, and S. R. Palumbi. 1996. Molecular genetic identification of whale and dolphin products from commercial markets in Korea and Japan. Molecular Ecology 5:67–685.

Baker, C. S., and Palumbi, S. R. 1994. Which whales are hunted? A molecular genetic approach to monitoring whaling. Science 265:1538–1539.

Convention on International Trade in Endangered Species of Wild Fauna and Flora (CITES). 1973. [Text of the Convention.] CITES Secretariat, Châtelaine-Genève, Switzerland. (For more information, go to *www.wcmc.org.uk/CITES*.)

Evett, I. W., and B. S. Weir. 1998. Interpreting DNA evidence: Statistical genetics for forensic scientists. Sinauer Associates, Sunderland, Massachusetts.

Harvey, W. D. 1990. Electrophoretic techniques in forensics and law enforcement. Pages 313–321 *in* D.H. Whitmore, editor. Electrophoretic and isoelectric focussing techniques in fisheries management. CRC Press, Boca Raton, Florida.

IUBMBNC (International Union of Biochemistry and Molecular Biology, Nomenclature Committee). 1992. Enzyme nomenclature, 1992. Academic Press, San Diego, California.

Kirby, L. T. 1990. DNA fingerprinting: an introduction. Stockton Press, New York.

Lapedes, D. N. 1978. Dictionary of scientific and technical terms. McGraw-Hill, New York.

Seeb, J. E., G. H. Kruse, L. W. Seeb, and R. G. Weck. 1990. Genetic structure of red king crab populations in Alaska facilitates enforcement of fishing regulations. Alaska Sea Grant College Program Report 90-04:91–502.

Seeb, J. E., and L. W. Seeb. 1986. Genetic stock identification of confiscated high seas chum salmon. Summary Report to the U.S. Department of Commerce, National Marine Fisheries Service, Law Enforcement, Seattle.

Seeb, L. W. 1986. Biochemical systematics and evolution of the scorpaenid genus *Sebastes*. Doctoral dissertation. University of Washington, Seattle.

Thomasen, H. V., M. J. Thomson, G. G. Shutler, and L. T. Kirby. 1989. Development of DNA fingerprints for use in wildlife forensic science. Wildlife Society Bulletin 17:321–326.

Utter, F. 1991. Biochemical genetics and fishery management: an historical perspective. Journal of Fish Biology 39A:1–20.

18

Population Viability Analysis

Eric Hallerman

18.1 Introduction

Maintaining a population's **viability**—its ability to sustain itself demographically—is a fundamental goal of conservation biology and increasingly is becoming so for fisheries management. The justifications for doing so are several. Wild fish populations have intrinsic value (Naess 1986); fish populations play important ecological roles in the structure and function of aquatic communities; and many fish populations have economic or esthetic value. Further, key environmental laws mandate maintaining the viability of populations.

The National Forest Management Act of 1976 (16 U.S.C. §§ 1600 to 1640) mandates that viable populations of all native vertebrates be maintained in our national forests. Regulations call for identification and monitoring of **management indicator species,** which include threatened or endangered species; species commonly fished, trapped, or hunted; economically important species; and species sensitive to human activities.

As amended in 1978, the U.S. Endangered Species Act of 1973 (ESA) (16 U.S.C. §§ 1531 to 1544; ESA Amendments of 1978, Pub. L. 95-632, 92 Stat. 3751, and Pub. L. 97-304) provides "a means whereby the ecosystems upon which endangered species and threatened species depend may be conserved." As defined in the ESA, the term **species** includes "any subspecies of fish or wildlife or plants, and any distinct population segment of any species of vertebrate fish or wildlife which interbreeds when mature" (Section 3[15]). The ESA, however, does not offer guidance regarding how distinct population segments are to be delineated, and evolutionary biologists have struggled to define defensible guidelines for the purpose (Box 18.1). The issue of defining appropriate and defensible management units is the subject of a growing body of the scientific literature (Nielsen 1995).

Hence, there is considerable impetus for designing fisheries management plans aimed at protecting the viability of fish populations. Once the management unit has been defined, the issue turns to one of how best to practice management in order to ensure that management unit's continuing viability, providing the context for this chapter. We must begin by establishing a working understanding of the principles of population viability analysis.

■ Box 18.1
Evolutionarily Significant Units

The U.S. Endangered Species Act of 1973 (ESA; 16 U.S.C. §§ 1531 to 1544) provides legal authority for the protection of threatened or endangered species, subspecies, or distinct population segments of any species of fish or wildlife or plants. It does not explain, however, how distinct population segments might be evaluated or measured. The U.S. Fish and Wildlife Service and the National Marine Fisheries Service adopted a joint policy (*Federal Register* 61 [7 February 1996]:4722–4725) clarifying interpretation of the term **distinct population segment** for the purpose of listing, delisting, or reclassifying vertebrates by the Fish and Wildlife Service and National Marine Fisheries Service. Three elements are considered in a decision regarding the status of a population segment as endangered or threatened under the ESA: (1) discreteness of the population segment in relation to the remainder of the species to which it belongs, (2) significance of the population segment to the species to which it belongs, and (3) the population segment's conservation status in relation to the ESA's standards for listing.

The discreteness of a fish stock can be quantified by analyzing the distribution of genetic diversity in the species (Chapter 13). From deliberations regarding protection of salmonid stocks under the ESA came the concept of the evolutionarily significant unit (ESU). Waples (1991) proposed that a population warrants protection under the ESA if it represents an ESU of the biological species. To qualify as an **evolutionarily significant unit,** a population must satisfy two conditions.

— It must be *reproductively isolated* from other conspecific units, although it does not have to be absolutely isolated reproductively.
— It must represent an important component of the *evolutionary legacy* of the species.

To be considered an important component of the evolutionary legacy, a population must satisfy at least one of the following criteria.

— Is the population *genetically distinct*?
— Does it occupy *unique habitat*?
— Does it exhibit *unique adaptation* to its environment?
— If it went extinct, would this be a significant loss to the *ecological or genetic diversity* of the species?

The ESU concept can support rational decision making regarding protection of populations or genetic stocks, whether or not these decisions are being made in the context of status determinations under the ESA. The concept and implementation of the ESU framework have been the subject of much discussion (Nielsen 1995).

18.2 Principles Underlying Population Viability Analysis

18.2.1 *Maintenance of Population Viability and Evolutionary Adaptivity*

From the viewpoint of population genetics, we might define our management goal as the stewardship of genetic variability in managed populations for purposes of maintaining both the viability and adaptivity of those populations. These are short- and long-term aspects of the same issue. We manage populations in order to support their ability to maintain their viability over an ecological time scale and their potential for continuing adaptation over an evolutionary time scale. Assessing the viability of individual fish populations can be a difficult problem for fisheries biologists. There is little quantitative support or guidance for such assessments. Extinction risks for fish populations are influenced by complex and interacting factors that often are difficult to identify and measure. However, understanding the nature of extinction processes can lead to management prescriptions that minimize risks to local populations (Rieman et al. 1993). Insights drawn from the developing field of conservation biology can provide useful guidance to fisheries managers.

What, then, are the conditions for maintaining the viability of natural populations? Community ecologists have discussed the issue of what minimum habitat areas might be set aside. Population geneticists, however, have approached the issue in terms of minimum population sizes (Soulé and Simberloff 1986). These are not really separate issues; rather, one approach complements the other. Extensive discussions of viability issues led to consideration of the minimum viable population size. **Minimum viable population size** has been defined (Soulé 1987) as the population size that provides a given probability of persistence of the population for a given amount of time, for example, a 95% expectation of persistence without loss of fitness for several centuries. The mention of probability is an explicit recognition that we cannot definitively ensure the viability of a population. Explicit mention of a long planning horizon follows from our desire to maintain not only the short-term viability but also the long-term adaptive capability of populations. Objections have been raised to managing for *minimally* viable populations (e.g., Conner 1988), because it suggests brinksmanship and a high level of precision in estimates of viability. Hence, minimum has been dropped in many later considerations and publications. In this chapter, I will use the term viable population number.

18.2.2 *Factors that Affect Population Persistence*

To estimate the population size needed to avoid extinction, we must consider the various factors that could cause extinction. Four classes of factors affecting the persistence of a population were discussed by Shaffer (1981) and many other authors. There are two classes of **extrinsic factors,** that is, factors outside the biology of the population at issue. **Environmental uncertainty** is posed by normal variation in ecological conditions. For example, conditions of streamflow and temperature affect habitat and food availability, which, in turn, affect the size of the population and, hence, its viability. Larger populations are better buffered against the effects of environmental uncertainty than are smaller populations. **Natural catastrophe,** as the name suggests, is posed by natural phenomena outside the normal range or variation. For example, the eruption of Mount St. Helens in the northwestern United States blocked the access of certain salmonid stocks to their spawning grounds, a catastrophe leading directly to extinction of those stocks. Other rare events, such as droughts and debris torrents, can be regarded as natural catastrophes. Larger populations are more likely to have some members survive a catastrophe than are small populations. Clearly, human-induced catastrophes, for example, dam construction or toxic discharges, also can impact the viability of populations. There are two classes of **intrinsic factors,** those that operate within the biology of the population at issue. **Demographic uncertainty** refers to processes affecting age structure and rate of population growth. Fluctuations in population size arise because populations contain a discrete number of individuals, and population changes are caused by a progression of individually unpredictable births and deaths (Nisbet and Gurney 1982). Demographic stochasticity produces sampling variances of survival and reproductive rates inversely proportional to population size (Lande 1988). Fluctuations in sex ratio also can affect the ability of a population to propagate itself. Demographic uncertainty is regarded as becoming important when population sizes become small, on the order of 20 adults (Leigh 1981; Shaffer and Samson 1985; Quinn and Hastings 1987; Shaffer 1987). **Genetic uncertainty** is the result of processes affecting the fitness of members of the population. The critical processes in this context are inbreeding (Chapter 10) and loss of variation due to random genetic drift (Chapter 9). Some authors add mention of loss of gene combinations critical for local adaptation (Chapter 11).

Other authors (e.g., Leigh 1981; Gilpin and Soulé 1986; Shaffer 1987) characterize processes contributing to extinction as deterministic, stochastic, or genetic. **Deterministic extinctions** occur with the cumulative loss or permanent change of a critical component in the species' environment (Gilpin and Soulé 1986), resulting in progressive increases in the population death rate, decreases in birth rate, or both. For example, salmon populations have declined due to increasing mortality at dams in the migration corridor. **Stochastic processes** are those associated with chance events resulting in unpredictable fluctuations in population numbers and include those mediated by demographic stochasticity and environmental stochasticity. Although cogent arguments can be offered supporting one or another framework for characterizing the risks faced by a population, the underlying principles are strikingly similar.

18.2.3 *Interaction of Factors Affecting Extinction Probability*

Whatever the framework for characterizing factors affecting vulnerability to extinction, it is important to recognize that these factors do not operate independently. For example, loss of genetic variability may compromise a population's ability to respond to environmental variation. Environmental variability may isolate suitable patches of habitat, increasing a population's vulnerability to demographic and genetic risks. Clearly, the relative importance of the factors posing extinction of a population varies among ecological contexts. For large natural populations, demographic and genetic factors are unlikely to pose risk, leaving environmental uncertainty and natural catastrophe as the major factors posing extinction. However, for small natural populations, all four types of extinction factors may be important. We can buffer captive populations from risks of environmental uncertainty and natural catastrophe, but because of small population size, demographic and genetic uncertainty may prove important.

Study of the interaction of factors affecting extinction or persistence has given rise to an approach called population viability analysis (PVA). These interactions have not been integrated into a widely applicable model of population persistence, although there are useful analytic (e.g., Shaffer and Samson 1985; Dennis et al. 1991; Stacy and Taper 1992) and simulation (e.g., Vortex, Lacy 2000) models. In a typical application, variance in population growth rate resulting from environmental or demographic factors interacts with population size to determine the probability of extinction in a given time period. The expected time of persistence decreases as population size decreases and the variation in population growth rate increases. Small populations are more likely to experience high temporal variation and lower genetic and phenotypic diversity and to have fewer refuges available (Gilpin and Soulé 1986; Shaffer 1987; Saunders et al. 1990).

Gilpin and Soulé (1986) proposed that environmental change can set up positive feedback loops of biological and environmental interactions that have further negative effects on a population, possibly leading to its extinction. They referred to these event trains as **extinction vortices,** or self-reinforcing processes that drive a population toward extinction. They described four types of vortices: variance in reproductive rate (the R vortex), patchiness in species distribution (the D vortex), inbreeding (the F vortex), and loss of adaptive potential due to random genetic drift (the A vortex). The action and interaction of these extinction processes is modeled in the Vortex PVA software package (Lacy 2000). Use of Vortex has proven popular for modeling possible extinction processes for organisms for which demographic data exist and for which the data fit within the model's constraints. However, the utility of the modeling package is limited for many aquatic organisms. Age-specific demographic information is lacking for many aquatic organisms. Fecundities (or litter sizes) for aquatic organisms frequently exceed 49, the maximum value accepted by the program.

Results for a variety of species suggest that populations isolated from other populations face higher risk. For example, Rieman and McIntyre's (1993) viability analysis of populations of bull trout *Salvelinus confluentus* indicated that few have high probability of persisting 100 years in isolation. The analyses also suggested that extinction risks for the local, isolated populations increase as population sizes drop below 1,000 to 2,000 individuals (Figure 18.1). The issue of connectedness of local demes within metapopulations is considered below (section 18.4.1), but first we must gain a better sense of how genetic risk has been considered in PVAs.

18.3 Estimation of Viable Population Number

One cannot define a general guideline for the minimum number of individuals needed to protect a population against genetic risks to viability. However, fisheries and wildlife managers desire guidance as to the magnitude of the number of individuals at issue. Although no process for estimating viable population number, or VPN, with any degree of precision has been put forward, a classical formulation for estimating the magnitude of VPN is cited frequently. The approach considers a management unit, either a captive stock or a population in the wild, as an isolated population (i.e., there is no in- or out-migration). Let's consider the classical process for estimating VPN, especially noting embedded assumptions.

18.3.1 Genetic and Demographic Population Size

Viability of a population classically has been modeled as a function of its heterozygosity. A viable population can maintain its heterozygosity and, therefore, its adaptive potential through the planning horizon. Heterozygosity should be regarded as a surrogate measure for adaptability over the short term. Why hasn't viability been modeled as a function of allelic diversity, which arguably would provide a better measure of adap-

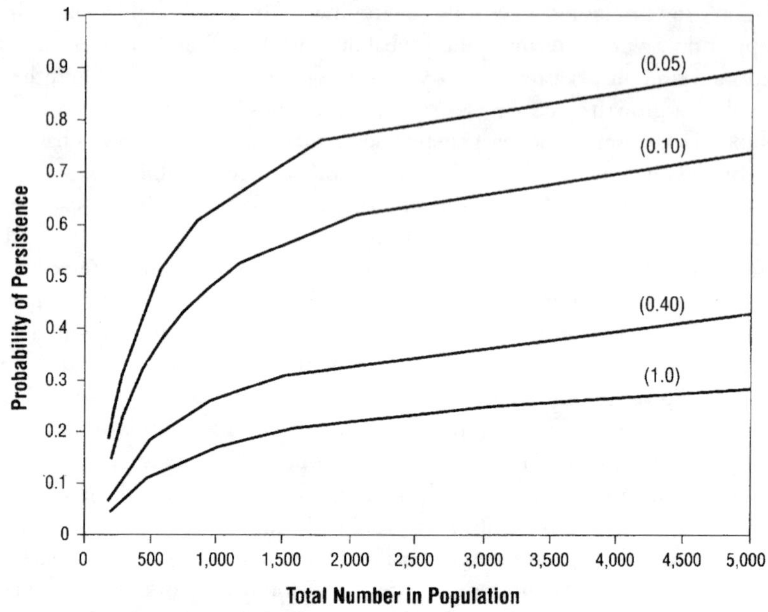

Figure 18.1 Estimated probability that a population will persist above a threshold of 20 adults for 100 years. Based on initial population size and variance (in parentheses) in annual population growth rate, probability curves are estimated using the method of Dennis et al. (1991). The threshold of 20 was adopted because it represents a point at which risks from demographic or genetic processes make recovery unlikely without migration from other populations. The estimated probability of persistence is low for populations smaller than 1,000 individuals and for populations with high variance in population growth rate. Figure adapted from Rieman et al. 1993.

tivity over the long term? Heterozygosity, which varies from zero to one, is more given to modeling than is allelic diversity, which varies from one to no identifiable upper bound. Hence, the viability of a population has been modeled in terms of its loss of heterozygosity. What, then, is the sequence of steps used for estimating the magnitude of the VPN?

1. Identify the maximum rate of heterozygosity loss that is acceptable for the population within the planning horizon.
2. Determine what effective population size, N_e, corresponds to that maximum rate of heterozygosity loss.
3. Determine what population size corresponds to that N_e. Let's term this the genetic *N*.
4. Determine what population size is needed for the population to maintain its demographic viability. Let's term this the demographic *N*.
5. Determine whether the genetic *N* or demographic *N* is larger. Adopt the larger as the VPN for the population, that is, as the minimum population size allowable, recognizing that risk of genetic- or demographic-mediated extinction processes will decline as the population size increases.

Let's examine each of these considerations in greater detail.

Franklin (1980) estimated the magnitude of N_e that must be maintained so that the maximum acceptable rate of loss of heterozygosity (denoted *H*) is not exceeded. The rate of heterozygosity loss is related to N_e as $\delta H = 1/2N_e$. Thus, if 100 adults contribute equally to propagation of the next generation, there will be a 0.5% loss of *H*. Embedded in the consideration of the acceptable *rate* of loss of heterozygosity is the important issue of the acceptable *cumulative* loss of heterozygosity. There are no firm guidelines for how much heterozygosity may be lost before the short-term adaptivity of a population is compromised. For purposes of reaching an order-of-magnitude estimate of VPN, the threshold of survival for a species was assumed to be a 40–50% loss of heterozygosity. Assuming that 40–50% of heterozygosity may be lost, the rate at which the loss will occur becomes a function of the length of the planning horizon. Under the National Forest Management Act, a planning horizon of 50 years was mandated. To population geneticists, this period of time is quite short! Recognizing that in the context of public policy 50 years is a long time, let's take the U.S. Congress at its intent that a long-term planning horizon should be adopted. Whatever the planning horizon in years, we must account for the generation time of the species at issue. Hence, a planning horizon of 50 generations for short-term conservation, or 500 for long-term conservation, may seem appropriate.

What N_e corresponds to the maximum allowable rate of heterozygosity loss, H_{critical}? Effective population size and heterozygosity loss can be related as

$$H_{\text{critical}} = 1 - (1 - 1/N_e)^{50}. \tag{18.1}$$

Assuming that a loss of 40% of heterozygosity is acceptable, and substituting into the equation above $0.60 = 1 - (1 - 1/2N_e)^{50}$, we obtain an N_e of 49.19, or approximately 50. This yields a target N_e, or $N_{e\ \text{critical}}$, of 50 for short-term conservation. What target N_e is appropriate for long-term conservation? It has been suggested that multiplication of the short-term $N_{e\ \text{critical}}$ by 10, yielding an $N_{e\ \text{critical}}$ of 500, will ensure enough genetic variability over the long term to maintain the evolutionary potential of the conserved population. This conservation guideline, sometimes termed the **50/500 rule**, has been widely cited. I emphasize that this guideline represents nothing more than an order-of-magnitude estimation of what size a population must be for conservation purposes, not a precise recommendation suitable for ready inclusion in any particular species management plan.

18.3.2 Target Effective Population Size

What census population size, *N*, corresponds to $N_{e\ \text{critical}}$? Using the classical population genetic models relating *N* and N_e (see Kimura and Crow 1963) proves impractical in most fisheries applications because many key parameters will not be known. Hence, a number of models relating *N* and N_e, models such as those of Ryman et al. (1981), Lacava and Hughes (1984), Reed et al. (1986), Lande and Barrowclough (1987), Lemkuhl (1984), Wood (1987), and Crow and Denniston (1988), subsequently have been developed to address different aspects of demographics and departures of populations from "ideal" conditions (Chapter 9). Certain of these models will prove appropriate for application to a given species and others not. Some species, notably including aquatic species, may not be represented well by any of these models. Harris and Allendorf (1989) presented an assessment of estimators of N_e for large mammals, for which, like aquatic species, key demographic parameters frequently are unknown. Using grizzly bear *Ursus arctos* as a case study, they estimated N_e by means of nine different analytic methods and a computer simulation model. The ratio of estimated N_e/N varied considerably among methods. Using the simulation-based method as a reference point, three models (Ryman et al. 1981; Hill 1972; Reed et al. 1986) yielded close agreement. Certain models (e.g., Lacava and Hughes 1984) yielded an overestimation of N_e, which could provide a false sense of security regarding the viability of managed populations. Findings demonstrated the utility of simulation modeling—data requirements were less and results could be used to identify sensitive demographic parameters and thereby serve as a means for focusing future research effort. Although the study focused on grizzly bear, the lesson is more general.

Models that require extensive demographic data often are problematic for application to aquatic species. Because demographic data frequently are lacking for aquatic species, genetic methods for estimating N_e (e.g., Hill 1981; Nei and Takajima 1981; Pollack 1983; Waples 1989; 1990a, 1990b; Waples and Teel 1990, Jorde and Ryman 1995) are likely to prove most useful. A more thorough discussion of genetic methods for estimation of N_e is presented in Chapter 9.

18.3.3 Validity of the 50/500 Rule

Given that the 50/500 rule often is cited in species or population management plans, what evidence supports it validity? One line of supporting evidence is empirical. In a retrospective study of population viability, Newmark (1986) examined historical records of census population sizes of mammals in U.S. national parks. Knowing whether a particular species had persisted from the founding of the park until the time of his study, Newmark found (Table 18.1) that the best predictor of the persistence of mammalian species in the U.S. national parks is the estimated population size at the time the park was established. The mean *N* for most surviving populations was one hundred to several thousand, depending on taxon and trophic level. The required population numbers for lagomorphs (rabbits and related evolutionary lineages) were higher. This is because they have boom–crash cycles of population size and, hence, are highly vulnerable to demographic factors causing extinction. The population sizes in Table 18.1 should be regarded as minimal estimates of VPN. It has been only 100 years since most of these parks were established, and these are census population sizes, not N_e.

The other line of evidence supporting the 50/500 rule is theoretical. On the basis of theory used for estimating N_e, one would expect that a target N_e of 500 would correspond to a census *N* of a few thousand individuals (Chapter 9). Still, a PVA should be performed for each population of planning interest (Box 18.2). Each case is different. There are some species that, because of unusual features of their biology, have VPNs that are known to depart from the 50/500 rule. For example, species that have density-dependent reproduction and live in highly stable environments, such as pupfishes *Cyprinodon* spp., may have lower VPN. Species exhibiting wide, cyclical population fluctuations, such as many small mammals and lagomorphs, will have higher VPNs.

Table 18.1 *Comparison of estimated initial median population sizes of extinct and extant mammals in 24 western North American national parks and park assemblages. Data from Newmark (1986) as presented in Soulé (1987).*

	Extinct populations			Extant populations		
Mammal group	Median size	95% confidence interval	*N*	Median size	95% confidence interval	*N*
Lagomorphs	3,276	702–56,952	9	70,889	34,720–173,150	60
Artiodactyls	241	3–1,273	7	792	429–1,504	88
Large carnivores	24	14–68	16	108	70–146	153
Small carnivores	156	112–880	26	1,203	908–1,704	127

■ Box 18.2
Overview of Population Viability Analysis

Process

In general the population viability assessment process includes the following steps.

— Determine the census population size, N.

— Estimate the effective population size, N_e, using the most appropriate model(s) available.

— Use the estimated N_e and the rules of thumb below to make a series of qualitative statements about the short- or long-term viability of the population. Bear in mind that extinctions can be caused by environmental stochasticity, natural or human-induced catastrophes, demographic stochasticity, and genetic uncertainties.

— Determine critical research needs for better quantifying key unknowns for populations identified as at risk.

— Determine and implement key management actions for enhancing the viability of populations at risk.

Rules of Thumb

If the estimated N_e for the population at issue is greater than 500, risks due to demographic and genetic factors tend to be low. Consider the possible effects of environmental stochasticity and natural or human-induced catastrophe on population viability.

If the estimated N_e for the population at issue is less than 500 but greater than approximately 50, consider the following.

— There may be some risk of extinction due to environmental and demographic stochasticity and to catastrophe.

— If immigration is low (e.g., less than one immigrant per generation), the population may become subject to loss of genetic variability due to random genetic drift and expression of inbreeding depression, presenting a long-term genetic risk.

If the estimated N_e for the population at issue is less than approximately 50, consider the following.

— The population is likely to be at risk from environmental and demographic stochasticity and catastrophe.

— If the immigration rate is low, the population may be subject to loss of genetic variability through random genetic drift and to expression of inbreeding depression.

Points to Remember

Keep in mind that these rules of thumb provide only preliminary indications of risks to populations. No rule of thumb is a substitute for detailed information about a population's ecology, demography, and genetics.

Be aware that some populations serve as sources and others serve as sinks for individuals. Local populations may or may not be linked by migration into functional metapopulations. These considerations affect assessment of a local population's viability.

Be resourceful about finding information (e.g., published information and personal communications) and assessment tools (e.g., simulation models such as Vortex [Lacy 2000]). Often, relevant local knowledge may prove a useful supplement to systematically acquired empirical data.

18.3.4 *Incorporation of Population Viability Analysis Theory into Practice*

One of the major contributors to the development of PVA, Soulé (1987), poignantly posed the question, "Are the theories any good?" Going on to offer an answer, he pointed out that we cannot plug species-specific parameters into a formula to obtain a reliable estimate of a VPN. However, we can make defensible order-of-magnitude estimates. Notwithstanding factors specific to species and locations, for a 95% expectation of persistence for several centuries, VPN is on the order of several thousand individuals. He pointed out that this answer is an improvement over the case ten or more years before. Then posing the question, "Where do we go from here?" Soulé (1987) identified two key theoretical inputs needing research: the effects of inbreeding and random genetic drift on *r*, the intrinsic rate of population increase, and the effects of age structure and overlapping generations on N_e. He identified four key gaps in data needed for deepening our understanding of population viability processes:

— retrospective studies examining the likelihood of extinction of isolated populations as a function of variables such as carrying capacity and climatic variability,
— long-term studies of extant populations with monitoring of parameters such as *N*, *r*, and N_e,
— comparative genetic and demographic studies of very small and very large populations of the same species to detect the effects of inbreeding, and
— studies of metapopulation dynamics.

Because most population viability work carried out to date has focused on terrestrial species, I add that the methods and findings need to be extended to aquatic species.

The difficulty of incorporating a variety of factors into a realistic model of extinction prompted conservation biologists to suggest VPNs largely based on genetic factors. Although widely cited, the 50/500 rule has proven controversial. For example, Lande (1988) argued that demography is usually of more immediate importance than is population genetics for determining the minimum viable sizes of wild populations. He asserted that management plans based primarily on population genetics threatened the existence of populations of northern spotted owl *Strix occidentalis caurina* and red-cockaded woodpecker *Picoides borealis* because basic demographic factors were ignored. He concluded that management of particular species should incorporate details of the species' ecology, especially its life history and demography. Incorporating demographics may require maintaining larger populations than has been suggested on genetic grounds alone. Applying such an approach to chinook salmon *Oncorhynchus tshawytscha,* Ratner et al. (1997) developed an age-structured, density-dependent model of the population dynamics that incorporated both demographic and environmental stochasticity.

18.4 Incorporation of Population Viability Analysis into Management Practice

In recent years, PVA has become incorporated into many fisheries and wildlife management programs. It has affected implementation of ecosystem management, development of metapopulation theory, and prioritization of populations for conservation.

18.4.1 *Metapopulation Structure*

Among the factors that strongly influence risks of local and regional extinction (Rieman et al. 1993) are replication of and dispersal and connection among the populations that collectively constitute a metapopulation. A **metapopulation** is a collection of local populations, or **demes**, that interact through the exchange of individuals, that is, through migration (Chapter 7). The process of extinction of a local population and founding of a new population in that habitat by migration from nearby populations is a key part of metapopulation dynamics. Metapopulations are associated with large watersheds, lakes, or river basins, depending on the degree of connection among streams and the dispersal rates typical of each species. When habitat for a species is lost or migration routes are blocked, metapopulations can become fragmented into local populations.

The effect of metapopulation structure on maintenance of genetic variability and the associated conservation implications have been investigated using simulation modeling. Boecklen (1986) compared rates of genetic drift and fixation of alleles among intact populations and subdivided populations with the same total numbers of individuals. Intact populations typically exhibited a short-term (through 50 generations) advantage over subdivided populations in maintaining heterozygosity. However, subdivided populations with occasional inter-subpopulation migrants typically exhibited more heterozygotes over the long term and preserved alleles longer than did intact populations. Boecklen concluded that archipelagos of refuges with occasional inter-refuge migration appeared to be the optimal strategy for genetic conservation. Lacy (1987) examined the interacting effects of genetic drift, mutation, migration, selection, and population subdivision on loss of genetic variability in small, managed populations. Stochastic events were simulated with a pseudo-random number generator, and expected heterozygosity was modeled in 25 populations for 100 generations. A system of subdivided populations rapidly lost variability from within each subpopulation but retained more variation across subpopulations than did large panmictic populations. The strongest population genetic process in this context, drift, could be countered by the occasional natural or human-mediated introduction of migrants.

18.4.2 *Ecosystem Management*

Among the important implications emerging from considerations of population viability is the finding that VPNs are large, in most cases on the order of thousands of individuals. Often, there won't be enough critical habitat for the population in any one reserve or other management unit. In the case of migratory species, populations will move between management units defined for other purposes. The implications are that in some cases fisheries and wildlife managers will have to maintain a population, or system of populations, in several reserves or other management units. Project planning and assessment will have to address habitat disruption and population responses at both the local and regional scales. There is a general consensus that preserving genetic and phenotypic diversity requires maintaining populations across a wide geographic range in a variety of habitats. Maintaining viable populations in intact habitat throughout the landscape and preserving metapopulation structure and function may be the most effective conservation strategy (Rieman et al. 1993). Some authors argue that conservation goals which include maintaining spatially dispersed, quality habitats may prove more effective than goals based on estimates of VPNs (see Boyce 1992). The concept of management at the regional scale has been applied to a number of aquatic environments and fish populations (Sheldon 1988; Moyle and Sato 1991; Reeves and Sedell 1992).

The demands of planning recovery of the northern spotted owl brought issues of interjurisdictional and landscape-level conservation planning into sharpened focus. The realization that conservation of not only spotted owl, but also anadromous salmonids and marbled murrelets (*Brachyramphus marmoratus*, a small, marine bird that nests in old-growth forests) posed similar issues and led to a drive to plan land use across the entire Pacific Northwest coastal ecoregion. Embedded within the emerging landscape-scale land use planning paradigm is the notion that species-by-species conservation ultimately proves untenable; hence, the development of the concept of **ecosystem management,** an approach aimed at providing habitat for all organisms in an ecosystem. The paradigmatic expression of the ecosystem management concept was the report of the Forest Ecosystem Management Assessment Team (FEMAT 1993), which presented options for timber harvest and the anticipated consequences not only for spotted owl, anadromous salmonids, and marbled murrelets but also for a wide range of other species. Landscape-level planning subsequently was initiated for the "eastside" ecoregion east of the Cascade Mountains, the Tongass National Forest in Alaska, and the southern Appalachian ecological region. While focusing primarily on viability at the species level, these plans to various degrees also reflect genetic aspects of conservation, particularly for salmonids. Superceding management unit-level planning, ecosystem management represents a new paradigm in fisheries and wildlife management. Many uncertainties are faced as agencies go forward with implementing ecosystem management. The degree to which implementation actions protect and maintain genetic diversity at the population level will become understood only with the passage of time.

18.4.3 Prioritization of Populations for Conservation

Many natural populations face threats to their viability (Box 18.3). With only limited resources available for conservation and recovery programs, prioritization of populations or stocks (Chapter 13) may become necessary if programs of meaningful scope are to be implemented. A prominent example of the need for prioritization is provided by stocks of Pacific salmonids (*Onchorhynchus* spp.). Native stocks of Pacific salmonids are threatened with extinction over vast portions of their range. Nehlson et al. (1991) reported that 214 native stocks of Pacific salmon, steelhead *O. mykiss*, and coastal cutthroat trout *O. clarki clarki* are at risk of extinction in California, Oregon, Washington, and Idaho. The FEMAT report (FEMAT 1993) included the results of subsequent assessments to bring the total to 314 native stocks. Given that these assessments cover only a part of the ranges of the respective species, it is clear that a large number of native Pacific salmon stocks need conservation attention. Regardless of what recovery actions are contemplated, limited resources may require prioritizing stocks for future action.

Several authors have developed biologically based methods for prioritizing at-risk stocks. For example, Mace and Lande (1991) ranked species on the basis of extinction threat alone, whereas others (reviewed by Given and Norton 1993) also considered biological, social, and other aspects in reaching a prioritization. Allendorf et al. (1997) proposed criteria aimed at prioritizing at-risk Pacific salmonid stocks. Based on their method, stocks first are ranked by risk of extinction, as assessed either by PVA or by a set of surrogate measures. Stocks then are ranked regarding biological consequences of extinction, based on sets of questions designed to establish the genetic, evolutionary, and ecological consequences if a stock were to become extinct. Allendorf et al. found that the prioritization process was most likely to work successfully when applied to stocks on which data exist, when several experts carry out the prioritization, and when the results are peer reviewed. Recovery of priority stocks should begin with actions having immediate benefit or having the highest chance of success. Management actions that prevent a threatened stock from descending near to or below the threshold of viability and that preserve the remaining habitat for priority stocks are important first steps in conservation.

■ Box 18.3
Viability Assessment and the Red List

The International Union for Conservation of Nature and Natural Resources (IUCN) is an international conservation organization that has collected endangered species status data for over 35 years. It recently has revised its criteria for listing species as endangered based on objective criteria, in part reflecting thinking on population viability analysis. The IUCN classifies species into three categories, critically endangered, endangered, and vulnerable. A species classified as endangered will meet any of the following criteria.

— Its population has fallen at least 50% in the last 10 years or three generations.
— The population is projected to halve in the next 10 years or three generations.
— The population is less than 250 mature individuals.
— The population is less than 2,500 individuals and is expected to decline at least 20% in the next 5 years or two generations, whichever is longer.
— Analysis shows that the probability of extinction in the wild is at least 20% within 20 years or five generations.

Critically endangered species have an 80% and vulnerable species a 20% chance of extinction within 10 years or three generations. Use of the new criteria (Baillie and Groombridge 1996) suggests that previous estimates of the number of endangered species worldwide may have been too low.

Below are some highlights of the revised *IUCN Red List of Threatened Animals* (Baillie and Groombridge 1996).

— In total, 5,205 species fall within one of the three endangered categories, including 11% of bird species, 20% of reptiles, 25% of amphibians, and 34% of fishes, most of them freshwater fishes.
— More than 100 species of marine fishes were added in 1996, including some sharks, tunas, coral reef fishes, and sea horses.
— Forty-four percent of crocodile species are endangered, an improvement since conservation efforts were initiated in 1971.

The "Red List" is recognized to be incomplete because the viability of so few species has been assessed.

■ 18.5 References

Allendorf, F. W., and nine coauthors. 1997. Prioritizing Pacific salmon stocks for conservation. Conservation Biology 11:140–152.

Baillie, J., and B. Groombridge (compilers and editors). 1996. IUCN red list of threatened animals. International Union for Conservation of Nature and Natural Resources, Gland, Switzerland.

Boecklen, W. J. 1986. Optimal design of nature reserves: consequences of genetic drift. Biological Conservation 38:323–338.

Boyce, M. S. 1992. Population viability analysis. Annual Review of Ecology and Systematics 23:481–506.

Conner, R. 1988. Wildlife populations: minimally viable or ecologically functional? Wildlife Society Bulletin 16:80–84.

Crow, J. F., and C. Denniston. 1988. Inbreeding and variance effective population numbers. Evolution 42:482–495.

Dennis, B., P. L. Munholland, and J. L. Scott. 1991. Estimation of growth and extinction parameters for endangered species. Ecological Monographs 6:115–143.

FEMAT (Forest Ecosystem Management and Assessment Team). 1993. Forest ecosystem management: an ecological, economic, and social assessment. U.S. Forest Service, Portland, Oregon.

Franklin, I. R. 1980. Evolutionary change in small populations. Pages 135–150 *in* M. E. Soulé and B. A. Wilcox, editors. Conservation biology: an evolutionary-ecological perspective. Sinauer Associates, Sunderland, Massachusetts.

Gilpin, M. E., and M. E. Soulé. 1986. Minimum viable populations: processes of species extinctions. Pages 19–34 *in* M. E. Soulé, editor. Conservation biology: the science of scarcity and diversity. Sinauer Associates, Sunderland, Massachusetts.

Given, D. R., and D. A. Norton. 1993. A multivariate approach to assessing threat and for priority setting in threatened species conservation. Biological Conservation 64:57–66.

Harris, R. B., and F. W. Allendorf. 1989. Genetically effective population size of large mammals: an assessment of estimators. Conservation Biology 3:181–191.

Hill, W. G. 1972. Effective population size with overlapping generations. Theoretical Population Biology 3:278–289.

Hill, W. G. 1981. Estimation of effective population size from data on linkage disequilibrium. Genetical Research 38:209–216.

Jorde, P. E., and N. Ryman. 1995. Temporal allele frequency change and estimation of effective size in populations with overlapping generations. Genetics 139:1077–1090.

Kimura, M., and J. F. Crow. 1963. The measurement of effective population number. Evolution 17:279–288.

Lacava, J., and J. Hughes. 1984. Determining minimum viable population levels. Wildlife Society Bulletin 12:370–376.

Lacy, R. C. 1987. Loss of genetic diversity from managed populations: interacting effects of drift, mutation, immigration, selection, and population subdivision. Conservation Biology 1:143–158.

Lacy, R. C. 2000. VORTEX population viability analysis software, version 8.3. *http://pw1.netcom.com/~rlacy/vortex.html.*

Lande, R. 1988. Genetics and demography in biological conservation. Science 241:1455–1460.

Lande, R., and G. F. Barrowclough. 1987. Effective population size, genetic variation, and their use in population management. Pages 87–123 *in* M. E. Soulé, editor. Viable populations for conservation. Cambridge University Press, Cambridge, UK.

Leigh, E. D. 1981. The average lifetime of a population in a varying environment. Journal of Theoretical Biology 90:213–239.

Lemkuhl, J. F. 1984. Determining size and dispersion of minimum viable populations for land management planning and species conservation. Environmental Management 8:167–176.

Mace, G. M., and R. Lande. 1991. Assessing extinction threats: toward a reevaluation of IUCN threatened species categories. Conservation Biology 5:148–157.

Moyle, P. B., and G. M. Sato. 1991. On the design of preserves to protect native fishes. Pages 155–173 *in* W. L. Minckley and J. E. Deacon, editors. Battle against extinction: native fish management in the American West. University of Arizona Press, Tucson.

Naess, A. 1986. Intrinsic value: will the defenders of nature please rise? Pages 504–515 *in* M. E. Soulé, editor. Conservation biology: the science of scarcity and diversity. Sinauer Associates, Sunderland, Massachusetts.

Nehlson, W., J. E. Williams, and J. A. Lichatowich. 1991. Pacific salmon at the crossroads: stocks at risk from California, Oregon, Idaho, and Washington. Fisheries 16(2):4–21.

Nei, M., and F. Takajima. 1981. Genetic drift and estimation of effective population size. Genetics 98:625–640.

Newmark, W. D. 1986. Mammalian richness, colonization, and extinction in western North American national parks. Doctoral dissertation. University of Michigan, Ann Arbor.

Nielsen, J. L., editor. 1995. Evolution and the aquatic ecosystem: defining unique units in population conservation. American Fisheries Society, Symposium 17, Bethesda, Maryland.

Nisbet, R. M., and W. S. C. Gurney. 1982. Modeling fluctuating populations. Wiley, New York.

Pollack, E. 1983. A new method for estimating the effective population size from allele frequency changes. Genetics 104:531–548.

Quinn, J. F., and A. Hastings. 1987. Extinction in subdivided habitats. Conservation Biology 1:198–207.

Ratner, S., R. Lande, and B. B. Roper. 1997. Population viability analysis of spring chinook salmon in the South Umpqua River, Oregon. Conservation Biology 11:879–889.

Reed, J. M., P. D. Doerr, and J. R. Walters. 1986. Determining minimum population sizes for birds and mammals. Wildlife Society Bulletin 14:255–261.

Reeves, G. H., and J. R. Sedell. 1992. An ecosystem approach to the conservation and management of freshwater habitat for anadromous salmonids of the Pacific Northwest. Transactions of the 57th North American Wildlife and Natural Resources Conference 57:408–415. Wildlife Management Institute, Washington, D.C.

Rieman, B., D. Lee, J. McIntyre, K. Overton, and R. Thurow. 1993. Consideration of extinction risks for salmonids. U.S. Forest Service Fish Habitat Relationships Technical Bulletin 14 (1993-583-068). U.S. Government Printing Office, Washington, D.C.

Rieman, B. E., and J. D. McIntyre. 1993. Demographic and habitat requirements for conservation of bull trout. U.S. Forest Service Intermountain Research Station, General Technical Report INT-302, Ogden, Utah.

Ryman, N., R. Baccus, C. Reuterwall, and M. H. Smith. 1981. Effective population size, generation interval, and potential loss of genetic variability in game species under different hunting regimes. Oikos 36:257–266.

Saunders, H. R., R. J. Hobbs, and C. R. Margules. 1990. Biological consequences of ecosystem fragmentation: a review. Conservation Biology 5:18–32.

Shaffer, M. L. 1981. Minimum population sizes for species conservation. BioScience 31:131–134.

Shaffer, M. L. 1987. Minimum viable populations: coping with uncertainty. Pages 69–86 *in* M. E. Soulé, editor. Viable populations for conservation. Cambridge University Press, New York.

Shaffer, M. L., and F. B. Samson. 1985. Population size and extinction: a note on determining critical population sizes. American Naturalist 125:144–151.

Sheldon, A. L. 1988. Conservation of stream fishes: patterns of diversity, rarity, and risk. Conservation Biology 2:149–156.

Soulé, M. E. 1987. Where do we go from here? Pages 175–183 *in* M. E. Soulé, editor. Managing for viable populations. Cambridge University Press, New York.

Soulé, M. E., and D. Simberloff. 1986. What do genetics and ecology tell us about the design of nature reserves? Biological Conservation 35:19–40.

Stacy, P. B., and M. Taper. 1992. Environmental variation and the persistence of small populations. Ecological Applications 2:18–29.

Waples, R. S. 1989. A generalized method for estimating population size from temporal changes in allele frequency. Genetics 121:379–391.

Waples, R. S. 1990a. Conservation genetics of Pacific salmon. II. Effective population size and the rate of loss of genetic diversity. Journal of Heredity 81:267–276.

Waples, R. S. 1990b. Conservation genetics of Pacific salmon. III. Estimating effective population size. Journal of Heredity 81:277–289.

Waples, R. S. 1991. Pacific salmon, *Oncorhynchus* spp., and the definition of "species" under the Endangered Species Act. U.S. National Marine Fisheries Service Marine Fisheries Review 53(3):11–22.

Waples, R. S., and D. J. Teel. 1990. Conservation genetics of Pacific salmon. I. Temporal changes in allele frequency. Conservation Biology 4:144–156.

Wood, J. W. 1987. The genetic demography of the Gainj of Papua New Guinea. 2. Determinants of effective population size. American Nauralist 129:165–187.

Glossary

Abbreviations used in this glossary are base pair (bp), adenine (A), cytosine (C), guanine (G), thymine (T), mitochondrial DNA (mtDNA), messenger RNA (mRNA), ribosomal RNA (rRNA), transfer RNA (tRNA), and polymerase chain reaction (PCR).

absolute gene diversity (D_{ST}) the between-subpopulation component of total heterozygosity in a population.

absolute viability the probability of an individual surviving from fertilization to maturity, ranging from 0 to 1; contrast with **relative viability**.

acrocentric referring to a chromosome with one short arm.

adaptation genetic change between consecutive generations that increases a population's fitness in a given environment.

adaptive referring to a character conferring fitness upon a carrier in a given environment.

adaptive landscape a model of fitness of populations depicted by allele frequencies at fitness-related loci *A* and *B* as the two planar axes of the graph and fitness as a third, perpendicular dimension, much as elevation is depicted on a conventional topographic map (Wright 1932).

adaptive management explicit design and implementation of management actions or programs as "experiments" that require regular monitoring for obtaining valid experimental data and for tracking progress toward program goals and objectives. Adaptive management also requires systematic evaluation of actions and making adaptive changes or midcourse corrections based on conclusions drawn from such evaluations.

adaptive norm the mean value for the range of phenotypes of a particular trait exhibited by a population adapted to a particular environment.

adaptivity the ability of a population to respond to changes in its environment.

additive gene effect a mode of allelic interaction in which dominance is absent—the phenotype of the heterozygote is intermediate between those of the homozygotes. The cumulative effect of all loci at which additive allelic interaction occurs upon expression of a polygenic quantitative trait.

additive variance (V_A) the fraction of phenotypic variance attributable to loci for which phenotypic expression of the quantitative trait is determined by the summation of individual allelic effects; that is, phenotypic expression is *not* affected by interactions of alleles.

alike in state referring to alleles that are identical but are not recent copies of an ancestral allele; contrast with **identical by descent**.

allele an alternative form of a given gene that differs from other alleles in DNA sequence or in effect on phenotype.

allele frequency the percentage of all alleles at a particular locus represented by a particular allele in the gene pool of a given population.

allopatric literally, "different countries." Referring to processes depending upon reproductive isolation among populations; for example, allopatric speciation is a mode of speciation requiring reproductive isolation among differentiating populations. Contrast with **sympatric**.

allopolyploidy the state of having four sets of chromosomes resulting from hybridization between two species followed by duplication of the sets of chromomsomes; contrast with **autopolyploidy**.

allozyme allelic form of an enzyme encoded at a given locus. Allozymes usually are distinguished by electrophoresis and histochemical staining and are observed as a difference in electrophoretic mobility due to differences in net electric charge or molecular weight.

allozyme electrophoresis an analytical technique examining variation among that class of proteins which controls much of cell metabolism—bidirectional catalytic enzymes—by means of electrophoresis and histochemical staining for enzyme activity.

altruism the expression of behaviors that increase the fitness of other individuals at the expense of one's own fitness.

amplification primers short, single-stranded oligonucleotides, generally ranging from 10–20 bp in length, that anneal to both ends of a DNA segment under investigation and support use of the PCR to amplify that DNA segment.

anadromous referring to a migratory life cycle in which spawning occurs in the freshwater and juvenile growth in the sea. Anadromy is exhibited, for example, by many salmonids; contrast with **catadromous**.

analysis of molecular variance (AMOVA) a data analytic procedure used for partitioning molecular genetic variance into components.

aneuploidy the condition of a cell having one, two, or several chromosomes more or less than the normal chromosome number for the species (e.g., $2n - 1$ is monosomy, and $2n + 1$ is trisomy).

anticodon the triplet of nucleotides in a tRNA that pairs with a specific triplet codon on a DNA template or on mRNA.

artificial selection the process of intentionally choosing or inadvertently retaining parents for mating on the basis of one or more heritable phenotypic traits, leading to genetic and phenotypic change of the trait(s) in the progeny generation.

associative overdominance high fitness of the heterozygote resulting from variation at loci associated with the marker locus; contrast with **direct overdominance**.

assortative mating nonrandom mating, with individuals preferentially mating with individuals with like (positive) or unlike (negative) phenotypes.

augmentation rehabilitation of a demographically depressed population by integrating naturally and hatchery reproduced fish to maximize seeding of habitat by naturally reproducing adults.

autopolyploidy the condition in which one set of diploid chromosomes in a given individual is duplicated; contrast with **allopolyploidy**.

autoradiography the making of an image by exposing X-ray film to a gel or Southern blot containing radioactively labeled DNA fragments in a light-proof container.

autosomes chromosomes other than sex chromosomes; that is, they have no role in sex determination. Autosomal refers to genes or genetic markers located on the autosomes; contrast with **gonosomes** and **sex chromosomes**.

autozygous describing individuals that bear alleles that are identical by descent from a common ancestor. Autozygosity, then, is a measure of the frequency (or expectation) that two alleles are identical by descent.

B chromosomes "extra," or supernumerary, chromosomes that are not duplicates of normal chromosomes as in aneuploidy. These chromosomes are usually heterochromatic and sometimes are so small that they can be classified as microchromosomes. Many are derived from hybridization events and often display irregular, non-Mendelian inheritance.

back mutation the mutation of a DNA sequence back to the sequence from which it originally mutated. In some cases, a change in a mutant gene will restore its ability to produce a functional protein. Sometimes referred to as reverse mutation.

backcross a cross between an individual and one of its parents or a cross between an F_1 hybrid and one of the parental lines.

balancing selection a type of natural selection in which a diversity of alleles is maintained by the changing action of selection in a dynamic environment.

band on a chromosome, a dark striation on a condensed, chemically stained chromosome during metaphase. On a gel, the location where material migrates based on size (nucleic acids), net electrical charge (proteins), or both. On an autoradiograph, a darkened area where a labeled probe exposed the X-ray film, indicating a hybridized, homologous DNA sequence.

binding site a site in the genome where a regulatory molecule becomes attached to the DNA, affecting transcription at a gene nearby. A site on a protein where a molecule becomes attached, affecting the biological activity of the protein.

bootstrapping a statistical procedure for creating pseudoreplicate data sets in cases in which it is impossible to obtain true replicate data to evaluate whether the results of analysis can be explained by chance alone. Computationally, bootstrapping generates a large number of pseudoreplicate data sets by random sampling (with replacement) of the original data set. Each of these data sets is analyzed, and the results are used to calculate the frequency of occurrence of a given branch in a dendrogram.

bottleneck effect loss of genetic variation consequent to a restricted number of individuals reproducing within an existing population.

broad-sense heritability (h^2_b) or **degree of genetic determination** the extent to which phenotypic variation within a population is determined by genetic factors, measured as the ratio of genetic to phenotypic variation; contrast with **narrow-sense heritability**.

broodstock captive adult fish used to propagate a stock.

C bands highly repetitive DNA that typically is condensed and seen under the light microscope as darkly staining heterochromatin. Represents tandemly arranged highly repetitive DNA, with the typical repeat unit being 100–500 bp in length. Most genomes contain several subclasses of these sequences, which vary in the base sequence of the repeating unit.

canalized referring to the case in which phenotypic expression of a trait is well buffered through ontogeny.

captive broodstock program the collection of individuals (or gametes) from a natural population, the rearing of obtained individuals to maturity in captivity, and, generally, the release of their progeny into the wild to aid recovery of the natural population.

catadromous referring to a migratory life cycle in which spawning occurs in the ocean and juvenile growth in freshwater. Catadromy is exhibited, for example, by American eel *Anguilla rostrata;* contrast with **anadromous**.

centiMorgan a unit of relative distance along a chromosome, 1/100 of a Morgan. One Morgan, M, of distance represents the length of a chromosome with a meiotic crossover frequency of 100%. Named after T. H. Morgan, a geneticist who performed linkage mapping in fruit fly *Drosophila melanogaster.*

centric transposition a shift in the position of a centromere. These multiple-break rearrangements are rare and probably the result of sequential rearrangements of chromosomal material.

centromere region of chromosome, generally bounded by heterochromatin-containing repetitive DNA, where spindle fibers attach during meiosis and mitosis.

chimera in eukaryotic organisms, an individual composed of a mixture of genetically different cell or tissue types, for example, exhibiting different nuclear, mitochondrial, or plastid genotypes. At molecular level, a splicing of DNA or RNA molecules from different sources. Chimeric DNA is a DNA construct composed of fragments from different sources.

chromatids the two arms of a chromosome, attached by a centromere.

chromatin the complex of nucleic acid (RNA or DNA) and protein (e.g., histones and nonhistones) in eukaryotic chromosomes. It is a fiber 11 nm in width that is formed by one linear, double-stranded DNA molecule running the length of each chromosome and wound onto histone complexes.

chromosomal aberration karyotypic change resulting from loss, duplication, or rearrangement of genetic material within one chromosome or across chromosomes; see **duplication, inversion,** and **translocation.**

chromosomal linkage the occurrence of alleles or genes in proximity on the same chromosome.

chromosome literally, "colored body," referring to the appearance of a stained chromosome under a microscope. In the nucleus of a eukaryotic cell, one of the threadlike structures composed of chromatin; that is, DNA encoding genetically determined traits and associated proteins. In prokaryotic cells, the circular DNA molecule encoding genetically determined traits of the organism.

cladogenic referring to a branching pattern of evolutionary differentiation of lineages or referring to an evolutionary event giving rise to differentiation among lineages.

classical, or **Mendelian, genetics** the subset of genetics concerning the transmission of discrete traits from generation to generation and the mode of their expression, following laws deduced by Gregor Mendel.

cline a gradual spatial variation in allele frequencies exhibited by a species along a line of environmental or geographic transition.

cloning a molecular genetic procedure involving insertion of foreign DNA into a vector such as a plasmid or phage, which then carries the foreign DNA into bacteria in which the chimeric DNA is replicated.

cluster analysis a group of data analytic procedures that use explicit sets of rules to classify individuals into groups. The most commonly used ones for genetic stock identification are agglomerative, hierarchical techniques that sequentially join individual samples into similar groups, or clusters, based the differences (or similarities) in the genetic distance matrix.

coadaptation the process by which favorable allele combinations arise by chance and become spread through the population and maintained by selection.

coadapted gene complexes the groups of alleles produced and maintained via coadaptation.

codominant referring to the situation in which two alleles at a locus are phenotypically expressed in the heterozygote; therefore carriers of the genotypes *AA*, *Aa*, and *aa* can be distinguished. In a gel, both alleles at a single locus can be visualized.

codon nucleotide triplet in DNA or mRNA that specifies the amino acid to be inserted in a specific position of a polypeptide; see **genetic code**.

cofactor a coenzyme (e.g., nicotinamide adenine dinucleotide, NAD) or metal ion (e.g., Mg ion) required in addition to an enzyme for a given biochemical reaction to occur.

complete dominance a mode of allelic interaction in which the expression of the dominant allele fully masks that of the recessive allele; that is, the phenotype of the heterozygote is indistinguishable from that of the homozygous dominant.

complete-linkage method a clustering procedure in which the distance between two clusters is the distance between the two least-similar pairs of samples.

continuous traits characters exhibiting a continuum of phenotypes that can be measured. There are infinitely many possible phenotypes, among which discrimination is limited only by our ability to measure them. Examples of continuous traits include growth rate, yield, and morphometric traits (e.g., head length:standard length; eye diameter:head length; and body depth:standard length).

correlation of alleles (*r*) the frequency at which alleles occur together. Contrast with linkage, by which alleles occur together because they are expressed at loci on the same chromosome segment. In small populations, correlation of alleles leads to their occurring together without their being linked.

crossing-over the exchange of genetic material among homologous chromosomes at meiosis. This is an important mechanism for creation of new genetic variability.

C-value a measure of the amount of DNA in a haploid genome for a given species. Estimates of the amount of nuclear DNA in typical piscine cells are in the range of 1 to 3 pg DNA.

C-value paradox the often observed negative correlation between the range of C-values reported for a species and the evolutionary complexity of that species.

cytogenetics the science that combines the methods and findings of cytology and genetics and that focuses on the structure, function, and evolution of chromosomes.

cytology the study of the structure, function, and life history of the cell.

cytophotometry a cytogenetic technique that uses an instrument combining microscopy with spectrophotometry to analyze Feulgen-stained tissue homogenates in order to study the changes of DNA content in cells throughout their life cycles.

dams females used in a mating design, for example, mated in a hierarchical mating experimental design; contrast with **sires**.

deamination the oxidative removal of amine (NH_2) groups from amino acids.

degree of dominance the degree to which a dominant allele determines the phenotype, and hence, the fitness (W) of the heterozygote. Several cases can be distinguished: no dominance (W_{Aa} = midpoint between W_{AA} and W_{aa}, often termed an additive gene effect); partial dominance (W_{Aa} is above the midpoint between W_{AA} and W_{aa}); complete dominance ($W_{Aa} = W_{AA}$); and overdominance ($W_{Aa} > W_{AA}$).

deletion the mutational loss of a nucleotide (or sequence of nucleotides) or chromosomal segment.

deliberate genetic marking the propagation of a cultured stock in order to produce an array of allele frequencies differing from that of a receiving population into which the cultured stock is released.

deme a local population or isolated reproductive unit.

demographic uncertainty processes affecting the likelihood of a population's persistence that act through effects upon age structure and rate of population growth.

dendrogram the result of cluster analysis, displayed as a tree showing hierarchical structure in the data set.

depurination the loss of purines, at acidic pH, from the polysaccharide–phosphate backbone of DNA.

detection observation of enzyme activity by tying the specific chemical reaction of a given enzyme to a colorimetric or fluorescent detection system.

deterministic extinctions extinctions that occur with the cumulative loss or permanent change of a critical component in the species' environment, resulting in progressive increases in the population death rate, decreases in birth rate, or both.

dimer an enzyme molecule that is functional when composed or two identical or near-identical subunits; contrast with **monomer** and **tetramer**.

diploid referring to an organism having the (usual) two sets of chromosomes, one from each parent.

diploid chromosome number (2*n*) the number of chromosomes corresponding to the condition of having paired chromosomes, as is typical of eukaryotic organisms.

diploidization the evolutionary process of reversion to the normal pairing of chromosomes following a polyploidization event.

direct overdominance generally regarded as the exceptional case, the situation in which high fitness of the heterozygote is the consequence of variation at a single locus, a genetic marker locus; contrast with **associative overdominance**.

directional selection a mode of selection favoring phenotypes at *one* end of the population's phenotypic distribution.

discrete phenotypes alternative phenotypes for a particular trait that are readily distinguished from one another; for example, in pea plants dwarf or normal height, yellow or red flowers, and wrinkled or smooth seeds.

disomic the situation in which each gene is encoded on two homologous chromosomes, for example, disomic inheritance of a gene is a mode of inheritance consistent with the gene being encoded on two homologous chromosomes. Contrast with **tetrasomic (inheritance)**.

dispersed gene family a set of genes that descended by duplication and variation from some ancestral gene and are presently dispersed on different chromosomes. Examples include actins, myosin, and histones, which are repeated and spread throughout the genome.

disruptive selection a mode of selection that favors extreme phenotypes *at both ends* of the phenotypic distribution.

distinct population segment three criteria are considered regarding the status of a population segment as distinct for the purpose of listing, delisting, or reclassifying vertebrates under the U.S. Endangered Species Act: (1) discreteness of the population segment in relation to the remainder of the species to which it belongs, (2) significance of the population segment to the species to which it belongs, and (3) the population segment's conservation status in relation to the act's standards for listing.

DNA deoxyribonucleic acid, the molecule that encodes genetic information needed for normal embryological development, homeostasis, and reproduction of an organism.

DNA amplification fingerprinting a molecular genetic analytic technique that involves DNA amplification by means of a single primer of random sequence of approximately 10 bases, followed by analysis of amplification products on polyacrylamide gels.

DNA library a collection of fragments of the DNA of a species (e.g., DNA complements of cell-specific mRNA, restriction-digested total genomic DNA, or DNA that has been selectively enriched for repeat sequences) that have been inserted into a vector (such as plasmid or phage) and cloned in a host organism (e.g., *Escherichia coli*).

DNA ligase repair enzyme that seals single-stranded nicks in DNA by catalyzing formation of a phosphodiester bond between adjacent 3′ and 5′ DNA termini.

DNA microsatellite array a sequence of tandem repeats of a short motif of 2–8 nucleotides at a particular locus; see **microsatellite**.

DNA sequencing a collection of procedures used to determine the exact nucleotide sequence of a target fragment of DNA.

DNA shuffling a process by which transposons capture genes and move them to other parts of the genome.

domesticated genetically adapted to conditions of artificial propagation, for example, domesticated fish stocks adapted to production under aquacultural conditions.

domestication selection selection that operates during artificial propagation to produce adaptations to the captive environment. Domestication selection eliminates individuals that are not adapted to the captive environment, which may include some individuals that are well adapted to the natural environment.

dominance the situation in which an allele will fully express its phenotype even when present in the heterozygous state.

dominance hypothesis the hypothesis that loss of fitness due to inbreeding results from the heightened frequency of expression of deleterious recessive alleles across the genome; contrast with **overdominance hypothesis**.

dominance variance (V_D) the proportion of phenotypic variance attributable to loci for which phenotypic expression of a quantitative trait is affected by interactions of alleles *at a locus*, that is, by dominance allelic interactions.

dominant refers to alleles that are fully manifested regardless of the other allele occurring at the locus.

dot blot a molecular genetic technique based on hybridization between homologous DNA sequences that is used for showing the presence of or estimating the relative abundance of nucleic acid fragments that are bound to a nylon or nitrocellulose membrane. Multiple samples of test DNAs are spotted on a membrane in equal amounts and then allowed to hybridize to a labeled probe (e.g., total genomic DNA or a specific sequence of interest). The intensity of signal from a dot indicates the extent of hybridization, which is taken to indicate the extent of similarity between the bound and probe sequences.

double helix spiral DNA structure based on hydrogen bonding between two polynucleotide strands spiraling around each other. The nucleotides are linked side by side in a direction known as 5′ to 3′ along the length of a strand. The 5′ and 3′ refer to the numbered carbons of the pentose deoxyribose.

duplication in molecular genetics, an increase in the number of nucleotides or repeat units. In classical genetics, an increase in the number of loci borne by a chromosome. In cytogenetics, an intrachromosomal rearrangement resulting in a change in chromosome length. Those rearrangements involving euchromatin tend to be lethal unless only small segments are involved. When heterochromatin is involved, these types of chromosomal rearrangements are typically selectively neutral, although they may result in alteration of recombination patterns.

dysmetric load the load of fitness-reducing alleles in a population maintained by mismatch between allele frequencies and environment.

ecological interaction the overlapping of resource needs and uses (e.g., food, cover, and spawning habitat) among sympatric populations or species that may result in indirect genetic change if competition results in loss or alteration of genetic resources in the populations at issue.

ecosystem management a conservation approach aimed at providing habitat for all organisms in an ecosystem, as opposed to managing species (or populations) separately.

effective number of alleles (n_e) the number of equally frequent alleles in an ideal population subjected to mutation pressure under the infinite alleles model.

effective number of genes (n_E) the effective, or minimum, number of loci contributing to the difference in a quantitative trait between two inbred lines grown in a common environment. Estimates are obtained by comparing the phenotypic means and variances of the character in parental, F_1, F_2, and backcross populations.

effective population number or size (N_e) the number of reproducing individuals in an ideal population that would lose genetic variation due to genetic drift or inbreeding at the same rate as the number of reproducing adults in the real population under consideration. Typically, N_e is less than either a population's total number of sexually mature adults present or the total number of adults that reproduced. Effective population number can be defined either in terms of the amount of increase in homozygosity (inbreeding effective number) or the amount of allele frequency drift (variance effective number).

eigenvalue effective number the effective size of a population defined in terms of the rate of loss of heterozygosity.

electrophoresis a procedure for separation of charged molecules in an electric field, for example, for screening of allozyme variation.

end-labeling the molecular genetic process in which radioactively labeled nucleotides are attached to the end of a DNA fragment by means of a polymerase enzyme.

enhancement the management practice of using habitat restoration and stocking of hatchery-derived individuals to increase the size of a managed fish or shellfish population.

environmental component of variance the proportion of phenotypic variation resulting from a variety of nongenetic factors.

environmental uncertainty a class of factors operating through normal variation in ecological conditions that affect the likelihood that a population will persist for a given period of time.

enzyme a molecule that converts one substrate to another without itself being affected.

epigamic selection a type of selection in which the female is the selective agent, acting by choosing a mate from a field of genetically variable males of her species.

epistasis interaction of nonallelic genes, usually manifested as the masking of expression of one gene by another gene.

epistatic, or **interaction, variance (V_I)** the proportion of phenotypic variance attributable to loci for which phenotypic expression of the quantitative trait is affected by interactions of alleles among loci.

erythrocyte a hemoglobin-containing blood cell, that is, a red blood cell.

euchromatin the chromatin that is uncoiled during interphase and condenses during mitosis. It produces the staining pattern of the majority of an organism's chromosomal complement.

eukaryotic referring to the superkingdom that contains organisms whose cells contain membrane-bound nuclei and mitochondria (Protista, Fungi, Animalia, and Plantae). Sexual cells of these organisms undergo meiosis, and somatic cell division is by mitosis.

evolutionarily significant unit a population or assemblage of populations that is the object of management concern. This unit must be *reproductively isolated* from other conspecific units, although it does not have to be absolutely isolated, and it must represent an important component of the *evolutionary legacy* of the species.

evolutionary potential the long-term ability of a population to evolve and thereby persist in the face of environmental change. Maintenance of evolutionary potential requires existence of genetic variation in the population.

exon in a eukaryotic gene, the portion of the DNA sequence that is transcribed in the nucleus of the cell and remains after splicing ultimately to become part of structural RNA in the cytoplasm.

exonuclease any of several types of enzymes that digest nucleic acids.

expression of a gene the timing of production, cellular or tissue specificity, and level of production of the product of a gene.

extinction loss of all members of a population or species.

extinction vortices self-reinforcing processes affecting a population's probability of extinction, respectively relating to variance in reproductive rate (the R vortex), patchiness in species distribution (the D vortex), inbreeding (the F vortex), and loss of adaptive potential due to random genetic drift (the A vortex).

extirpated extinct in a portion of a species' historic range.

extraction the procedure in which cell membranes are broken to release soluble enzymes into solution.

extrinsic factors factors affecting extinction probability that are outside the biology of the population at issue; contrast with **intrinsic factors**.

factorial mating division of an individual's gametes into aliquots and fertilization of those gametes with gametes from a number of mates. This practice maximizes the genetic diversity in an individual's offspring.

family size (k) the number of progeny of one female or one male parent that survive to reproduce. For a given population, there is an average family size ($\bar{k}$) and an associated variance in family size (V_k).

50/500 rule a rule of thumb suggesting a target effective population size (N_e), or $N_{e\ \mathrm{critical}}$, of 50 for short-term conservation and an $N_{e\ \mathrm{critical}}$ of 500 for long-term conservation to maintain the evolutionary potential of the conserved population.

first filial generation (F_1) the offspring of a cross of interest. Often used to refer to a hybrid individual from the cross of two distinct species, strains, or lines. The individual will be heterozygous at all loci for which the two parental species, strains, or lines were fixed for alternate alleles.

fission the mutational breaking of a chromosome arm, resulting in a shorter chromosome arm and a fragment that may become a distinct chromosome; contrast with **fusion**.

fitness (W) the ability to survive to reproductive age and leave viable offspring. The frequency distribution of reproductive success for a population of sexually mature adults.

fixation the state of a locus having just one allele, with a frequency of 1.00. All other alleles have become lost, with frequencies of 0.00.

fixation index (F_{ST}) the proportion of the variation at a locus attributable to divergence among populations. Measures the reduction in total expected heterozygosity of the entire system of populations due to random drift and other differentiating processes among isolated populations. See ***F*-statistics**.

fixed referring to population- or species-specific diagnostic genetic markers; referring to the frequency of an allele becoming 1.00.

flow cytometer an instrument that sorts and measures fluorescence of dyed particles as they pass single-file through a narrow laser beam.

fluctuating asymmetry the occurrence of a difference in a character between the left and right sides of an individual. At the population level, fluctuating asymmetries are normally distributed about a mean of 0.

fluorescence in situ hybridization (FISH) a cytogenetic technique involving denaturation of a DNA probe and chromosomes, hybridization of the probe to chromosomal DNA, post-hybridization washes, and detection and visualization of the fluorescent probe. The technique shows the chromosomal location of DNA sequences complementary to the probe.

forensic science use of scientific methods to make inferences regarding past events, especially with regard to questions relevant to fisheries, wildlife, and conservation law enforcement

forward mutation given that the mutation rate per generation from allele *A* to allele *a* is μ, and from *a* to *A* is ν, when μ is greater than ν, μ is regarded as the rate of the forward mutation.

founder effect the phenomenon by which a group newly split off from a larger group, especially if small, does not exhibit all the heritable variations, or not in the same proportions, as the original group. Loss of variability occurring consequent to a restricted number of individuals colonizing new habitat.

frame-shift mutation addition or subtraction of a nucleotide, resulting in improper translation of the message after the point of addition or loss; that is, message is read in triplets, but from the mutation site onward, codons specify the wrong amino acids or possibly chain termination.

frequency-dependent selection a type of selection in which the frequency of a phenotype determines the fitness of the carrier.

***F*-statistics** a suite of hierarchical tests designed to estimate genetic population structure using Wright's theoretical relationship $1 - F_{IT} = (1 - F_{IS})(1 - F_{ST})$, where F_{IT}, the overall inbreeding coefficient, estimates the reduction of heterozygosity in the total sample; F_{IS}, the inbreeding coefficient, is an estimation of the reduction of heterozygosity due to nonrandom mating in subpopulations; and F_{ST}, the fixation index, estimates the reduction in heterozygosity due to random genetic drift among subpopulations.

functional gene a DNA sequence that encodes a product or regulates expression of a coding sequence.

functional repetitive DNA sequences repetitive DNA sequences encoding a product or regulating expression of a coding sequence, including dispersed gene families, tandem gene families, and noncoding telomeric sequences.

fundamental number (FN) the total number of chromosome arms in a karyotype.

fusion the mutational joining of chromosomes, chromosome arms, or chromosome fragments; contrast with **fission**.

G bands bands appearing upon Geimsa staining of a metaphase chromosome spread. Each band indicates an alternation in the base content and relative condensation of chromatin along the chromosome that was enhanced by staining, so that each chromosome can be identified by its specific banding pattern. The darkly staining regions are AT-rich and replicate late in the cell cycle, and the lightly staining regions are GC-rich and replicate early in the cell cycle.

gametic disequlibrium a state that exists in a population when alleles at different loci are not distributed independently in the population's gamete pool, often because the loci are linked.

gametic phase disequilibrium transmission of given combinations of alleles by chance as opposed to by linkage or by migration.

gene a hereditary unit that occupies a specific location (locus) on a chromosome, the physical entity that is transmitted from parent to offspring. Often, the term is used to refer to a unit encoding a specific trait or protein.

gene differentiation (G_{ST}) is determined by partitioning the genetic variation (heterozygosity) observed in the system of populations into two components, the relative proportion existing *within* populations and the proportion attributable to differences *among* the populations.

gene family a collection of genes within a genome sharing homology. Many gene families evolve by gene duplication, after which the genes may remain together in clusters (tandem arrays) or be dispersed throughout the genome following chromosomal rearrangements.

gene flow genetically effective migration; the movement of genes among populations of a species.

gene migration migration-mediated movement of genes among populations of a species.

gene pool the combination of all allelic variation at all loci summed across all members of a population or discrete breeding group.

genetic architecture the number, relative strengths, and modes of expression for genes affecting expression of a trait.

genetic code the nearly universal system whereby nucleotide triplets, or codons, specify the sequence of amino acids for protein synthesis. Nuclear DNA and mtDNA have slightly different codes.

genetic component of variance the proportion of phenotypic variation with an underlying genetic basis.

genetic distance in population genetics, an estimation of the number of allelic substitutions per locus that have occurred since separation of a population pair.

genetic distance matrix a tabular presentation of pairwise genetic distance values for a group of populations.

genetic diversity all of the genetic variation within a species. Genetic diversity includes both within- and between-population components.

genetic drift random changes in allelic frequencies that occur in each generation due to natural sampling errors. The rate of genetic drift increases as effective population size decreases.

genetic hazard as a result of hatchery propagation activities, adverse genetic consequences due to (1) extinction, (2) loss of within-population genetic variation, (3) loss of between-population genetic variation (population identity), or (4) domestication (Busack and Currens 1995).

genetic interaction direct interbreeding between hatchery fish and wild or natural fish.

genetic load the proportional amount by which the average fitness of the population is depressed for genetic reasons below that of the genotype with maximum fitness (Crow 1958).

genetic marking the purposeful change of allele frequencies in a group of fish or shellfish in order to identify them or their descendants within a larger collection of conspecifics.

genetic resources genetic diversity considered as the raw material for adaptation or exploitation.

genetic risk the probability of a given genetic hazard, that is, the probability of extinction, loss of population identity (among-population genetic diversity), loss of within-population genetic variability, or artificial or domestication selection (Busack and Currens 1995).

genetic stock a group of organisms, sharing a gene pool, that is sufficiently discrete and nominally identifiable that it warrants management as such (Larkin 1972). Panmixia is the case within, but not between, genetic stocks.

genetic stock identification (GSI) in the fisheries science literature, use of any of an array of genetically based methods for species or population identification. These approaches are based on the assumption that because genetic differences underlie species and population differentiation, genetic data and analyses can be used to identify and characterize these entities.

genetic stock structure a type of assortative mating, in which the gene pool of a species is composed of a group of subpopulations, or stocks, that mate panmictically within themselves.

genetic uncertainty processes affecting viability of a population acting through impacts upon the fitness of members of the population.

genetic variance the proportion of phenotypic variance that is attributable to genetic factors.

genetic variation all the variation due to different alleles and genes in an individual, population, or species, including variation in alleles and genes influencing both qualitative traits (under single gene determination) and quantitative traits (under polygenic and environmental determination).

genetics the scientific study of inheritance. Genetics can be regarded as having five interrelated branches: classical or Mendelian genetics, cytogenetics, molecular genetics, population genetics, and quantitative genetics.

genic coadaptation a mode of coadaptation that results from interactions among alleles at different loci that confer fitness on their carriers. The simplest case would involve two loci, A and B, where A_1B_1 and A_2B_2 are well-adapted allele combinations, and A_1B_2 and A_2B_1 are not.

genome all of the genetic information carried on one copy of all the chromosomes of a species of interest.

genome mapping construction of an ordered listing of the genes or genetic markers occurring on the respective chromosomes of a species of interest.

genotype the set of alleles for one or more genes in an organism. The entire set of genes carried by an individual.

genotype × environmental interaction the case in which phenotypic variation is affected by interaction between genotype and environmental variation.

glycolysis the metabolic pathway by which glucose is broken down.

gonosomes the sex-determining chromosomes; contrast with **autosomes**.

grilse an alternative reproductive strategy (of Atlantic salmon *Salmo salar*), analogous to the jack strategy of Pacific salmonids, in which males return to the spawning grounds after only 1 year at sea.

group-average method a clustering algorithm in which the distance between two clusters is defined by an average of the distances between samples.

haploid referring to the number of chromosomes or amount of genetic material in one chromosome set; referring to an individual bearing one set of chromosomes.

haploid chromosome number (*n*) the number of chromosome pairs in a diploid species. Gametes contain one member of each pair, that is, one haploid set.

haplotype a contraction of the term haploid genotype, used to represent a specific mtDNA pattern or a collection of coinherited nuclear DNA alleles or markers.

hard selection a type of selection in which genotype is the absolute criterion determining the fitness of an individual, independent of environmental or other genetic factors.

Hardy–Weinberg law or **equilibrium** a model that predicts that for genetic characters (1) the frequency of alleles remains constant from one generation to the next (i.e., at equilibrium) so long as the organisms are diploid, reproduce sexually, and mate randomly and the population is not subject to mutation, migration, or selection, and (2) the frequencies of genotypes and alleles can be related as $(p + q)^2 = p^2 + 2pq + q^2$, where p and q are the frequencies of alleles A and a at a locus, p^2 is the frequency of the AA genotype, $2pq$ is the frequency of the Aa genotype, and q^2 is the frequency of the aa genotype.

hatchery (population) a group of interbreeding fish that are artificially propagated in a hatchery setting. The breeding history of their ancestors may or may not be known.

heritability the proportion of phenotypic variance in a population attributable to genetic factors (broad-sense heritability—contrast with narrow-sense heritability).

heterochromatin chromosomal regions, predominantly composed of repetitive DNA, that are maximally condensed during interphase. Heterochromatin is principally composed of repetitive DNA and generally is presumed to be genetically inactive.

heterochrony temporal differences in gene expression. An ontogenic change that occurs during evolution, altering the growth rate or development of an organ or morphological feature.

heterogametic sex the sex that bears two types of gametes, one giving rise to males, the other to females. For example, in salmonids, XY gives rise to males, the heterogametic sex, and XX gives rise to females, the homogametic sex.

heteroplasmy the presence in an individual of mtDNA molecules of more than one size (length heterosplasmy) or with different occurrences of restriction sites (restriction site heteroplasmy).

heterozygosity the proportion of individuals in a population that are heterozygous at a particular locus, loci, or the entire genome.

hierarchical mating a mating design in which gametes of individuals of one sex are used in crosses with multiple individuals of the other sex. Hierarchical mating of dams nested within sires is the mating design in which males are each mated to several different females. Individuals are randomly chosen and paired for mating. The progeny so measured represent a collection of full-sib families nested within half-sib families. Hierarchical mating of sires nested within dams is the mating design in which females are each mated to several different males. Individuals are randomly chosen and paired for mating. The progeny so measured represent a collection of full-sib families nested within half-sib families. Often, phenotypic data from hierarchical matings are used for estimation of heritability.

highly repetitive centromeric DNA DNA sequences that repeated 10^5–10^9 times, generally tandemly arranged, and often localized in constitutive heterochromatin of the centromere of a chromosome.

histones small, evolutionarily conserved, DNA-binding proteins that act as spools around which chromosomal DNA wraps, resulting in structures known as nucleosomes.

hitchhiking (allele) the spread of a selectively neutral or deleterious allele through a population because it is tightly linked to an adaptive allele favored by selection.

homing the propensity for returning to the natal area to spawn.

homologous referring to structures, functions, or protein or DNA sequences that have the same evolutionary origin.

homologous chromosomes chromosomes that pair during meiosis and mitosis and that contain the same sequence of genes. In sexually reproducing species, one member of each pair was inherited from each parent.

homology the quality of showing a fundamental similarity in structure, function, or protein or DNA sequence due to the same evolutionary origin.

hooknose males male salmon that exhibit the reproductive strategy of maturing at the large size typical of their stock and securing matings through aggressive defense of breeding territory; contrast with alternative male reproductive strategies such as **grilse**, **jacks**, or **precocious parr**.

hybrid vigor increase in the phenotypic value of a trait, or of fitness of a hybridized population, relative to the distinct, parental populations that were interbred. Hybrid vigor is attributed to overdominance or to increased heterozygosity (also called heterosis).

ideal population a population displaying certain qualities pertinent to transmission of genetic information; that is, unlimited number of breeders, equal numbers of males and females breeding, equal probability of reproductive success for each breeder (i.e., random variation in family size), non-overlapping generations, and absence of migration, mutation, and selection.

identical by descent referring to alleles that are identical because they are inherited from a recent common ancestor; contrast with **alike in state**.

immigration the movement of individuals into a population followed by successful reproduction and transmission of genes.

inadvertent selection unintended change in allele frequencies in a cultured population resulting from nonrandom collection of hatchery broodstock from the wild or from unintentional selection occurring in the hatchery environment.

inbreeding the mating of related individuals, a particular case of positive assortative mating; see also **inbreeding depression**.

inbreeding coefficient (*F*) the estimated level of inbreeding. Inbreeding is quantified in a relative fashion by estimating the increase in homozygosity, or decrease in heterozygosity, relative to some ancestral population. See ***F*-statistics**.

inbreeding coefficient (F_{IS}) a fixation index that measures the average departure from Hardy–Weinberg expectations among populations in terms of the average deficit of heterozygotes within populations.

inbreeding depression reduction in the fitness or vigor of individuals in a lineage due to inbreeding, resulting from increased homozygosity or expression of recessive deleterious alleles.

inbreeding effective number the genetically effective size of an idealized population exhibiting the same increase in homozygosity as a population under consideration.

inclusive fitness the aspect of fitness that is manifested by an altruist who increases the likelihood of survival or reproduction of alleles carried by relatives as opposed to those same alleles carried by his or her self.

indirect genetic effects genetic changes within or among populations that are induced without actual infiltration of exogenous genomes (i.e., introgressive hybridization) into native populations.

infinite alleles model a model of mutation developed by Kimura and Crow (1964) concerning the interplay of mutation and drift. It is assumed that every mutation creates a new allele, that is, one that does not already exist in the population.

infinitesimal model regarding the expression of quantitative traits, the assumption that a large number of loci each contribute a small and equal amount to expression of the trait.

insertional mutagenesis the mutation of a DNA sequence or gene by insertion of a transposon, virus, injected DNA, or other DNA sequence.

insertional translocation insertion of a portion of chromatin from one locus into another on the same or a different chromosome. Centric shifts are essentially intrachromosomal translocations.

interdeme selection a population genetics process in which interplay of migration and selection can give rise to fusion of demes, loss of low-fitness populations, and spread of high-fitness populations, a controversial concept put forward by Sewall Wright.

interphase the eukaryotic cell cycle stage between mitoses when most of a chromosome is not condensed around histones and does not display the conventional "X" shape. Some of the DNA, however, may remain highly coiled.

intrasexual selection a mechanism of sexual selection described by Darwin involving adaptation(s) conferring the ability of a male to dominate other males in battle.

intrinsic coadaptation a mode of coadaptation relating to factors internal to the gene pool of the population. Intrinsic coadaptation occurs when a karyotype or gene complex evolves in response to the state of other chromosomes or genes. Contrast with **local adaptation**.

intrinsic factors factors affecting extinction probability that are within the biology of the population at issue; contrast with **extrinsic factors**.

introgression repeated backcrossing of hybrid descendants with a parental line, population, or species, resulting in the incorporation of genes from one gene pool into the other.

introgressive hybridization the infiltration of the genome of one species or population into that of another.

intron noncoding DNA that interrupts exons in a eukayote; a segment that is transcribed into nuclear RNA but does not survive processing in the nucleus, is removed from the transcript, and does not occur in the cytoplasm.

inversion a change in the arrangement, but not the total number, of nucleotides, loci, or chromosome segments, in which the DNA sequence or chromosomal segment is excised, rotated 180°, and reinserted into the same chromosome. The fitness effects of heterozygosity for an inversion depend on the effects on gene expression of affected chromosomes, meiotic pairing configurations, and frequency of crossing-over in heterozygotes.

island model a model of migration that considers populations as discrete, spatially separated units analogous to a number of islands, among which there are exchanges of individuals. The likelihood of migration among all islands is equal.

isoelectric point the pH at which the positive and negative charges on a protein are equal and a charged molecule no longer migrates in an electric field.

isoenzymes multiple forms of an enzyme that catalyze the same reaction but are encoded at different loci. Isoenzymes may differ in optimum reaction properties such as temperature, pH, or substrate concentration.

isolation by distance a model of migration (Wright 1943) in which populations are distributed continuously over space and the only barrier to migration is distance itself.

iteroparity the life history characteristic of an individual or species capable of reproducing more than once in the lifetime; contrast with semelparity.

jack a male salmon that matures before passing through the usual number of years at sea, sometimes after just 6 months at sea.

karyotype the number, relative sizes, and morphologies of the chromosomes in an individual or species. A photomicrograph of paired metaphase chromosomes arranged in order of size.

karyotypic coadaptation a type of intrinsic coadaptation resulting from certain combinations of chromosomal variants conferring fitness on their carriers whereas other combinations do not.

Kreb's (or **citric acid**) **cycle** the biochemical pathway for energy generation in mitochondria.

law of independent assortment the Mendelian law that states that members of different pairs of characters associate randomly, or independently; that is, alleles at different loci are incorporated independently into gametes at meoisis provided they reside on different chromosomes.

law of segregation the Mendelian law that states that the two members of a pair for a character in a parental organism segregate; that is, the two alleles at a locus in a parental organism segregate into different gametes at meiosis.

length heteroplasmy presence in an individual of mtDNA molecules of more than one size.

lethal gene a gene whose expression in the homozygous state results in death of the carrier. Lethal genes persist in a gene pool because only the homozygous recessive genotype is lethal. If the recessive allele is indeed a lethal allele, then all the offspring with the homozygous recessive genotype die. By noting that in the system of matings there are families whose survival conforms to the ratio 1:2:0, a geneticist can infer the segregation of a lethal allele.

level of inbreeding the probability that the two alleles at a given locus are identical copies of a single ancestral allele.

life history the significant features of the life cycle through which an organism passes, with particular reference to strategies influencing survival and reproduction.

LINEs see **long interspersed elements**.

linkage disequilibrium the situation that occurs when alleles of genes that reside on the same chromosome are distributed nonrandomly into gametes of individuals in a pedigree or population.

linkage equilibrium the situation in which the respective loci are on different chromosomes, and their alleles are in random association.

linkage group a group of genes or genetic markers occurring in sufficient proximity on the same chromosome that they are inherited together.

load of mutations the negative effect of mutant alleles on the fitness of a population (Muller 1950).

local adaptation the process in which allele frequencies at fitness-related loci of a population are subjected to selection by extrinsic, or environmental, factors, thereby increasing the fitness of the population in that environment. The array of allele frequencies approaches and then tracks the optimum for that local environment. Local adaptation is one mode by which coadaptation can occur. Contrast with **intrinsic coadaptation**.

locus (plural, **loci**) the site that a gene or molecular sequence of interest occupies on a chromosome.

long interspersed elements (LINEs) a family of dispersed highly repetitive DNA sequences, generally greater than 5 kilobase pairs in length and present in at least 10^4 copies per genome; includes retrotransposons.

lymphocyte a class of leukocytes (white blood cells) that function in tissue repair and production of antibodies.

major gene a gene that plays a large role in determining expression of a quantitative trait.

maladaptive (alleles) alleles conferring reduced fitness upon the carrier.

management indicator species species whose occurrence and abundance provide an indication of the success of fisheries or wildlife management actions, including threatened or endangered species; species commonly fished, trapped, or hunted; economically important species; and species sensitive to human activities.

marking effort the ratio of the number of fish screened for genetic markers relative to the number of fish subsequently spawned.

marking strategy the approach of selecting for or against marker alleles at one or more loci in a deliberate genetic marking program.

mass selection type of selection that operates upon individuals as opposed to families. In the context of selective breeding, a geneticist has the choice of the unit of selection. Also applied in the context of the shifting balance theory.

maternal effects a source of phenotypic variation stemming from nongenetic effects of the mother on her young, regarded as an environmental source of variation. For example, maternal size and age may have nongenetic effects upon the size of eggs and of fry at hatch.

maximum likelihood method a data analytic approach for evaluating the evolutionary history of groups, as represented by a particular tree topology, in terms of a proposed model of evolution. The investigator proposes a specific model for genetic change and then calculates the probability that the observed genetic data would have generated a particular tree.

maxisatellites tandem repeats of motifs that are hundreds to thousands of nucleotides in length, found clustered in the centromeres of chromosomes.

megasatellites repetitive monomers; tandem repeats of motifs hundreds to thousands of nucleotide base pairs in length, often found clustered in the centromeres of chromosomes.

meiosis the process of gamete formation. A chromosomal duplication occurs, followed by two successive nuclear divisions, resulting in haploid gametes. It is at meiosis that crossing-over occurs, creating new genetic variation.

meiotic drive any meiotic mechanism underlying preferential recovery of a specific allele in gametes of a heterozygote.

Mendelian traits phenotypic characters for which substitution of one allele for another at a given locus yields a large change in phenotype. These traits include discrete phenotypes, visible mutations, and lethal genes.

meristic traits characters exhibiting phenotypes expressed in discrete elements that are counted to form integral classes. An individual's phenotype is characterized as the number of elements of the trait it exhibits. Examples of meristic traits in fishes are number of scales along the lateral line, number of fin rays, and number of pyloric caecae.

messenger RNA (mRNA) RNA that is the transcript of expressed DNA sequences and is translated to yield proteins.

metacentric describing a chromosome in which the centromere is approximately at the center of the chromosome, so that the chromosome arms are similar in length.

metaphase one of four phases of mitosis, during which the chromosomes are condensed and become arranged at the equatorial region of the spindle, then to be separated toward the poles before cell division occurs.

metapopulation a system of local demes that are linked by migration. The genetic composition of individual populations linked by gene flow may coevolve as one complex unit, the metapopulation.

Metazoa all multicellular animals except the Protozoa.

methylation addition of a methyl (CH_3) group to DNA.

microchromosomes supernumerary chromosomes that are very small.

microsatellites genomic DNA sequences containing tandem arrays of short (2–5 bp) motifs of nucelotides, for example, poly(GT), poly(GATA), and so on.

migration the movement of animals between populations, followed by successful reproduction and resulting gene flow.

minimum spanning tree a depiction of population genetic relatedness in which a network shows the shortest path among all the objects or samples. A computer algorithm uses the pairwise genetic distances from a genetic distance matrix to find the tree that connects all the different samples and has the shortest total length.

minimum viable population size the population size that provides a given probability of persistence of a given population for a given amount of time; for example, a 95% expectation of persistence without loss of fitness for several centuries (Soulé 1987).

minisatellites repeating motifs of nucleotides, generally 9–65 bp in length, dispersed throughout the genome.

missense mutations three kinds of mutations resulting from substitution of one base pair for another.

mitochondrial DNA (mtDNA) a small proportion (<1%) of the DNA of eukaryotic cells that is nonnuclear, located within organelles in the cytoplasm called mitochondria, and appears to have been endosymbiotic in early eukaryotic cells, as its genetic code differs from the "universal" genetic code. Animal mtDNA is a double-stranded, circular molecule usually ranging in size from 14,000 to 26,000 bp.

mitosis the process of nuclear division during the cell cycle that is composed of four stages: prophase, metaphase, anaphase, and telophase.

mixed-stock analysis (MSA) the statistical estimation of the composition of an assemblage of organisms, using prior knowledge of genetic or morphological characters, or both, that effectively discriminates among populations whose members are represented in the mixture.

mixed-stock fishery a harvested group of aquatic or marine organisms that comprises multiple, component populations.

mobile genetic element a member of a class of DNA sequences that can move from one place in the genome to another.

mobile repetitive DNA sequences repeated DNA sequences within the genome that have no known coding or regulatory function and can move from one place in the genome to another.

modes of selection the three fundamental ways in which selection can act upon the range of phenotypes in a population, that is, directional, stabilizing, and disruptive selection.

molecular evolutionary clock hypothesis the hypothesis that the rate of nucleotide substitutions in DNA is sufficiently constant that, given an observed degree of genetic divergence among lineages, the time since their divergence can be estimated.

molecular genetics the study of the structure, function, and dynamics of genes at the molecular level.

monomer a functional enzyme molecule composed of one subunit; contrast with **dimer** and **tetramer**.

monoploid referring to cells having only one set of chromosomes (n = the haploid number).

Monte Carlo simulation a class of techniques that involve using random numbers generated by a computer to simulate stochastic processes.

mRNA see **messenger RNA**.

mtDNA see **mitochondrial DNA**.

multidimensional scaling techniques that provide means by which the many pairwise genetic distances between samples are represented by points along a line, in a plane, or in three dimensions, such that the Euclidean distance between the points reflects as closely as possible the same relationship as the distances in the original matrix.

multi-locus DNA fingerprinting a set of molecular genetic procedures by which genotypes at multiple loci are observed simultaneously. Often, it is not possible to infer allelic relationships among the bands so observed.

multiple arbitrary amplicon profiling a general class of PCR-based techniques used to detect anonymous, or arbitrary, sequences (the specific locations and functions of many nuclear DNA sequences are unknown).

multivalent referring to an association of more than two chromosomes whose homologous regions are synapsed by pairs (as in autotetraploids) (King and Stansfield 1990).

mutability enhancer gene a gene that regulates the rate of mutation in a genomic region.

mutagen a physical or chemical agent that causes a structural change in DNA or chromosomes.

mutant an individual bearing one or more mutations.

mutation the group of processes through which an organisms acquires heritable genetic changes.

mutation load the collection of fitness-reducing alleles maintained in a population by mutation.

mutation pressure the tendency for allele frequencies to change as a result of recurrent mutation.

narrow-sense heritability (h^2_n) the proportion of phenotypic variance attributable to additive genetic variance (expressed as the ratio V_A / V_P); contrast with **broad-sense heritability**.

natural catastrophe the process of extinction posed by natural phenomena outside the normal range of environmental variation.

natural population a group of interbreeding fish that reproduce without the aid of humans and whose ancestors probably include hatchery-propagated fish. The degree of contribution from hatchery fish may be known or unknown.

natural selection the differential perpetuation of genes and gene complexes (Dobzhansky 1970:97); that is, the evolutionary force mediating adaptation of populations to their environment. The process is mediated by individuals with more favorable (adaptive) phenotypes having higher survival and reproduction rates than those with less favorable phenotypes, eventually leading to heightened frequencies of favored phenotypes and underlying genotypes and elimination of less favorable phenotypes and underlying genotypes.

negative assortative mating nonrandom mating that occurs when unlike individuals mate more frequently than expected by chance, as in the rare-male advantage of *Drosophila*; contrast with **positive assortative mating**.

neighborhood the geographic range within which individuals move between birth and reproduction.

neighborhood models models of migration that assume that individuals are distributed over a continuum and that genetic structure within the continuum arises if the average distance that individuals move between birth and reproduction is small relative to the species' range.

nested mating design a mating design involving overlapping full-sib and half-sib families. See **hierarchical mating** for examples.

neutralist the viewpoint that the great majority of evolutionary changes, at least at the DNA level, result from random fixation of neutral or nearly neutral mutants through nonselective processes, such as random genetic drift; contrast with **selectionist**.

neutrality the property of a genetic variant whose phenotypic expression has no effect on the fitness of the carrier.

noncoding region a region of DNA that does not contain structural genes, that is, those coding for expressed proteins.

noncoding telomeric sequences short, simple sequences at the tips of chromosomes that fold back on themselves in a hairpin structure so that lagging 5′ ends can have 3′ ends for priming of polymerization.

nonfunctional repetitive DNA sequences satellite DNA found in the heterochromatin that flanks centromeres, including repetitive monomers (sometimes called maxisatellites) and variable number of tandem repeat sequences (VNTRs), which include minisatellites and microsatellites.

nonintrogressive hybridization the case of matings among species for which sterility or reduced fitness precludes reproductive success of hybrids.

nonmetric multidimensional scaling (MDS) a data analytic technique in which the goal is to maintain the original pattern in the genetic distance matrix by using the rank order of the genetic distance values. If we imagine the points in a three-dimensional space as a structure being joined by elastic bands that equal the genetic distance between samples, the computations in MDS attempt to find the minimum amount of distortion (bending, stretching, and squeezing of the rubber bands) required to flatten the structure to two dimensions while maintaining the rank order of magnitudes among the samples that existed in the original distance matrix.

nonmobile repetitive DNA sequences repeated DNA sequences not associated with a mechanism to move to another site in the genome.

nonrecombinant gametes gametes bearing genotypes observed in the parents. In a cross of inbred lines bearing *AB* and *ab* genotypes, respectively, if two loci are linked, in the gametes of the F_2 there will be higher and equal frequencies of the two parental (nonrecombinant) types, *AB* and *ab*, than of the recombinant types *Ab* and *aB*; contrast with **recombinant gametes**.

nonsense mutations mutations resulting in production of a termination codon for translation of a protein.

nucleolar organizer region (NOR) the RNA-rich region of the chromosome in an organism that contains the rRNA genes, the products of those genes, their associated proteins, and a variety of enzymes associated with gene replication.

nucleosome eukaryotic chromosome structure consisting of a core of eight histones wrapped by a DNA segment (ca. 150 bp in length) and separated from adjacent nucleosomes by DNA linkers (ca. 50 bp in length).

nucleotide the basic molecular unit of which DNA (and RNA) is composed. Each nucleotide is composed of a purine (or pyrimidine) base, a pentose sugar, and phosphoric acid. The nucleotides of DNA often are referred to by the purine or pyrimidine base of which each is composed (or its first letter), adenine (A), guanine (G), cytosine (C), and thymine (T) (uracil [U] replaces thymine in RNA).

null allele in molecular genetics, an allele that produces no functional product, for example, at an isozyme locus. In PCR-based analyses, a sequence that is not amplified because a variation in the DNA sequence annealing to the 3′ end of a primer results in nonamplification of the expected segment.

objective referring to analyses of population genetic or phylogenetic data, those analyses that anyone repeating the analysis using the same data and sequence of methods would produce the same classification.

oligonucleotide a synthetic sequence of nucleotides, often used as a primer for PCR or as a hybridization probe.

oncogene a gene, viral or cellular in nature, that can induce uncontrolled cell proliferation.

ordination techniques data analytic methods that provide means by which the many pairwise genetic distances between the samples are represented by points along a line, in a plane, or in three dimensions, such that the Euclidean distance between the points reflects as closely as possible the same relationship as the distances in the original matrix.

outbreeding depression the phenomenon of reduction in fitness following intraspecific hybridization (matings between individuals from different populations), either in the immediate hybrids or perhaps delayed until the backcross or later generations.

outcross the breeding of unrelated individuals.

outgroup a taxon that is not part of the group or groups of interest, but shares a common ancestor with that group or groups.

overdominance a mode of gene expression in which the phenotype of the heterozygote is outside the range of those for the respective homozygotes; for example, fitness of the heterozygote, W_{Aa}, is greater than W_{AA}.

overdominance hypothesis the hypothesis that fitness is determined by genome-wide heterozygosity levels. Under this hypothesis, high heterozygosity is inherently advantageous; hence, it sometimes is called the heterozygous advantage hypothesis. Inbreeding depression, then, is the consequence of reduced heterozygosity throughout the genome due to matings between related individuals. Contrast with **dominance hypothesis**.

panmictic well mixed, referring to the set of genotypes in a population.

panmixia the situation in which all individuals in a unit of reference, most often a population, are randomly interbreeding.

paracentric inversion a chromosomal mutation in which the inverted segment does not include the centromere; contrast with **pericentric inversion**.

parental male a male that exhibits territoriality and nest defense as part of normal spawning behavior, as in salmonids and centrarchids. Contrast with **grilse, jack, precocious parr, satellite male**, and **sneaker male**.

parthenogenesis reproduction of a virgin female in a species that normally exhibits sexual reproduction. In the absence of fertilization, the haploid egg divides mitotically, and the cleavage nuclei fuse to form a diploid nucleus, which then initiates embryogenesis.

partial dominance a mode of gene expression in which the phenotype of the heterozygote is distinguishable from that of the homozygote for the dominant allele.

PCR see **polymerase chain reaction**.

pericentric inversion a chromosomal mutation in which the inverted segment includes the centromere; contrast with **paracentric inversion**.

phage a bacterial virus.

phenetic referring to population genetic analyses that are based on similarity.

phenotype the physical manifestation of a genotype, for example, the coloration pattern of a fish or the banding pattern for lactate dehydrogenase on a stained gel slab.

phylogenetic relating to the hypothesized evolutionary relationships of individuals, populations, or species.

phylogenetic similarity the degree to which populations or species exhibit similar characters due to their evolutionary relatedness.

phylogeography analysis of genealogy, population genetics, or evolution within a geographical context.

plasmid double-stranded, closed, extra-chromosomal DNA molecule found in some bacterial species, often conferring an adaptive advantage to the strain such as antibiotic resistance.

pleiotropy the situation in which a particular gene affects multiple traits.

polygenic refers to a trait or phenotype whose expression is the result of the interaction of numerous genes.

polymerase any enzyme that catalyzes formation of DNA (from deoxyribonucleotides) or RNA (from ribonucleotides).

polymerase chain reaction (PCR) a molecular genetic technique used to amplify the number of copies of a target DNA sequence. The reaction requires oligonucleotide primers, deoxyribonucleotides, a thermostable polymerase, and an instrument that cycles between 95°C (to separate complementary strands), various annealing temperatures (to allow binding of primers), and extension temperatures (to allow polymerase-mediated generation of complementary strands). The process results in 2^n copies of the template, where n is the number of replication cycles.

polypeptide chain an amino acid sequence that is produced as a mRNA is translated; specific tRNAs bring their amino acids, and each added amino acid is bound by a peptide bond to the preceding amino acid.

polyploidization the process whereby the entire chromosomal complement of a species is duplicated.

polyploidy the situation of cells or individuals having additional complete sets of chromosomes (e.g., $3n$ is triploidy and $4n$ is tetraploidy).

population a group of organisms that belong to the same species and freely interbreed.

population genetics the study of the genetic composition and evolution of populations. The scope of population genetics cuts across numerous disciplines, including other branches of genetics, ecology, systematics, natural history, and statistics.

population identity the uniqueness of a population as indicated by genetic differences between it and other populations.

population viability analysis the process of assessing the likelihood that a population will persist for a specified period of time.

positive assortative mating the situation by which mates are chosen on the basis of greater than average phenotypic resemblance, for example, regarding morphological traits; contrast with **negative assortative mating**.

posttranslational modification change in the structure of a protein following translation from mRNA. For example, the protein may be folded, portions may be excised, or functional groups such as glucose rings may be added.

precocious parr in Atlantic salmon *Salmo salar*, males that never migrate to sea, instead wintering in their natal freshwater stream and becoming reproductively mature.

primary structure the specific sequence of monomeric subunits (i.e., nucleotides or amino acids) in DNA or a protein, respectively.

primer a short, single-stranded oligonucleotide (ranging from 10–50 bp in length) that anneals to one or both ends of a DNA segment under investigation. Often, primers are used in PCR.

principal components analysis (PCA) a data analytic procedure that uses the covariance matrix of allele frequencies to ordinate populations.

principal coordinates analysis (PCO) a data analytic procedure that produces a pattern of points that represents the original dissimilarities or distances in the genetic distance matrix as closely as possible. The PCO is a generalized form of principal components analysis. If the original distances were Euclidean, then the results of the PCO would be equivalent to the results from the principal components analysis.

probe a radioactively or chemically labeled DNA fragment that hybridizes to homologous target DNA sequences, permitting their visualization.

probe × restriction enzyme combination in production of an autoradiograph of a Southern blot, the use of a particular restriction enzyme to digest an organism's DNA into fragments and the use of a particular probe to visualize DNA fragments bearing homologous DNA sequences. Because variation in DNA sequence can affect the presence or absence of a restriction site at a particular locus, and because the choice of probe determines which sequences are visualized on the autoradiograph, the choice of probe × restriction enzyme combination is a major determinant of the results of a Southern-blot-based screening for genetic variation.

promoter element a DNA sequence proximal to a gene that controls its expression. It is here that RNA polymerase binds to initiate transcription.

Protista the kingdom of living organisms that includes eukaryotic microorganisms, such as algae, slime and water molds, and protozoans.

purine an adenosine or guanine nucleotide.

pyramidine a cytosine or thymidine nucleotide.

Q bands in analysis of a metaphase chromosome spread, the G-banding pattern obtained by staining with fluorochromes that are specific for AT-rich DNA, such as quinacrine.

quadrivalent a meiotic association of four homologous chromosomes, synonymous with tetravalent.

quantitative genetics the study of traits exhibiting phenotypes that are quantitative in nature.

quantitative trait a phenotypic trait that is described by a measurement (e.g., weight at maturity) and whose expression results from actions of genes at many loci, environmental factors, and interactions of genetic and environmental factors.

quantitative trait loci (QTLs) genes determining expression of quantitative traits.

quaternary structure assembly of nuclear DNA into chromatin or chromosomes through the complex association of DNA with histones and other proteins.

R bands in analysis of a metaphase chromosome spread, a reverse pattern (to G bands) obtained by staining with chromomycin A_3 or mithromycin, which stain GC-rich DNA.

random genetic drift chance-mediated change in allele frequencies due to genetic sampling errors.

random sampling selection of individuals independently from a population in a manner such that each individual is equally likely to be chosen.

randomly amplified polymorphic DNA (RAPD) a method of producing PCR products from genomic DNA template using one or a combination of oligonucleotide primers (typically ca. 10 bp). Amplification products are separated by gel electrophoresis, and variation is scored as presence or absence of bands, typically more than 500 bp in length. The visualized amplification products, often referred to as RAPDS, are a dominant genetic character.

rare male advantage a frequency-dependent selective mechanism in which females select as their mates males with an uncommon phenotype.

realized heritability (h^2) a heritability estimated by imposing a selection differential, *S*, upon a population and observing the response to selection, *R*; that is, $h^2 = R/S$.

recessive alleles alleles whose phenotypic expression is masked by dominant alleles.

reciprocal translocations the exchange of chromosomal segments between two nonhomologous chromosomes. These exchanges may involve only a portion of a chromosome or may involve the swapping of entire chromosome arms (whole-arm translocations).

recombinant gametes gametes bearing genotypes not observed in the parents. In a cross of inbred lines bearing *AB* and *ab* genotypes, respectively, if two loci are linked, in the gametes of the F_1 there will be higher and equal frequencies of the two parental (nonrecombinant) types, *AB* and *ab*, than of the recombinant types *Ab* and *aB*; contrast with **nonrecombinant gametes**.

recombination the process by which new combinations of alleles are created by independent assortment or crossing-over at meiosis.

recombination fraction the proportion of recombinant gametes produced by an individual. Parental-type *AB* and *ab* gametes each are produced with frequency $(1 - r)/2$, where *r* equals the recombination frequency (see below). Recombinant-type Ab and aB gametes each are produced with frequencies of $r/2$. The recombination fraction depends on the relative positions of two genes of interest on a chromosome.

recombination frequency (*r*) the frequency at which meiotic crossing-over produces new, nonparental types of gametes.

reference or **base population** for estimation of the inbreeding coefficient, the reference population is assumed to be noninbred, that is, its inbreeding coefficient (*F*) equals 0. The reference population may be the original source population or an intervening past generation.

regulatory element a DNA sequence that influences the expression of a structural gene, most often located immediately downstream.

regulatory gene a type of gene whose primary function is to control the expression (i.e., the rate of synthesis of the product) of other (often distantly located) genes, usually as a part of an inducible system including a regulator, repressors, and operator genes.

relative viability the absolute viability of a particular genotype or phenotype relative to that of the fittest genotype or phenotype, which is assigned the value of 1.00.

repetitive DNA functional and nonfunctional nucleotide sequences that occur repeatedly in the genome—tens, hundreds, thousands, or millions of times.

repetitive monomers adjacent repeats of motifs hundreds to thousands of nucleotide base pairs in length found clustered in the centromeres.

replication slippage a mutational mechanism in which a newly replicated DNA strand may slip relative to the strand being copied, so that the new strand ends up shorter or longer than the original, depending on whether it slipped forward or backward.

reproductive fitness the aspect of Darwinian fitness relating to reproductive success. Reproductive fitness may be quantified in either absolute or relative terms; however, because reproductive success can vary from zero to an undefined upper bound, the theory and associated modeling become more complex.

reproductive success the number of viable offspring parented by an individual, which can vary from zero to an undefined upper bound.

response to selection following an episode of selection, the change in the mean phenotype observed in the progeny generation relative to the parental generation or to a contemporary, randomly bred stock from the same base population.

restoration the re-establishment of a self-sustaining populations in historical habitat no longer occupied by the species.

restriction enzyme (endonuclease) an enzyme that cleaves DNA molecules at specific recognition sequences; for example, the enzyme *Alu* I cuts DNA at the sequence AG↓CT. In the producing organism, restriction enzymes are a defense against foreign DNA. In molecular genetics, they are utilized in cloning and population genetic analysis.

restriction fragment length polymorphism (RFLP) variation in the length of DNA fragments generated within a species when treating longer DNA segments with a restriction enzyme. The variants may be due to differences in DNA sequence at the recognition site where cleavage occurs (i.e., the enzyme does or does not cut) or to variations in length of the cleaved segment (often due to presence of varying numbers of tandem repeat motifs).

restriction site heteroplasmy presence in an individual of more than one type of mtDNA molecule, which have different restriction sites for an endonuclease.

restriction site a nucleotide sequence at which a particular restriction enzyme will cleave the DNA molecule, for example, AG↓CT for the restriction enzyme *Alu* I.

retroposon a mobile element (e.g., a retrovirus) that transposes via an RNA intermediate. Transpositions occur following reverse transcription of RNA into DNA, and the copies of the retroposon can reintegrate into the genome at numerous locations.

retrotransposon a mobile element that occurs following reverse transcription of RNA into DNA, the copies of which reintegrate into the chromosome at numerous locations and are flanked by long terminal direct repeats. These are structurally related to retroviruses.

retroviruses a group of RNA-bearing viruses that utilize reverse transcriptase during their life cycle. The term *retro*virus refers to the "backwards" transcription. The transcribed viral DNA becomes integrated into the genome of the host cell.

RFLP see **restriction fragment length polymorphism**.

ribosomal DNA (rDNA) any DNA region that encodes rRNA, generally found as a cluster of tandemly arranged rRNA genes; see **ribosomal RNA genes**.

ribosomal RNA (rRNA) the transcripts of ribosomal genes that mediate translation of mRNA into protein.

ribosomal RNA genes genes encoding rRNA that reside as tandem repeats (each unit separated from the next by a nontranscribed spacer, NTS) in the nucleolar organizer regions of eukaryotic chromosomes. In fishes, each unit contains three cistrons that encode 28S, 18S, and 5.8 S rRNA.

ribosome a protein–RNA structure upon which the translation of mRNA into proteins takes place.

RNA polymerase an enzyme that transcribes an RNA molecule from the antisense strand of the DNA molecule.

Robertsonian (centric) fusion interchromosomal rearrangement whereby the arms of two chromosomes break and fuse to form a single chromosome or two single-armed chromosomes fuse to form a biarmed metacentric or submetacentric chromosome. This is a fundamental mechanism of reducing the number of chromosomes without changing the total DNA content of a cell.

Robertsonian fission interchromosomal rearrangement whereby a biarmed chromosome splits at the centromere producing two uniarmed chromosomes.

rRNA see **ribosomal RNA**.

Ryman and Laikre effect reduction of effective population number that may result from breeding–release programs aimed at supporting natural populations demographically.

sampling of breeders a mechanism of random genetic drift consequent to the "indiscriminate destruction or failure to propagate" of part of a population (Gulick 1905:209).

sampling of gametes a mechanism of genetic drift occurring because heterozygous organisms may leave homozygous offspring, by chance they may leave only offspring homozygous for one of the alleles in question, or an allele may become lost.

satellite a small segment separated from the main portion of the chromosome by secondary constrictions.

satellite DNA the fraction of eukaryotic DNA, usually highly repetitive, that differs enough in base composition (either lighter because it is A + T rich or heavier because it is C + G rich) to separate from the remaining DNA as a distinct band(s) following centrifugation in a CsCl gradient. Satellite DNA often is found in the heterochromatin flanking centromeres.

satellite male an alternative male reproductive strategy in which a male expressing the coloration pattern and outward behavior of a female will wait near the water surface near an active nest. When a spawning female begins to release eggs, the satellite will enter the nest and release milt.

second-order (secondary) structure the helical, or twisting, appearance of the DNA (or protein) molecule.

segregation load the collection of fitness-reducing alleles in a population maintained by selection on alleles that are favored in heterozygotes but not in homozygotes.

selection the natural or artificial process by which breeding individuals have differential reproduction due to individual differences in fitness. See **artificial selection**, **domestication selection**, and **natural selection**.

selection coefficient the degree of selection against a genotype.

selection differential (*S*) the difference in the mean phenotypes of the selected subpopulation (μ_s) and the population overall (μ); that is, $S = \mu_s - \mu$. Selection differential is a key parameter in estimation of realized heritability.

selection plateau the weak response or absence of response of a population to further selection following sustained directional selection on a quantitative trait. This lack of response is due to fixation of genes affecting expression of the trait and negative pleiotropies among genes affecting the trait.

selectionist the viewpoint that the key driving force of evolution is Darwinian natural selection operating upon advantageous mutations; contrast with **neutralist**.

selective breeding the genetic improvement strategy in which humans choose the matings to be produced in order to approach a breeding goal.

selective equilibria the persistence of allele frequencies at variable loci; allele frequencies tend toward equilibrium values, which prove stable for generations.

selective neutrality the situation in which a particular genotype confers no selective advantage or disadvantage upon its carrier.

semelparity the life history characteristic of an individual or species reproducing just once in the lifetime; contrast with iteroparity.

sequence characterized amplified region (SCAR) a single-locus PCR-based genomic DNA marker that is identified using a pair of locus-specific oligonucleotide primers typically 16–24 nucleotides in length.

sex chromosomes (gonosomes) chromosomes that differ between males and females and that contain the sex-determining loci.

sex-linked the special case of linkage that occurs when a gene is located on the chromosome near the sex-determining locus.

sexual dichromatism the life history characteristic that the sexes differ in coloration.

sexual dimorphism the life history characteristic that the sexes differ in size or morphology.

sexual selection type of selection in which there is competition among males for mates, and characters enhancing the reproductive success of the carriers are perpetuated irrespective of their survival value.

shifting balance theory the theory proposed by Sewall Wright that demes, or isolated reproductive units, "explore" the adaptive topography seeking to maximize their fitness.

short interspersed elements (SINES) a family of dispersed highly repetitive DNA sequences, generally less than 500 bp in length and present in at least 10^5–10^6 copies per genome.

silent substitutions substitutions of one nucleotide for another that do not change the amino acid that the codon specifies. Many third-position base changes are silent substitutions.

SINEs see **short interspersed elements**.

single-copy nuclear (scn) DNA nonrepetitive nuclear sequences that occur with a frequency of one per haploid genome.

single-linkage method a clustering algorithm in which the similarity between two clusters is defined as the distance between the two most similar pairs of samples.

sires males used in a mating design, for example, males mated to several females in hierarchical mating experimental design; contrast with **dams**. See **hierarchical mating**.

slippage a molecular mechanism considered most likely to cause polymorphism in copy number of tandem repeats in repetitive DNA. Strand-slippage replication is speculated to occur in a polymerase complex, primarily during lagging strand (discontinuous) synthesis. If a newly synthesized DNA strand slips upon dissociation, the transient bulge that is created is subsequently repaired, leading to shortening of the repeat (if the bulge is removed) or elongation of the repeat (if the complementary nucleotides are attached to the bulge).

sneaker male an individual exhibiting the alternative reproductive strategy in which he hides near an active nest of a parental male and waits for a pair to start spawning. When the eggs are released, he enters the larger male's nest and releases milt, thereby obtaining fertilizations.

soft selection a type of selection in which the fitness of a phenotype is related to the kind, frequency, and distribution of phenotypes in the population; contrast with **hard selection**.

somatic referring to eukaryotic cells other than reproductive cells.

Southern blot a membrane bearing denatured DNA fragments that were transferred from an electrophoretic gel by capillary action. Particular fragments will be visualized by hybridization of a DNA probe.

species groups of actually or potentially interbreeding populations that are reproductively isolated from other such groups (biological species concept, Mayr 1940).

spectrophotometer an instrument that utilizes narrowly focused optics to compare the intensity of a specific wave length beam before and after passing through cells or other light-absorbing particles.

spindle the collection of microtubules responsible for the movement of eukaryotic chromosomes during mitosis (King and Stansfield 1990).

spontaneous lesion a naturally occurring mutation in DNA sequence, for example, through biochemically mediated changes in the condition of DNA (depurination and deamination), leading to pairing with normally noncomplementary bases.

stabilizing, or **normalizing, selection** a mode of selection favoring intermediate phenotypes.

stable referring to a population genetic or phylogenetic classification that does not change if more individuals or more descriptive traits are added to the analysis.

stable equilibrium the type of equilibrium in which, once displaced, an allele frequency converges once again to its equilibrium frequency.

stepping-stone model a model of migration (Kimura and Weiss 1964) in which populations are spatially distributed such that the primary sources of migrants into a population are the adjacent populations.

stepwise mutation model a model of mutation that assumes that new alleles arise in a stepwise fashion by the gain or loss of repeated units.

stochastic processes processes associated with chance events. In the context of assessing the viability of populations, stochastic processes include those mediated by demographic and environmental stochasticity, and result in unpredictable fluctuations in population numbers.

stock a group of organisms that shares the same demographic parameters, as for a fished stock; contrast with **genetic stock**.

stock concept the concept that maintaining the productivity and evolutionary potential of a species is dependent upon maintaining the abundance and diversity of its component stocks.

stratified random sampling random sampling within distinct segments or strata. Strata are chosen so that all members within a stratum are similar for some trait (e.g., run timing or size) but substantially different from members of all other strata. An equal proportion is sampled within each stratum.

straying the behavior of anadromous salmonids or other migratory species by which some fail to home in order to reproduce.

structural genes DNA sequences that encode the amino acid sequence of a protein.

sublethal allele an allele segregating in a population that reduces the survival rate of its carriers but that is not always lethal. If in a system of matings there are families whose survival conforms to the ratio 1:2:0, a geneticist can infer the segregation of a lethal allele. Should the last term in the observed ratio prove to be less than one, the recessive allele is termed sublethal.

submetacentric referring to chromosomes that have the centromere located toward one end, so that the arms are unequal in size (termed long and short arms).

subpopulation breeding groups within a larger population between which migration is significantly restricted.

substitution the mutational replacement of a nucleotide, DNA sequence, or chromosomal segment with another.

subtelocentric referring to chromosomes with very short arms.

supernumerary referring to extra chromosomes that are not duplicates of the normal-sized chromosomes.

supplementation the management practice of using hatchery-propagated fish to increase the number of naturally spawning fish of a targeted stock available for harvest in commercial or sport fisheries while conserving genetic resources.

supportive breeding the use of breeding–release programs to support demographically natural populations.

sympatric literally, "same country." Refers to populations or species occurring in the same ecosystem; contrast with **allopatric**.

synaptonemal complex a linear proteinaceous structure that forms between homologous chromosomes during meiotic prophase I. Synaptonemal complexes provide increased resolution for examining both the meiotic pairing behavior and chromosomal morphology.

synonymous mutations transitional mutations that do not change the amino acid sequence of the expressed protein.

taking under the U.S. Endangered Species Act of 1973 (16 U.S.C. §§ 1531 to 1544), "to harass, harm, pursue, hunt, shoot, wound, kill, trap, capture, or collect" an individual of a protected species, "or attempt to engage in such conduct."

tandem duplications adjacent repeated units of short DNA sequences.

tandem fusions end-to-end fusion of two chromosomes into a single element, typically involving the loss or "silencing" of one centromere.

tandem gene families groups of genes that have arisen via duplication, occur in arrays of repeated, adjacent copies, and whose products are needed in large quantities (e.g., rRNA, tRNA, and histones).

tautomeric (shifts) in molecular genetics, a reversible change in the location of a hydrogen atom on a nucleotide that allows it to join with a noncomplementary base (A joins C, T joins G).

telocentric referring to a chromosome that has only one visible arm, with the centromere at the end of the chromosome.

telomere specialized tandemly repeated DNA sequence at the ends of eukaryotic chromosomes (e.g., $[A_2C_4]_{60}$).

tetramer an enzyme molecule that is functional when composed of four identical or near-identical subunits; contrast with **monomer** and **dimer**.

tetrasomic (inheritance) the transmission of four gene copies at a single locus on four homologous chromosomes, two maternal and two paternal.

third-order (tertiary) structure the three-dimensional conformation of nucleic acid chains, such as folding and formation of intrastrand complementary base pairing.

threshold trait a discrete trait that is either present or absent in an individual. Generally, both genetic and environmental factors affect expression of the trait. Observation of the trait in a particular individual implies that it has a liability over the threshold for its expression. This liability is not always directly observable.

transcription the copying of the DNA template for a structural gene into its complementary mRNA product.

transfer RNA (tRNA) an amino-acid-specific RNA molecule that helps build proteins by having on one end the specific anticodon that matches a codon on a mRNA molecule and having on the other end a binding site for the amino acid.

transition a mutation in which a purine (A or G) is replaced by a different purine or a pyrimidine (C or T) is replaced by a different pyrimidine.

translation the tRNA-mediated conversion of the triplet code of nucleotides along an mRNA molecule into an amino acid (protein) sequence.

translocation for DNA, relocation of a portion of a chromosome (either within or between chromosomes). For RNA, the movement of mRNA through a ribosome during translation.

transposable elements DNA sequences that have the ability to excise themselves from one point in the genome and insert themselves elsewhere in the genome, including transposons and retroposons.

transposed sequences DNA sequences that are excised and then reintegrated elsewhere in the genome.

transposition the process of movement of a chromosomal segment to a new location on the same or a different chromosome, without reciprocal exchange. The process of insertion of a replica of a transposable element at a second site, with the original copy remaining at the first site (King and Stansfield 1990).

transposon a DNA sequence with the ability to move from one chromosomal position to another. This type of transposable element is flanked by inverted repeats, which are in turn flanked by direct repeats.

transversion the replacement of a purine (A or G) by a pyramidine (C or T).

trivalent an association of three homologous chromosomes during meiosis.

tRNA see **transfer RNA**.

truncation selection a mode of selection that favors all individuals with phenotypes above or below the truncation point, *T*.

unequal crossing-over during meiosis, the unequal exchange of genetic material among homologous chromosomes. This may result in duplication of a locus or chromosome segment in one chromatid and its loss in the other. Also, a molecular mechanism proposed as a cause of polymorphism in copy number of tandem repeats in repetitive DNA. The mechanism involves improper pairing across a repetitive DNA segment followed by crossing-over, resulting in one chromatid having more repeat copies than the other.

unlinked loci loci that do not occur on the same chromosome, or at least not in the same linkage group on one chromosome. The loci segregate independently at meiosis and are not inherited as a group.

unstable equilibrium the situation in which, once displaced, an allele frequency progresses away from its equilibrium value.

variable number of tandem repeats (VNTRs) tandemly repeated DNA sequences dispersed throughout the genome. The number of repeats at a particular locus often varies among individuals. VNTRs include minisatellites and microsatellites.

variance effective number the effective size of a population defined in terms of the amount of random genetic drift of allele frequency.

vector a type of DNA such as a plasmid or phage that is self-replicating and that can be used to transfer DNA segments among host cells.

viability a population's ability to sustain itself demographically.

viable population a population that is expected to persist for a specified length of time.

viable population number the size of a population that is expected to result in persistence of the population for a specified length of time.

visible mutation a mutation that gives rise to a readily observable change in the phenotype of the carrier, for example, eye color, wing shape, and abdominal bristle number in the fruit fly *Drosophila melanogaster.*

VNTR see **variable number tandem repeats.**

Wahlund effect the deviation of genotype frequencies from Hardy–Weinberg expectations in a sample due to the mixing of genetically differentiated subpopulations within that sample. When a population is subdivided into many breeding units and some inbreeding or drift occurs within each subgroup, the frequency of homozygotes in the aggregate sample is higher and the frequency of heterozygotes is smaller than the expected Hardy–Weinberg proportion. The effect also is characteristic of populations that receive immigrants.

wild (population) a self-sustaining population that has maintained successful natural reproduction and that is known to have had little or no supplementation from hatcheries in past generations.

wild fish fish with no hatchery-produced ancestors.

within-population genetic variation genetic differences among individuals of the same population.

References

Busack, C. A., and K. P. Currens. 1995. Genetic risks and hazards in hatchery operations: fundamental concepts and issues. Pages 71–80 *in* H. L. Schramm, Jr., and R. G. Piper, editors. Uses and effects of cultured fishes in aquatic ecosystems. American Fisheries Society, Symposium 15, Bethesda, Maryland.

Crow, J. F. 1958. Some possibilities for measuring selection intensities in man. Human Biology 30:1–13.

Dobzhansky, T. 1970. Genetics of the evolutionary process. Columbia University Press, New York.

Gulick, J. T. 1905. Evolution, racial and habitudinal. Carnegie Institution of Washington, Washington, D.C.

Kimura, M., and J. F. Crow. 1964. The number of alleles that can be maintained in a finite population. Genetics 49:725–738.

Kimura, M., and G. H. Weiss. 1964. The stepping stone model of population structure and the decrease of genetic correlation with distance. Genetics 49:561–576.

King, R. C., and W. D. Stansfield. 1990. A dictionary of genetics, 4th edition. Oxford University Press, New York.

Larkin, P. A. 1972. The stock concept and management of Pacific salmon. Pages 11–15 *in* R. C. Simon and P. A. Larkin, editors. The stock concept in Pacific salmon. H. R. McMillan Lectures in Fisheries, University of British Columbia, Vancouver.

Mayr, E. 1940. Speciation phenomena in birds. American Naturalist 74:249–278.

Muller, F. 1950. Our load of mutations. American Journal of Human Genetics 2:111–176.

Soulé, M. E. 1987. Where do we go from here? Pages 175–183 *in* M. E. Soulé, editor. Managing for viable populations. Cambridge University Press, New York.

Wright, S. 1932. The role of mutation, inbreeding, crossbreeding, and selection on evolution. Proceedings of the sixth international congress of genetics 1:356–366. Reprinted, pages 161–171 *in* W. B. Provine. 1986. Sewall Wright: evolution, selected papers. University of Chicago Press, Chicago.

Wright, S. 1943. Isolation by distance. Genetics 28:114–138.

Index

Note: italics indicate illustrations

H

W